Organometallic Compounds in the Environment

Organometallic Compounds in the Environment

Second edition

Edited by P.J. Craig
School of Molecular Sciences,
De Montfort University,
Leicester, UK

First Edition published by Longman, 1986. ISBN:0582 46361 0

Copyright © 2003   John Wiley & Sons Ltd, The Atrium, Southern Gate, Chichester,
West Sussex PO19 8SQ, England
Telephone (+44) 1243 779777

Email (for orders and customer service enquiries): cs-books@wiley.co.uk
Visit our Home Page on www.wileyeurope.com or www.wiley.com

Other Wiley Editorial Offices

John Wiley & Sons Inc., 111 River Street,
Hoboken, NJ 07030, USA

Jossey-Bass, 989 Market Street,
San Francisco, CA 94103–1741, USA

Wiley-VCH Verlag GmbH, Boschstr. 12, D-69469
Weinheim, Germany

John Wiley & Sons Australia Ltd, 33 Park Road,
Milton, Queensland 4064, Australia

John Wiley & Sons (Asia) Pte Ltd, 2 Clementi Loop #02–01,
Jin Xing Distripark, Singapore 129809

John Wiley & Sons Canada Ltd, 22 Worcester Road,
Etobicoke, Ontario, Canada M9W 1L1

Library of Congress Cataloging-in-Publication Data

Organometallic compounds in the environment/edited by P.J. Craig.–2nd ed.
    p. cm.
Includes bibliographical references and index.
ISBN 0-471-89993-3 (cloth: alk. paper)
    1. Organometallic compounds. 2. Environmental chemistry. I. Craig, P.J., 1944–

QD411 .O719 2003
547′.05–dc21

2002191061

British Library Cataloguing in Publication Data

A catalogue record for this book is available from the British Library

ISBN 0471 89993 3

Typeset in 10/12pt Times by Kolam Information Services Pvt Ltd, Pondicherry, India.
Printed and bound in Great Britain by Biddles Limited, Guildford, UK.
This book is printed on acid-free paper responsibly manufactured from sustainable forestry in which
at least two trees are planted for each one used for paper production.

DEDICATION

This book is dedicated to my wife Carole Ruth Craig

CONTENTS

CONTENTS

CONTRIBUTORS

Paul Andrewes
Environmental Carcinogenesis Division, United States Environmental Protection Agency, Research Triangle Park, NC 27711, USA

Janina Benoit
Chesapeake Biological Laboratory, University of Maryland, Solomons, MD20688, USA

F. Cima
Dipartimento di Biologia, Universita di Padova, Via Ugo Bassi 58B, 35121, Padova, Italy

P.J. Craig
School of Molecular Sciences, De Montfort University, Leicester LE1 9BH, UK

William R. Cullen
Department of Chemistry, University of British Columbia, Vancouver, British Columbia, Canada V6T 1Z1

J.S. Edmonds
National Institute for Environmental Studies, 16–2 Onogawa Tsukuba, Ibaraki 305–8506, Japan

G. Eng
Department of Chemistry and Physics, University of the District of Columbia, 4200 Connecticut Ave, Washington, DC 20008, USA

Jörg Feldmann
Department of Chemistry, University of Aberdeen, Aberdeen AB24 3UE, Scotland

O. Flassbeck
Institut für Umweltanalytik, University of Essen, D45117, Essen, Germany

K.A. Francesconi
Institute of Chemistry, Analytical Chemistry, Karl-Franzens-University Graz, Universitaetsplatz 1, 8010 Graz, Austria

Walter Goessler
Institute of Chemistry, Analytical Chemistry, Karl-Franzens-University Graz, Universitaetsplatz 1, 8010 Graz, Austria

B. Gruemping
GFA, Otto-Hahn Str. 22, D48161, Muenster, Germany

C.F. Harrington
School of Molecular Sciences, De Montfort University, Leicester, LE1 9BH

A.V. Hirner
Institut für Umweltanalytik, University of Essen, D45117, Essen, Germany

R.O. Jenkins
School of Molecular Sciences, De Montfort University, Leicester, LE1 9BH UK

Doris Kuehnelt
Institute of Chemistry, Analytical Chemistry, Karl-Franzens-University Graz, Universitaetsplatz 1, 8010 Graz, Austria

W.A. Maher
Ecochemistry Laboratory, Division of Science and Design, University of Canberra, Australia, ACT 2616

Robert P. Mason
Chesapeake Biological Laboratory, University of Maryland, Solomons, MD20688, USA

Jun Yoshinaga
Institute of Environmental Studies, The University of Tokyo, Hongo 7–3–1, Bunkyo, Tokyo 113–0033, Japan

PREFACE TO SECOND EDITION

The aims of this second edition are similar to those of the first, published in 1986. As well as the normal requirements of updating, the fundamental purpose of this edition is the same as that of the previous one.

Commercially, organometallic compounds have been used extensively over the past 50 years and in many of these uses there is a direct interaction with the natural environment. Some examples include their use as pesticides (organo-mercury or organotin compounds), gasoline additives (methyl- and ethylleads), polymers (organosicons) and other additives and catalysts. In addition there is an increasing realization that organometallics also exist as natural products in the environment (e.g. arsenic species). This has led to intensive research into their biological properties, toxicities, pathways and transformations in the environment and to their ultimate fate and disposal.

Although much material now exists in the literature, it is still often scattered or in large-scale form and can be rather inaccessible to the reader who needs to form an understanding of the topic as a whole. The present volume is designed as a single-volume source of information for this area.

The use of the term 'organometallic' is discussed in Chapter 1, but it is generally taken here for compounds with a direct metal to carbon sigma bond. The work also emphasizes the importance of as full an identification of the chemical substances present as possible, i.e. 'speciation'. Uses, chemical reactions in the environment, environmental pathways (biogeochemical cycles), biological properties and toxicities are covered.

It is intended that the book is read either as a whole, or as a series of stand-alone chapters which cover organometallic compounds of a single element together. The purpose of Chapter 1 is to cover the underlying properties of organometallic molecules that are relevant to their environmental behavior. The reader can consider these aspects first and then go on to consider the role of compounds of individual elements in the environment. Euqually, s/he may consider the elements separately and then draw the topic area together by using Chapter 1.

In order to allow this flexible use of the book a certain amount of redundancy is built in to allow chapters to be meaningful on their own. This element of repetition is intentional and it is hoped that it is helpful.

Finally, some other useful sources of information are given at the end of the book.

EDITOR'S ACKNOWLEDGEMENTS

I am delighted to thank the following, with whom I had helpful discussions or who have read parts of the text in the manuscript: Drs Richard Jenkins and Chris Harrington of De Montfort University, Dr Francesca Cima of the University of Padova, Italy and Dr Karl Ryder of Loughborough University of Technology. Any mistakes which remain are of course my own responsibility. I also thank Mrs Lesley Winder of De Montfort University for expert typing in connection with this work.

The Editor can testify personally that a great effort has been made to obtain copyright permission for anything we have reproduced here from other works. If there are any omissions we sincerely apologize.

1 Occurrence and Pathways of Organometallic Compounds in the Environment—General Considerations

P.J. CRAIG

School of Molecular Sciences, De Montfort University, Leicester, UK

GEORGE ENG

Department of Chemistry, University of the District of Columbia, Washington, DC, USA

and

R.O. JENKINS

School of Molecular Sciences, De Montfort University, Leicester, UK

1.1 SCOPE OF THIS WORK

The compounds considered in this work are those having metal–carbon (normally metal–alkyl) bonds, and which have environmental implications or properties. There is limited reference to metal–carbon π systems (e.g. $Mo(CO)_6$, $W(CO)_6(\eta\text{-}CH_3C_5H_4Mn(CO)_3)$) and mechanistic discussion of metal hydrides and ethene (e.g. decomposition by β elimination). In terms of formation of organometallics, methyl groups predominate but there is also reference to other metal hydrocarbon compounds (e.g. ethyl or phenyl mercury, ethyl leads, butyl tins). However, much of this work refers to metal methyl compounds as these are formed naturally in the environment (biomethylation)

The thrust of the work involves a good deal of analytical chemistry, but that is not the prime focus of the book. However, without the modern developments in analytical chemistry of the past 50 years, knowledge of most of the chemistry described in this book would barely exist. The analytical work that has led to this chemistry is described in the appropriate chapter but it is not the main theme. Several recent and comprehensive works that focus on the analytical chemistry of the environment have recently appeared and the reader is referred to those for the technical details of the analytical chemistry (see Standard Reference Sources and References at the end of the book).

Organometallic Compounds in the Environment
Edited by P.J. Craig © 2003 John Wiley & Sons Ltd

Amongst these, the recent comprehensive work by Crompton [1] focuses almost exclusively on the analysis of metal cations with minor consideration of organometallic compounds, and as its title denotes, is concerned with analysis from aqueous media. Similarly the work of Ure and Davidson [2] is mainly directed towards metal cations. The recent work edited by Ebdon *et al.* [3] also focuses more on analytical chemistry but takes full account of the complete molecular identity of the metallic compounds present (not focusing exclusively on organometallic compounds), i.e., speciation. (The importance of speciation is discussed later.)

Given that the stability, transport and toxicities of organometallic compounds depends on the number and type of the metal alkyl or aryl groups present, and that different compounds of the same or different metals may coexist at the same location in the environment, then separate detection of each species (speciation) is necessary. Separation and detection go together in so-called interfaced or hyphenated analytical systems. This is of particular importance because such species, although having real environmental and/or toxicity *effects*, often occur at very low concentrations in the environment (ppb, ppm— see later for definitions).

Nevertheless a broad statement on analysis needs to be made. There are two considerations:

(i) The metal, organometallic fragment or full compound needs to be *detected* by a sufficiently sensitive method (e.g. Hg, CH_3Hg^+, CH_3HgCl respectively), and

(ii) as a variety of organometallic compounds of the same element may be present together in the same matrix (e.g. butyltins, butyl/methyltins) and they each have different toxicity and environmental properties, then they must be *separated* before individual detection.

The main methods of detection are as follows:

(i) Atomic absorption spectroscopy (flame, graphite furnace, Zeeman, hydride generation/quartz furnace),

(ii) atomic fluorescence spectroscopy (alone or via hydride generation),

(iii) atomic emission spectroscopy (usually inductively coupled plasma),

(iv) voltammetry,

(v) mass spectrometry (conventional or chemical ionization, electrospray, tandem, isotope dilution, plasma),

(vi) X-ray and neutron methods.

The main methods of separation are:

(i) Gas chromatography (conventional or capillary),

(ii) thermal desorption methods (which depend on boiling points),

(iii) high performance liquid chromatography,

(iv) flow injection methods,

(v) ion exchange chromatography,
(vi) ion chromatography.

Many organometallic compounds, or cations are insufficiently volatile to undergo gas chromatography, but may be induced to do so by *derivatization*.

This is generally achieved by (formally) SN_2 attack by hydride (from $NaBH_4$), ethyl ($NaB(C_2H_5)_4$) or other alkyl group (e.g. from a Grignard reagent), e.g. Equation (1.1)

$$As(CH_3)_2X \xrightarrow{NaBH_4} As(CH_3)_2H \qquad (1.1)$$

X = environmental counter ion (in this case not riboside—see Chapter 5)

This has been widely achieved, even in the case of mercury where it had been thought that mercury hydrides were too unstable for use in analysis [4].

Coupling of these separation and detection techniques is now ubiquitous and provides an intensive battery of techniques for analytical work, well described in Reference [3]. Without these, little knowledge of organometallic compounds in the environment would be possible.

The present work is also not primarily a work of toxicology although the toxicity properties of the compounds are discussed in the chapters element by element. The reader is referred to several excellent works specifically dedicated to toxicity studies of organometallic compounds [5].

The present work is a consideration of the inputs (natural and anthropogenic) and/or formation of organometallic compounds in the natural environment (sediment, water and atmosphere), their properties and behaviour there, and their ultimate fate. Although much of our understanding in this field is derived from analytical chemistry and the methods are described where needed, the theme of the work is the overall behaviour of organometallics in the environment, not their analysis. Compounds are covered by chapters on an element-by-element basis.

Organometallic species (i.e. compounds, complexes or ions) may be found in the natural environment either because they are *formed* there or because they are *introduced* there. To date, the behaviour of the latter group is better understood, and their environmental impact has been assessed by studies of their direct toxicities, their stabilities and routes to decay, and by toxicity studies of their decay products. Organometallic compounds entering the environment may be deliberately introduced as products whose properties relate to the environment (e.g. biocides) or they may enter peripherally to a separate, main function (e.g. gasoline additives, polymer stabilizers). Compounds of arsenic, mercury, tin and lead have important uses as organometallic compounds. Their role and behaviour in the environment are covered in the appropriate chapters of this work (Chapters 2 to 5). The behaviour of other organometallic species in the natural environment is also covered (Chapters 6 to 10). However, not all organometallics found in the environment are introduced—some are *formed* after entry as inorganic species and constitute the organometallic components of global biogeochemical cycles. This process of environmental methylation is usually termed biomethylation and is, as the name implies, almost exclusively concerned with formation of metal–methyl bonds (although ethyl mercury has been found in the environment in circumstances removed from likely input as a product). Recent work has demonstrated the occurrence of transition metal carbonyls, likely to have been formed in the environment (see below).

1.2 GENERAL APPROACH: SPECIATION, CONCENTRATIONS AND TERMINOLOGY

This work considers those organometallic compounds that have relevance to the natural environment. It is concerned with compounds that are found there, or which may be formed there, or which may react or be transported within the environment. Accordingly we discuss inputs, formation, transportation and decay. The approach is to consider these processes element by element in each chapter. The present chapter links the work by considering those fundamental aspects of organometallic chemistry that are relevant to the environmental chemistry of the elements discussed in each chapter, including stabilities, and mechanisms of environmental formation and decay.

It is at this point that further consideration of the term 'speciation' should be made. A generation ago, and indeed much legislation concerning pollution, etc. still occurs in this context, chemists had to be content with discussing a contaminant by its defining *element* (e.g. total arsenic, mercury concentrations, etc.). In parallel with many chemists becoming environmental chemists, technology and necessity prompted further identification into partial or complete molecular identification of the contaminant (e.g. methylmercury, CH_3Hg^+, arsenobetaine, $(CH_3)_3^+AsCH_2COO^-$). Such full operational identification of a compound within a larger matrix is now commonly termed 'speciation'. Where possible, to accord with a speciation approach, we will discuss the chemistry in terms of compounds.

Additionally of course, speciation is now not only more possible, it is essential. The main toxicity and environmental properties depend markedly on what compound is present, not on what metal. Some arsenic compounds are notoriously toxic (e.g. As_2O_3), but some are effectively non-toxic (e.g. arsenobetaine). Toxicity also depends on the degree of alkyl substitution of a metal and the identity of the organic (alkyl) group, and it varies also for the same compound towards different (biological) species. Residence times may also vary with species (e.g. CH_3Hg^+ (long) and Hg^{2+} (shorter) in biological tissue), and this can determine toxic impact. In addition to toxicity, transportation parameters also vary for the same element with its speciation, e.g. partition to atmosphere or water. Organometallic cations (e.g. Bu_3Sn^+) tend to be more water soluble and non-volatile, but saturated compounds (e.g. $(CH_3)_4Pb$) are hydrophobic and volatile.

Additionally, a generation ago analytical chemists had generally to accept quantitative limits for their work of parts per thousand (ppt or mg g^{-1}) e.g. for arsenic in a matrix (containing medium). At the time of the first edition of the present work (1986) parts per million or billion (ppm, ppb or μg g^{-1}, ng g^{-1}) were being achieved. The standard now is commonly ppb (or ng g^{-1}; the quantity present in 10^9 parts of the matrix), but parts per 10^{-12} (ppt or pico grams per gram (fg g^{-1})) are now commonly reported. It should always be borne in mind how relevant in practical terms such extreme measures of dilution might be, and analytical and environmental chemists should pause on occasion to consider which chemical species may *not* be present in a matrix at fg g^{-1} or more dilute levels. Chemical analysis is usually targeted towards the species of interest and much else present may be missed or ignored. The question is, 'if the level of a certain species is of the order of 10^{-12} parts per gram, does it *matter* and if so, to *whom?*'

The above considerations also bring forward another point of terminology. Laboratory chemists usually express concentrations in molar terms, i.e. mol dm^{-3}. At greater levels of dilution parts per million (ppm) and similar terms are often used. These terms are less precise, often because the matrix in which the species of interest is present is not water or a similar solvent. It is often a wet, amorphous sediment. Hence 'ppm' can mean one of the following:

(i) grams of the relevant atom present in 10^6 grams of the matrix,
(ii) grams of a defined part of the molecule in 10^6 grams of the matrix,
(iii) grams of the whole molecule in 10^6 grams of the matrix.

In some published work, 'ppm' is not even defined as above. To add to the imprecision, the matrix (which may be a sediment or biological tissue) may be taken as wet (heavier) or dried (lighter)—giving two possible figures for the same measurement. Clarity of definition is not always present in quantitative work in this field, and the matrix is rarely the simple defined volume of known solvent that occurs in laboratory chemistry.

With regard to atmospheric or liquid measurements, terms such as ppm could mean:

(i) grams of the molecule (or relevant atom) present in 10^6 cm^3 of atmosphere (at STP?) or water, or
(ii) volume (cm^3) of the molecule (or relevant atom) present in 10^6 cm^3 of atmosphere or water.

Care therefore needs to be taken when results from different laboratories or groups are compared. Consistency is often absent (even orthodox molar concentrations are sometimes used, even at extremes of dilution).

To put this field into perspective—although ppm, ppb and similar concentrations can be of major physiological, toxicological or environmental significance, it will do little harm to repeat the comparison given in the first edition of this work—a ppm is equivalent to a needle in a haystack; a ppb is equivalent to a grain of sand in an Olympic-sized swimming pool (in checking the calculation the reader is also invited to consider this as a not completely outlandish example of the use of imprecise concepts to register concentrations).

Within the present work, full standardization of terms is not possible owing to wide differences in practice, methods, matrices and analytical feasibility. To overcome this as far as possible, a basic attempt at standardization has been made and cross-referencing will then be used to clarify detailed points.

1.3   TYPES OF ORGANOMETALLIC COMPOUND

Most, but not all, organometallic compounds of environmental interest are covalent, bound by a σ bond from a single carbon atom, to a main group element. The term 'organometallic' is generally defined as a compound with a bond $(M—C)$ polarized $M^{\delta+}—C^{\delta-}$ i.e. the metal is less electronegative than carbon. A compound containing carbon atoms, but where the bonds to the metal are not directly to the carbon atom (but may be via oxygen, nitrogen or halogen atoms instead), is not considered to be organometallic, although such a compound may be referred to as 'metal organic'. In general then, the compounds discussed in this book will involve carbon bound to a main group metal via a single carbon atom—these are referred to as 'monohapto' compounds. Despite the polarization above, in this work metal–carbon bonds are usually shown as R_nM to accord with common practice.

A vast organometallic chemistry of interest to synthetic and mechanistic organometallic chemists exists outside the definition above, which is of little significance (so far as is known) in an environmental context, other than by way of input from manufacture or use, in which case the pollutant is a decomposition product that is usually non-organometallic. (Transition metal organometallics have hugely important uses as synthetic and catalytic intermediates, but there is usually little pollution owing to the high cost of the metals concerned, e.g. rhodium, ruthenium, etc.) Within the context of these transition-metal compounds there are actually some instances of environmental significance, for example, $(\lambda\text{-}CH_3C_5H_4Mn(CO)_3)$ is used as an anti-knock agent in gasoline (Chapter 9) and $Mo(CO)_6$ and $W(CO)_6$ have been detected in the environment (Chapter 9).

Most of the 'metals' in the present work are clear cut main group elements (mercury, tin, lead, etc.) but certain 'metalloidal' elements are included because their environmental properties have so much in common (arsenic, antimony, etc.). Some elements generally considered to be non-metallic have strong interactions with this area of work and are mentioned as needed (e.g. sulphur, selenium). Polymeric organometallic compounds are covered in the chapters relating to siloxanes (Chapter 8) and tin (Chapter 3).

1.4   THERMODYNAMIC STABILITIES OF ORGANOMETALLIC COMPOUNDS

As noted above, most of the organometallic compounds in the present work involve covalent bonds between a carbon atom (usually in a methyl, CH_3, group) and a main group (non-transition element). Such bonds may have little or much polarization (polarization is a measure of the drift in electronic charge in a covalent bond from, in this case, the metallic element to the carbon atom, viz. $M^{\delta+}$—$C^{\delta-}$). Polarizations vary and affect stabilities considerably. In general the M—C bonds can be considered as localized (albeit polarized) with conventional σ electron pair bonds similar to those occurring in organic chemistry.

In considering stability, we have to ask stability with respect to *what* phenomenon (hydrolysis, oxidation, thermal). We consider stability of a compound with respect to decomposition into its own elements (this doesn't usually happen) and towards external chemical attack (the norm), e.g. by atmospheric oxygen, water and microbial-mediated decay. We consider inherent (thermodynamic) stability first because the general approach is helpful. However, in general it can be said that where the compounds are unstable it is not because of weak M—C bonds (thermodynamic) but because there are low energy pathways to decomposition (kinetic). A component is thermodynamically unstable with respect to decay to elements if the standard Gibbs free energy for the process is negative. In certain equilibrium reactions where this equals zero, there can still be a driving force for the reaction.

Thermodynamic stabilities depend mainly on the strength of the bonds from metal to carbon. They can be estimated by bond enthalpy ($\bar{E}_m$ M—C) measurements; compounds with strong M—C bonds are, not surprisingly, more stable than those with weaker M—C bonds. Although not inherently weak, M—C bonds are less strong than M—N, M—O and M—X (X = halogen) bonds. (Table 1.1). Free energies of formation (ΔG_f^θ) are not usually known for organometallics as standard entropies are rarely known and so enthalpies of formation (ΔH_f^θ) are usually used in comparing thermodynamic stabilities. Hence organometallic reactions are exothermic or endothermic (ΔH_f^θ is negative or positive), but free energies (ΔG_f^θ) are not known other than that they are negative and hence, in a formal sense, neither are thermodynamic stabilities. For alkyl organometallics mean bond enthalpies for the whole molecule ($\bar{E}$ (M—C)) are usually quoted, but there is a problem here in that *stepwise* bond dissociation energies ($D_1 - D_n$) may deviate a lot from the mean values, viz

$$\bar{E} = \frac{1}{n}\sum_1^n D_1. \quad \text{(see Me}_2\text{Hg below, Equation 1.2, 1.3)}$$

$$(CH_3)_2Hg \longrightarrow CH_3Hg + CH_3 \quad D_1(Hg\text{—}C) = 214\,kJ\,mol^{-1} \quad (1.2)$$

$$CH_3Hg \longrightarrow Hg + CH_3 \quad D_2(Hg\text{—}C) = 29\,kJ\,mol^{-1} \quad (1.3)$$

Table 1.1   Bond dissociation energies[a] of diatomic molecules (in kJ mol^{-1}). From Kerr JA, 1983, Strengths of chemical bonds, *Handbook of Chemistry and Physics*, Chemical Rubber Company, 65 edn, F171–181 Reprinted with permission. Copyright CRC Press Inc, Boca Raton, Fla, USA

Main group elements[b]

Gp 13	M-C	M-O	Gp 14	M-C	M-O	Gp 15	M-C	M-O	Gp 16	M-C	M-O
B	449	806	C	605	1080	N	773	—	O	1080	500
Al	—	449	Si	437	790	P	517	596	S	701	521
Ga	—	299	Ge	462	660	As	—	479	Se	584	424
Ln	—	323	Sn	—	525	Sb	—	391	Te	—	382
Tl	—	—	Pb	—	416	Bi	—	302	Po	—	—

[a] D(M—X) = M$^{\bullet}$ + X$^{\bullet}$, i.e. bond dissociation enthalpy; loosely, bond strength
[b] In kJ mol^{-1} measured spectroscopically (mass spectrum) from transient molecules as above, at 25 °C

Hence $\bar{E} = 121.5$ KJ mol^{-1}, but this does not reflect the difficulty of breaking the *first* Hg—C bond [6].

What matters here is that some organometallic compounds are exothermic towards decomposition to their elements, and some are endothermic, but not all of the exothermic compounds decompose (kinetics again!). Also, when they do decompose on heating they usually do so to a mixture of hydrocarbons, hydrogen and the metal, not to the elements. Nevertheless, quantification information from such measurements is a useful guideline and, is usually all that is available. Data is given in Tables 1.2 to 1.4

The information in the tables relates to decomposition to the elements in their standard states, but this is not what usually happens. Taking $(CH_3)_4Pb$ for example (Table 1.5) it can be seen that there are other routes than those to elements. On this basis route 4 is apparently the most favoured.

Despite the apparent thermodynamics (we don't know the entropy values), $(CH_3)_4Pb$ is an important, readily available and commercial compound. The reason is, again, kinetic stability. The reaction pathway at room temperature does not have an activation energy low enough for it to happen at a measurable rate. The activation energy in such cases may depend on the inherent M—C bond strength and this appears to be strong enough to allow $(CH_3)_4Pb$ stored in an inert atmosphere to be indefinitely stable.

In practice, in the environment, decay mediated by oxygen, water, free radicals and biology is much more relevant. Many kinetically stable organometallics in these terms may be very unstable in the environment, including $(CH_3)_4Pb$ which decays in days in the atmosphere (Chapter 4).

So while all carbon compounds (including organometallics) are thermodynamically unstable owing to the stability of the products (again assuming entropy values), many are kinetically stable because there is no low energy route to decomposition. This may be associated with there being a closed shell of electrons, often of spherical symmetry, around the metal atoms, i.e. full use of

Table 1.2 Standard enthalpies of formation ΔH_8^0 (kJ mol^{-1}) and *mean* bond enthalpies, $\bar{E}$ (M—C) (kJ mol^{-1}) of CH_3 derivatives in the gas phase[a]

MMe$_2$			MMe$_3$			MMe$_4$			MMe$_3$		
M	ΔH_8^0	$\bar{E}$	M	ΔH_8^0	$\bar{E}$	M	ΔH_8^0	$\bar{E}$	M	ΔH_8^0	$\bar{E}$
Zn	50	177	B	−123	365	C	−167	358	N	−24	314
Cd	106	139	Al	−81	274	Si	−245	311	P	−101	276
Hg	94	121	Ga	−42	247	Ge	−71	249	As	13	229
			Ln	173	160	Sn	−19	217	Sb	32	214
			Tl	—	—	Ph	136	152	Bi	194	141

[a] mean 'bond strengths'. $n\bar{E}(M—CH_3) = \Delta H_8^\theta M_{(g)} + n\Delta H_8^\theta(CH_3{}^0)_g - \Delta H_8^\theta(M(CH_3)_{n(g)}$

Notes
(i) Some compounds are exothermic, some are endothermic with respect to decomposition to elements n the absence of air/water.
(ii) $\bar{E}$ (M—C) decreases with increasing atomic number for main group elements (it is the opposite for transition metal–methyl compounds) due to increasing orbital dimensions on M and poorer covalent overlap.
(iii) These bonds are weak compared to metal–oxygen, metal–halogen, carbon–oxygen bonds in weakness of organometallic to oxidation.
(iv) Mean bond enthalpies can be misleading (see text).
(v) Analogous values of $\bar{E}$ for some transition metal carbonyls are $Cr(CO)_6$, $\bar{E} = 107$; $Mo(CO)_6$, $\bar{E} = 152$; $W(CO)_6$, $\bar{E} = 180$. Note *increasing* bond strength down the group.
(vi) Mode of decomposition is via homolytic breakage of the M—C bond to produce radical species.
Reprinted with permission from Ref. 7

Table 1.3 E_{298}^0 Values for the first M—C bond in some polyatomic molecules (kJ mol^{-1})

$CH_3—Ge(CH_3)_3$	346 ± 17
$CH_3—Sn(CH_3)_3$	297 ± 17
$CH—Pb(CH_3)_3$	238 ± 17
$CH_3—As(CH_3)_2$	280 ± 17
$CH_3—Sb(CH_3)_2$	255 ± 17
$CH_3—Bi(CH_3)_2$	218 ± 17
$CH_3—CdCH_3$	251 ± 17
$CH_3—HgCH_3$	255 ± 17
$CH_3—SH$	312 ± 4.2
$CH_3—SCH_3$	272 ± 3.8

Notes
(i) Compare with Table 1.2. The above bonds are $> \bar{E}$ (M—C).
(ii) Measurement of bond strengths in polyatomic molecules is not straightforward, being hard to measure (usually by kinetic methods). Some can be calculated at 298° from the following equation:

$$E^0(R—X) = \Delta_8 H^0(R^0) + \Delta_8 H^0(X) - \Delta_8 H^0(RX) \quad \text{or}$$
$$E^0(R—R) = 2\Delta_8 H^0(R^0) - \Delta_8 H^0(R—R)$$

Reprinted with permission from Ref. 8

Table 1.4 Mean bond enthalpies ($\bar{E}$ (M—C)) for oxides and halides

B—O	526	Si—O	452	As—O	301
B—Cl	456	Si—Cl	381	Bi—Cl	274
Al—O	500	Si—F	565		
Al—Cl	420	Sn—Cl	(tin)323		

Taken with permission from Ref. 9

Table 1.5 Decomposition of $(CH_3)_4Pb$

Decomposition route	ΔH^0	ΔHper M—C$^+$
(1) $Pb(CH_3)_{4(g)} \rightarrow Pb_{(s)} + 4C_{(s)} + 6H_{2(g)}$	-136	-34
(2) $Pb(CH_3)_{4(g)} \rightarrow Pb_{(s)} + 2C_2H_{6(g)}$	-307	-76.75
(3) $Pb(CH_3)_{4(g)} \rightarrow Pb_{(s)} + 2CH_{4(g)} + C_2H_{4(g)}$	-235	-58.75
(4) $Pb(CH_3)_{4(g)} \rightarrow Pb_{(s)} + 2H_{2(g)} + 2C_2H_{4(g)}$	-33	-8.25

$+kJ\ mol^{-1}$

Taken with permission from Ref. 6

the metal orbitals allowing no easy access to attacking reagents. Where available empty orbitals do exist, the compound may still be kinetically stable unless the metal–carbon (M—C) bonds are strongly polarized allowing, for example, nucleophilic attack by an external reagent.

Organometallics are also thermodynamically unstable with respect to oxidation to MO_n, H_2O and CO_2. Again kinetic reasons may render such compounds inert. Very reactive are compounds with free electron pairs, low-lying empty valence orbitals and highly polar M—C bonds.

We now consider kinetic stability for organometallic compounds.

1.5   KINETIC STABILITY OF ORGANOMETALLIC COMPOUNDS

In order to decompose, homolytic breakage of the M—C bond must first occur (Equation 1.4)

$$MR_{n+1} \rightarrow MR_n + R \qquad (1.4)$$

In isolation this process is thermodynamically controlled by bond strengths (or enthalpies)—see above. However, once the short lived and reactive radicals are formed, further rapid reaction takes place to produce thermodynamically stable final products. None the less an energy input is required to break the metal–carbon bond. If the input is large (i.e. for a strong M—CH$_3$ bond) then a thermodynamically unstable compound (ΔG^θ – ve) may be stable at room temperature. The required energy input is known as the activation energy ($G^{\ddagger}$). This is illustrated in Figure 1.1.

However, even when ΔG^θ is negative, $G^{\ddagger}$ may be large if the metal–carbon bond is strong. So thermodynamically unstable molecules may be kinetically stable, but when exposed to external attack, say by oxygen, water or microbes, they can soon decay. Strength of metal–carbon bonds is only a tendency towards environmental stability.

Some stability is dependent on molecular architecture other than metal–carbon bond strengths. Although many of the organometallics observed in the natural environment are metal methyls, others such as ethylleads and butyltins also exist and are observable in the environment. Organometallics

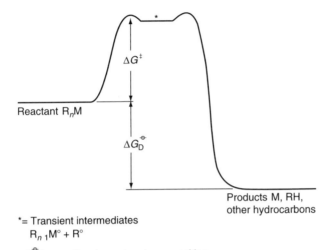

*= Transient intermediates
$R_{n-1}M^{\circ} + R^{\circ}$

$\Delta G_D^{\ominus}$ is negative, favouring decomposition

If $\Delta G^{\ddagger}$ is large there is no low energy pathway for decomposition and the compound is kinetically stable under normal conditions

Figure 1.1   Activation energy and decay of compounds.
Taken from this work, first edition, with permission

with metal alkyl groups other than methyl are susceptible to an important route to decomposition and on this ground alone should be less stable and less observed than metal methyls. This route is termed β elimination and occurs by migration of a hydrogen atom attached to a carbon atom at a remove of one other carbon atom from the metal (e.g. in an M—CH_2—CH_3 grouping but not limited to ethyl). The products are a metal hydride and ethenes (ethylene). That ethene is frequently observed in the natural environment may be related (Equation 1.5). Blockage of β elimination by alkyl groups not having a β hydrogen atom (e.g. —$CH_2C(CH_3)_3$, —CH_2CF_3, etc.) is not relevant to the environment.

$$CH_3CH_2MR_n^1 \rightleftharpoons HMR_n^1 \, CH_2{=}CH_2 \qquad (1.5)$$

β elimination proceeds more rapidly down the groups in the periodic table because at the intermediate stage the metal is, in effect, increasing its coordination number. It may be assumed that β elimination could be involved (even when accompanied by hydroxide attack) in decay routes for compounds where decay is known to occur (e.g. butyltins) in compounds inputted into the environment, and why few ethyls and no higher alkyl species appear to be *formed* and stable in the environment. β elimination requires an empty valence metal orbital on M to interact with the electron pair on the C_β—H bond. Hence β elimination is more important for groups 1, 2 and 13 than for groups 14, 15 and 16. Empty orbitals can be blocked by ligands to increase stability. Two other processes are feasible but have never been investigated as environmental routes.

These are α hydrogen elimination (clearly possible with metal–methyl group-ings) and orthometallation (where a nearby *ortho* aromatic hydrogen is trans-ferred to the metal). Methyl and phenyl mercury species do decay in the environment and may do so by these routes. However in the presence of air, water and microbes, the above routes are likely to be minor ones.

1.6 STABILITY OF ORGANOMETALLIC COMPOUNDS TO ATMOSPHERIC OXIDATION

All organometallic compounds are thermodynamically unstable to oxidation because of the much lower free energy of the products of oxidation (metal oxide, carbon dioxide, water). Equation 1.6 gives an example.

$$(CH_3)_4Sn(g) + 8O_2 = SnO_{2(s)} + 4CO_{2(g)} + 6H_2O_{(g)}$$
$$\Delta H_m^\theta = -3591 \text{ kJ mol}^{-1}$$
$$\Delta H_m^\theta = \text{enthalpy change in reaction}$$

(1.6)

Here again, we rely on ΔH not ΔG values because of lack of knowledge of ΔS; however, in view of the liberation of gaseous molecules, $T\Delta S$ will be an overall positive input into $\Delta G = \Delta H - T\Delta S$. This is very exothermic, but $(CH_3)_4Sn$ is quite stable in air. Some, but not all, organometallic compounds however are spontaneously inflammable. Others, although thermodynamically unstable $((CH_3)_4Sn)$ still do not oxidize in this way for kinetic reasons, as discussed in Section 1.5 above. Interestingly and relevant, compounds which *in bulk* may spontaneously oxidize (burn; ignite) may be stable or decay much more slowly when they are attenuated (in dilute form, e.g. ppm or ppb in air). $(CH_3)_3Sb$ has been noted in this respect.

Greater stability on attenuation follows from a consideration of the collision theory of gases. The rate constant is related to a collision number Z (the number of reactant molecules colliding per unit time) and the activation energy E, the Arrhenius equation (Equation 1.7),

$$k = Z \exp(-E/RT)$$

(1.7)

where k = rate constant, E = activation energy, R = gas constant and T = absolute temperature.

Z can be derived from classical gas kinetic theory as (Equation 1.8)

$$Z = \sigma_{AB}^2 \left[8\pi kT \frac{m_A + m_B}{m_A m_B} \right]^{\frac{1}{2}} n_A.n_B.$$

(1.8)

σ_{AB} is the collision cross section of A and B molecules with masses m_A and m_B and, crucially, concentrations n_A and n_B atoms as molecules per unit volume. Where reaction depends on collision between two molecules then it is quicker as concentration increases. This aspect is well discussed in Reference [10].

For these reasons, organometallics which, by their pyrophoric nature, alarm the laboratory chemist, may be much more stable in the environment.

The initial process of oxidation of organometallics by O_2 is a rapid charge transfer interaction that occurs, involving electron donation from the organometallic to oxygen [11]. This is shown in Equations 1.9–1.11.

$$R_nM + O_2 \rightarrow R_nM^+.O_2^- \rightarrow R^0 + R_{n-1}^+MO_2^- \tag{1.9}$$

$$R^0 + O_2 \rightarrow RO_2^0 \tag{1.10}$$

$$RO_2^0 + R_nM \rightarrow RO_2R_{n-1}M^0 + R^0 \tag{1.11}$$

The species $R_nM^+.O_2^-$ may decay by various routes, and peroxides may be formed or coupling of the alkyl ligands may occur. Organocarbon and -mercury compounds are oxidized by radical chain SH_2 processes as exemplified below, following the initial charge transfer processes, (Equations 1.12 and 1.13); [11]

$$R_3B + RO_2^0 \longrightarrow R_2BOOR + R^0 \tag{1.12}$$

$$R^0 + O_2 \longrightarrow RO_2^0, \text{ etc} \tag{1.13}$$

The comparative susceptibilities of some metal–carbon bonds to oxidation to metal oxides are demonstrated in Tables 1.6 and 1.7. For main group elements they suggest increasing liability to oxidation as the group is descended.

Atmospheric oxidation will tend to occur rapidly:

(i) Where metal–carbon bonds are very polar, the partial charges on $M(\delta+)$ or $C(\delta-)$ may facilitate attack by external reagents including oxygen. This can be observed indirectly by considering electronegativity values (Table 1.8). Electronegativity is the ability of an atom in a molecule to attract a shared

Table 1.6 Comparative bond enthalpy terms for metal–carbon and metal–oxygen bonds[a] (in kJ mol^{-1}). Data gives an assessment of comparative bond strengths. (Reproduced with permission Johnson DA 1982 *Some Thermodynamic Aspects of Inorganic Chemistry* (2nd edn). Cambridge University Press, pp 201–2).

Group 14	M—C	M—O	Group 15	M—C	M—O	Group 16	M—C	M—O
C	347	358	N	314	214	O	358	144
					632(NO)			
Si	320	466	P	276	360	S	289	522(S=O)
Ge	247	385	As	230	326	Se	247	
Sn	218		Sb	218		Te		
Pb	155		Bi	141		Po		

B—C = 364   C—H = 413   O—H = 464   N—H = 391
B—O = 520   H—H = 436   Cl—H = 432   C—Cl = 346

[a] From thermodynamic data on the decomposition of the molecules, e.g. C_2H_6. Calculated from thermodynamic cycles
Definitions as in Table 1.3. M = elements listed

Table 1.7 Stability of methylmetals to oxygen[a]

Stable	Unstable[b]
$(CH_3)_2Hg$	CH_3PbX_3
$(CH_3)_4Si$, $[(CH_3)_2SiO]_n$, $(CH_3)_nSi^{(4-n)+}$,	CH_3Tl^+
$(CH_3)_6Si_2$	
$(CH_3)_4Ge$, $(CH_3)_4Ge^{(4-n)+}$, $(CH_3)_6Ge_2$	$(CH_3)_2Zn(CH_3Zn^+$ also$)$
$(CH_3)_4Sn$	$(CH_3)_2Cd(CH_3Cd^+$ also$)$
$(CH_3)_4Pb^c$	$(CH_3)_3B$
$CH_3HgX(C_6H_5$ and C_2H_5 also stable$)$	$(CH_3)_3Al$
$(CH_3)_{4-n}SnX_n$	$(CH_3)_3Ga$
$(CH_3)_3PbX$	$(CH_3)_3In$
$(CH_3)_2PbX_2$	$(CH_3)_3Tl$
$\pi(CH_3C_5H_4Mn(CO)_3{}^c$	$(CH_3)5As$
$CH_3Mn(CO)_4L^d$	$(CH3)3As^e$
$(CH_3)_2AsO(OH)$	$(CH_3)_3Sb^e$
$CH_3As(O)(OH)_2$	$(CH_3)_3Bi$
$(CH_3)_2S$	$(CH_3)_2AsH$
$(CH_3)_2Se$	CH_3AsX_2
$CH_3HgSeCH_3$	CH_3SbX_2
CH_3COB_{12} (solid state)	$(CH_3)_{4-n}SnH_n{}^e$
$(CH_3)_3SbO$	$(CH_3)_6Sn_2$ (At RT gives $[(CH_3)_3Sn]_2O)$
$(CH_3)_2SbO(OH)$	$(CH_3)_6Pb_2$ (to methyl lead products)
$CH_3SbO(OH)_2$	$(CH_3)_5Sb$
$(CH_3)_2Tl^+$, $(CH_3)_2Ga^+$	$(CH_3)_3AsO$
$(CH_3)_3S^+$	$(CH_3)_3P$
$(CH_3)_3Se^+$	$(CH_3)_4SiH_{4-n}$
$(CH_3)_3PO$	$(CH_3)_4GeH_{4-n}$

[a] At room temperature in bulk (assume similar but lesser environmental stability for ethyls). As against rapid (seconds, minutes) oxidation. Table to be read in conjunction with Table 1.6. Not necessarily stable against water (see Table 1.9)

[b] Variously unstable because of empty low lying orbitals on the metal, polar metal–carbon bonds and/or lone electron pairs on the metal.

[c] Gasoline additive.

[d] To exemplify ligand-complexed transition metal organometallics. Many of these synthetic compounds are oxygen stable but none have been found in the natural environment.

Reprinted with permission from this work, first edition

[e] But stable in dilute form and detected in the environment (Ch.3).

(ii) electron pair to itself, forming a polar covalent bond. The values of electro-negativity computed by Pauling (Table 1.8) still remain the best indicator of polarity, as dipole moments apply to the molecular as a whole, not just to the bond (whose dipole may therefore only be estimated indirectly).

(ii) Where empty low lying (valence) orbitals on the metal exist, or where the metal has a lone electron pair (non-bonding pair), then this also facilitates kinetic instability to oxidation. The oxygen molecule (a diradical) can attack empty orbitals on M, e.g. d orbitals.

(iii) Where the compound is thermally (thermodynamically) unstable anyway.

(iv) Where full coordination complexation with stabilizing ligands is not oc-curring.

Table 1.8 Some electronegativity values relative to carbon

Group 13	Group 14	Group 15	Group 16	Group 17
B 2.0				F 4.0
Al 1.5	Si 1.8	P 2.1	S 2.5	Cl 3.0
Ga 1.6	Ge 1.8	As 2.0	Se 2.4	Br 2.8
Ln 1.7	Sn 1.8	Sb 1.9	Te 2.1	I 2.5
Tl 1.8	Pb 1.8	Bi 1.9	Po 2.0	Al 2.2

Notes
1. The C—S bond is non-polar. The electrons are distributed midway between C and S.
2. For the other environmental main group elements (E) the electrons are closer in the covalent bond to carbon (i.e. $M^{\delta-}$—$C^{\delta-}$).
Taken from standard data in texts.

Conversely complexation requires available empty orbitals on the metal. It is a competition between available ligands and atmospheric oxygen. $(CH_3)_4Sn$ is quite stable in air (fully coordinated) whereas $(CH_3)_2Zn$ is pyrophoric (empty orbitals, coordinatively unsaturated).

1.7   STABILITY OF ORGANOMETALLICS TO WATER

The first step in the hydrolysis of an organometallic compound is usually nucleophile attack of the lone electron pair on the water oxygen atom to an empty metal orbital on the organometallic. Hence hydrolytic instability is also connected with empty low-lying orbitals on the metal and on the ability to expand the metal coordination number. The rate of hydrolysis is connected with the polarity of the metal–carbon bond; strongly polarized ($M^{\delta-}$—$C^{\delta-}$) bonds are unstable to water. These are found for example in groups I and II organometals and for those of zinc and cadmium. The influence of polarity is shown for alkylboron compounds which have low polarity and are water stable although unstable to air. Low polarity compounds which cannot easily expand their coordination number are expected to be water stable. Most metal alkyls and aryls are thermodynamically unstable to hydrolysis to metal hydroxide and hydrocarbon (Equation 1.14). Many, though, are kinetically stable. Examples are given in Table 1.9.

$$R_nM + nH_2O \longrightarrow M(OH)_n + nRH \qquad (1.14)$$

To illustrate, $SiCl_4$ is easily attacked by water, owing to the low lying 3d orbitals on silicon being polarized ($\delta+$) by the electron attracting chloride ligands. $(CH_3)_4Si$ is kinetically inert to hydrolysis at room temperature because the Si—C bonds are less polarized than the Si—Cl bonds.

Table 1.9 Stability of organometallic species to water

Organometallic	Stability, comments
R_2Hg, R_4Sn, R_4Pb	Only slightly soluble, stable, diffuse to atmosphere. Higher alkyls less stable and less volatile. Species generally hydrophobic and variously volatile
CH_3HgX	Stable, slightly soluble depending on X
$(CH_3)_n Sn^{(4-n)+}$	Soluble, methyltin units stable but made hexa- and penta-coordinate by H_2O, OH^-. Species are solvated, partly hydrolysed to various hydroxo species. At high pH poly-nuclear bridged hydroxo species form for $(CH_3)_2 Sn^{2+}$
$(CH_3)_3 Pb^+$	Soluble, hydrolysis as methyltins above. Also dismutates to $(CH_3)_4 Pb$ and $(CH_3)_2 Pb^{2+}$ at 20 °C
$(CH_3)_2 Pb^{2+}$	Soluble as for $(CH_3)_3 Pb^+$ above. Disproportionates to $(CH_3)_3 Pb^+$ and CH_3^+ slowly. These reactions cause eventual total loss of $(CH_3)_3 Pb^+$ and $(CH_3)_2 Pb^{2+}$ from water.
$(CH_3)_2 As^+$	Hydrolyses to $(CH_3)_2$ AsOH then to slightly soluble $[(CH_3)_2 As]_2 O$
$CH_3 As^{2+}$	Hydrolyses to $CH_3 As(OH)_2$, then to soluble $(CH_3 AsO)_n$
$(CH_3)_2 AsO(OH)$	Stable and soluble ($330\,g\,dm^{-3}$). Acidic p$Ka = 6.27$, i.e. cacodylic acid, dimethlyarsonic acid. Detected in oceans
$CH_3 AsO(OH)_2$	Stable and soluble. Strong acid p$K_1 = 3.6$, p$K_2 = 8.3$—methylarsinic acid. Detected in oceans
$(CH_3)_3 S^+$, $(CH_3)_3 Se^+$	Stable and slightly soluble
$(CH_3)_n SiCl_{4-n}$	Hydrolyse and condense but methylsilicon groupings retained
$(CH_3)_n Ge^{(4-n)+}$	Stable, soluble, have been discovered in oceans. Hydrolyse but $(CH_3)_n$ Ge moiety preserved
$(CH_3)_2 Tl^+$	Very stable, soluble, but not been detected as a natural environment product
$(CH_3)_3 AsO$, $(CH_3)_3 SbO$	Stable and soluble
$(CH_3)AsH$, $CH_3 AsH_2$	Insoluble, diffuse to atmosphere, air unstable
$(CH_3)SbO(OH)$	Stable and soluble. Detected in oceans
$CH_3 SbO(OH)_2$	Stable and soluble. Detected in oceans

Other species

Stable and insoluble: R_4Si, $(R_2SiO)_n$, $CH_3 \ldots H_3HgSeCH$, most C_6H_5Hg derivatives, $(CH_3)_2S$, $(CH_3)_2Se$, $(CH_3)_4Ge$, $(CH_3)_3B$

Unstable: CH_3Pb^+ (has not been detected in the environment), R_2Zn, R_2Cd, R_3Al, R_3Ga, $(CH_3)_6Sn_2$, $(CH_3)_6Pb_2$, $(CH_3)_5Sb$, CH_3Tl^{2+}, CH_3Cd^+, $(CH_3)_2Cd$, $(CH_3)_2Sb^+$, CH_3Sb^{2+}.

Solubility here refers to *air-free* distilled water, no complexing ligands. Range of solubilities is from mg dm^{-3} to g dm^{-3}. Data from references.

Reprinted with permission from this work, first edition.

1.8 STABILITY OF ORGANOMETALLICS TO LIGHT AND ATMOSPHERIC REAGENTS

The primary radiolytic decomposition process for organometallic compounds is electronic absorption leading to organic radical formation. The absorption may lead to d–d electronic transitions in the case of transition elements, or to charge transfer to or from metal orbitals. The former often causes the dissociation of metal–ligand bonds, and hence coordinative unsaturation, and the latter may facilitate nucleophilic attack at the metal. For organometallic compounds photo properties are more dependent on the wavelength of excitation radiation than is the case for organic compounds. Light stability is more relevant for volatile, i.e. R_nM, species as it is they that enter the atmosphere, and not the organometallic cationic derivatives that are complexed in sediments, water, etc. The most important of these are $(CH_3)_2Hg$, R_4Pb, R_4Sn, $(CH_3)_3As$, $(CH_3)_3 Sb$, and $(CH_3)_2Se$.

Photolysis of $(CH_3)_2Hg$ at 254 nm in the gas phase produces CH_3Hg^0 and CH_3^0 radicals, further reactions producing ethane and methane by hydrogen abstractions. At normal temperatures ethene is formed; methane occurs at higher temperatures. CH_3HgI in organic solvents at 313 nm forms CH_3^0 by breaking of the mercury–carbon bond. In the gas phase, however, the mercury–halogen bond breaks. Diphenylmercury in organic solvents is photolysed to $C_6H_5^0$ and also decomposes thermally [12].

Degradation of methyltin halides in water at about 200 nm was observed to produce inorganic tin via sequential degradation, [13] and irradiation of alkyllead compounds at 254 nm also leads to breakdown [14]. These wavelengths exist in the homosphere (see below) and hence these materials would be expected to decay if they volatilize to the atmosphere. Atmospheric fates are discussed in detail in the appropriate chapters.

Processes in the real atmosphere are more complex and, in general, lead to much reduced stability from that suggested by laboratory experiments. There is the additional presence of oxygen, other free radicals and surfaces on which enhanced decomposition will take place. Where this has been measured the lifetime of organometallics in the atmosphere may be in terms of hours or days, rather than years. The lifetime of $(CH_3)_4Pb$ in the atmosphere, for example, has been estimated as several days [15].

From Figure 1.2, laboratory processes using wavelengths shorter than 340 nm might be thought to be less relevant under normal conditions for the lower atmosphere as radiation of these wavelengths hardly penetrates to the earth's surface [16]. However, at up to 85 km the atmosphere is homogeneous (homosphere) [17] and volatile materials released into it will, if stable, eventually circulate to that height and be subject to interaction with radiation penetrating to levels below 85 km. A wavelength of 120 nm is equivalent to 998 kJ mol^{-1}; 240 nm is equivalent to 499 kJ mol^{-1} and 340 nm equals 352 kJ mol^{-1}. These are generally sufficient to homolyse metal–carbon bonds. The main absorbing medium at wavelengths below 340 nm is ozone, whose importance to biology is clear in view of the toxicity of short wavelength radiation. It might, therefore, be inferred that $(CH_3)_2Hg$, R_4Pb and R_4Sn, etc. will photolyse in the atmosphere to methyl radicals and $(CH_3)_nM^0$, and that further reactions to produce methane, ethane and other hydrocarbons will occur. However this is not the main decomposition process.

In addition to direct photolysis, reactions with other species produced by atmospheric photochemistry will also occur (e.g. $OH^\bullet$, O^3P and O_3). These

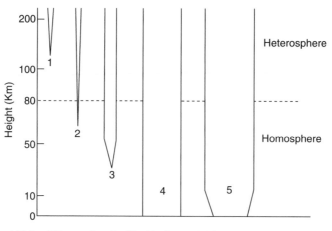

1 UVλ <120 nm, absorbed by N_2, O_2 2 UVλ = 120 180 nm, absorbed by O_2
3 UVλ =180 340 nm, absorbed by O_3, 4 UV-visλ = 340 700 nm
5 IRλ => 700 nm, slight absorption by O_2, H_2O (near IR)

Figure 1.2   Absorption of light in the atmosphere
Reprinted from this work, first edition, with permission

occur faster than alternative heterogeneous processes available after adsorption
on particles. $(CH_3)_4Pb$ has been most thoroughly investigated [18] from this
point of view (details are given in Chapter 4) but hydroxyl radical attack seems
to be the most important first step reaction [19] here, viz. Equations (1.15–1.17).
These chemical decomposition routes are much faster, in fact, than the photolytic decomposition processes described above when the real atmosphere is
considered.

$$(CH_3)_4Pb + OH^0 \longrightarrow (CH_3)_3PbCH_2^0 + H_2O \qquad (1.15)$$

$$(CH_3)_4Pb + O_3 \longrightarrow (CH_3)_3PbOOH + CH_2O \qquad (1.16)$$

$$(CH_3)_4Pb + O(^3P) \longrightarrow (CH_3)_3PbCH_2^0 + OH^0 \qquad (1.17)$$

In view of these being reactions of a general nature, i.e. abstraction of hydrogen from a methyl group; insertion of ozone into a metal–carbon bond, it is likely
that other volatile organometallics are decomposed chemically in the atmosphere
this way, and similar mechanisms are likely to exist for R_4Sn, etc.

Thermal homolysis of metal alkyls leading initially to radical species
$(R_nM^0 + R^0)$ produces the radicals in their ground energy states. Photolysis
produces products in higher excited energy states and, therefore, more energy is
required to decompose compounds than is needed by the thermal route.

Photochemistry is, however, more effective than thermal decomposition in
promoting permanent decomposition as the excited product fragments are less
likely to recombine. The energy required for photochemical dissociation varies

but for relevant organometallic molecules it will be of the order of 100–120 kJ mol^{-1} [20]. Taking D_1 values (Table 1.10) for mercury, tin and lead, we should expect energies around 320, 370 and 300 kJ mol^{-1} to be required for the photochemical decompositions of $(CH_3)_2Hg$, $(CH_3)_4Sn$ and $(CH_3)_4Pb$ respectively. This is easily accomplished by, for example, 240 nm solar radiation (equivalent to 499 kJ mol^{-1}) which penetrates down to below 50 km altitude.

The volatile organometallics considered here are heavy molecules and their mixing ability in the atmosphere and consequent exposure to solar radiation must be considered. Higher diffusion into the atmosphere will give greater exposure to stronger intensities of radiation and faster photolysis. Higher energy radiation at greater height will also produced higher energy fragments, which are less likely to recombine. Mixing to cloud base level (varying from 0.5 to 12 km) is a rapid but complex phenomenon dependent on seasonal factors, temperature, inversions, maritime influence and the amount of atmospheric pollution in the area. Diffusion to greater heights is a slower process. For a fairly similar mass molecule CCl_3F (MW 137.5 versus 178.7 for $(CH_3)_4Sn$), it has been calculated that 50 to 80 years elapse between the release of CCl_3F at ground level and photochemical destruction at over 25 km altitude. (Boltzman distribution is assumed, leading to varied lifetimes for molecules.)

It can be concluded that the photochemical processes alone would decompose organometallics into radical species and the metals but that this might take decades. However, it should also be considered that other processes are occurring in the atmosphere, particularly where it is polluted. Atmospheric chemistry is complex. Species involved include O_3, O (triplet ^3P), OH°, ClO_x°, HO_x°, NO_x°. All of these are highly reactive radicals particularly able to abstract hydrogen atoms from hydrocarbons or organometallics. In addition, particle-based heterogeneous decomposition may occur. Radical or singlet oxygen attack on organometallic species in polluted atmospheres will actually decompose

Table 1.10 Bond energies in alkylmetal compounds (in kJ mol^{-1})

Compound	Mean bond enthalpy, $\bar{D}^a$	D_1	D_2
$(CH_3)_2Hg$	123	218	29
$(C_2H_5)_2Hg$	101	179	25
$(iC_3H_7)_2Hg$	88	113	63
$(CH_2{=}CH)_2Hg$	141	202	80
$Hg(CN)_2$	302	517	94
	180	252	97^b
$(CH_3)_4Sn$	218	270	
$(CH_3)_4Pb$	155	205	
$(CH_3)_2Zn$	172	197	147^c
$(CH_3)_2Cd$	141	189	92^c

^{a} $\bar{D} = \frac{D_1 + D_2}{2}$ etc
^{b} Separate estimates
^{c} But note environmental instability despite $\bar{D}$ and D_1 values compared to $(CH_3)_2Hg$
Reprinted from this work, first edition, with permission.

these molecules rather than direct photochemical decomposition alone. It can be concluded that the lifetime of organometallic species in the atmosphere is of the order of hours or days rather than decades, with metals and hydrocarbons being produced. Actual measurements and estimates reinforce these conclusions (see, for example, Chapter 4).

Actual decay rates will vary widely and have been studied under environmental conditions, e.g. for lead where rates vary from 3 to 4% per hour in winter to 16 to 29% in summer for $(CH_3)_4Pb$ [15, 18, 19]. For $(C_2H_5)_4Pb$ the corresponding rates are 17–23 and 67–93%. These are upper limits but it can be seen that organometallics are not persistent in the atmosphere, with direct photolysis accounting for little of the decay. Even in the area of 340 nm radiation, where the atmosphere is fully transparent to light, the energy associated with the wavelength, (352 kJ mol^{-1}), is sufficient to break most metal-carbon bonds (Table 1.2) to produce ground energy state products. Shorter wavelengths are needed to produce higher energy state products, but normally *chemical* decomposition processes predominate [21].

For $(CH_3)_2Hg$ also, a relatively fast degradation by hydroxyl attack has been deduced, with elemental mercury being produced and the rate constant for hydroxyl attack calculated at 2×10^{11} cm^3 mol^{-1} s^{-1} [22]. Overall it should be assumed that the common organometallic species emitted to atmosphere degrade rapidly and cannot be considered a serious environmental contaminant in most locations owing to their low concentrations (see relevant chapters) and relatively rapid rates of decay. The role and fate of their non-organometallic products merges into the field of organic and metallic atmospheric chemistry in general. That organometallics have a source and a transport role for metals in the atmosphere is clear.

1.9 COORDINATION PREFERENCES OF ORGANOMETALLIC SPECIES IN THE ENVIRONMENT

The general coordination preferences for metal cations has been understood by the Hard and Soft Acid and Base Principle. Hard acid metals are small, have higher oxidation states and are not easily polarized—they bond preferably to hard donor bases. Soft acid metals are larger, have lower oxidation states and are more polarizable, and prefer to coordinate to soft donor bases.

Inorganic metals in the environment tend to follow this principle—hard acid metals such as magnesium, calcium, tin(IV), etc. are found in the earth's crust as oxides or carbonates, while softer metals such as mercury or lead are found in sulphides. Methyl groups are more electronegative than metals and generally withdraw electron density from metals (increasing hardness). Larger aliphatic groups are more electron releasing and polarizable than methyl groups and tend towards producing softer organometallic molecules overall, though the effect is small. Table 1.11 gives details.

Table 1.11 Hard and soft acids and bases

Hard acids	Borderline acids	Soft acids
$(CH_3)_n As^{(3-n)^+}$	$(CH_3)_n Pb^{(4-n)^+}$	CH_3Hg^+
$(CH_3)_n Ge^{(4-n)^+}$	$(CH_3)_n Sb^{(3-n)^+}$	$(CH_3)_3 Ti^{(3-n)^+}$
$(CH_3)_n Sn^{(4-n)^+}$	$R_n Sn^{(4-n)^+} (R > CH_3)$	

Hard bases (ligands) include $-NH_2$, $-OR$, $-OH$, OA_2, Cl^- etc.; soft bases include higher alkyl groups, $-SH$, $-SR$, I^- etc. and so $HgCH_3^+$ will be found bonded to sulphur where possible but chloride where necessary (sea water).

Taken from standard data in texts

Organometallic compounds in the aqueous and sediment environment will usually be stabilized by coordination to sediments and suspended particulate matter containing natural sulphur, by oxidation, or via nitrogen donor ligands which, by occupying coordination positions, will stabilize the metallic atom from attack. The fact that these ligands are often multidentate further increases stability owing to increased loss of entropy in the forward reaction and hence a more negative ΔG.

It should be pointed out that certain coordination reactions with oxygen and sulphur ligands [23–25] may, however, lead to *labilization*, not *stabilization*, through dismutation processes (see Chapters 2, 3 and 4). Examples are shown in Equations 1.18 and 1.19.

$$2CH_3Hg^+ + S^{2-} \longrightarrow (CH_3Hg)_2S \xrightarrow{\Delta,\ h\nu} (CH_3)_2Hg + HgS \qquad (1.18)$$

$$6(CH_3)_3Sn^+ + 3S^{2-} \longrightarrow 3((CH_3)_3Sn)_2S \xrightarrow{\Delta,\ h\nu} 3(CH_3)_4Sn + c((CH_3)_2SnS)_3 \qquad (1.19)$$

The driving force for these labilizations is the formation of stable products, e.g. HgS or $c((CH_3)_2SnS)_3$. In practice these processes have been shown to reduce the stability of the above organometallic species in the natural environment (see particularly Chapter 2), albeit increasing transport ability via the atmosphere.

1.10 STABILITY OF ORGANOMETALLIC COMPOUNDS IN BIOLOGICAL SYSTEMS

Details of the toxicity and fate of organometallic compounds in organisms are given in the appropriate chapters. Where toxicity exists, numerous routes for breaking down organometallic species are available to organisms. These vary in rate; methylmercury has a half-life in man of about 70 days, triethyllead of around 35 days in one human compartment and 100 days in another. Half-lives vary between species and organisms but mechanisms for the decomposition of the metal–carbon bond do exist. There are several common routes linking decay processes for a variety of organometallics in various organisms.

Decay usually takes place by dealkylation. Methylmercury may be decomposed by numerous bacteria, first to methane and mercury(II), and finally to the mercury(0) state. Arylmercury salts produce the arene and mercury(0). Full saturated tetramethyl organometallics (e.g. $(CH_3)_4M$, $M = Sn$, Pb) are easily absorbed but rapid decay by loss of methyl group to the more stable trimethyl metal form takes place. For tetramethyllead, this conversion occurs rapidly in the liver—in rat liver there is a half-life of minutes. The trialkyls decay at various rates, eventually giving the metallic ions in various yields. In physiological saline, $(CH_3)_2PbCl_2$ is broken down to lead(II) with a half-life of about 12 days. Half-lives for methyl species are usually longer than for ethyl and higher alkyl analogues, owing to the presence in the latter of facile routes to decay [22]. The main route to dealkylation for higher alkyl metals, however, does not lie in direct metal–carbon bond cleavage but usually occurs after initial hydroxylation of the by the liver [23].

Enzymatic dealkylation of metals has been widely reported [24]. It occurs for example in the dealkylation of butyl- and cyclohexyltin compounds (Chapter 3) and in the decay of the ethyllead moiety (Chapter 4). Decay is usually reported at the β carbon but it also occurs at the α carbon; attachment at the metal may also occur, for example, in the case of silicon where the initial silanol product condenses to a silane. Phenyl groups do not usually undergo enzymatic hydroxylation but it is known for the phenylsilicon grouping [25].

For lower alkyl groups the organic products have been reported as alkanes and alkenes; for ethyllead species the product is reported to be ethanol. The gasoline additive λ-$CH_3C_5H_4Mn(CO)_3$ is also attacked at the methyl position [26]. Hydroxylation is mainly carried out by liver microsomal monooxygenase enzyme systems [27]. From the above it can be seen that the importance of hydroxylation is the labilization it produces in the metal–carbon bond, allowing eventual dealkylation.

By contrast, it has been shown that organic arsenic compounds occurring naturally in seafood are transferred through the body with little change . The common organoarsenic compounds found in the environment, aquatic $(CH_3)_2As(O)(OH)$ and $CH_3AsO(OH)_2$, are very stable against biological decay routes in plants and animals $(CH_3)_3As^+CH_2COO^-$ is also very stable. Soil bacteria have the ability to demethylate methylarsenic compounds, giving carbon dioxide by an oxidative route. A lack of arsenic demethylation in animals may be seen in the context of organoarsenic compounds being less toxic than their inorganic counterparts. The reverse is the case for most organometals (e.g. those of mercury, lead or tin).

1.11 GENERAL COMMENTS ON THE TOXICITIES OF ORGANOMETALLIC COMPOUNDS

The toxicities of the important environmental organometallic compounds are dealt with in the relevant chapters. There is also a detailed discussion of the toxicity of organometallics and their interaction with living organisms in a recent book [28]. The toxicities of individual groups of organometallic compounds are separately discussed in a number of sources [29–32].

This section will present a brief summary of the main toxic mechanisms for environmental organometallics. In general, organometallic compounds are more toxic than the inorganic metal compounds from which they derive, often substantially so. Mercury, lead and tin obey this general rule, however, as noted, arsenic is an exception. Usually the toxic effects are at a maximum for the formal monopositive cations, i.e. the species derived by loss of one organic group from the neutral, fully saturated, organometallic, viz. R_3Pb^+, R_3Sn^+, CH_3Hg^+. It should be noted that the toxic effects of the saturated organometallics (R_4Sn, R_4Pb, etc.) usually derive from conversion to R_3M^+, etc., in organisms. In general alkyl groups are more toxic than aryl groups when attached to metals. The most toxic alkyl groups (attached to a common metal) vary from organism to organism, but methyl, ethyl and propyl groups tend to be the most toxic. For tin compounds toxicity of the organotin cations is at a maximum for the trialkyl series. For trialkyltins methyl and ethyl groups are the most toxic to mammals, with higher alkyl groups being more toxic to bacteria, invertebrates and fish; longer chain alkyl compounds are of lower toxicity to mammals.

Mechanisms of toxicity are varied, but good coordination to base atoms (e.g. S, O, N) on enzyme sites seems to be the main one. Coordination of the organometallic species (usually R_nM^{m+}) to the enzyme blocks the sites and prevents reaction with the biological substrate. Enzymes blocked in this way include lipoic acid (sulphur site), acetylcholinesterase (ester site) and aminolaevulinic acid dehydrase.

The other main result of organometal introduction into vertebrates is a diminution of the myelin coating of nerve fibres. In addition, water accumulation and oedema in the central nervous system may occur. Other toxic effects are due to coordination to non-enzyme sites (e.g. thiols, histidine residues of proteins, haemoglobin, cytochrome P_{450}, cerebral receptors). Haematopoietic bone marrow, immune and essential trace metal systems in organisms may be affected by organometallic compounds. Fundamental interferences with DNA, protein synthesis, mutagenicity and genotoxicity have also been reported [33].

The reason for the enhanced toxicity of organometallics over the inorganic derivatives lies with the existence of lipophilic or hydrophobic groups (R) on the same species also having a hydrophilic dipole. This allows transport in aqueous body fluids, and also solubility and transport through fatty tissue and cell walls by diffusion. As Lewis acids (Table 1.11) similar to inorganic metal species, there is good bonding to Lewis base coordination sites within the organism (e.g. thiol groups).

The main result of organometallic poisoning in vertebrates is damage to the central nervous system, leading to varied symptoms including coma, ataxia, hyperactivity, varied motor difficulties, speech problems and psychological–attitudinal changes. Within the most prevalent series of environmental organometallic compounds, triorganotin toxicity, for example, arises through disruption of calcium (Ca^{2+}) and mitochondrial functions, membrane damage, disruption of ion transport and inhibition of adenosine triphosphate (ATP) synthesis. Dialkyltins also inhibit oxygen uptake in mitochondria and inhibit α-ketoacid oxidation. Acute organic lead poisoning is chiefly focused on the central nervous system, with pathological changes to the brain being found, including neuron destruction and degeneration of nerve tracts. In general, serious but non-specific damage to the nervous system occurs in acute lead poisoning with organic lead compounds, with nerve cells in the hippocampus, reticular formation and cerebellum being particularly sensitive. At lower levels of poisoning incipient anaemia effects have been noted. Trialkyllead and -tin compounds, together with monomethylmercury, all bind to sulphide residues. Both trialkyllead and -tin compounds may destroy the normal pH gradient across mitochondrial membranes, thereby uncoupling oxidative phosphorylation.

The results of organomercury poisoning with respect to brain function are quite similar to those considered above. Rapid penetration of the blood–brain barrier leads to sensory disturbance, tremor, ataxia, visual and hearing difficulties. Methylmercury is lipid soluble, rapidly diffuses through cell membranes, and once it enters the cell is quickly bound by sulphydryl groups. It rapidly penetrates the blood–brain barrier. Methylmercury inhibits protein synthesis and RNA synthesis and causes particular damage to the developing brain.

A final general point is that the most subtle and early results of organometal poisoning cannot always be detected, particularly for lead and mercury. Here behavioural and psychological effects are shown and these are difficult to evaluate, especially in children. For instance, one of the earliest and most sensitive indices of methylmercury poisoning, paraesthesia, is very hard to measure and evaluate in children. In the case of lead, there is of course the debated and incompletely resolved case of low-level intellectual effects in children following (mainly) inorganic lead accumulation from organic lead gasoline additives. A particularly subtle effect of such cases is the difficulty in separating genuine toxic effects from other disease or socio-economic symptoms. These second level (or chronic) effects cause different problems in terms of evaluation and action than do acute high-concentration poisonings, and are an integral feature in the consideration of the environmental effects of organometallic compounds.

Discussion of the role of microorganisms in the oxidation, reduction and biomethylation of metals is discussed in greater detail in Section 1.13.

1.12 GENERAL CONSIDERATIONS ON ENVIRONMENTAL REACTIVITY OF ORGANOMETALLIC COMPOUNDS

A compound may be reactive if:

(i) the free energy of formation of products is negative,

(ii) the M—C bonds are polar,

(iii) the metal has empty low energy orbitals which a reagent may attack,

(iv) the metal has lone electron pairs (which may attach external environmental reagents or be used for coordination expansion),

(v) the metal—carbon bonds are weak (low bond energies),

(vi) the metal is coordinatively unsaturated, i.e. it can coordinate further ligands perhaps by using lone electron pairs. This coordinatively expanded species may be unstable, say, for steric reasons and lead to the decay of the entire assembly.

(vii) hydrogen atoms are present on the carbon atom β to the metal.

In general terms these effects produce low (activation) energy routes to decay.

1.13   MICROBIAL BIOTRANSFORMATION OF METALS AND METALLOIDS

1.13.1   INTRODUCTION

All microorganisms, whether prokaryotic[†] or eukaryotic[‡], interact with the various classes of metals. The alkaline earth metals (Ca and Mg) have structural and catalytic functions, whereas some other metals (vanadium, chromium, manganese, iron, cobalt, nickel, copper, zinc, molybdenum, tungsten) also have a catalytic function. The metalloid selenium is also known to have a catalytic function in certain microorganisms. In these roles, the concentrations of the elements required are very low and such interactions have biological rather than general environmental significance. Microorganisms can also transform metal and metalloid species by oxidation, reduction, methylation (alkylation) and demethylation (dealkylation). Many of these biotransformations have great environmental significance since changed chemical forms of metals can alter the physical and chemical properties, mobility and toxicity of an element. Some microbial biotransformations also have potential for bioremediation, e.g. of environmental sites contaminated with metal(loid)s.

In the following sections, some of the microbial transformations of metal (loid)s in each of four categories—reduction, oxidation, methylation and demethylation—are considered and summarized on an element by element basis. The intention is to illustrate the great diversity of metal(loid) biotransformations and their microbial catalysts. Where appropriate, comments relating to the physiological role, biochemical mechanism, environmental significance and bioremediation potential of the microbial biotransformations are included.

[†] Microorganisms lacking a nucleus, and other membrane-enclosed organelles, i.e. bacteria and archaea.

[‡] Organisms with membrane-bound nucleus, such as fungi, algae and protozoa.

1.13.2 REDUCTION AND OXIDATION

In cellular respiration, electron transport occurs from a growth substrate (e.g. glucose) to an electron acceptor. In this process protons are pumped across a membrane to create a proton motive force, which is then used to drive cellular processes directly or to generate biologically useful energy in the form of adenosine triphosphate (ATP) (Figure 1.3). In eukaryotic microorganisms, such as fungi and algae, the electron acceptor molecule is oxygen and the process is termed aerobic respiration. Some fungi can also ferment—a process by which ATP is generated by substrate level phosphorylation—and this does not involve molecular oxygen. It is the prokaryotes (bacteria and archaea), however, that are well adapted to survival and growth in anaerobic environments, with anaerobic respiration being the major mode of ATP generation. In this mode of metabolism, a variety of inorganic and organic compounds may serve as alternatives to oxygen as terminal electron acceptors in respiration.

Only prokaryotes can use metal(loid)s as electron donors and conserve energy for growth from the oxidation. Similarly, it is only prokaryotes that have the capability of conserving energy for growth through reduction of metal(loid)s. In this role the metal(loid) serves as a terminal electron acceptor in anaerobic respiration, with energy being generated via coupling of an electron transfer chain to oxidative phosphorylation. Since relatively high concentrations of the metal(loid) are required to support microbial growth, either through oxidation or reduction, such transformation occurs on a large scale in the environment. For some microbial/metal(loid) interactions, oxidation or reduction of the metal (loid) occurs but the organism does not conserve energy from the transformation, i.e. it is not linked to growth. In many such cases, the transformations are thought to serve as a detoxification reaction for the metal (loid).

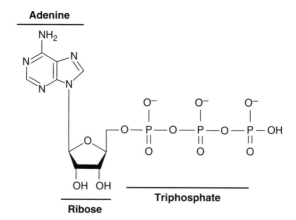

Figure 1.3 Adenosine triphosphate (ATP)

Several metal(loid)s are less soluble in the reduced state and microbial reduction can convert soluble metal(loid) species to insoluble forms that can more readily be removed, for example from contaminated waters. This offers the potential for bioremediation of metal(loid) contamination from polluted waters or waste streams. In the case of mercury, reduction of soluble Hg(II) forms volatile Hg(0), which is also regarded as a potential bioremediation strategy, albeit in a sense of transporting a problem elsewhere.

Some prokaryotes are also capable of producing volatile hydrides of metal (loid)s, such as arsine (AsH$_3$) and stibine (SbH$_3$). These transformations, although essentially bioreductions, will be considered alongside biomethylation since this process also generates volatile derivatives of metal(loid)s, including methyl hydrides

We now consider redox reactions on an element by element basis, as the oxidation state of a metal is relevant to the stability and transport of its organometallic forms.

Chromium

A wide range of bacteria have been shown to reduce Cr(VI) to Cr(III) [34, 35] enzymatically, including the facultative anaerobe *Enterbacter cloacae*, the anaerobe *Desulfovibrio vulgaris* and a cyanobacterium [36]. The reduction is thought to be at least a two-step reaction involving Cr(V). Although chromium-reducing bacteria are widespread within the natural environment the extent of Cr(VI) reduction is determined largely by environmental factors, such as availability of electron donors, pH, redox conditions and temperature. Chromium bioreduction has potential for remediation of waters contaminated with Cr(VI) since Cr(III) is much less toxic [37]. Cr(III) also becomes immobilized in anaerobic sediments by adsorption [38], which can be viewed as a form of bioremediation.

A variety of carbon growth substrates have been shown to provide electrons for Cr(VI) reduction, including a range of sugars (e.g. glucose, mannose, sucrose, lactose), amino acids, aromatic compounds and fermentation end products such as acetate, butyrate and propionate [39–41]. The type of carbon substrate utilized is highly dependent upon both the nature of the organism and the prevailing aerobic/anaerobic growth conditions. In addition, *D. vulgaris* has been shown to reduce Cr(VI) under anaerobic conditions utilising hydrogen as the electron donor [42].

The rate of Cr(VI) reduction by microbial systems is influenced by the initial concentration of the ion, although differing trends have been observed. In the case of *E. cloacae*, high initial Cr(VI) concentrations showed relatively low rates of reduction [43, 44], whereas for *E. coli* the highest rates have been obtained at high Cr(VI) concentrations [45]. With regards to both pH and temperature, the highest rates of Cr(VI) reduction coincide with the optima for growth rate of the organisms [41, 43]. The influence of redox potential on Cr(VI) is also organism dependent. For *Bacillus* spp, Cr(VI) reduction has been shown to occur over a wide range of redox potentials from about $+250\,mV$ to $-500\,mV$ [46], whereas no reduction of Cr(VI) by *E. coli* occurs at initial redox potentials greater than $-140\,mV$ [47].

Cr(VI) bioreduction can occur under either aerobic or anaerobic conditions, or under both, depending upon the organism [34, 35, 44]. Under aerobic conditions, reduction of Cr(VI) occurs through the action of a soluble reductase protein, utilizing NADH as electron donor [42, 48]. Under anaerobic conditions, Cr(VI) is thought to serve as a terminal electron acceptor, with electrons being passed via a soluble and/or membrane reductase [42, 49]. Cytochromes are also thought to be involved; viz. cytochromes *b* and *d* in *E. coli* [48] and cytochrome *c* in *Enterobacter cloacae* [50]. There is no firm evidence, however, that electron transport to Cr(VI) is linked to energy generation (i.e. growth) [34]. The physiological role of Cr(VI) reduction to Cr(III) has not been established and may be a detoxification mechanism or a fortuitous reaction [51].

Agrobacterium radiobacter and *E. coli* have been shown to reduce chromium in liquid media under either aerobic or anaerobic conditions, although lower rates of Cr(VI) reduction are observed in the presence of oxygen [48]. Similarly, Cr(VI) reduction by *E. cloacae* under anaerobic conditions is lower in the presence of sulphate and nitrate; this inhibitory effect was not observed for *E. coli* [43, 45].

There appear to be no reports of microbial oxidation of Cr(III) to Cr(VI).

Selenium

Bacteria can reduce both SeO_4^{2-} and SeO_3^{2-} to Se(0), but only the former reduction has been shown to support growth [34]. SeO_4^{2-}-respiring bacteria are thought to be ubiquitous in the natural environment [52]. Dissimilatory[†] SeO_4^{2-} reduction has been reported to be inhibited by NO_3^- but not by SO_4^{2-}, which suggests a role for nitrate reductase in dissimilatory selenium reduction [52]. However, the bacterium *Thauera selenatis* has been shown to grow anaerobically by reducing SeO_4^{2-} via specific reductases that are distinct from those involved in NO_3^- reduction [53–55]. Similarly, the facultatively anaerobic bacterium *Enterobacter cloacae* respires SeO_4^{2-} under anaerobic conditions and reduces NO_3^- concurrently [56, 57]. Washed cell suspensions of *E. cloacae* have also been shown to reduce SeO_3^{2-}, but this did not support growth. Interestingly, reduction of SeO_3^{2-} by this organism is inhibited by nitrate and nitrite but *stimulated* by sulphate. Bacterial reduction of SeO_4^{2-} to Se(0) is thought to be the major biological transformation for reduction of selenium in anaerobic sediment. Reduction of Se(0) to Se(II) (selenide) has been reported for *Thiobacillus ferrioxidans* [58].

[†] Reactions that are directly or indirectly linked to energy formation but the element is *not* incorporated into biomass

Oxidation of Se(0) has been shown to occur in soils, although the organisms responsible have not been identified [59, 60]. Losi and Frankenberger [61] showed that the oxidation occurred at relatively slow rates, was mainly biotic in nature, and produced SeO_3^{2-} alone or with SeO_4^{2-}. Oxidation of Se(0) by soils was enhanced by prior exposure to selenium. In liquid enrichment culture experiments, in the presence or absence of glucose or carbonate as carbon source, only SeO_3^{2-} was detected as a product of Se(0) oxidation. Growth was detected in treatments containing Se(0) as the sole energy source, which suggests that some bacteria may conserve energy from the oxidation of selenium.

Uranium

Certain bacteria can grow by using U(VI) as a terminal electron acceptor in respiration, forming U(IV). *Shewanella putrefaciens* couples oxidation of hydrogen to U(VI) reduction, while *Geobacter metallireducens* uses acetate as electron donor [62]. Both bacteria are known iron reducers [63] and U(VI) is thought to replace Fe(III) as the terminal electron acceptor in respiration. Conversely, several *Desulfovibrio* species reduce U(VI) but this is not coupled to growth [64, 65], i.e. energy is not conserved in this reductive transformation. Cytochrome c_3 has been shown to be involved in U(VI) reduction by *D. vulgaris* [66].

Abdelouas *et al.* [67] have reported on the biocatalysed reduction of U(VI) in groundwater in the presence of high concentrations of nitrate, sulphate and carbonate. Microbially mediated reactions were sequential in order of decreasing redox potential, with U(VI) reduction accompanying that of sulfate and sulfide. U(VI) precipitated as a U(IV) solid adhering to bacteria.

Arsenic

Microbial reduction of arsenate (As (V) to arsenite As III) has been demonstrated for several bacterial genera [68], some of which are known to reduce arsenite further to form arsine or dimethylarsine (see Section 1.13.3). *Micrococcus aerogenes* and cell extracts of *M. lactilyticus* reduce arsenate to arsenite [69], while *Methanobacterium* strain MoH can reduce arsenate to dimethylarsine [70]. Certain *Alcaligenes* spp and *Pseudomonas* spp are known to reduce arsenate and arsenite to arsine (AsH_3) [71]. Since arsenate reduction in these strains has not been linked to growth, the physiological role of arsenate-to-arsenite reduction is likely to be detoxification only. Indeed, the only known means of microbial redox arsenate detoxification is reduction to arsenite followed by its expulsion. Studies on bacterial arsenate detoxification have mainly involved *Escherichia coli* and *Staphylococcus aureus*. The reductive detoxification enzymes (ArsC enzymes) are small (15 kDa molecular mass) monomeric protein being located in the cytosol[†] [72, 73]; the enzymes are inducible[‡] and encoded by plasmid[§] located genes [72, 74]. Reductive detoxification occurs in both aerobic and anaerobic environments, although the extent to which this mechanism contributes to overall arsenate reduction in relevant natural environments has not been estimated.

Arsenic respiration is now thought to be widespread in the environment and can be performed by diverse bacteria [75]. Bacteria from a reed bed [76], pristine freshwater and saline environments [77], surface sediments of a lake [78] and marsh-land [79] have been shown to grow by reducing arsenate. Isolated arsenic respirers include the Gram-positive *Desulfotomaculum auripigmentum*, capable also of respiring sulphate but not nitrate [80], and the Gram-negative *Chrysiogenes arsenatis*, capable of respiring nitrate but not sulphate [76].

Bacteria able to oxidize arsenite to arsenate have been isolated, including *Bacillus* and *Pseudomonas* spp [68]. The oxidation is thought to be a detoxification mechanism since bacteria are an order of magnitude more resistant to arsenate than to arsenite. Studies on a strain of *Alcaligenes faecalis* able to oxidize arsenite indicate that the transformation consumes oxygen, is induced by arsenite and does not yield utilizable energy [81]. Oxygen was shown to be the terminal electron acceptor, since respiratory inhibitors prevented further oxidation of arsenite. Heterotrophic bacteria are thought to play a major role in detoxifying the environment, catalysing up to 90 % of arsenite transformation to arsenate [82].

Mercury

A variety of bacterial genera can reduce inorganic mercury (Hg(II)) to metallic mercury (Hg(O)), including *Cryptococcus, Pseudomonas, Staphyloccus* spp and a range of enteric bacteria [83–85]. The reduction occurs during aerobic growth and is thought to be a detoxification mechanism; the ability to reduce Hg(II) is often correlated with mercury resistance. The enzyme involved in mercury resistance— mercuric reductase (MR)—is a NADPH-dependent flavoprotein located in the cytosol [84, 86], which transfers two electrons to Hg(II) (Equation 1.20)

$$NADPH + RS—Hg—SR + H^+ \longrightarrow NADP^+ + Hg(0) + 2RSH^-$$ (1.20)

Activity of the enzyme is enhanced by addition of thiols (R—SH) such as dithiothreitol and glutathione. The product Hg(0) is volatile and is released from the cell; it is much less toxic to the microorganism and to humans as compared to Hg(II).

Mercury resistance has been intensively studied in the bacterium *Pseudomonas aeruginosa*, where the genes for mercury resistance (*mer* genes) are located on a plasmid. The *mer* genes are arranged in an operon under the control of a regulatory protein. Hg(II) forms a complex with MerR, which activates

[†] Cellular content inside the cytoplasm, excluding membrane bound organelles.
[‡] Synthesis of enzymes stimulated by presence of ions/molecules (inducers) such as arsenate.
[§] An extrachromosomal genetic element that is not essential for growth and has no extracellular form.

transcription of the *mer* operon,[†] leading to synthesis of other Mer proteins, including mercuric reductase and a membrane transport protein for Hg(II). The role of the *mer* operon in Hg(II) resistance is illustrated in Figure 1.4.

Iron

Microbial dissimilatory reduction of Fe(III) to Fe(II) is thought to catalyse most Fe(III) reduction that occurs in sedimentary environments. A diverse range of bacterial genera have been associated with Fe(III) reduction, including *Desulfovibrio, Desulfobacter, Geobacter, Aeromonas, Thiobacillus, Rhodobacillus, Clostridium, Shewanella* and *Bacillus* [87]. This form of anaerobic respiration can be coupled to the oxidation of a wide range of both organic (fatty acids, aromatic compounds, some amino acids) and inorganic electron donors [87], a feature arising from the high reduction potential of the Fe(III)/Fe(II) couple (0.77 mV). Since Fe(II) is a more soluble form of iron, dissimilatory bacterial reduction solubilizes iron from rocks and soils where Fe(III) is found as one of the most common metals.

A few bacteria can obtain energy from the oxidation of Fe(II) to Fe(III); the so called 'iron bacteria' [88]. Since only a small amount of energy is conserved from this oxidation, the iron bacteria must oxidize large amounts of iron to grow and the ferric iron forms insoluble ferric hydroxide. *Thiobacillus ferrooxidans* is the best-known species and grows aerobically in acid mine drainage waters where the low pH stabilizes Fe(II) to chemical oxidation. Because the reduction potential of Fe(III)/Fe(II) is high (+0.77 V at pH 3) the path of electrons to oxygen is very short and the iron bacteria exploit the pre-existing proton gradients of their environment to generate ATP. The respiratory chain of *T. ferrooxidans* includes a copper-containing protein (rusticyanin) that accepts electrons from Fe(II) and donates electrons to membrane bound cytochrome *c*, with subsequent transfer via cytochrome a_1 to oxygen. The natural proton motive force across the *T. ferrooxidans* membrane replenishes the protons used to reduce $0.5O_2$ via ATPase, generating ATP. Some aerobic iron-oxidizing bacteria—such as *Gallionella ferruginea* and *Sphaerotilus natans*—exist at near-neutral pH, but only at the interfaces between anoxic and oxic conditions where Fe(II) and O_2 coexist. In entirely anoxygenic environments Fe(II) is not oxidized abiotically at neutral pH, but can be oxidized by the purple bacteria. In these phototrophic bacteria, Fe(II) is used as an electron donor to reduce cytochrome *c* in the photosystem.

Manganese

Microorganisms able to reduce manganese are ubiquitous in the natural environment. They include fungi and both anaerobic and aerobic bacteria

[†]   A cluster of genes whose expression is controlled by a single operator.

(a)

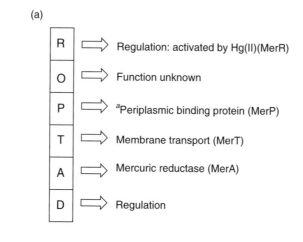

(b)

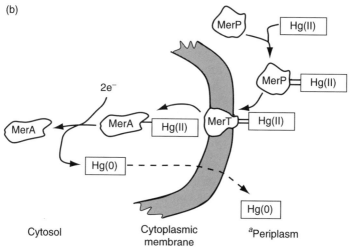

Figure 1.4 Hg(II) reduction to Hg(0) in *Pseudomonas aeruginosa*. (a) The *mer* operon governing expression of Mer proteins. (b) Involvement of Mer proteins in transport and reduction of Hg(II)

[a] Area between the cytoplasmic membrane and the cell wall in certain (Gram negative) bacteria
Adapted from Brock, *Biology of Microorganism*, 9th edn, Madigan, M. T., Martinko, J. M., and Parker, J., Prentice Hall Int. Inc. with permission.

[89, 90]. Many different genera of bacteria have been shown to reduce manganese, including *Pseudomonas, Bacillus, Corynebacterium* and *Acinetobacter* [89].

Manganese can be reduced indirectly by spontaneous chemical reactions arising from bacterial metabolism. For example, manganese oxides are reduced by sulfur-reducing bacteria, such as *Desulfovibrio* spp, which generate sulfide as an end-product of dissimilatory sulfur reduction [91, 92]. Similarly, iron-reducing bacteria can reduce Mn(IV) oxides through the formation of Fe(II) [93, 94]. Methane oxidation by methylotrophic bacteria is also coupled to manganese oxide reduction [95].

Direct enzymatic reduction of Mn(IV) by cell extracts *of Acinetobacter calcoaceticus* has been demonstrated [96], although involvement of an assimilatory nitrate reductase or a distinct Mn(IV) reductase has not been resolved. Dissimilatory Mn(IV) reduction is known to occur in certain facultatively anaerobic bacteria, such as *Geobacter metallireductens* [97]. In this process, manganese reduction is coupled to the oxidation of a non-fermentable carbon source (e.g. lactate), growth indicating that Mn(IV) serves as a terminal electron acceptor in anaerobic respiration. Inhibitors of electron transport have also been used to demonstrate the involvement of Mn(IV) in anaerobic respiration [89]. Studies using mutants lacking ferric and nitrate reductases demonstrate the involvement of a specific manganese reductase [90].

Various bacteria, algae and fungi oxidize Mn(II) [98]. The range of Mn(II) oxidizing bacteria is particularly diverse and includes common Gram-negative (e.g. *Arthrobacter, Bacillus*) and Gram-positive (e.g. *Pseudomonas, Vibrio*), sheathed bacteria (e.g. *Lepothrix, Vibrio*) and budding and appendaged bacteria (e.g *Caulobacter, Pedomicrobium*). Unequivocal evidence that bacteria can gain energy for growth through Mn(II) oxidation has not yet been obtained and the role of biotic Mn(II) oxidation is thought to be a protection against the toxicity of Mn(II) or that of the oxidant. Both indirect and direct mechanisms of manganese biooxidation are thought to exist. Indirect oxidation may occur through biological generation of hydrogen peroxide or oxygen free radicals [98, 99]. These indirect mechanisms of manganese oxidation are thought to protect aerobic bacteria, such as *Leptothrix* and *Arthrobacter*, from the toxic effects of strong oxidants produced during aerobic respiration. Direct oxidation of manganese involves binding of the element to oxidizing proteins. An intracellular protein is involved in *Pseudomonas* spp [100], whereas in *Leptothrix* spp [91] and *Bacillus* spp [101] an extracellular protein and a spore[†] protein have been implicated respectively. Other cellular components, such as the cell wall [102] and exopolysaccharides [103] are thought to be nucleating sites for manganese oxidation. Direct and indirect oxidation of soluble Mn(II) forms relatively insoluble Mn(III) or Mn(IV), which can precipitate around microbial cells.

1.13.3 METHYLATION AND ALKYLATION

Bioalkylation refers to the process whereby living organisms cause direct linkage of alkyl groups to metals or metalloids, thus forming metall(oid)–carbon bonds. The methyl group is the most common alkyl group transferred in a process termed biomethylation. This process has been extensively studied

---

[†] Resistant resting structure.

in nature and biomethylation activity has been found in soil, but mainly occurs in sediments from environmental waters such as estuaries, harbours, rivers, lakes and oceans. The attachment of a methyl group to a metall(oid) changes the chemical and physical properties of that element, which in turn influences toxicity, mobility and geological cycling. The organisms responsible for metall(oid) biomethylation are almost exclusively microorganisms. Anaerobic bacteria are thought to be the main agents of biomethylation in sediments and other anoxic environments. Some aerobic and facultatively anaerobic bacteria, as well as certain fungi and lower algae, have also been shown to be capable of metall(oid) biomethylation. With the exception of their cobalt content as vitamin B_{12}, *higher organisms* do not seem to be capable of biomethylating true metals. The situation is different, however, for the metalloids arsenic, selenium and tellurium, and many higher organisms have been shown to form methyl derivatives of these elements, for instance methylarsenicals are formed in a wide range of organisms, including marine biota and mammals (including man).

Volatile methyl and hydride derivatives of metal(loid)s have been found in gases released from natural and anthropogenic environments [104–107] (e.g. geothermal gases, sewage treatment plants, marine sediments, landfill deposits), and biological production of volatile compounds is thought to be a significant part of the biogeological cycles of metal(loid)s such as arsenic, mercury, selenium and tin. With the exception of arsenic and selenium (and perhaps antimony), biomethylation increases toxicity because methyl derivatives are more lipophilic and therefore more biologically active.

Bioalkylation (the addition of a carbon bonded counter ion, e.g. riboside) is much rarer than biomethylation and has only been found for those metalloids capable of being biomethylated by higher organisms, i.e. arsenic, selenium and tellurium. Bioalkylation of arsenic has been studied most extensively and occurs mainly in marine species, with up to 96% of cellular arsenic being present in alkylarsenical (riboside) forms in certain algae. Conversion of inorganic arsenic to other arsenic–carbon containing organoarsenicals, such as arsenolipids and arsenosugars, is thought to be a detoxification mechanism (see Chapter 5, Section 5.6). Bioalkylation may exist for mercury, lead, tin, etc., but no systematic search for a carbon containing counter ion has been made.

1.13.4 MECHANISMS OF BIOMETHYLATION

Biomethylation (i.e. methyl transfer) in organic molecules—such as proteins, nucleic acid bases, polysaccharides and fatty acids—occurs in all living cells and is an essential part of the normal intracellular metabolism. The three main biological methylating agents for organic molecules—*S*-adenosylmethionine (SAM), methylcobalamine and *N*-methyltetrahydrofolate—have all been shown to be involved in the biomethylation of metall(oids) (Figure 1.5). All three methylating agents have an extensive chemistry and have been studied intensively in relation to methylation of organic molecules. SAM is considered a ubiquitous methylating agent. It is synthesized by the transfer of an adenosyl group from ATP to the sulphur atom of methionine (Figure 1.6).

The positive charge on the sulphur atom activates the methyl group of methionine, making SAM a potent methyl carbonium ion donor. The methyl group in biochemistry is transferred as an intermediate radical ($CH_3^{\bullet}$) or as a carbonium ion (CH_3^+) and thus the atom receiving the methyl group must be a nucleophile, which requires an available pair of electrons in the valency shell. In this oxidative addition process there is alternation of M^{n+} and $M^{(n+2)+}$ oxidation states. The reaction is often referred to as the Challenger mechanism, after Frederick Challenger who first formulated it for methylation of arsenic [108, 109] (see Figure 1.7 for the Challenger scheme). SAM has subsequently been also shown to be the methylating agent for the metall(oids) selenium, tellurium, phosphorus [110] and antimony [111], all of which have an available lone pair of electrons. N-Methyltetrahydrofolate (Figure 1.5), like SAM, is thought to transfer methyl groups as carbonium ions or as an intermediate radical, but its transfer potential is not as high as SAM. Conversely, for N-methylcobalamin (Figure 1.5)—a derivative of vitamin B_{12}—the methyl group is transferred as a carbanion (CH_3^-) and the recipient atoms must be electrophilic. Methylcobalamin is well established as a methylating agent for mercury and is also involved in methylation of lead, tin, palladium, platinum, gold and thallium [110]. Although methylcobalamin appears to be the sole carbanion-donating natural methylating agent, in the natural environment the carbanion can also be transferred to metals from other organometallic species that may be present, such as $(CH_3)_3Pb^+$ or $(CH_3)_3Sn^+$. Regardless of the methylating agent, only one methyl group is transferred to metal(loids) in each step, although further methyl groups may then be transferred to the same receiving atom.

(a)

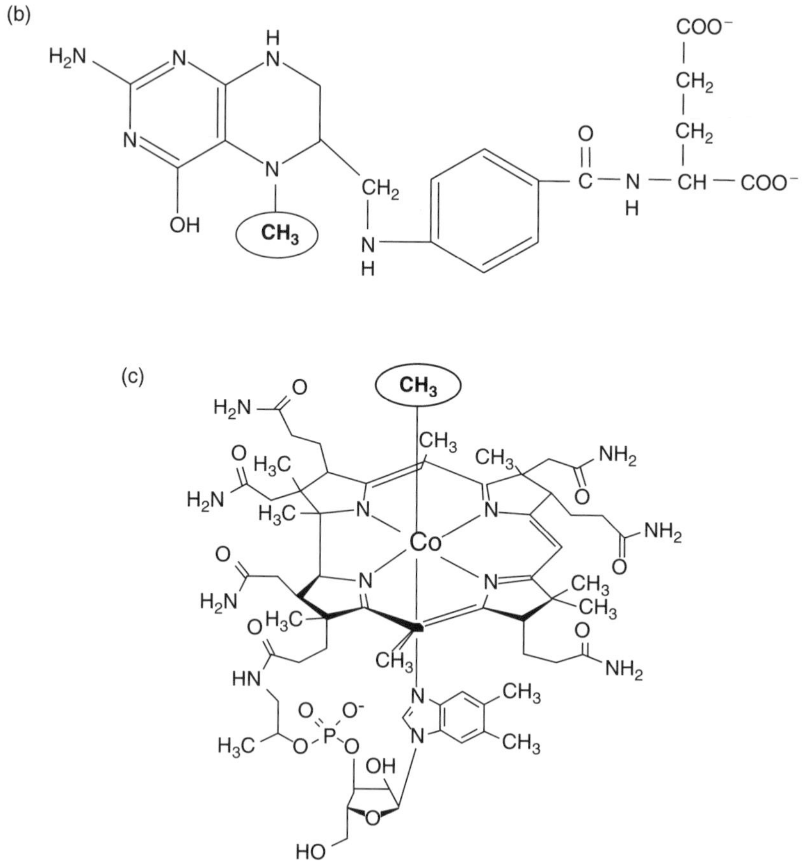

Figure 1.5 Biological methylating agents involved in methylation of metall(oids). (a) *S*-Adenosylmethionine. (b) *N*-Methyltetrahydrofolate. (c) Methylcobalamin. The donated methyl groups are circled

Mercury

Biomethylation of mercury has been studied extensively. One reason for this is that methylation of mercury enhances lipid solubility and thus enhances bioaccumulation in living organisms. This has led to mercury entering the food chain and to severe mercury poisoning incidents, e.g. in Japan [112]. Many different bacteria in aerobic and anaerobic environments are known to catalyse mercury biomethylation, although anaerobic sediments are the main sites of environmental methylmercury formation [110]. Sulphur-reducing bacteria, such as *Desulfovibrio desulfuricans*, are thought to be the most important methylators of mercury [113–115]; both low pH and high sulfate concentrations promote mercury biomethylation activity within environmental sediments [116, 117].

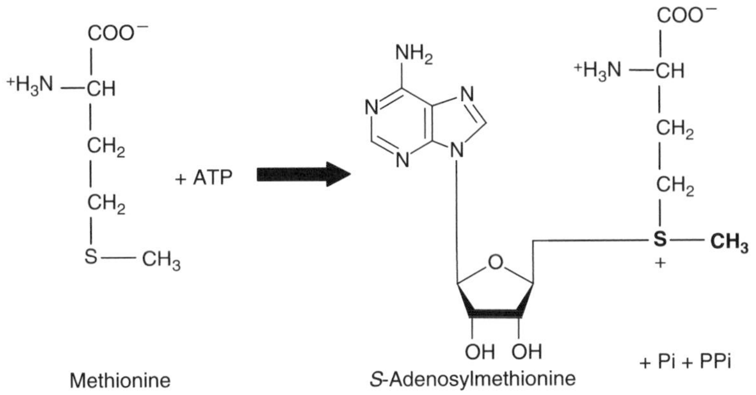

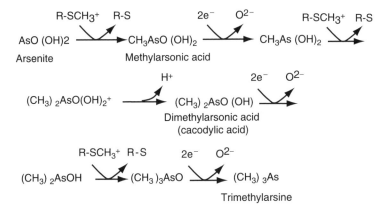

Figure 1.6   Biological synthesis of S-adenosylmethionine, involving ATP

Figure 1.7   Scheme for biomethylation of arsenic. The methyl donor, S-adenosylmethionine, is illustrated as $R\!-\!SCH_3^+$.

As mentioned earlier, methylcobalamin is the methyl donor, giving rise to monomethylmercury and (in a slower step) dimethylmercury (volatile) products (Equations 1.21, 1.22).

$$CH_3CoB_{12} + Hg^{2+} \longrightarrow CH_3Hg^+ + H_2OCoB_{12}^+ \qquad (1.21)$$

$$CH_3CoB_{12} + CH_3Hg^+ \longrightarrow (CH_3)_2Hg^+ + H_2OCoB_{12}^+ \qquad (1.22)$$

Further reactions of methylmercury can occur via various abiotic mechanisms (e.g. disproportionation reaction involving H_2S) also giving rise to dimethylmercury (see Section 1.9).

The key factor governing the concentration of mercury in biota is the concentration of methyl mercury in environmental waters, which is determined by the relative efficiencies of the methylation and demethylation processes. Methyl mercury in the environment is discussed in detail in Chapter 2.

Arsenic

A great diversity of microorganisms can methylate arsenic and in the natural environment methylarsenic compounds are widely produced, under both aerobic and anaerobic conditions [118]. The importance of fungal methylation of arsenic dates back to the second half of the nineteenth century, when several poisoning incidents in England and Germany were associated with arsenic-containing wall coverings. Gosio, in 1901, reported that a garlic-smelling, methylated arsenic compound was released from moulds growing in the presence of inorganic arsenic. Challenger and coworkers [108], in 1933, identified the gas as trimethylarsine $((CH_3)_3As)$. Several fungi have been shown to methylate arsenic under aerobic conditions, including filamentous *Scopulariopsis brevicaulis, Penicillium* spp, *Gliocladium roseum* and the yeast *Cryptococcus humicola* [109, 119, 120]. As mentioned earlier, the mechanism of arsenic methylation in fungi was established by Challenger [109] and involves a series of reductions and methylations, with SAM as methylating agent (Figure 1.7).

Certain bacteria have also been shown to methylate arsenic under aerobic conditions, including *Flavobacterium* spp, *Escherichia coli* and *Aeromonas* spp. Such bacteria-catalysed transformations can lead to di- and trimethylarsine derivatives, which have been widely found in environmental waters [121].

Arsenic biomethylation can also be catalysed by obligately anaerobic bacteria. *Methanobacterium* spp. can reduce AsO_3^- to AsO_2^-, with subsequent methylation to methylarsonic acid, dimethylarsenic acid and dimethylarsine [121]. Methylcobalamin is the methyl donor for arsenic methylation by the methanogenic archaea [70]. Until recently, bacterial methylation of arsenic under anaerobic conditions was thought to stop at dimethylarsine $((CH_3)_2AsH)$ [121]. However, Michalke *et al.* [122] reported trimethylarsine in the gas phase above anaerobic cultures of *Methanobacterium formicicum, Clostridium collagenovorans* and two *Desulfovibrio* spp. For *C. collagenovorans* and *D. vulgaris*, trimethylarsine was the sole volatile arsenic species detected. Conversely, *M. formicicum* volatilized arsenic as mono-, di- and trimethylarsine and as arsine (AsH_3), while *Methanobacterium thermoautotrophicum* produced only arsine. These data serve to illustrate the high dependence of organism type on speciation of metal(loid) microbial transformations.

The biomethylation of arsenic is thought to serve as a detoxification process, since methylated forms of arsenic are far less toxic than arsenic(V) or its salts, etc. Further, methylation facilitates arsenic removal from living terrestrial organisms by excretion as water-soluble forms, such as methylarsonic acid and cacodylic acid, or by volatilization as methylarsine derivatives [110, 121]. Higher animals, including mammals, are thought to be able to methylate arsenic enzymatically [123], although the contribution from intestinal bacteria to formation of methylated arsenic compounds and thus to arsenic detoxification in higher animals is unclear. These aspects, and the extensive chemistry of natural product arsenoriboside in the marine and terrestrial environments, are discussed in Chapters 5 and 6.

Antimony

Mono- and dimethylantimony species have been found in both marine and terrestrial natural waters [124, 125]. Freshwater plants from two Canadian lakes influenced by mine effluent, and plant material from an abandoned antimony mine in the UK, have also been shown to contain methylantimony species [126, 127].

Biomethylation of inorganic antimony by the aerobic fungus *Scopulariopsis brevicaulis* is well documented [128–133] and is thought to involve methyl transfer via SAM [111]. Recently, the wood rotting fungus *Phaeolus schweinitzii* has also been shown to biomethylate antimony [134]. Undefined mixed cultures of bacteria growing under anaerobic conditions have been shown to generate volatile trimethylantimony [$(CH_3)_3Sb$] as the sole volatilized antimony species [135–137]. Volatilization of antimony from environmental sediments and municipal waste dumps also suggests that certain anaerobic and/or facultatively anaerobic bacteria can biomethylate [138, 139] this element. Recently, Michalke *et al.* [122] reported on the biomethylation of inorganic antimony by pure cultures of anaerobic bacteria. *C. collagenovorans, D. vulgaris* and three species of methanogenic archaea, were shown to produce trimethylantimony in culture headspace gases. Interestingly, *M. formicicum* was shown to generate stibine (SbH_3) into culture headspace gases, together with mono-, di and trimethyl antimony; previously only trimethyl antimony was detected as a biovolatilization product of antimony in headspaces of soil enrichment cultures and of pure cultures of *S. brevicaulis*. These authors [122] also reported the formation of trimethyl bismuth by pure cultures of *M. formicicum*, which may provide a microbial basis for the presence of this volatile compound in gases from municipal waste deposits and sewage gases [104, 105]. Environmental properties of organoantimony and-bismuth species are discussed in Chapters 7 and 8.

Selenium

The chemistry of arsenic and selenium are similar in many regards, and biomethylations of these two elements have much in common. Several fungi, including *S. brevicaulis* and a *Penicillium* sp, biomethylate selenium using

SAM as a methyl donor [140, 141]. Selenite, selenate and inorganic selenide have been shown to be biomethylated to the trimethyl selenium cation, via a pathway involving dimethylselenone [142] (Figure 1.8).

Some obligately aerobic bacteria have also been shown to methylate selenenium and, in the natural environment, anaerobic bacteria are likely to contribute to this process [142]. The capacity to biomethylate selenium in microorganisms appears to be related to inorganic selenium resistance. The methylated forms of selenium are highly volatile and, like arsenic, biomethylation of selenium facilitates removal of selenium either as a water-soluble form or through (hydrophobic) volatilization. Microbial selenium methylation has been used successfully for *in situ* bioremediation of water and soil, resulting in removal of inorganic selenium (e.g. SeO_4^-) through volatilization [143].

Microbial demethylation/dealkylation

For some metals, such as mercury and tin, microbial demethylation or dealkylation of organometallic forms are important reactions in detoxification mechanisms. Bacterial Hg(II) resistance, for example, involves reduction of Hg(II) to Hg(0) via mercuric reductase (MR). The reduced form of the metal is less toxic, more volatile, and is rapidly removed to the atmospheric compartment. Detoxification of organomercurials proceeds via organomercurial lyase (OL), a product of the *Mer* B gene, that enzymatically cleaves the Hg—C bond to form Hg(II), which is then removed via mercuric reductase [142, 144] (Equations 1.23 and 1.24). For example:

$$CH_3Hg^+ \xrightarrow{OL} CH_4 + Hg(II) \xrightarrow{MR} Hg(O) \qquad (1.23)$$

$$C_2H_5Hg^+ \xrightarrow{OL} C_2H_6 + Hg(II) \xrightarrow{MR} Hg(O) \qquad (1.24)$$

Bacterial oxidative demethylation of methylmercury, involving liberation of CO_2, also occurs. The mechanism is thought to involve enzymes associated with bacterial metabolism of C-1 compounds [145] and is believed to occur

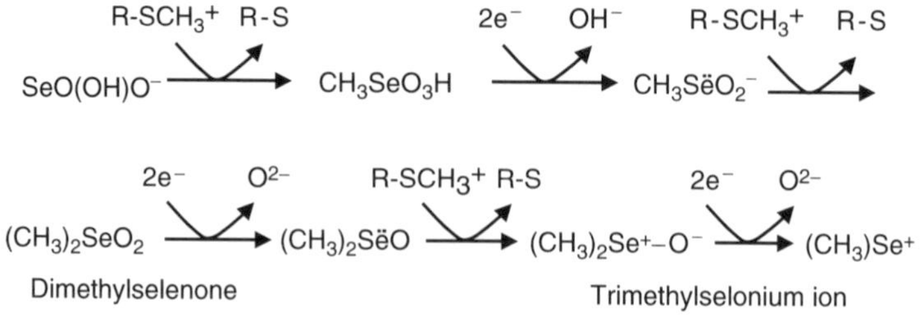

Figure 1.8 Scheme for biomethylation of selenium. The methyl donor, *S*-adenosylmethionine, is illustrated as $R\!-\!SCH_3^+$

widely in freshwater and aqueous environments, under both aerobic and anaerobic conditions.

Degradation of organotin compounds has been demonstrated for a wide range of microorganisms, including bacteria, fungi and algae [146]. Organotin degradation (Equation 1.25) is thought to involve sequential cleavage of tin–carbon bonds, resulting in removal of organic groups and a general reduction in toxicity [147–150]. Cleavage is initiated by hydroxo attack.

$$R_4Sn \longrightarrow R_3SnX \longrightarrow R_2SnX_2 \longrightarrow RSnX_3 \longrightarrow SnX_4 \qquad (1.25)$$

The wood preservatives tributyltin oxide (TBTO) and tributyltin napthenate (TBTN) have been shown to be degraded by fungal action to di- and monobutyltins. Certain Gram-negative bacteria and the green alga *Ankistrodesmus falcatus* can also dealkylate tributyltins, giving rise to dibutyl, monobutyl and inorganic tin products; the alga was able to metabolize around 50% of the accumulated tributyltin to less toxic dimethyltin(IV) over a 4-week period [151]. Similar end-products are formed by the action of soil microorganisms on triphenyltin acetate [152]. Tin–carbon bonds are also cleaved abiotically, for example by UV light, and it has proved difficult to establish the relative importance of abiotic and biotic mechanisms of organotin degradation in the natural environment [146]. In some circumstances, environmental conditions (such as pH and redox potential), established by microbial activity, greatly influence the extent of abiotic degradation of organotins [146]. These comments relating to abiotic and biotic mechanisms of organotin degradation also apply to other organometal(-loids). Environmental organotin chemistry is discussed in Chapter 3.

Bacterial demethylation of methylarsenicals is known to occur in aerobic aqueous and terrestrial environments, giving rise to CO_2 and arsenate [153]. Several commmon soil bacteria, such as *Achromobacter, Flavobacterium* and *Pseudomonas*, have been shown to possess organoarsenical demethylation ability. Bacterial demethylation of methylarsenic acids excreted by marine algae is an important part of the biogeological cycling of arsenic.

Dimethylselenide has been shown to be demethylated in anaerobic environments by methanogenic and sulfate reducing bacteria, with the liberation of CO_2 and CH_4 [78]. Several bacterial isolates form aerobic soils, including members of the genera *Pseudomonas, Xanthomonas* and *Corynebacterium*, have been shown to utilize methylselenides as their sole source of carbon [154].

1.14 CORRELATION OF QUANTITATIVE STRUCTURE–ACTIVITY RELATIONSHIPS WITH ORGANOTIN MOLECULAR TOPOLOGY AND BIOACTIVITY

The concept of correlating the activity of a compound to one or more descriptors of the molecule is not new. Earlier correlations have mainly focused on organic systems. These types of correlation have been used extensively by the drug industry. However, these types of relationship are also applicable to organomet-allic molecules, and organotin systems will be used to illustrate the concept here.

In the past several decades, there has been a dramatic rise in the production of organotin chemicals. The production of organotin compounds was 2000 tons in 1960 [155] and rose to more than 50 000 tons in 1983 [156]. In fact, there are more commercial uses of organotins than of any other organometallic system [157]. The increased usage of organotins is undoubtedly related to the diverse biocidal properties possessed by organotin compounds, discussed in more detail in Chapter 3. It is well established that the toxicity of organotin compounds is dependent on both the nature and the extent of alkylation or arylation of the tin atom (see Chapter 3).

The increased production of these chemicals as well as the formulation of new organotin compounds has led to increased concern about the fate of these com-pounds and their degradation products as environmental pollutants. Thus, the development of reliable correlations between the toxicities of organotins and some descriptor or descriptors of the molecule is of great value. It will allow the predic-tion of toxicities of new or untested compounds, and promote quantitative mo-lecular designed prospects for safer industrial and medical applications.

A common technique used for relating structures of molecules to their toxicological activities is to develop a quantitative structure–activity relation-ship (QSAR). The quantitative structure–activity relationship is a regression equation that relates some measurable biological activity of a compound to some molecular descriptor or descriptors of the chemical. Molecular descriptors can be classified into four groups: physicochemical, topological, geometrical and electronic.

The various chapters of this work show that liquid chromatography provides a basis for both theoretical and practical assessments of fragmental or additive substituent contributions to molecular solution behaviour, including organo-tins as trace cations or neutral species. However, the introduction of a more complete approach using the molecule's entire geometrical or electronic config-uration was even more appealing [158]. This type of approach has been used in determining biological response in organic drug design [159]. With the rapid advances in current computer technology, the calculations of various topo-logical descriptors for organometallic compounds is now commonplace. Among the leading topology descriptors are:

(i) Molecular connectivity. This is a bond-centred summation of electronega-tivity and electron density that is related to the molecular volume and valence state [160].
(ii) Branching index. This is a perturbational analysis of central and adjacent atoms that is related to the molecular refraction and volume [161].
(iii) Total surface area (TSA). This is computed using the Van der Waals radii of the atoms assuming a specific conformation for the molecule [162]. It is

related to the molecular volume and the hydrophobic solvent cavity or exclusion volume in polar solvents [161, 163].

The topological descriptors (molecular connectivity and TSA) correlate well with physicochemical properties of organic hydrophobic molecules that influence biological activity such as solubility [161, 164] partition coefficient [165], or solvent polarity [166]. In addition, estimates of simple molecular size or shape have proven accurate in predicting chromatographic retention for homologous series of toxic polycyclic aromatics [167], while refined connectivity indices correlate well with chromatographic retention of highly polar aliphatic compounds [168] or barbiturates [169]. For all those properties reported in common with the three molecular topology methods listed, correlation matrices indicate good agreement.

Bioconcentration factors (K_β) for aquatic organisms have been explained [170] in terms of the log P or aqueous solubility (S) for the cellular uptake of dilute solute molecules by organisms. Thermodynamic equilibrium is presumed and these techniques will be invalid if there is significant degradation of the molecule by any of the chemical or biological side reactions in the environment. However, the bioaccumulation of a wide range of organic compounds, including aliphatic and aromatic hydrocarbons, halides, acids and phosphate pesticides in fish has been linearly correlated [171] with the solubility (S) or log P value of the compounds and is given in Equations 1.26 and 1.27.

$$\log K_\beta = m(\log S) + \text{constant} \tag{1.26}$$
$$\log K_\beta = m(\log P) + \text{constant} \tag{1.27}$$

However, the specific uptake sites or route of entry of the chemical into the test organism cannot be stated with certainty based on these predictions. Lipophilic transmission is suggested due to the observed strong dependence of log P. For large molecules, however, membrane permeation resistance is expected [170, 172].

Consistent with this picture, Wong *et al.* showed a direct correlation between the organic substituent (R), partition coefficient and the toxicity of triorganotin compounds towards algae [173]. Prior to the work of Brinckman *et al.*, none of the topological descriptors had been applied to organometallic systems for predicting chromatographic retention or toxicity [174].

1.14.1   APPLICATION OF TOTAL SURFACE AREA (TSA) AS A PREDICATOR FOR ORGANOTIN COMPOUNDS

1.14.1.1   Additivity of Total Surface Area for Aqueous Solubility and Chromatographic Parameters

Total surface area (TSA) as a descriptor has been used because of its relevance and simplicity to stereochemical constructs of mono- and polynuclear core structures as well as to the diverse number of organic substituents that is common to organometallic molecules. Organotin TSA values are easily calculated since a broad data base of bonding distances, angles and Van der Waals radii are readily available in the literature. Using these values, Brinckman et al. [175] were able to calculate the TSA values of various individual organotin molecules. Calculations for the individual molecules involved various degrees of coordination, charge or likely conformations (Figure 1.9) of the tin molecules. In addition to the calculation of the individual organotin molecules, the mean fragment TSA values for several organic groups (R) and labile inorganic ligands (Table 1.12) were also evaluated [174]. The mean fragment TSA values were calculated from either full tetra- or pentacoordinated triorganotin(IV) configurations. The use of mean fragment values as additive substituent TSA values along with a tin core structure TSA value may be subject to error due to the various conformations in organotin structures. This is because the TSA parameter has been shown to be sensitive to conformational changes [176]. In spite of this limitation, surprisingly good agreement with complete structure computer TSA calculations was obtained.

A good correlation, in the form of Equation 1.26 [177] was found between the TSA values for a series of methyl elements, $(CH_3)_n E$, where $E = Br, I, Hg, C, Sb$ and Sn, and their solubilities. Figure 1.10 shows that a good correlation was obtained between the solubility data for a series of tetraalkyl derivatives of group IV elements [177, 178] and their TSA values.

Brinckman et al. [175] were able to apply the TSA values to estimate the chromatographic retentivity for two series of mixed tetraorganotins. Two excellent linear correlations were obtained for the two classes of tetraalkyltin and arylalkyltin compounds. However, two distinct patterns of solvophobic behavior were observed with the conformationally flexible alkyl groups exhibiting far greater retentivity in the C_{18} reverse bonded-phase column than the more rigid planar phenyl substituents.

Since the TSA parameter has been shown to be sensitive to conformational changes [176], the calculation of the TSA of the entire molecule would eliminate the problems encountered with fragmental values. To account fully for the various possible conformers of molecules, a 'holistic' approach was used by Eng et al. [179] for the calculation of the TSA values to estimate the capacity factors for a homologous series of group VA phenyl derivatives, $(C_6H_5)_3M$. This approach would preserve the molecule's entire geometric and/or electronic conformation. The term 'holistic' refers to a particular fixed conformation for the complete molecule and can distinguish the local carbon and M heteroatom geometries and their contributions to the overall TSA predictor. It does not necessarily imply that the TSA value for the whole molecule is more than the sum of its parts. This representative approach would distinguish the local carbon and the M heteroatom geometries. Using this approach, the authors were able to obtain a significant linear correlation between the capacity factors ($\ln k'$) and the TSA values for only one particular set of conformers, implying preferred solute-column chemistry based on preferred C—M bond rotations. A

TSA = 343 Å²

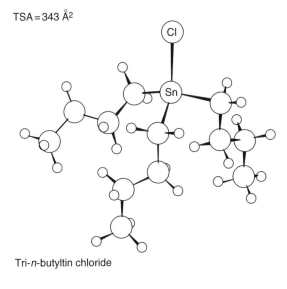

Tri-*n*-butyltin chloride

TSA = 334 Å²   1 Å

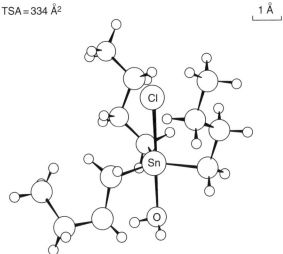

Tri-*n*-butyltin chloride hydrate

Figure 1.9   Structures for tributyltin chloride species
Reproduced with permission from this work, first edition

later study by Tierney *et al.* [180], using both mean fragment and holistic TSA calculations, found good correlations between TSA values and the natural logarithms of the capacity factors for both fluxional and rigid organotin systems as defined by summed carbon hybridization. Tie lines could be drawn

Table 1.12   Additive group/atom TSA values[a]

	TSA (Å^2)	Standard deviation
Organic group		
CH_2—	20.7	1.5
CH_3	32.7	0.02
C_2H_5	55.4	0.01
n—C_3H_7	74.4	1.6
i—C_3H_7	75.6	2.2
n—C_4H_9	95.4	1.8
i—C_4H_9	95.6	1.9
c—C_6H_{11}	117.9	2.2
C_6H_5	92.3	0.1
Inorganic group		
Sn	17.7	0.9
Cl	27.6	2.3
CO_3	45.4	3.3
OH	20.2	0.05
OH_2	23.4	2.2

[a] Mean TSA derived from 4- and 5-coordinate R_3Sn structures (cf. Figure 1.9)
Reproduced with permission from this work, first edition

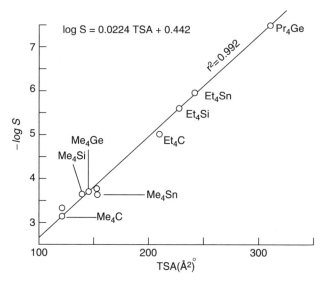

Figure 1.10   Solubility data for tetraalkyl derivatives of group IVA
Reproduced with permission from this work, first edition

between the two systems. The use of these column-dependent tie lines allowed the prediction of unreported compounds with mixed ligands.

Recently, Radecka [181] used the relative change of the trans-membrane current to study the interaction of several organotin compounds with a gramicidin ion-conducting channel. It was found that this parameter was relatable to the TSA of the compounds and the model presented may be applicable to the estimation of the toxicity of the compounds as well as other organometallic systems.

1.14.1.2 Molecular Topology Prediction of Organotin Toxicities in Aquatic Organisms

Several groups of mud crab larvae, *Rhithropanopeus harrisii* were exposed to seawater containing eight structurally distinct triorganotins for the duration of their zoeal development by Brinckman *et al.* [175]. The results of this study indicated that there was a strong correlation between the LC_{50} values (the concentration that causes 50% mortality) and the TSA values for the eight triorganotin compounds studied. A similar study was conducted for the analogous diorganotins [174]. The principal conclusions derived from the correlations of the LC_{50} values with TSA values computed for each organotin toxicant were:

(i) High linear correlations existed between LC_{50} values and the individually calculated TSA values [175] for either the tetra- or pentacoordinated triorganotin solvates or the chloroaquates of $R_2SnCl(H_2O)^+$
(ii) Only the mononuclear di- or triorganotin moiety appeared to be the rate-determining toxic agent.
(iii) A slightly better fit was obtained for the neutral triorganotin pentacoordinated hydrate forms over the tetracoordinated chloride, however, it was not significant at the 95% confidence level.

Using the additive TSA values in Table 1.12, Brinckman *et al.* were able to apply this approach to the study by Wong *et al.* [173] who determined toxicity data for freshwater algal species (*Ankistrodesmus falcatus*) from Lake Ontario. As before, only the neutral four- and five-coordinated triorganotin species gave excellent correlations between ln LC_{50} and TSA. The chlorotin species is unlikely to form in freshwater, but from the values in Table 1.12, the distinction between –OH, –Cl or H_2O probably affects the quality of the line fit far less than the Sn—C bond angles and the accompanying number of labile inorganic groups in the tin's inner coordination sphere. In a similar evaluation [174], the toxicity of a trialkyltin series towards spheroplasts obtained from *Escherichia coli* [182] was evaluated, and again it was found that TSA values from Table 1.12 correlated with the extent of bacterial organotin uptake.

A later study by Eng *et al.* [183] found a high correlation between TSA values and diorganotin toxicity towards several distinct types of organisms. The high correlation between the toxicities and the TSAs, which is independent of the cell type, suggests that it is the hydrophobic behaviour of the organotin species that governs the toxicity process in all cell types studied. The use of TSA as a predictor of toxicity can be expanded to other organometallic systems. For example, Eng *et al.* [184] found a high correlation between the TSA values of triorganolead chlorides and their toxicity towards a bacterium (*Escherichia coli*) and an alga (*Selenastrum capricornutum*). A similar high correlation was obtained for the triorganotin chlorides. However, attempts to correlate other Group IVA organometals incorporating silicon or germanium as the metal center were unsuccessful due to the solubility of the compounds. A study on the inhibitory effects of alkytins on anaerobic bacteria by Belay *et al.* [185] revealed that the TSA relationships observed with aerobic systems may not be valid for anaerobic systems. In a study on the organotin inhibition of methanogenic bacteria, Boopathy *et al.* [186] found that the inhibition of the bacteria increased with a decrease of the TSA value of the compound, which is in contrast to earlier studies. A similar negative correlation was reported by Lascourréges *et al.* [187] in their study involving the toxicity of a series of organotins and three pure strains of sulphate-reducing bacteria.

A topological descriptor closely related to the total surface area of a molecule is the molecular volume. Since both of these descriptors are related to the radius of the molecule, Luedke *et al.* [188] were able to show that both the total surface area and molecular volume of a series of organotins can be used as a descriptor in QSAR studies with an equal level of confidence.

Brinckman *et al.* [175] were able to show in their aquatic toxicity studies that a high correlation existed between the TSA value of the organotin compounds and their LC_{50} values. Attempts to correlate the toxicity with descriptors that assign electronic and steric contributions of the molecule such as the ionic substituent constant (σ^φ) were unsuccessful [175]. These types of descriptor do not provide recognition of either geometric or configuration details that might influence the actual mode of transfer of the organotin species from the bulk solvent into the cell. The transfer most likely involves a reorganization zone, where only one *n*-alkyl group undergoes transition while the three stereochemically rigid R groups remain unaffected during the uptake process.

It should be noted that not all uptake processes on cells result in the transmission of the organotin. It was also shown that when the tri-*n*-butyltin cation was rapidly adsorbed on to the outer membranes of Gram-negative heterotrophic bacteria isolated from the Chesapeake Bay, essential metabolic processes continued unabated. Also, the bulk of the aquated tri-*n*-butyltin cation could be recovered unchanged by washing live cells with methanol. Such passive resistance to toxic metal-containing species by metal sorption is well known and has great environmental importance in metal biogeochemical cycles [189, 190]. It now remains to be seen whether such non-transmission can also be predicted by molecular topology methods, or whether some special geometrical features of the tin core and/or conformation of certain organic groups on the tin atom regulate the path and rates of these processes.

1.15 PRACTICAL CONSIDERATIONS

This section brings together into an environmental context the theoretical and experimental aspects of fundamental organometallic chemistry discussed above. Certain broad conclusions may be considered as follows.

(i) If an organometallic compound is detected in the *atmosphere* it must still be being inputted to the atmosphere (by a natural or anthropogenic process). So many routes to decay occur in the atmosphere that no organometallic compound would be detected if one was contemporarily being inputted.

(ii) The same, ultimately, holds true for organometallics detected in water, organisms or sediments. So many routes to decay exist that, ultimately, the organometallic would be eliminated. However, routes to decay here are shown (CH_3Hg in humans is 70 days, butytins are still being detected and high levels in and nearby to marinas long after they had been introduced).

(iii) Sediments in particular (anoxic, no light penetration) might be expected to stabilize organometallic compounds for longest—particularly as stabilizing coordinating ligands are amply present.

(iv) Non-compliance with the disposal regulations might account for confining presence of organometallics in water and sediments (e.g. butyltins) but inherently there is little citable evidence for this.

1.16 REFERENCES

1. Crompton, T.R., *Determination of Metals in Natural and Treated Waters*, Spon Press (Taylor and Francis) London, 2002.
2. Ure, A.M., and Davidson, C.M. (Eds), *Chemical Speciation in the Environment*, Blackie Academic and Professional, Glasgow, 1995.
3. L., Ebdon, L., Pitts, R., Cornelis, H., Crews, O.F.X., Donard and Ph. Quevauviller (Eds), *Trace Element Speciation for Environment, Food and Health*, Royal Society of Chemistry, London, 2001.
4. Craig, P.J., Needham, M.I., Ostah, N., Stojak, G., Symons, M., and Teesdale-Spittle, P., *J. Chem. Soc. Dalton*, 1996, 153.
5. See, for example, Sigel, H., and Sigel, A. (Eds), *Metal Ions in Biological Systems*, Vol. 29, Marcel Dekker, New York, 1993.
6. Erlschenbroich, Ch., and Salzer, A., *Organometallics, A Concise Introduction*, VCH Weinheim, 1992, pp. 10–11.
7. Abel, E.W., Stone, F.G.A., and Wilkinson, G. (Eds), *Comprehensive Organometallic Chemistry*, Pergamon, Oxford 1982.
8. *Handbook of Chemical Data*, CRC Press, Boca Raton, Florida, USA, 1997–98.
9. Huheey, J.E., *Inorganic Chemistry*, 3rd edn.
10. Alcock, C.B., *Thermochemical Processes: Principles and Models*, Butterworth Heinemann, Oxford, 2001, pp. 45–46.
11. Kochi, J.K., *Organometallic Mechanisms and Catalysis*, Academic Press, New York, 1978, pp. 230–232.
12. Balzoni, V., and Carasiti, V., *Photochemistry of Coordination Compounds*, Academic Press, New York, 1970, p. 281.
13. Blunden, S.J., *J. Organomet Chem.*, 1983, **248**, 155.

14. Gilroy, S.M., Price, S.J., and Webster, N.J., *Can. J. Chem.*, 1972, **50**, 2639.
15. Harrison, R.M., and Laxon, D.P.H., *Environ. Sci. Technol.*, 1978, **12**, 1384.
16. Manahan, S.E., *Environmental Chemistry*, 5th edn, Lewis, Chelsea MI, USA, 1991, pp. 219–249.
17. Harrison, R.E. (Ed.), *Pollution: Causes, Effects and Controls*, Royal Society of Chemistry, London, 1990, pp. 1–20.
18. Neilson, O.J., O'Farrell, D.J., Treacy, J.J., and Sidebottom, H.J., *Environ. Sci. Technol.*, 1991, **25**, 1098.
19. Harrison, R.M., and Laxon, D.P., *Environ. Sci. Technol.*, 1978, **12**, 1384.
20. Fox, M.F., De Montfort University, personal communication.
21. Rowland, F.S. In J.D. Coyle, R.R. Hill and D.R. Roberts (Eds), *Light Chemical Change and Life*, Open University Press, 1982, pp. 142–148.
22. Lindqvist, O., Jernelov, A., Johnsson, K., and Rodhe, H., *Mercury in the Swedish Environment*, Nat. Swed. Environ. Prot. Board, SNV PM 1816, 1982.
23. Craig, P.J., and Moreton, P., *Marine Pollut. Bulletin*, 1984, **15**, 406.
24. Craig, P.J., and Bartlett, P.D., *Nature*, 1978, **275**, 635.
25. Craig, P.J., and Rapsomanikis, S., *J. Chem. Soc. Chem. Commun.*, 1982, 114.
26. Cremer, J.E., *Occup. Health Rev.*, 1965, **17**, 14.
27. Jensen, A.A. In *Biological Effects of Organolead Compounds*, P. Grandjean (Ed.), CRC Press, Boca Raton, 1984, pp. 97–115.
28. See Ref 5.
29. Ref. 5, pp. 50–52.
30. Ref. 5, pp. 71–73.
31. Ref. 5, pp. 82–88.
32. Ref. 5, pp. 90–95.
33. Hamasaki, T., Sato, T., Nagase, T., and Kito, H., *Mutat. Res.*, 1993, **300**, 265.
34. Lovley, D.R., *Ann. Rev. Microbiol.*, 1993, **47**, 263.
35. Wang, Y.-T., and Shen, H., *J. Ind. Microbiol.*, 1995, **14**, 159.
36. Garnham, G.W., and Green, M., *J. Indust. Microbiol.*, 1995, **14**, 247.
37. Palmer, C.D., and Wittbrodt, P.R., *Environ. Health Perspectives*, 1991, **92**, 25.
38. Moore, J.W., *Inorganic Contaminants of Surface Water*, Springer-Verlag, New York, 1990.
39. Shen, H., Pritchard, H., and Sewell, G.W., *Environ. Sci. Technol.*, 1996, **30**, 1667.
40. Shen, H., Pritchard, H., and Sewell, G.W., *Biotechnol. Bioeng.*, 1996, **30**, 1667.
41. Wang, Y.-T., and Xiao, C.S., *Water Research*, 1995, **29**, 2467.
42. Lovley, D.R., and Phillips, E.J.P. *Appl. Environ. Microbiol.*, 1994, **60**, 726.
43. Komori, K., Rivas, A., Toda, K., and Ohtake, H., *Biotechnol. Bioeng.*, 1990, **35**, 951.
44. Komori, K., Wang, P.C., Toda, K., and Ohtake, H., *Appl. Microbiol. Biotechnol.*, 1990, **31**, 567.
45. Shen, H., and Wang, Y.-T., *J. Environ. Eng.*, 1994, **120**, 560.
46. Enterline, P.E., *J. Occup. Med.*, 1974, **16**, 523.
47. Gvozdyak, P.I., Mogilevich, N.F., Rylskii, A.F., and Grishchenko, N.I., *Mikrobiologiya*, 1986, **55**, 962.
48. Shen, H., and Wang, Y.-T., *H Appl. Environ. Microbiol.*, 1993, **50**, 3771.
49. Wang, P.C., Toda, K., Ohtake, H., Kusaka, I., and Yabe, I., *FEMS Microbiol. Letters*, 1991, **78**, 11.
50. Wang, P.C., Mori, T., Komori, K., Sasatsu, K., Toda, K., and Ohtake, H., *Appl. Environ. Microbiol.*, 1989, **55**, 1665.
51. Cervantes, C. *Antonie van Leeuwenhoek*, 1991, **59**, 229.
52. Oremland, R.S., Steinberg, N.A., Presser, T.S., and Miller, L.G., *Appl. Environ. Microbiol.*, 1991, **57**, 615.
53. Rech, S.A., and Macy, J.M., *J. Bacteriol.*, 1992, **174**, 7316.

54. Macy, J.M., and Lawson, S., *Arch. Microbiol.*, 1993, **160**, 295.
55. Macy, J.M., Rexch, S., Auling, G., Dorsch, M., Stackebrandt, E., and Sli, L.I., *Int. J. Systematic Bacteriol.*, 1993, **43**, 135.
56. Losi, M.E., and Frankenberger, W.T. Jr, In *Environmental Chemistry of Selenium*, W.T. Frankenburger Jr and R.A. Engberg (Eds), Marcel Dekker, New York, 1997, pp. 515–544.
57. Losi, M.E., and Frankenberger, W.T. Jr, *Environ. Toxicol. Chem.*, 1997, **16**, 1851.
58. Bacon, M., and Ingledew, W.J., *FEMS Microbiol. Letters*, 1989, **58**, 189.
59. Zawislanski, P.T., and Zavarin, M., *Soil Sci. Soc. Am. J.* 1996, **60**, 791.
60. Tokunaga, T.K., Pickering, I.J., and Brown, G.E. Jr, *Soil Sci. Soc. Am. J.*, 1996, **60**, 781.
61. Losi, M.E., and Frankenberger, W.T. Jr, *J. Environ. Quality*, 1998, **27**, 836.
62. Lovley, D.R., Phillips, E.L.P., Gorby, Y.A., and Landa, E.R., *Nature*, 1991, **350**, 413.
63. Lovley, D.R. In *Advances in Agronomy*, Vol. 54, L. Donald and L. Sparks (Eds), Academic Press, New York, 1995, pp. 175–231.
64. Lovley, D.R., and Phillips, E.J.P., *Appl. Environ. Microbiol.*, 1992, **58**, 850.
65. Lovley, D.R., Roden, E.E., Phillips, E.J.P., and Woodward, J.C., *Marine Geol.*, 1993, **113**, 41.
66. Lovley, D.R., Widman, P.K., Woodward, J.C., and Phillips, J.P., *Appl. Environ. Microbiol.*, 1993, **59**, 3572.
67. Abdelouas, A., Yongming, Lu., Lutze, W., and Nuttall, H.E., *J. Contam. Hydrol.*, 1998, **35**, 217.
68. Tamaki, S., and Frankenberger, W.T. Jr, *Rev. Environ. Contam. Toxicol.*, 1992, **124**, 81.
69. Woolfolk, C.A., and Whitely, H.R., *J. Bacteriol.*, 1962, **84**, 647.
70. McBride, B.C., and Wolfe, R.S., *Biochemistry*, 1971, **10**, 4312.
71. Chen, C.-M. and Focht, D., *Appl. Environ. Microbiol.*, 1979, **38**, 494.
72. Ji, G., Garber, E.A.E., Armes, L.G., Chen, C.-M., Fuchs, J.A., and Silver, S., *Biochemistry*, 1994, **33**, 7294.
73. Chen, C.-M., Mistra, T.K., Silver, S., and Rosen, B.P., *J. Biol. Chem.*, 1986, **261**, 15030.
74. Gladysheva, T.B., Oden, K.L., and Rosen, B.P., *Biochemistry*, 1994, **33**, 7288.
75. Newman, D.K., Ahmann, D., and Morel, F.M.M., *Geomicrobiology*, 1998, **15**, 255.
76. Macy, J.M., Nunan, K., Hagen, K.D., Dixon, D.R., Harbour, P.F., Cahill, M., and Sly, L.I., *Int. J. Sys. Bacteriol.*, 1996, **46**, 1153.
77. Dowell, P.R., Laverman, A.M., and Oremland, R.S., *Appl. Environ. Microbiol.*, 1996, **62**, 1664.
78. Newman, D.K., Kennedy, E.K., Coates, J.D., Ahmann, D., Ellis, D.J., Lovley, D.R., and Morel, F.M.M., *Arch. Microbiol.*, 1997, **168**, 380.
79. Laverman, A.M., Switzer Blum, J., Schaefer, J.K., Phillips, E.J.P., Lovley, D.R., and Oremland, R.S., *Appl. Environ. Microbiol.*, 1995, **61**, 3556.
80. Newman, D.K., Beveridge, T.J., and Morel, F.M.M., *Appl. Environ. Microbiol.*, 1997, **63**, 2022.
81. Phillips, S.E., and Taylor, M.L., *Appl. Environ. Microbiol.*, 1996, **32**, 392.
82. Wakao, N., Koyatsu, H., Komai, Y., Shimokawara, H., Sakurai, Y., and Shiota, H., *Geomicrobiol. J.*, 1988, **6**, 11.
83. Furukawa, K., and Tonomura, K., *Agric. Biol. Chem.*, 1972, **36**, 2441.
84. Rinderle, S.J., Booth, J.E., and Williams, J.W., *Biochemistry*, 1983, **22**, 869.
85. Summers, A.O., and Silver, S., *J. Bacteriol.*, 1972, **112**, 1228.
86. Fox, B., and Walsh, C.T., *J. Biol. Chem.*, 1982, **257**, 2498.
87. Lovley, D.R., *FEMS Microbiol. Rev.*, 1997, **20**, 305.
88. Madigan, M.T., Martinko, J.M., and Parker, J. *Brock: Biology of Microorganisms*, Prentice Hall International, London, 1997.

89. Nealson, K.H., and Myers, C.R., *Appl. Environ. Microbiol.*, 1992, **58**, 439.
90. Nealson, K.H., and Little, B. In *Advances in Applied Microbiology*, Vol. 45, S.L. Neidleman and A.I. Laskin (Eds), Academic Press, London, 1997, pp. 213–237.
91. King, G.M., *FEMS Microbial Ecology*, 1990, **73**, 131.
92. Nealson, K.H., Rosson, R.A., and Myers, C.R. In *Metal Ions and Bacteria*, T.J. Beveridge and R.J. Doyle (Eds), Wiley-Interscience, New York, 1989, pp. 383–411.
93. Lovley, D.R., and Phillips, E.J.P., *Geomicrobiol. J.*, 1988, **6**, 145.
94. Golden, D.C., Chen, C.C., Dixon, J.B., and Tokashiki, Y. *Geoderma*, 1988, **42**, 199.
95. Zehnder, A.J.B., and Brock, T.D., *Appl. Environ. Microbiol.*, 1980, **39**, 194.
96. Karavaiko, G.I., Yurchenko, V., Reimizov, V.I., and Klyushnikova, T.M., *Microbiology*, 1987, **55**, 553.
97. Lovley, D.R., Giovannoni, S.J., White, D.C., Champine, J.E., Phillips, E.J.P., Gorby, Y.A., and Goodwin, S. *Arch. Microbiol.*, 1993, **159**, 336.
98. Ghiorse, W.C., *Ann. Rev. Microbiol.*, 1984, **38**, 515.
99. Pingitore N.E. Jr, Eastman, M.P., Sandidge, M., Oden, K., and Freiha, B., *Marine Chem.*, 1988, **25**, 107.
100. Jung, W.K., and Scheiwsfurth, R., *Z. Allg. Mikobiol.*, 1979, **19**, 107.
101. Adams, L.F., and Ghiorse, W.C., *J. Bacteriol.*, 1987, **169**, 1279.
102. Beveridge, T.J. In *Metal Ions and Bacteria*, T.J. Beveridge and R.J. Doyle (Eds), Wiley-Interscience, New York, 1988, pp. 1–29.
103. Nealson, K.H., Tebo, B.M., and Rosson, R.A., *Adv. Appl. Microbiol.*, 1988, **33**, 279.
104. Feldmann, J., and Hirner, A.V., *Intern. J. Environ. Anal. Chem.*, 1995, **60**, 339.
105. Feldmann, J., Krupp, E.M., Glindemann, D., Hirner, A.V., and Cullen, W.R., *Appl. Organomet. Chem.*, 1999, **13**, 1.
106. Hirner, A.V., Feldmann, E., Krupp, E., Grümping, R., Goguel, R., and Cullen, W.R., *Org. Geochem.*, 1998, **29**, 1765.
107. Feldman, J, Grumping, R., and Hirner, A.V., *Fresenius J. Anal. Chem.*, 1994, **350**, 228.
108. Challenger, F., Higginbottom, C., and Ellis, L., *J. Chem. Soc. Transactions*, 1933 (32), 95.
109. Challenger, F. In *Organometals and Organmetalloids. Occurrence and Fate in the Environment*, F.E. Brinckman and J.M. Bellama (Eds), American Chemical Society, Washington, DC, 1978, pp. 1–20.
110. Fatoki, O.S., *South Afr. J. Sci.*, 1997, **93**, 366.
111. Andrewes, P., Cullen, W.R., Feldmann, J., Koch, I., and Polishchuk, E., *Appl. Organometal. Chem.*, 1999, **13**, 681.
112. Tsubaki, T. In *Minamata Disease*, T. Tsubaki and K. Irukayama (Eds), Kodansha, Tokyo, 1977, pp. 57–78.
113. Compeau, G.C., and Bartha, R., *Appl. Environ. Microbiol.*, 1985, **5**, 498.
114. Compeau, G.C., and Bartha, R., *Appl. Environ. Microbiol.*, 1987, **53**, 261.
115. Kerry, A., Welbour, P.M., Prucha, B., and Mierle, G., *Water, Air Soil Pollut.*, 1991, **56**, 565.
116. Winfery, M.R., and Rudd, J.M.W., *Environ. Toxicol. Chem.*, 1990, **9**, 853.
117. Gilmour, C.C., and Henry, E.A., *Environ. Pollut.*, 1991, **71**, 131.
118. Cullen, W.R., and Reimer, K.J., *Chem. Rev.*, 1989, **89**, 731.
119. Thayer, J.S., *Organometallic Compounds and Living Organisms*, Academic Press, New York, 1984.
120. Andreae, M.O. In: *Organometallic Compounds in the Environment*, P.J. Craig (Ed.), Longman, Harlow, UK, 1986, pp. 198–228.
121. Takahashi, K., Yamauchi, H., Mashiko, M., and Yamamura, Y., *Chem. Abstr.* 1990, **114**, 96350k.
122. Michalke, K., Wickenheiser, M., Mehring, A., Hirner, A.V., and Hensel, R., *Appl. Environ. Microbiol.* 2000, **66**, 2791.

123. Healy, S.M., Casarez, E.A., Ayalo-Fierra, F., and Vasken Aposhiam, H., *Toxicol. Appl. Pharmacol.*, 1998, **148**, 65.
124. Andreae, M.O., Asmode, J.-F., Foster, P., and Van't Dack, L., *Anal. Chem.*, 1981, **53**, 1766.
125. Andreae, M.O., and Froelich, P.N., *Tellus*, 1984, **36B**, 101.
126. Craig, P.J., Forster, S.N., Jenkins, R.O. and Miller, D.M., *Analyst*, 1999, **124**, 1243.
127. Dodd, M.S., Pergantis, A., Cullen, W.R., Li, H., Eigendorf, G.K., and Reimer, K.J., *Analyst*, 1996, **121**, 223.
128. Craig, P.J., Jenkins, R.O., Dewick, R., and Miller, D.P., *Sci. Total Environ.*, 1999, **228**, 83.
129. Jenkins, R.O., Craig, P.J., Goessler, W., and Irgolic, K.J., *Human Experimental Toxicol.*, 1998, **17**, 231.
130. Jenkins, R.O., Craig, P.J., Goessler, W., Miller, D., Ostah, N., and Irgolic, K.J., *Environ. Sci. Tech.*, 1998, **32**, 882.
131. Andrewes, P., Cullen, W.R., Feldmann, J., Koch, I., Elena, P., and Reimer, K.J., *Appl. Organomet. Chem.*, 1998, **12**, 827.
132. Andrewes, P., Cullen, W.R., and Polishchuk, E., *Appl. Organomet. Chem.*, 1999, **13**, 659.
133. Andrewes, P., Cullen, W.R., and Polishchuk, E., *Environ. Sci. Technol.*, 2000, **34**, 2249.
134. Andrewes, P., Cullen, W.R., Polishchuk, E., and Reimer, K.J., *Appl. Organomet. Chem.*, 2001, **6**, 473.
135. Gates, P.N., Harrop, H.A., Pridham, J.B., and Smethurst, B., *Sci. Total Environ.*, 1997, **205**, 215.
136. Gürleyük, H., Van Fleet-Stalder, V., and Chasteen, T.G., *Appl. Organomet. Chem.*, 1997, **11**, 471.
137. Jenkins, R.O., Craig, P.J., Miller, D.P., Stoop, L.C.A.M., Ostah, N., and Morris, T.-A., *Appl. Organomet. Chem.*, 1998, **12**, 449.
138. Hirner, A., Feldmann, J., Goguel, R., Rapsomanikis, S., Fischer, R., and Andreae, M., *Appl. Organomet. Chem.*, 1994, **8**, 65.
139. Feldmann, J., Gruemping, R., and Hirner, A.V., *Fresenius J. Anal. Chem.*, 1994, **350**, 228.
140. Fleming, R.W., and Alexander, M., *Appl. Microbiol.*, 1972, **24**, 424.
141. Doran, J.W. In *Advances in Microbial Ecology*, Vol. 6, G.F. Nordberg (Ed.), Plenum Press, New York, 1982, p. 1ff.
142. Gadd, G.M., *FEMS Microbiol. Rev.*, 1993, **11**, 297.
143. Losi, M.E., and Frankenburger, W.T., *Soil Sci.*, 1997, **162**, 692.
144. Bellivean, B.H., and Trevors, J.T., *Appl. Organomet. Chem.*, 1986, **3**, 283.
145. Oremland, R.S., Culbertson, C.W., and Winfrey, M.R., *Applied and Environ. Microbiol.*, 1991, **57**, 130.
146. Gadd, G.M., *Sci. Total Environ.*, 2000, **258**, 119.
147. Cooney, J.J., and Wuertz, S., *J. Ind. Microbiol.*, 1989, **4**, 375.
148. Thayer, J.S., *Organometallic Compounds and Living Organisms*, Academic Press, New York, 1984.
149. Cooney, J.J., *J. Ind. Microbiol.*, 1988, **3**, 195.
150. Cooney, J.J. In *The Biological Alkylation of Heavy Metals*, P.J. Craig and F. Glockling (Eds), Royal Society of Chemistry, London, 1988, pp. 92–104.
151. Maguire, R.J., *Environ. Sci. Technol.*, 1984, **18**, 291.
152. Barnes, R.D., Bull, A.T., and Poller, R.C., *Pestic. Sci.*, 1973, **4**, 305.
153. Andreae, M.O. In *Organometallic Compounds in the Environment*, P.J. Craig (Ed.) Longman, Harlow, UK, 1986, pp. 198–228.
154. Doran, J.W., and Alexander, M. *Appl. Environ. Microbiol.*, 1977, **33**, 31.

155. Van der Kerk, G.J.M. *Organotin Compounds: New Chemistry and Applications*, J.J. Zuckerman (Ed.), ACS Advances in Chemistry Series No. 157, Washington, DC, 1976, pp. 1–25.
156. Bennett, R.F. *Indust. Chem. Bull.*, 1983, **2**, 171.
157. Blunden, S.J., Cusack, P.A., and Hill, R., *The Industrial Uses of Tin Chemicals*, Burlington House, London, 1985, pp. 1–8.
158. Hansch, C., *Acc. Chem. Rev.*, 1969, **2**, 232; see also Dagani, R. *Chem. Eng. News.*, 1981, (March 9), 26.
159. Kier, L.B., and Hall, L.H., *Molecular Connectivity in Chemistry and Drug Research*, Academic Press, New York, 1976. See also Froimowitz, M., *J. Med. Chem.*, 1982, **25**, 689 and *J. Med. Chem.*, 1982, **25**, 1127.
160. Kier, L.B., Hall, L.H., *J. Pharm. Sci.*, 1981, **70**, 583.
161. Cammarata, A., *J. Pharm. Sci.*, 1979, **68**, 839.
162. Pearlman, R.S., *Partition Coefficient Determination and Estimation*, W.J. Dunn III, J.H. Block and R.S. Pearlman (Eds), Pergamon Press, New York, 1986, pp. 3–20. For an interesting example of a triorganotin aquate, see Davies, A.G., Goddard, J.P., Hursthouse, M.B., and Walker, N.P.C., *J. Chem. Soc., Chem. Commun.* 1983, 597.
163. Max, N.L., *J. Mol. Graphics*, 1984, **2**, 8.
164. Hine, J., and Mookerjee, P.K., *J. Org. Chem.*, 1975, **40**, 292.
165. Yalkowsky, S.H., and Valvani, S.C., *J. Med. Chem.*, 1976, **19**, 727.
166. Kier, L.B., *J. Pharm. Sci.*, 1981, **70**, 930.
167. Wise, S.A., Bonnett, W.J., Guenther, F.R., and May, W.E., *J. Chromatogr. Sci.*, 1981, **19**, 457.
168. Kier, L.B., and Hall, L.H., *J. Pharm. Sci.*, 1979, **68**, 120.
169. Wells, M.J.M., Clark, C.R., and Patterson, R.M., *J. Chromatogr. Sci.*, 1981, **19**, 573.
170. Mackay, D., *Environ. Sci. Technol.*, 1982, **16**, 274.
171. Neely, W.B., Branson, D.R., and Blau, G.E., *Environ. Sci. Technol.*, 1974, **8**, 1113. See also Chiou, C.T., Freed, V.H., Schmedding, D.W., and Kohnert, R.L., *Environ. Sci. Technol.*, 1977, **11**, 475.
172. Fendler, J.H., *Acc. Chem. Res.*, 1980, **13**, 7.
173. Wong, P.T.S., Chau, Y.K., Kramar, O., and Bengert, G.A., *Can. J. Fish Aquat. Sci.*, 1982, **39**, 483.
174. Laughlin, R., Johannesen, R.B., French, W., Guard, H.E., and Brinckman, F.E., *Environ. Toxicol. Chem.*, 1985, **4**, 343.
175. Johannesen, R.B., and Brinckman, F.E. 1981, unpublished results. See also Laughlin, R.B., French, W., Johannesen, R.B., Guard, H.E., and Brinckman, F.E. *Chemosphere*, 1984, **13**, 575.
176. Funasaki, N., Hada, S., and Neya, S. *J. Phys. Chem.*, 1985, **89**, 3046.
177. Brinckman, F.E., Olson, G.J., and Iverson, W.P. In *Atmospheric Chemistry*, E.D. Goldberg (Ed.), Springer-Verlag, Berlin, 1982, pp. 231–49.
178. De Ligny, C.L., and Van der Veen, N.G., *Rec. Trav. Chim. Pays Bas*, 1971, **90**, 984.
179. Eng, G., Johannesen, R.B., Tierney, E.J., Bellama, J.M., and Brinckman, F.E., *J. Chromatogr.*, 1987, **403**, 1.
180. Tierney, E.J., Bellama, J.M., Eng, G., Brinckman, F.E., and Johannesen, R. B., *J. Chromatogr.*, 1988, **441**, 229.
181. Radecka, H., *Pol. J. Environ. Stud.*, 2000, **9**, 419.
182. Yamada, J., *Agric. Biol. Chem.*, 1981, **45**, 997.
183. Eng, G., Tierney, E.J., Bellama, J.M., and Brinckman, F.E., *Appl. Organomet. Chem.*, 1988, **2**, 171.
184. Eng, G., Tierney, E.J., Olson, G.J., Brinckman, F.E., and Bellama, J.M., *Appl. Organomet. Chem.*, 1991, **5**, 33.
185. Belay, N., Rajagopal, B.S., and Daniels, L., *Curr. Microbiol.*, 1990, **20**, 329.

186. Boopathy, R., and Daniels, L., *Appl. Environ. Microbiol.*, 1991, **57**, 1189.
187. Lascourréges, J.F., Caumette, P., and Donard, O.F.X., *Appl. Organomet. Chem.*, 2000, **14**, 98.
188. Luedke, E., Lucero, E., and Eng, G., *Main Group Metal Chem.*, 1991, **14**, 59.
189. Blair, W.R., Olson, G.J., Brinckman, F.E., and Iverson, W.P., *Microb. Ecol.*, 1982, **8**, 241.
190. Sterritt, R.M., and Lester, J.N., *Sci. Total Environ.*, 1980, **16**, 55.

2 Organomercury Compounds in the Environment

ROBERT P. MASON and JANINA M. BENOIT

Chesapeake Biological Laboratory,
University of Maryland, USA

2.1 INTRODUCTION

Interest in the impact of mercury (Hg) released into the environment, from both natural but primarily anthropogenic sources, has been driven by the human health concerns associated with Hg ingestion, mainly through the consumption of seafood. More recently, similar concerns have been raised in terms of wildlife, especially with regard to piscivorous mammals and birds. Two large-scale poisonings, in Minamata, Japan and in Iraq, between 1950 and 1975, focused attention on the health risks associated with Hg consumption, especially in its methylated form. In Minimata, methylmercury (MMHg) from a vinyl chloride–acetic acid plant was the source of contamination. Inorganic Hg, used as a catalyst, was primarily methylated to MMHg within the plant and subsequently discharged in the wastewater. Historically, fish levels as high as 20 ppm and sediment levels above 500 ppm were reported [1, 2]. Patient hair concentrations were similarly elevated—as high as 700 ppm. More than 1000 people died, and about 2250 patients were recognized [3], and Hg poisoning is now often referred to as Minamata disease. Dredging began in 1977 [1], and since the removal of contamination, levels in fish have decreased to values below 0.5 ppm [2]. Levels in human hair have also decreased to values below 10 ppm, which is the currently considered a 'safe' level [4]. However, a recent investigation of fishermen from the region found neurological effects even though the patients current hair concentrations were below 10 ppm [5].

The poisoning incident in Iraq was the result of the consumption of seed grain preserved with MMHg, a commonly used preservative at the time [6]. Mercury, as inorganic Hg or as organomercurials, has been used in a wide variety of industrial applications: as a catalyst; in the chlor-alkali industry; in the pulp and paper industry; and in batteries, paint and pharmaceuticals (as discussed by Craig) [7]. Many of these uses of Hg and organomercurials are now banned in the USA, Europe and elsewhere, and industry has sought

Organometallic Compounds in the Environment
Edited by P.J. Craig © 2003 John Wiley & Sons Ltd

alternative methods that do not rely on Hg use. This practice has resulted in an overall decreased direct discharge of Hg to the aquatic environment. However, since these early cases, there have been many other sites identified where local Hg contamination has resulted from discharge associated with industrial usage (see, for example, Ebinghaus *et al.* [8]) and historical local industrial contamination is a continuing problem. In many direct discharge incidents, contrary to Minamata, elevated levels of Hg in fish, as MMHg, resulted from methylation of the Hg in the aquatic environment.

Elimination of major point source inputs to aquatic systems has refocused attention on Hg inputs into the atmosphere by man's activities [9–15]. Lake sediments and other historical records document an increase in atmospheric Hg of the order of three to five in the last century [12, 16–18]. Currently, for example, coal burning and waste incineration have been identified as the primary anthropogenic atmospheric Hg sources within the USA [19]. Regulation of the waste incineration industry and changes in Hg usage are resulting in reductions in Hg emissions from these sources and, in the USA, coal burning, unless regulated, is projected in the next 10 years to become the main anthropogenic Hg source. Similarly, changes in usage and regulation have lead to a decrease in Hg emissions in Scandinavia and other European countries. However, atmospheric input of Hg from man's activities may continue to increase as many Third World countries are rapidly industrialized, with concomitant increased coal usage, and with little environmental oversight. Other sources of Hg to the atmosphere include biomass burning and mercury use in gold recovery via Hg amalgamation. The crude methods which are often used to mine gold result in large losses of Hg to aquatic systems and to the atmosphere [20, 21].

While sources are often localized or in urban environments, dispersion of Hg through the atmosphere, because of its volatility and relatively slow reactivity in the atmosphere, has resulted in the contamination of aquatic systems that are remote from point source inputs [9, 11, 12, 22, 23]. These remote aquatic environments have received an insult of inorganic Hg via wet and dry deposition processes and, as Hg is effectively methylated within the water bodies, MMHg is bioaccumulated into fish, and other wildlife [9, 24, 25]. The processes leading to Hg methylation and bioaccumulation will be discussed in detail later in the chapter (Section 2.6), but the accumulated evidence suggests that sulfate-reducing bacteria (SRB) are the primary methylators of Hg in aquatic systems [26, 27]. Overall, the accumulation of MMHg into fish provides the major vehicle for human and wildlife exposure.

2.2 HUMAN HEALTH CONCERNS RELATED TO METHYLMERCURY

Mercury, particularly as MMHg, is a neurotoxin. Exposure of humans to elemental and ionic Hg(II) has occurred throughout history because of the use of Hg compounds in various application such as in hat making (hence the term 'mad as a hatter'), in medicine, and in industrial activities including a recent widespread use in the chlor-alkali industry. In these cases, exposure was primarily related to the inhalation of elemental Hg. Elemental Hg is taken up via gas exchange in the lungs and is oxidized within the blood to ionic Hg in the presence of catalase. Severe poisoning incidents from inorganic Hg are now rare [28].

The same is not true for MMHg (CH_3Hg^+) as there is a continuing and widespread exposure of the human population to MMHg, primarily from consumption of fish and other aquatic organisms [28–30]. Methylmercury selectively damages the brain as it readily transverse the blood–brain barrier. MMHg can also transverse the placenta and disrupt the normal development of the fetal brain, therefore fetal exposure is a major concern as this represents the most susceptible stage in the life cycle [28]. Estimations of toxicity related to consumption of fish has lead to consumption advisories in many countries, especially in Europe and North America. For example, in the USA, more than 40 states have issued fish consumption advisor warnings based on Hg [19]. Long-range atmospheric transport and deposition of Hg is the dominant source of this contamination in remote environments in the Northern Hemisphere [9, 11, 25]. In contrast, health effects and high fish levels in South America, Africa and other regions have resulted from the use of Hg in gold mining, and from other sources such as biomass burning and forest clearing [31, 32].

The widespread occurrence of elevated MMHg levels in fish has lead to a reevaluation of the toxic dose [33–36] especially as the current advisories are based primarily on the older data sets, such as the Iraq poisoning incident. Clarkson [37] concludes that at-risk populations are those that consume large quantities of freshwater fish, and pregnant mothers and their offspring. For example, while most of the population has exposure below the World Health Organization (WHO) limit, about 0.1% of the population are at risk based on the WHO guidelines (0.47 µg MMHg kg^{-1} body weight d^{-1}). Stern [34] has reevaluated the results of the early studies and additional studies of human fish consumption and laboratory animal exposure [38–41] with reference to prenatal exposure. Estimation of a no-observed-effects-level (NOEL) dose is limited by uncertainties in the estimation of the mass of fish consumed per day, the concentration and percent age of MMHg in the fish (95–100% for piscivores and marine mammals), and the fraction of the MMHg that is absorbed across the gastrointestinal tract (typically assumed to be 90–100%). Assuming the NOEL to be a tenth of the lowest-observed-effect-level (LOEL), Stern [34] estimated a value of 0.07 µg kg^{-1} day^{-1} for developmental effects. For an average fish consumption of around 30 g day^{-1}, this corresponds to a fish MMHg level of 0.16 ppm. While most ocean fish have MMHg levels comparable to this value, many freshwater fish have much higher concentrations, and marine mammals have values often exceeding 0.5 ppm, the WHO limit (see, for example, Wagemann et al. [42]). In marine mammals, Hg in muscle is essentially present as MMHg, while the percent MMHg in other tissues such as the liver is often lower, and blubber concentrations are typically lower than muscle. Two large epidemiological studies have further assessed human health impacts. A study in the Faroe Islands is investigating the impact of Hg exposure through consumption of marine mammals (whale meat) while a study in the Seychelles Islands is looking at the impact associated with the consumption of marine fish. As fish consumption is an important protein source, and because there are health benefits associated with fish consumption, it is crucial to weigh these positive aspects against the potential adverse impact of MMHg exposure [43]. Additionally, consumption of meat relative to fish could also have adverse health effects. Overall, it has been concluded that the Faroe Island population has suffered neurological effects, while the Seychelles population does not appear to be impacted by consuming fish with levels similar to those in commercial marine fish [44]. However, these studies differed in the criteria used to gauge exposure and assess effects. The Faroe Island study used more sensitive evaluation criteria [45].

In the Seychelles study, fish levels ranged from 0.05–0.2 ppm, while in the Faroe Island study, the levels in whale meat were much higher. Similarly, levels of MMHg in freshwater fish are often significantly higher than those of ocean fish. It is possible that consumption of lower amounts of fish with higher MMHg levels could lead to adverse effects because of the minimization of any health benefits associated with fish consumption [45]. Additionally, potential synergistic adverse effects of other pollutants needs to be assessed, as many waterbodies in the USA and elsewhere that have Hg advisories also have advisories for organic contaminants such as polychlorinated biphenyls (PCBs) and dioxins.

In the Seychelles study, a cohort of over 700 mother–child pairs was studied [46]. The median maternal hair Hg concentration during pregnancy was 5.8 ppm (range 0.5–26.7 ppm) and no association between exposure and achievement of development milestones (age of walking or talking) was found [47, 48] nor were any adverse outcomes to age-appropriate neurodevelopmental tests noted at 66 months at which time the children's average hair Hg concentration was 6.5 ppm. In the Faroe Islands a cohort of 1022 children, for whom information was available on prenatal exposure based on maternal hair concentration [4], has been followed since birth. The median maternal hair Hg concentration was 4.5 ppm, but 13 % of the mothers had hair concentrations exceeding 10 ppm. Examination of 917 children at 7 years of age did not show any clearcut Hg-related abnormalities based a range of neurophysical tests [49]. However, examination of those children whose mother's hair levels ranged from 10–20 ppm showed mild decrements in motor function, language and memory compared with the control group (mother's hair <3 ppm), suggesting that subtle effects are possible at what are often considered 'safe' levels of exposure [4]. Statistical treatment of the results to remove the influence of PCBs did not negate the relationships found [45, 49].

Other recent studies have found corroborating evidence. For example, evaluation of 149 children in a Madeiran fishing village (average maternal hair levels of 9.6 ppm) [50] suggested impact on children whose mother's hair exceeded 10 ppm. The results of this study are similar to those of the Faroe Island study. Overall, current knowledge suggests that levels greater than 10 ppm in maternal hair can be associated with neurological damage to children through prenatal exposure. However, the complicating factors of magnitude and duration of exposure associated with differences in fish concentration, and frequency and amount of consumption need to be further examined.

2.3   TOXICITY AND WILDLIFE

The accumulation, impact and storage of MMHg in mammals is similar in many ways to that of humans. There is evidence to suggest demethylation in the brain with half-lives of MMHg being from 1 to 2 months, depending on the mammal [51]. All forms of Hg cross the placenta but only MMHg tends to concentrate in the fetal brain of mammals and birds. At high levels, MMHg causes symptoms related to neurotoxicity but lower level chronic MMHg exposure has been little studied [51]. Most of the MMHg is excreted in the feces and the overall half-life for MMHg is estimated at around 70 days, which is nearly twice as long as that for inorganic Hg (40 days).

Studies with otters [52] suggest a NOEL of 2 ppm MMHg (0.09 mg MMHg kg^{-1} body mass). For mink, a NOEL of 0.046 mg MMHg kg^{-1} body mass (0.44 ppm in food) was determined. In a separate study, food levels of 1 ppm were shown to cause neurologic lesions [53–55]. These food levels are somewhat higher than those found generally in the environment, but there are lakes in continental North America and Europe that have fish containing mercury levels high enough to cause toxicological effects in piscivorous mammals.

Mercury impacts on birds are largely related to those that are piscivorous, and the impact is primarily on embryonic development and the level of reproduction success is the most sensitive response of birds to MMHg exposure, and egg levels provide a measure of this impact. Demethylation as a detoxification strategy is widespread amongst birds and the role of Se in alleviating toxicity has also been extensively studied [56]. Estimates of the half-life of MMHg in birds are in the range of 2 to 3 months [51].

The impact of MMHg accumulation in loons has received particular attention in the mid-west USA and in other regions of North America. Concentrations in loon blood and/or feathers appear to correlate with predicted relative amounts of atmospheric Hg deposition in the respective regions, although lake chemistry and hydrology also play a role [57]. Blood Hg levels are strongly correlated with those of forage fish in the respective lakes—as has been found for oceanic birds (see, for example, Monteiro *et al.* [58])—and fish Hg concentrations at levels associated with impact on loon reproduction (0.3–0.4 ppm MMHg) are widespread in lakes frequented in summer by these birds [57, 59, 60]. Male loons appear to have higher concentrations than females [57, 59]. Studies looking into reproductive success have found correlations between lake pH and loon blood Hg levels (both adults and juveniles) and hatching success correlated inversely with chick blood Hg levels. Levels of Hg in eggs was within the range associated with reduced hatching success (0.5–1.5 ppm [61, 62]) although other compounding factors associated with lake pH, other than Hg levels in fish, might also be impacting the productivity of loons. More research is clearly required. Similarly, in the Florida Everglades, egrets and other wading birds also have elevated Hg concentrations [63]. Concentrations in egret nestlings were approximately six-times higher in Everglades birds compared with birds of comparable age sampled elsewhere [64]. A detailed study of egret food consumption by fish species in the Everglades indicated that birds were being exposed to an average tissue burden of 0.4 ppm Hg. Exposure to MMHg is highest for chicks [65] and is comparable to levels that have caused effects in laboratory-raised chicks [63].

Individual tissue concentrations and Hg speciation studies show that Hg concentrations are high in liver and kidney, but most of the Hg is inorganic and high levels are correlated with high Se. This contrasts with muscle tissue where >80% of the Hg is MMHg [66]. A correlation between Hg and Se supports the notion of a role for Se in binding Hg as a Hg:Se complex in both birds and mammals [67–69]. Hoffman and coworkers [56, 70] have examined the impact of Hg and Se in both laboratory exposures with mallards and in wild populations of diving ducks, and have demonstrated a complex role for Se, in the form of selenomethionine, in protecting against MMHg intoxification. Results of studies with inorganic Se show differences to those of Hoffman and coworkers in that inorganic Se appears to act as a protective agent overall [71]. These complex interactions need further scrutiny.

Given the above, reported liver and kidney total Hg concentrations are not a strong indicator of toxic levels in birds but levels above about 5 ppm can be considered elevated [51, 72]. Feather concentrations, while attractive as a monitoring tool because of the potential non-invasive/non-destructive nature of sampling, should be interpreted cautiously. Feathers do not directly reflect exposure as deposition to feathers is a method of excretion at the time of moult [73–75]. Egg levels are indicative as Hg is a potent embryonic toxicant and dietary Hg is effectively transferred to avian eggs [51]. Methylmercury is the dominant form in eggs and levels between 0.5 and 1.5 ppm are considered to effect hatchability [61, 62] with higher levels resulting in more severe impacts [51]. Overall, therefore, there is mounting evidence to suggest that fish-eating birds and mammals are being exposed to concentrations of MMHg that are impacting their ability to survive and reproduce.

2.4 ANALYTICAL METHODS

Analytical methods for Hg and MMHg have been recently reviewed and the reader is referred to these more detailed publications [76, 77]. The stability of organomercury species in environmental samples depends to a degree on the sample matrix as adsorption to the container walls, volatilization and decomposition can occur [76]. In natural waters, especially those of higher DOC, MMHg is likely present as a DOC complex (Figure 2.1). In seawater, the

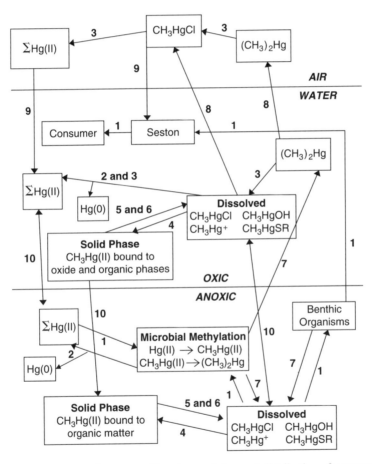

Figure 2.1 Principal reactions controlling the speciation and distribution of organomercury compounds in the environment. The arrows show the principal pathways and major net processes that occur. Numbers refer to the specific processes involved: 1, biological uptake; 2, biotic demethylation; 3, abiotic demethylation; 4, absorption to the solid phase; 5, desorption; 6, dissolution of the solid phase; 7, excretion from organisms; 8, gas exchange; 9, wet and dry deposition; 10, physical processes. Note that SR is used to denote a generic thiol association, i.e. binding to organic matter.

strength of the chloride complex and the low DOC concentrations, suggest that MMHg is present as the inorganic complex. In anoxic waters, MMHg is present as sulfide complexes. Samples are either kept frozen until analysis [78, 79] or acidified with 1 % HCl and refrigerated [76, 80]. It is likely that this acidification step converts organic complexes to the chloride complex, which is possibly more stable than the organic complexes which could degrade by photochemical or other processes. Biota and sediment samples are typically kept frozen until analysis, or are freeze dried. Heat and high acid levels can promote MMHg degradation in water and sediments [76, 81]. Low temperatures will hinder any potential methylation or demethylation that could occur in samples.

The original methods for the measurement of MMHg in biota and sediments were developed by Westoo [82] and involved an organic solvent extraction of MMHg followed by gas chromatography with electron capture detection (GC/ECD). These methods were suitable for tissue and sediment but were not sensitive enough for water analysis. Concentrations of MMHg and dimethyl-mercury (DMHg) in water are at the low to sub-pM (ppt; $\sim$ ng m^{-3}) level typically and thus most techniques rely on preconcentration prior to quantifi-cation. Bloom and Fitzgerald [83] developed the first method for the determin-ation of low levels of organomercurial species in air, based on cryo-trapping, chromatographic separation, and cold vapor atomic fluorescence detection (CVAFS). The sensitivity of CVAFS allowed the development of methods for water analysis, which were based on the method developed by Rapsomanikis et al. [84] for the measurement of alkyllead species. This method relies on the ethylation of the MMHg to form a volatile derivative, methylethylmercury. The derivative can then be stripped from solution, trapped on a solid absorbent column, and subsequently quantified by GC/CVAFS. Recently, isothermal GC methods have been devised so that cryo-trapping is no longer required. How-ever, DMHg, if present, must be stripped from solution prior to ethylation, as it is relatively unstable, especially in acidified medium, and cryo-trapping is required [85–87]. Connection of the derivatization/degassing system to an in-ductively coupled plasma mass spectrometer (ICP-MS) has allowed the deter-mination of samples of Hg species simultaneously with other methylated metals and metalloids [87].

The ethylation step, which uses sodium tetraethylborate as the reagent, was found to suffer from interference in terms of pH effects, and inhibition at high matrix concentrations of chloride [80], and dissolved organic matter [88]. Thus, a pre-separation step is routinely included in the procedures. The methods that have been used for separation include co-distillation of MMHg as CH$_3$HgCl with water [81, 89] and extraction into organic solvents with back-extraction into water [80, 86]. These extraction procedures have been intercompared [81, 90] and distillation was deemed a better method.

However, there has been concern that MMHg can be formed as an artifact of the distillation procedure [91, 92]. It is now apparent that the potential for formation of MMHg can occur with many of the extraction procedures—

distillation, acid and alkaline extraction [93–95]—and the potential increases with increasing inorganic Hg and organic content (particularly carboxylic acids and humic material). For water samples, tissue analysis, and typical environmental sediments, this is not usually a problem. Furthermore, the potential for artifact formation can be checked by suitably spiking the samples with inorganic Hg, although the complication of the reactivity of spiked versus unspiked Hg is an issue [76].

For a number of years, Hg methylation rates have been measured based on the conversion of ^{203}Hg(II) into MM^{203}Hg in water samples, sediment slurries, or intact cores after the method of Furutani and Rudd [96]. Recently, custom synthesis of high-purity ^{203}Hg has allowed its use at near-tracer levels [97, 98]. The method consists of introduction of the isotope, incubation, freezing to halt metabolic activity, extraction of MMHg from thawed subsamples, and gamma counting of the extract. Rate constants derived in this way—the fraction of spike converted per unit time—must be multiplied by some ambient Hg pool to estimate *in situ* methylation rates (see Gilmour *et al* [99]). Typically, total sediment Hg concentration has been used, however, the bioavailability of added Hg(II) may not be comparable to that of ambient Hg, so this calculation may overestimate MMHg production. Demethylation is measured using ^{14}C-labeled MMHg by counting the volatile ^{14}C [100]. This compound cannot yet be produced with high enough specific activity for tracer use, although custom synthesis has produced material that approaches that goal [101].

ICP-MS has been used to determine Hg stable isotope abundances [102]. The use of stable isotopes rather than radioisotopes for mechanistic studies has certain advantages, besides obvious safety issues. Mercury has a number of reasonably abundant stable isotopes, and use of stable isotopes can be advantageous, for example, in studying methylation and demethylation at natural levels by simultaneously using Hg and MMHg enriched with different isotopes [102]. A number of methods exist for GC separation with ICP-MS detection for the analysis of inorganic and methylated Hg species [87, 103–107] and the ability to work with stable isotopes allows tracking of Hg added in any form to experimental systems over the longer term, compared with ^{203}Hg, which has a 47-day half-life.

There is also growing interest in the use of hydride generation techniques. Early research discounted this method as it was found that the hydride of MMHg was unstable. However, a number of recent papers [108, 109] confirm the usefulness of this technique which is subject to less interference, and can, for example, be used to analyze salt water samples directly [109]. The addition of EDTA to the solution to be analyzed helps to stabilize the hydride of MMHg and also reduces any competitive hydride formation by high levels of other metals in the solution. Craig *et al.* [110] discuss the advantages and disadvantages of the hydride generation technique. This technique has been used to measure MMHg in salt marsh sediments and plants [77, 111], and in sediments and fish [108] although more work is required to demonstrate its general applicability to the low levels found in natural waters.

Solvent extraction is routinely used for analysis of samples by conventional GC/electron capture detection (ECD). In these techniques, initial extraction is into an organic phase, after acidification with HCl, HBr or HI [76, 110]. Copper(II) is added to sediment and tissue samples to aid in the release of MMHg from organic complexes. Back extraction into water is used as a purification step and cysteine is added to aid extraction. Finally, extraction into an organic solvent is required prior to injection into the GC. These techniques are suitable for high levels of MMHg, such as in biota samples. However, biota samples can also be analyzed by ethylation/GC/CVAFS after alkaline digestion as the concentrations are typically high enough that only a small amount of sample ($< 100 \mu L$) need be used and interferences are minimized [80]. Microwave-assisted digestion has been used by a number of investigators for rapid digestion of tissue and sediment samples [108, 112], and a number of investigators use the method of trapping MMHg on a sulfhydryl column, followed by elution with HCl [113]. This method is useful for low halide matrices.

Methods involving CVAFS detection have low detection limits in the region of 0.05 pM for MMHg in water; 0.05 pmol g^{-1} dry weight for sediments and tissue. The overall detection limit is a function of sample size for water samples, and sample mass to extraction volume for solid samples. In reality, only a small number of methods are routinely used for the determination of MMHg in environmental matrices. While this may appear advantageous, there is the potential that intercalibration and other quality assurance procedures will not highlight potential artifacts in the methods if all investigators are using similar methods. Strict quality assurance/quality control protocols are essential for the determination of the low levels of MMHg in environmental samples and the continued development of standard reference materials is required [76] to ensure the accuracy of reported results.

2.5 METHYLMERCURY CYCLING AT THE EARTH'S SURFACE

2.5.1 METHYLMERCURY IN WATER

A number of recent modeling studies have estimated the fluxes of Hg to and from the principal reservoirs, and compared natural versus anthropogenic Hg sources to the atmosphere [10, 11, 114]. This has lead to a reasonably coherent picture of the major sources and reservoirs for Hg. However, there has been little attempt to construct a similar budget for MMHg and this is partially a result of the lack of data on concentrations of MMHg in the various reservoirs and on fluxes between reservoirs. Concentrations of MMHg in surface waters are typically higher in freshwater environments compared with the open ocean.

The amount of MMHg in a system is to a degree a function of size (sediment surface area/water volume) because of the importance of methylation in sediments, but other factors also control MMHg formation. The role of sulfate in stimulating MMHg formation by SRBs at low sulfate concentrations has been documented [115], but there is also inhibition of methylation at higher sulfate concentrations through the complex interaction between Hg and sulfide, and inorganic Hg complexation appears to control the bioavailability of Hg to methylating bacteria [116, 117]. Wetlands are important sources of MMHg [118] both in temperate lakes and in the coastal zone, and seasonal anoxia plays a role in increasing MMHg concentrations in the water column and biota by enhancing the release of MMHg from sediments [119, 120]. Demethylation, both biotic and abiotic, is also substantial in most aquatic systems and the steady state concentrations therefore depend on the relative rate of formation compared with decomposition/demethylation. These processes, being largely biologically mediated, vary seasonally and spatially in aquatic systems.

Monomethylmercury appears to be the principal methylated Hg species in uncontaminated terrestrial systems [120–122] while dimethylmercury (DMHg) is more prevalent, and at times it is the dominant species in open ocean waters [78, 85, 86, 123–125]. In estuarine environments there have been reports of DMHg [126], but there is little information of the factors controlling its formation in preference to MMHg. Methylation of MMHg to form DMHg can also be biologically mediated, but it is not known whether sulfate-reducing bacteria or other organisms, are the primary methylators in open ocean sub-thermocline waters where the highest concentrations of DMHg and MMHg occur. While MMHg is likely to be produced directly, it is also the product of DMHg decomposition, and the stability of DMHg in ocean waters, even in the absence of light, is such that DMHg decomposition accounts for a large fraction of the MMHg [86, 124]. In general, therefore, concentrations of DMHg and MMHg in the ocean are low in surface waters (at or below detection limits (DLs); <0.01 pM for DMHg and <0.05 pM for MMHg), and are a maximum in sub-thermocline environments below productive regions (e.g. equatorial upwelling zones; up to 0.7 pM) [85]. In deep ocean waters, overall decomposition is greater than formation and concentrations are low in these environments.

In contrast, in the surface waters of temperate lakes, uncontaminated by point source inputs of Hg, MMHg concentrations range from 0.05 pM for clearwater, oligotrophic environments to 5 pM or more for high DOC environments [122, 127, 128]. The correlation between MMHg concentration and DOC in many aquatic systems reflects the complex role of DOC in mediating both MMHg concentration and distribution, and MMHg bioavailability. DOC complexation keeps MMHg in solution, and high DOC lakes tend to have a higher percentage of MMHg in water [128]. A high organic carbon pool probably stimulates the microbial community, and also leads to development of low oxygen conditions, with resultant increased MMHg production. However, DOC appears to hinder bioaccumulation [129]. The interactions of the microbial community (bacteria/phytoplankton/benthic algae) play a large role in determining food chain 'length' and a combination of all these factors leads to the complex interaction between DOC and MMHg burdens in fish discussed below.

The speciation of MMHg in oxic natural waters is a function of pH, chloride concentration and DOC (Figure 2.1). In the open ocean, where DOC concentrations are low, Cl complexation dominates (i.e. CH_3HgCl is >90% of the total) but DOC is likely to be the more dominant ligand in the coastal zone and in many freshwaters, outcompeting the chloride and hydroxide ligands. Formation constants for MMHg have been estimated [130] and the modeling studies of Hudson *et al.* [131] suggest that about 70% of the MMHg in northern Wisconsin lakes is bound up as organic complexes. While there is little information concerning the binding sites for MMHg with organic matter, consideration of Hg chemistry suggests that thiol groups are the most likely candidates and the estimated conditional constants for MMHg complexation to organic matter are consistent with this notion [131]. In freshwater, low pH will tend to favor chloride complexation over organic complexation. In the presence of sulfide, it appears that measurable sulfide concentrations ($> 10^{-7}$ M) are sufficient to outcompete the organic ligands in most natural waters, including porewaters [132]. Clearly, the extent of organic complexation will impact the bioavailability of the MMHg to microorganisms, especially for uptake via passive diffusion processes [131, 133]. Similarly, and as discussed in detail below, complexation of inorganic Hg plays a crucial role in the accumulation, and thus methylation, of Hg by microorganisms.

For sediments, based on equilibrium modeling studies using the estimated DOC complexation constants [130, 131] and constants estimated for oxide absorption (based on data in Dzomback and Morel) [134], absorption of MMHg to Fe oxide phases appears to be important in both oxic freshwater and estuarine systems, over a fairly wide range of organic content (<10% POC) for sediments of moderate Fe content (>0.5% Fe). These results fit with field data [135]. Clearly, while competition is between the oxide phases and the organic phases in the oxic environment (Figure 2.1), deeper in sediments, complexation to organic matter via thiol ligands dominates, as the relative concentration of thiol ligands increases in concert with dissolution of the oxide phases in reduced environments. However, the coexistence of reduced (especially as FeS/AVS) and oxidized solid Fe phases in surface sediments complicates the simple thermodynamic treatment considered above, and in many sediments the concentration of Fe, AVS and POC covary [136]. No attempt has been made to use selective extraction methods to isolate the phases for MMHg in sediment and it is likely that the harsh methods typically used would decompose the MMHg. However, understanding the binding of MMHg in sediments is crucial if the fact and mobility of MMHg in sediments is to be properly determined.

While MMHg concentrations are often a relatively small fraction of the total Hg in oxic waters, in seasonally or permanently anoxic environments, MMHg concentrations can be significantly higher. Values for dissolved MMHg in anoxic waters as high as 10 pM have been measured for both seasonally and permanently stratified lakes [121, 137], and permanently anoxic fjords [138] and estuaries [139]. Mixing of these seasonally stratified high MMHg waters at lake 'turnover' can be an important source of MMHg to aquatic organisms [140, 141].

Water column concentrations of MMHg in the Everglades [142], and pore water fluxes of MMHg from Lavaca Bay sediments [119] were seen to increase during periods of darkness. Diurnal mobilization of MMHg from sediments is thought to be linked to dissolution of Fe oxide phases in association with changes in the depth of the redox interface because of the lack of sediment photosynthesis in the dark [119]. This view is consistent with observed minimum K_d values for MMHg at sediment depths exhibiting maximum dissolved Fe concentration in Lavaca Bay sediments [143] and the porewater data of Gobeil and Cossa [135]. Our results from incubation of Baltimore Harbor sediments also shows an increased MMHg flux in association with that of Fe, ammonia, arsenic, and the appearance of measurable sulfide in the overlying water. These results are consistent with the proposal that the presence of a surficial oxic layer in sediments may prevent the efflux of dissolved MMHg that builds up in underlying anoxic pore waters [144]. More studies are required to elucidate the factors controlling MMHg flux from sediments, but clearly the extent of the flux is dependent on sediment redox status.

2.5.2 METHYLATED MERCURY IN THE ATMOSPHERE

Concentrations of MMHg in the atmosphere are small, typically a few percentage points of the total Hg in the gas and particulate phase, as they are in wet deposition. Methylmercury in wet deposition varies from typically <1% for open ocean rain to <1–7% for terrestrial precipitation [145–150]. Concentrations in Maryland, USA average around 0.3–0.4 pM, depending on location, for rural sites, and are somewhat higher in the urban Baltimore region (0.7 pM average) [79]. There does not appear to be a strong difference between urban and rural location in terms of the fraction of the total Hg in precipitation as MMHg [79] suggesting that urban sources, particularly anthropogenic sources such as coal burning and waste incineration, are not dominant sources for MMHg (e.g. Prestbo and Bloom [151]). Measurements in other locations of North America and Europe show similar or higher concentrations, up to 3 pM [147–149]. Similarly, in Sweden, MMHg concentration in precipitation ranged from 0.25 to 0.3 pM. Given these concentrations of MMHg in precipitation, wet deposition of MMHg is not a significant source of MMHg to estuaries and the open ocean [79, 85]. In lakes, direct deposition and indirect inputs of MMHg from watershed runoff can be a more important source of MMHg [150], but this is not the case for seepage lakes [121, 131].

Measurements of MMHg in atmospheric particulate and in the gas phase are essentially non-existent, and this is a function to a large degree of the difficulty of measuring the low concentrations present. Preliminary measurements of MMHg in atmospheric particulate in Wisconsin found values of 5–10 fmol m^{-3}; equivalent to 4–30% of the total particulate Hg [148]. These fractions are higher than in soils where the percentage of MMHg is generally <1% [152] which suggests other sources besides resuspension of soil particles, including the adsorption of gas-phase MMHg. While incineration is not a likely source of MMHg, there are other potential anthropogenic inputs to the atmosphere. Loss of MMHg from municipal sludge applied to soil has been suggested as an important source [153]. Other sources such as losses via gas exchange from sewage treatment plants, and other waters of high MMHg concentration, need further investigation. Bloom *et al.* [154] have suggested that one source of atmospheric MMHg to the west coast of North America (off Washington state, USA) is the upwelling of deeper ocean waters containing DMHg. Gas evasion of DMHg and subsequent decomposition in the atmosphere [155, 156] could be a source of MMHg. Other regions of the ocean where deep winter mixing (e.g. the far North Atlantic [78]) or substantial upwelling brings deeper ocean waters containing DMHg to the surface could also account for MMHg in the atmosphere. The stability of DMHg in the atmosphere [155, 156] is such that it will not persist far downwind of sources.

Given the relatively high solubility of MMHg, and its correspondingly low Henry's law coefficient, dry deposition of MMHg in the gas phase to surface waters is a potential sink for the atmosphere. Fitzgerald *et al.* [157] found gas-phase atmospheric concentrations to be below their DL of 25 fmol m^{-3} for studies in northern Wisconsin. In Washington state, Bloom *et al.* [154] measured higher values of between 10–100 fmol m^{-3}. Mason [158] found values to be < 10 fmol m^{-3} for the equatorial Pacific atmosphere. If MMHg were in the gas phase in the atmosphere as the complex CH_3HgCl in equilibrium with ocean surface waters at 25 fM MMHg, then its concentration would be around 0.5–1 fmol m^{-3} at 25 °C (Henry's law coefficient, $H = 2$–4×10^{-5}) [159, 160] a value which is at or below current detection limits, but not inconsistent with measurement. For lakes, assuming CH_3HgOH to be the volatile species ($H = 0.4 \times 10^{-5}$) [160] the lower H value compensates for the relatively higher surface water MMHg concentrations (2.5–5 pM) and equilibrium concentrations would be similar at 25 °C. If these concentrations occurred in the atmosphere, then equilibrium partitioning with gaseous phase MMHg could account for all the MMHg in rain. Overall, MMHg concentrations in the gas phase in the terrestrial atmosphere are generally in the region of 5–25 fmol m^{-3}, although higher concentrations are possible, and the concentration is probably controlled by removal by wet and dry deposition. For the ocean, the lack of detection of MMHg in rain (<50 fM) [161, 162] suggests that air concentrations are less than 1 fmol m^{-3}.

It is instructive to calculate a budget for MMHg for the ocean to illustrate the potential importance of the various fluxes on a larger scale. As the concentration in some media is not well known, it is assumed here that the ocean is at steady state with the atmosphere in terms of gas exchange of CH_3HgCl. This assumption is valid given the high gas phase deposition velocity of CH_3HgCl. An equilibrium atmospheric concentration of CH_3HgCl of 0.5 fmol m^{-3} is consistent with surface water values and measured rain concentrations, assuming equilibrium partitioning. A flux of 6×10^4 mol yr^{-1} is derived, assuming an average dry deposition velocity of 1 cm s^{-1} for CH_3HgCl at 25 °C (based on equations in Seinfeld and Pandis) [163]. If MMHg concentration in atmospheric particulates over the ocean is similar to that of Wisconsin, then the dry particulate flux would be a important contributor to the total atmospheric flux, even given the low particulate concentrations—typically < 10 fmol m^{-3} total particulate Hg. A value of around 1.1×10^4 mol yr^{-1} is calculated for 10% MMHg in particulates and a deposition velocity of 0.1 cm s^{-1}. The wet deposition flux is estimated at 5×10^4 mol yr^{-1} based on fluxes for total Hg in Mason et al. [11] and 0.5% MMHg in wet deposition. Overall, a total atmospheric flux to the ocean of 12.1×10^4 mol yr^{-1} is calculated.

The other major source to the ocean is riverine input. Assuming 3% MMHg in river water entering the ocean [79, 126], the flux is 3×10^4 mol yr^{-1}. Similarly, considering deep ocean sediment to contain about 0.3% MMHg on average, the burial flux for the deep ocean is around 3×10^3 mol yr^{-1}. The major sink for MMHg is accumulation into fish, and this value was estimated by Rolfhus and Fitzgerald [164] to be 22×10^4 mol yr^{-1}. Another sink is the potential loss of DMHg from the ocean. The concentrations in the surface ocean are mostly at the detection limit (<10 fM), except in regions of deep water mixing or upwelling. Given the rapid decomposition of DMHg in the atmosphere [155, 156] any DMHg evaded, except in coastal areas, is likely redeposited to the ocean as MMHg [123, 154]. Thus the maximum flux is probably less than that of atmospheric deposition. These estimations suggest that overall there is little net exchange of MMHg across the air–water interface and therefore, the main balance for uptake into fish must be supplied by in situ net formation of MMHg, and the difference between the fluxes identified above can be used to estimate the rate of net formation of methylated Hg species (DMHg and MMHg) in the ocean. Considering air–sea exchange to be in balance, the net formation required is around 19×10^4 mol yr^{-1}. Assuming that methylation only occurs in waters below productive areas, and in coastal zones (as suggested by extensive data) [86, 78, 124–126], the net rate of formation is estimated at about 2% of the available substrate per year (i.e. for 10% of the ocean area; 500 m of water depth where methylation occurs; 0.5 pM available Hg(II) substrate). These calculations demonstrate the potential importance of the atmosphere as a source of MMHg to the ocean; the potential for MMHg and DMHg cycling at the water surface; and the importance of *in situ* production as the source for MMHg in fish.

In lakes, a number of investigators have similarly concluded that *in situ* production is the main source of MMHg [121, 131, 157]. However, when considering the mass balance approaches above in freshwater systems, it is clear that, because of the lack of DMHg, there is no mechanism for significant exchange of methylated Hg between the water and atmosphere, except if MMHg is being lost from the lake surface. Given the higher concentrations of MMHg in epilimnetic waters, such a scenario is possible. Additionally, given the flux of MMHg estimated from wet deposition, and the apparent lack of MMHg in anthropogenic inputs, there appear to be missing sources of terrestrial atmospheric MMHg. One suggestion is that the ocean, via DMHg decomposition in the atmosphere, is a source; another that evasion of MMHg occurs from sewage sludge applied to land. Other similar sources are likely. Further investigations are needed to identify these terrestrial sources of MMHg to the atmosphere.

2.6   METHYLMERCURY PRODUCTION, DECOMPOSITION, AND BIOACCUMULATION

2.6.1   PATHWAYS OF MERCURY METHYLATION

Biological methylation of Hg in aquatic sediments was first demonstrated by Jensen and Jernelov [165]. Although humic subsances may act as methyl donors in abiotic methylation of Hg [166–168], the inability of 'killed' samples to methylate Hg has been observed in a number of studies [169–171], and it is generally accepted that Hg methylation in nature is principally a biological process. A number of organisms in pure culture can produce MMHg from added Hg(II) including: *Neurospora, Clostridium, Pseudomonas, Bacillus,* and *Escherichia coli,* to name only a few [172], and early evidence of Hg methylation by extract of methanogens [173] pointed to this group as the principal MMHg producers in sediments. However, several lines of evidence point specifically to sulfate reducing bacteria (SRB) as the most important Hg methylators in the environment.

Studies have used specific metabolic inhibitors, addition of sulfate, and coincident measurement of sulfate reduction rate and MMHg production to elucidate the role of these microorganisms in Hg methylation in natural sediments. Compeau and Bartha [26] isolated Hg-methylating *Desulfovibrio desulfuricans* strains from estuarine sediments and found that addition of bromoethane sulfuric acid (BES), a specific inhibitor of methanogens, increased Hg methylation while molybdate, a specific inhibitor of sulfate reduction, dramatically decreased MMHg production. Since that time, molybdate inhibition of Hg methylation has indicated the importance of SRB in estuarine [174], freshwater lake [27, 97, 175], saltmarsh [176], and Everglades [99] sediments. Additionally, in freshwater environments, addition of sulfate may stimulate Hg methylation [27, 97, 177]. Experiments with estuarine [178] and saltmarsh [176] sediments have demonstrated a strong correlation between sulfate reduction and Hg methylation rates. Devereaux *et al.* [179] used 16S ribosomal RNA-targeted probes to estimate SRB distributions in estuarine sediments, and found that sub-surface peaks in Hg methylation potential corresponded to peaks in some groups of SRB.

The ability of methylcobalamin to methylate Hg spontaneously *in vitro* [180–183], prompted investigations into the methyl carrier in sulfate reducers. Berman *et al.* [184] demonstrated light-reversible inhibition of Hg methylation in *D. desulfuricans* LS by propyl iodide, which indicated that methylation was mediated by a cobalt porphyrin compound in this organism. This corrinoid was identified by Choi and Bartha [185] to be cobalamin (vitamin B_{12}), and these authors suggested that Hg methylation is an enzymatically catalyzed process *in vivo*. Cobalamin is not unique to SRB, and has been identified in some aceto-gens and methanogens [186], so not all organisms that contain cobalamin are capable of methylating Hg.

Choi *et al.* [187] observed Michaelis–Menton kinetics, temperature and pH optima for Hg methylation, and a 600-fold higher rate of Hg methylation by cell extracts of *D. desulfuricans* LS compared to free methylcobalamin. They proposed a two-step, enzyme-mediated transfer of a methyl group from methyl tetrahydrofolate to a cobalamin, then from methylcobalamin to Hg(II). It had previously been shown that the methyl group in MMHg production by this organism originates from serine, a principal methyl donor to tetrahydrofolate [184]. The acetyl-CoA synthase pathway was shown to be a significant meta-bolic pathway of Hg methylation in *D. desulfuricans* LS [188]. In view of the presence of cobalamin and high levels of acetyl-CoA enzymes in methanogens and acetogens that do not methylate Hg, it has been suggested that some sulfate reducers possess a distinct or highly specific enzyme to catalyze this step [188]. The identity of the enzyme responsible for methyl transfer to Hg(II) is not known.

2.6.2 PATHWAYS OF MMHG DEMETHYLATION

Bacteria with broad-spectrum resistance to Hg are able to degrade a variety of organomercury compounds including methyl-and ethylmercury chloride [189]. Microbial degradation of methylmercury occurs through the cleavage of the carbon–Hg bond by the enzyme organomercurial lyase followed by reduction of Hg(II) by mercuric reductase to yield methane and Hg(0) [172]. These proteins are specified for by genes of the *mer* operon, and *mer*-mediated Hg resistance is widely distributed across microbial groups [190]. The physiology and genetics of Hg resistance have been extensively reviewed elsewhere [189–192].

In an early study using lake sediments, Spangler *et al.* [193] found methane and Hg(0) as the products of bacterially mediated Hg demethylation, which prompted the conclusion that MMHg demethylation in nature occurs primarily through the *mer* pathway (see Winfrey and Rudd [194]). However, CO_2 has also been observed as a major MMHg demethylation product by Oremland and coworkers [195, 196]. These authors suggested that MMHg degradation can occur through biochemical pathways used to derive energy from single C substrates, and they termed this process 'oxidative demethylation'. Marvin-DiPasquale and Oremland [101] further proposed a reaction analogous to microbial monomethylamine degradation:

$$4CH_3Hg^+ + 2H_2O + 4H^+ = 3CH_4 + CO_2 + 4Hg^{2+} + 2H_2.$$

Pak and Bartha [197] suggested another explanation for CO_2 production from MMHg, namely that anaerobic methanotrophs oxidized the methane released from MMHg after cleavage via organomercurial lyase. It is important to note that the Hg products should be different from the two pathways, in *mer*-mediated detoxification Hg(0) is produced, but Hg(II) is the likely product of the oxidative pathway [101]. It would be highly illuminating to investigate the Hg product in pure cultures or sediments that exhibit apparent oxidative demethylation.

The relative importance of *mer*-mediated versus oxidative demethylation in natural sediments is not known. A possible controlling mechanism may be the ambient MMHg concentrations to which resident bacteria are exposed. The *mer* pathway is inducible by exposure to high levels of Hg(II) or MMHg, although an absolute threshold concentration has not been established. The pathway of demethylation may also depend on the dominant microbial community present in a given environment. Oremland *et al.* [195] found that under aerobic conditions, MMHg degradation occurred via the organomercurial lyase system, but under anaerobic conditions sulfate reducers and methanogens used oxidative demethylation. Pak and Bartha [198] confirmed the ability of two sulfate reducer strains and one methanogen strain to demethylate Hg in pure culture, but, as discussed above, they disagree as to the source of the CO_2 endproduct [197].

A third pathway for MMHg degradation that is independent of lyase activity has been proposed. It has been shown that under laboratory conditions MMHg reacts with excess sulfide to produce dimethylmercury sulfide, which in turn decomposes to DMHg [199, 200]. However, the importance of this reaction in the natural environment has not been determined. Baldi *et al.* [201] reported MMHg conversion via this mechanism by two strains of *D. desulfuricans*. These experiments were, however, carried out at high sulfide and high MMHg concentrations, conditions that would thermodynamically favor the abiotic formation of DMHg via the reaction above.

Photodegradation of MMHg in lake surface waters was demonstrated by Sellers *et al* [202]. The kinetics of MMHg degradation were first order with respect to MMHg concentration and light intensity. Measured rates were sufficiently high to suggest that this process may represent a significant sink for MMHg in clear-water lakes; however, little collaborating data exits and the Hg end-product of photodemethylation was not determined. In experiments using surface waters from the Florida Everglades, Krabbenhoft *et al.* [142] tested the stability of total Hg and MMHg to light, and observed light-mediated loss of total Hg but no photodecomposition of MMHg. The authors suggested that the high DOC content of surface waters may limit light penetration and photodemethylation in the Everglades. However, further studies have revealed significant demethylation in the more oligotrophic, southern portions of the Everglades (Krabbenhoft, personal communication). Our studies with seawater suggest that MMHg is stable under typical light and pH conditions, and that both DMHg and MMHg become much more unstable at a pH below 5, and that the stability is enhanced in seawater [203]. This is the likely consequence of the greater stabilizing influence of chloride complexation of MMHg compared to that of the hydroxide [7, 204, 205]. Thus, the conditions of the experiment chosen by Sellers *et al.* [202] were probably most favorable for abiotic demethylation—oligotrophic freshwater with low pH and low DOC. The general applicability of their results requires further investigation, as does the mechanism. The nature of the product, Hg(II) or Hg^0, will determine the overall impact of this process on Hg cycling in aquatic ecosystems.

2.6.3   ENVIRONMENTAL FACTORS THAT AFFECT MMHg PRODUCTION

Environmental factors affecting Hg methylation and demethylation fall into two categories: (i) those that control the overall metabolic activity of the methylating organsims, and (ii) those that control the chemical form of the Hg in the matrix where methylation occurs. To date, the focus of most research has been on the chemical and physical characteristics that stimulate Hg methylation by affecting microbial activity, rather than on chemical forms of Hg that are available for uptake and methylation. For example, Bodaly *et al.* [206] found that the methylation rate (M) in epilimnetic sediments was positively related to temperature while the demethylation rate (D) was negatively related, so that net MMHg production (often expressed as the M/D ratio) increased dramatically with increasing temperature. They further explained an observed inverse relationship between Hg content in fish and lake size to greater net methylation in small lakes due to the higher water temperatures reached during the open-water season. Likewise, Ramlal *et al.* [207] found that temperature exerted a major control on both the location and timing of maximal M/D ratios in several Canadian Shield Lakes, and they concluded that the bulk of net MMHg production occurs in the epilimnetic sediments during the summer months. Similar seasonal variation in M/D were observed in sediments from Lake Clara in northern Wisconsin [208] and the Wisconsin River [209].

Since SRB are the important methylators of Hg, a general relationship between the percentage of MMHg in sediments and the sulfate reduction rate is expected. At relatively low sulfate concentrations, sulfate stimulates both sulfate reduction and Hg methylation [27], and increased sulfate deposition associated with acid rain may explain increased MMHg bioaccumulation in aquatic systems undergoing acidification [115]. At high sulfate concentrations, the build-up of dissolved sulfide inhibits methylation, as discussed below. The supply of labile carbon also exerts a positive control on methylation rates [96, 178, 210]. Therefore, the distribution of methylation activity is tied to the distribution of biodegradable organic matter. Accordingly, maximal M/D ratios are often observed in surface sediments [208, 209], where microbial activity is greatest. Systems with high levels of organic matter production, such as wetlands [118, 211–213] and recently flooded reservoirs [214–216] may exhibit extremely high rates of MMHg production.

An inverse correlation between lake water pH and Hg in fish tissues has been observed in a number of studies [217–221], and this relationship has led to investigation into the effect of pH on methylation and demethylation in aquatic ecosystems. In sediments from the Experimental Lakes Area [222], methylation was reduced when pH was lowered below pH 7 until it was virtually inhibited below pH 5, but a much less dramatic reduction in demethylation was observed. The decrease in methylation was attributed to loss of substrate Hg due to precipitation with sulfide released from AVS, as was consistent with observed decreases in pore water inorganic Hg with lowered pH. Similar results were obtained by Steffan et al [223], who observed a positive correlation between Hg methylation and pH in the range of 2.5 to 8.0 but no inhibition of demethylation between pH 4.5 and 8.0. These authors concluded that enhanced MMHg production in low pH sub-surface sediments could not explain elevated Hg content in fish from low pH lakes.

On the other hand, Xun et al. [224] found increasing rates of Hg methylation in epilimnetic lake waters and at the sediment surface with lowered pH, and Miskimmin et al. [225] observed an increase in MMHg production in lake water at pH 5 relative to pH 7. In surficial sediments from the experimentally acidified basin (pH 6) of Little Rock Lake, an increase in Hg methylation was observed relative to the reference basin [194]. Winfrey and Rudd [194] reviewed potential mechanisms for low pH effects on Hg methylation and suggested that changes in Hg binding could account for the seemingly conflicting results seen in all of these studies. They pointed out that lowering pH may lead to increased association of Hg with solid phases [222, 226, 227], decreased dissolved pore water Hg, and (presumably) lower availability of Hg(II) to bacteria. They argue that in the water column, bacterial particles may serve as important particulate binding sites for Hg, thus increasing the availability of Hg to methylating bacteria. Thus, it appears that the availability of Hg(II) to methylating organisms may play a key role in controlling MMHg production. This idea is consistent with the hypothesis of Fitzgerald *et al.* [157] who proposed that lower pH decreases loss of Hg(II) due to reduction and evasion from the lake surface, thereby increasing the overall substrate pool for methylation.

In many surface waters, DOC may be an important complexing agent for Hg as suggested by the association of Hg with DOC that has been observed in some surface lake waters [122, 228–230]. Miskimmin *et al.* [225] found that methylation rates in epilimnetic lake water were lower at pH 7 than at pH 5, but that DOC suppressed methylation regardless of pH. They concluded that DOC binds Hg thereby making it unavailable for methylation, but that lower pH leads to desorption of Hg and increased bioavailability. Demethylation was also greater at pH 7, but did not show any trend with DOC concentration [225]. DOC had no effect of methylation rate in sediment, although partitioning into the overlying water increased with DOC [231].

To summarize, acidification of lakes may increase MMHg production in the sediments due to sulfate fertilization and possibly due to lowered pH, which may stimulate MMHg production near the sediment–water interface. Although decreased pH may decrease methylation in deeper sediments, in many aquatic ecosystems the bulk of Hg methylation occurs near the sediment surface [194, 207–209]. The effects of DOC are less straightforward, and the relationship between DOC and fish Hg concentrations differs depending on lake type. These parameters are positively correlated in drainage lakes [232, 233] and negatively correlated in seepage lakes [220, 234]. A possible explanation for this difference is that in drainage lakes the bulk of the DOC is supplied from the watershed, which may bring in exogenous Hg and MMHg or provide labile carbon for increased microbial activity and MMHg production [228, 235, 236]. In seepage lakes, where DOC is mainly produced *in situ*, DOC may inhibit in-lake MMHg production [225]. Furthermore, low pH and high DOC tend to favor Hg(II) methylation relative to reduction and evasion [24, 237]. All of these studies on pH, DOC, and Hg methylation suggest that the effects of these water chemistry parameters on Hg concentrations in fish may only be partially explained by increased MMHg production within aquatic ecosystems.

The previous discussion brings up an important point—that the uptake mechanism for Hg(II) by methylating bacteria is not presently known. Although Hg-resistant bacteria possessing the *mer* operon have the ability actively to transport Hg (II), this operon is not present in many SRB that methylate Hg [237]. As discussed above, MMHg is produced in an accidental side reaction of a metabolic pathway, so it is not likely that active transport would be involved in Hg methylation, and a limited number of experiments with SRB suggest that Hg uptake is not an active process [237]. For these reasons, passive diffusion across the cell membrane been proposed as the important uptake mechanism by methylating bacteria [116, 237, 238]. This hypothesis is consistent with work by Gutknecht [239] and Mason *et al.* [133, 240] who demonstrated passive diffusion of neutral Hg complexes across artificial membranes and diatom cell membranes, respectively.

Another phenomenon that appears to be linked to the issue of bioavailability of Hg for methylation is the inhibition of MMHg production in high salinity sediments. While sulfate may stimulate methylation in freshwaters, low MMHg has been observed in wetland and estuarine sediments [26, 99, 116, 178] with high sulfate concentrations. The inhibition of Hg methylation in high sulfate environments has been attributed to the presence of sulfide, which is produced as a by-product of sulfate reduction. An inverse relationship between dissolved sulfide concentration and MMHg production and/or concentration has been observed in sediments from a number of aquatic ecosystems [99, 116, 174, 180, 194, 241]. The mechanism of sulfide inhibition is unknown, but it has been suggested that a decrease in the concentration of dissolved inorganic Hg, due to precipitation of mercuric sulfide (HgS(s)), leads to a decrease in the availability of substrate Hg to bacterial cells [174, 180, 194, 242]. However, while equilibrium calculations show that the solubility of mercuric sulfide increases in the presence of sulfide [243, 244], filterable Hg concentrations measured in pore waters from natural sediments may stay the same or increase [116, 143] with increasing sulfide; there is no correlation between dissolved Hg and MMHg in sediments [116, 117]; and mercuric sulfide added to SRB cultures is readily methylated [245]. An alternative explanation is that shifts in the complexation of Hg in pore waters may affect Hg bioavailability to bacteria.

A number of Hg complexes exist in solution in the presence of dissolved sulfide, including HgS^0, $Hg(SH)^+$, HgS_2^{2-}, $HgHS_2^-$, and $Hg(SH)_2^0$ [243, 244, 246, 247]. These complexes significantly increase the solubility of mercuric sulfide when excess sulfide is available. Of these complexes, HgS^0 and $Hg(SH)_2^0$ are expected to cross bacterial membrane most readily because they are small and uncharged. A chemical equilibrium model for dissolved Hg complexation in pore waters [116] suggested that as sulfide increases from μM to mM concentrations, Hg speciation shifts from predominantly HgS^0 toward disulfide complexes, mainly $HgHS_2^-$ near pH 7. The changes in octanol–water partitioning associated with this speciation shift was investigated [117], and it was determined that the overall partitioning coefficient (D_{ow}) for Hg decreased with increasing sulfide in a manner that was consistent with predicted decreases in the fraction of dissolved Hg present as HgS^0. Since D_{ow} is a measure of permeability of Hg to cell membranes by passive diffusion [133, 240] these results are consistent with decreased bioavailability of Hg with increasing pore water sulfide concentration.

2.6.4   FACTORS THAT AFFECT THE BIOACCUMULATION OF MMHG

High concentrations of Hg occur in fish from lakes with only atmospheric sources of Hg, especially in lakes with low acid-neutralizing capacity and pH, high DOC content, and in recently formed hydroelectric reservoirs (Wiener and Spry [248] and references therein). Studies addressing direct correlations between environmental factors and the Hg content of fish have shown: (i) an inverse relationship with lake water pH [217, 218, 249]; (ii) a positive relationship with eplilimnetic water temperatures [206], (iii) a positive relationship with lake water DOC [122, 232, 233]; a negative relationship with lake water DOC [220, 234]; (iv) a positive relationship with surface sediment total Hg concentration [220, 250], and (v) a negative relationship with calcium [232, 251]. As previously mentioned, chemical and physical characteristics of aquatic ecosystems including acidity, DOC, sediment organic matter content, and temperature may indirectly affect Hg accumulation in biota by influencing MMHg production by resident bacteria. In this section we will consider how post-production factors affect bioaccumulation of MMHg into aquatic food chains.

While mercury is unique among metals in that it biomagnifies through all levels of the food chain [9], the largest bioconcentration of Hg occurs between water and phytoplankton [9, 133, 252, 253]. Therefore, uptake at the base of the food chain may exert a primary control on Hg reaching higher trophic levels [131]. Mason et al. [133, 240] demonstrated passive uptake of neutral species of both inorganic Hg and MMHg by the diatom T. wesflogii. In these experiments, uptake was most efficient for the complex MMHgCl. At a given chloride concentration, the fraction of MMHg present as this complex decreases with increasing pH as it is replaced by MMHgOH. Because K_{ow} for MMHgCl is 24-times the K_{ow} for MMHgOH [133, 254], a decline in D_{ow} and cell permeability for MMHg with increasing pH is consistent with the observed relationship between pH and Hg in biota.

On the other hand, in several Wisconsin lakes [24, 121, 229], lake water DOC concentration was positively correlated with dissolved MMHg concentration but negatively correlated with bioaccumulation factors (BAF) for seston. Furthermore, while Hg concentration in fish from a number of Adirondack lakes increased with water column DOC, fish BAF decreased [122]. These findings suggest that organic matter complexation makes MMHg less bioavailable, so that positive relationships between lake DOC and Hg in fish cannot be explained by enhanced uptake of MMHg at the base of the food chain. Therefore, DOC effects on bioaccumulation are controlled by the conflicting effects of: (i) increased MMHg loading and/or bacterial production when DOC is supplied from the watershed; (ii) decreased bioavailability of DOC-bound Hg(II) to methylating organisms; (iii) decreased bioavailability of DOC-bound MMHg to phytoplankton, and (iv) changes in the food chain structure.

Although both inorganic Hg and MMHg enter diatom cells, more efficient transfer of MMHg from diatoms to copepods occurs due to greater sequestration of MMHg in the cytoplasm compared with binding to membranes [133]. Similarly, more MMHg is associated with the soft tissue of copepods, and this correlates with the higher assimilation of MMHg over inorganic Hg by fish feeding on these organisms [255]. A similar relationship between fractionation of an element in prey tissue and assimilation efficiency has been observed for other elements [256, 257]. These results were consistent with observations that trophic transfer is more efficient for MMHg than for inorganic Hg [258, 259], so that the percentage of MMHg tends to increase with increasing trophic levels [221, 229, 252, 260] until the bulk of Hg (>90%) in predatory fish tissues is in the form of MMHg [79, 221, 234, 261]. However, this is not true for herbivorous/detritivorous fish. For example, freshwater fish such as the brown bullhead (*Ameierus nebulosus*) have only 40–70% of the total Hg as MMHg [262]. It should be noted that many measurements of Hg in fish tissue focus only on muscle as this is the tissue fraction consumed by humans. In general, the percentage of MMHg in muscle tissue is much greater than in other tissues [262], and thus measurements of muscle tissue alone do not reflect the overall fraction of the Hg as MMHg for the whole organism. For example, in freshwater crayfish, muscle MMHg is >80% but the fraction was <60% for the hepatopancreas, gut and gonad, and <20% for the gill and carapace [79]. Overall, the MMHg content of the whole tissue is only 50–75% of the total Hg.

In summary, therefore, the fraction of body burden Hg as MMHg is a function of trophic, not taxonomic position. This is also true for invertebrates. Riisgård and Famme [262] observed 73% MMHg in carnivorous shrimps compared with 17% in suspension feeding mussels collected from the same area, and Mason *et al.* [263] found 60–100% MMHg in predatory insect larvae compared with <50% MMHg in herbivorous insects. A literature summary of the percentage of MMHg in tissues of a variety of invertebrate organisms, listed by trophic group, further illustrates this trend (Table 2.1).

The dominant uptake pathway, food versus water, of MMHg by zooplankton is not well known, but the MMHg content of these organisms may reflect water column MMHg concentrations. For example, a greater fraction of total Hg was present as MMHg in both water and zooplankton in the acidified basin of Little Rock Lake [252], and an experimental reservoir impoundment in Canada led to increased MMHg concentrations in zooplankton that were strongly correlated with MMHg concentrations in water [264]. Typical biomagnification factors (BMF = $[Hg]_{predator}/[Hg]_{prey}$) for MMHg are 3–5 per trophic level [9, 252, 260], but BMFs of less than unity have been observed [264]. A lack of relationship between pH and bioconcentration factors (BCF = $[Hg]_{organism}/[Hg]_{water}$) in seston, zooplankton, and fish [120, 121, 251], supports the idea that the lake acidification influences Hg bioaccumulation primarily through enhanced bacterial MMHg production and uptake at the level of phytoplankton, rather than through effects at higher trophic levels.

Concerns about human exposure to Hg via fish consumption has prompted a great deal of investigation into Hg bioaccumulation and toxicity in fish (see Wiener and Spry [248] for an excellent review). Bioenergetic models [265] and feeding experiments [266] indicate that diet accounts for more than 85% of MMHg uptake by fish, although a small amount may be produced in the gut [267] or taken up from water passed over the gill [248]. As a result, diet and

Table 2.1 Percentage of total Hg present at MMHg in tissues of invertebrate aquatic organisms

Organism	MMHg content (%)	Reference
Plants phytoplankton and macroalgae	< 10	268, 288
Deposit feeders		
Scobicularia plana (bivalve)	20	285
Nereis diversicolor (polychaete)	18	287
Diporeia (amphipods)	35 ± 12	288
Isopods (*Cyathura*)	30–55	136
Snail, (*Littorina*)	3–10	289
Annelids, echinoderms	4–36	289
Polychaetes, echinoderms	25–50	269
Polychaetes, Lavaca Bay	13 ± 13	281
Primary consumer/herbivores		
Cardium spp. (cockles)	30–90	286
Caddisfly, stonefly, mayfly, cranefly larvae	5–55	79
Zooplankton	18 ± 2	288
Zooplankton	35	131
Suspension feeders		
Mytilus edulis (mussel)	15	275
Clams (*Rangia, Macoma*)	5–25	136
Various bivalve molluscs, contaminated marsh	<1–35	289
Bivalves	18–30	269
Bivalves, Lavaca Bay, Texas	62 ± 43	281
Omnivores		
Dipteran and trichopteran larvae	30–60	274
Freshwater crayfish	50–75	79
Various crustaceans	65–90	269
Grass shrimp, Lavaca Bay, USA	87 ± 12	281
Blue crab, Lavaca Bay, USA	88 ± 16	281
Predators		
Stonefly, dragonfly, dobsonfly larvae	60–100	79
Crangon crangon (shrimp)	75	275
Mysis (zooplankton)	103 ± 36	288

trophic structure influence MMHg uptake in fish. In general, piscivorous fishes contain higher concentrations of Hg than fishes of lower trophic level from the same body of water [268–270], and the Hg concentration in fish is positively correlated with trophic level as measured by del ^{15}N [271]. Additionally, longer food chains can lead to higher Hg tissue concentrations in the top predators [272, 273]. Finally, Hg concentration in fish tends to increase with body size [234, 236], and with age as fish shift towards feeding at higher trophic levels [79, 268].

In benthic-based food chains, sedimentary invertebrates represent an important link between the sites of MMHg production and MMHg accumulation in organisms living in the water column. In addition, emerging insects may serve as an important transport vector for Hg transport out of the sediment [274]. Both MMHg and HgI are highly particle reactive, typically having sediment–water distribution coefficients ($K_d = $[ng Hg kg^{-1}]$_{sediment}$/[ng Hg L^{-1}]$_{water}$) of the order of 10^3–10^4 for MMHg and 10^4–10^5 for inorganic Hg [99, 144, 238] (see also Table 2.2). Experimental studies using benthic organisms have shown preferential accumulation of MMHg (compared with inorganic Hg) in insect nymphs [275], crustaceans [263], polychaetes [276], and amphipods [277]. In addition, inorganic Hg tends to be eliminated more rapidly than organic Hg, for example, Wang *et al.* [276] measured efflux rate constants of 0.027 d^{-1} for Hg and 0.014 d^{-1} for MMHg in the polychaete *Nereis succinea*. Both sediments and water represent potential exposure routes for benthic organisms, and the relative contribution of these two may depend on the MMHg content of pore water compared to sediment.

Organic matter (OM) has a particularly high affinity for Hg and MMHg, and the Hg concentration in sediments is often strongly correlated with OM content [136, 238, 278]. However, Hg accumulation in benthic fauna is inversely related to sediment organic matter content [136, 279, 280]. This difference may occur partly as a result of lower feeding rates in high organic matter sediments or differences in K_d across organic matter gradients [276]. Studies in Lavaca Bay, Texas [144, 281] and in the Chesapeake Bay [136] have shown a

Table 2.2 Average distribution coefficients for total Hg and MMHg

Site	log K_d		Reference
	Hg$_T$	MMHg	
Sediments			
Lavaca Bay, Texas, USA	4.9 ± 0.4^a	2.7 ± 0.8	143
Saguenay Fjord, Canada	4.3	3.3	144
St Lawrence Estuary, Canada	4.2		
Patuxent River Estuary, Maryland, USA	4.2–5.1	2.9–3.5	238
The Everglades, Florida, USA	4.3–5.0	2.7–4.1	99
Water Column			
Patuxent River Estuary, Maryland, USA	4.8–5.7	3.8–4.0	238
Scheldt Estuary, Belgium	5.3–6.0	3.6–4.5	290
Lake Michigan, USA	5.5–5.8	5.7	260
Wisconsin Rivers, USA	2.8–5.5	2.9–5.8	230
Anacostia, River, USA	3–6	3–5.5	288
Open ocean	5–6	—	78, 86
Compilation of river data	5.0–6.3	—	126 and references cited therein
Wisconsin Lakes, USA	5.3–6	5.6–6.2	121, 131
Finnish lakes	—	4.2–5.2	137

^{a} K_d for inorganic Hg (Hg$_I$ = Hg$_T$ – MMHg)

strong positive correlation between K_d and sediment OM for both Hg and MMHg, and a concomitant decrease in the bioaccumulation factor for benthic-organisms with increasing OM. These results point to the decreased bio-availability of MMHg from sediment with increasing OM. Laboratory studies have confirmed this trend [280] and have shown that the bioavailability of MMHg during deposit feeding—as measured by the *in vitro* solubilization of MMHg from sediment by the intestinal fluid of specific benthic invertebrates [280]—is also a function of sediment OM. Again, however, while more MMHg, on a percentage basis, is released into solution during incubation of sediment with invertebrate intestinal fluid with decreasing OM, the absolute amount released might not follow this trend as the overall concentration of MMHg in natural sediments tends to increase with increasing sediment OM [136].

Furthermore, a variety of studies looking at the bioaccumulation of MMHg by fish in the presence of different organic ligands, and studies looking at the uptake of MMHg from food [282, 283], suggest that the membrane transport across the gut is not the rate limiting step in accumulation. Indeed, studies, using perfusion techniques, of the factors controlling the accumulation of MMHg and Hg across the gut membrane of the blue crab [284] indicate that the rate of uptake of MMHg is of the same order as that of inorganic Hg, i.e. membrane transport *per se* cannot explain the higher accumulation of MMHg over inorganic Hg during trophic transfer. Rather, it appears from the results of the gut solubilization studies, perfusion experiments, and whole body accumulation studies, that the dissolution of MMHg from sediment or food during digestion is the main control over assimilation, and that factors influencing the extent to which MMHg is solubilized will determine the overall accumulation. For benthic deposit feeders, it is clearly the OM of the sediment that is the primary control over solubilization, although the importance of other factors such as AVS needs further study.

Overall, the bioaccumulation studies show the importance of the transfer of MMHg from water to microorganisms (phytoplankton/bacteria) and the accumulation from sediment into benthic invertebrates in controlling the net accumulation at higher trophic levels. For example, BAFs can range over two orders of magnitude, depending on sediment charateristics, and accumulation for the water can also be significantly influenced by water quality parameters. In contrast, transfer factors between trophic levels do not vary by more than an order of magnitude. Food chain length is also an important factor given the bioaccumulative nature of MMHg. While more studies are clearly warranted, current knowledge is sufficient for the bioaccumulation of MMHg into food chains to be adequately modelled. This is not so for the processes of formation and destruction of MMHg. While substantial advances have been made, there are still important gaps in our knowledge of the principal factors controlling the formation, fate and transport of MMHg in the environment.

2.7 REFERENCES

1. Fujiki, M. and Tajima, S., The pollution of Minamata Bay by mercury. *Wat. Sci. Technol.*, 1992, **25**, 133.
2. Kudo, A., Fujikawa, Y., Miyahara, S., Zheng, J., Takigami, H., Sugahara, M., and Muramatsu, T., Lessons from Minamata mercury pollution, Japan—After a contiunuous 22 years of observation, *Wat. Sci. Technol.*, 1998, **38**, (7), 187.
3. Harada, M., Minamata Disease—Methylmercury poisoning in Japan caused by environmental-pollution, *Crit. Rev. Toxicol.*, 1995, **25**, 1.
4. Grandjean, P., Weihe, P., White, R.F., and Debes, F., Cognitive performance of children prenatally exposed to 'safe' levels of methylmercury, *Environ. Res.*, 1998, **77**, 165.
5. Harada, M., Nakanishi, J., Konuma, S., Ohno, K., Kimura, T., Yamaguchi, H., Tsuruta, K., Kizaki, T., Ookawara, T., and Ohno, H., The present mercury contents of scalp hair and clinical symptoms in inhabitants of the Minamata area, *Environ. Res.*, 1998, **77**, 160.
6. Clarkson, T.W., Amin-Zaki, L. and Al-Tikriti, S.J., An outbreak of methylmercury poisoning due to the consumption of contaminated grain, *Fed. Proc.*, 1976, **35**, 2395.
7. Craig, P.J., Organomercury compounds in the environment. In *Organometallic Compounds in the Environment*, P.J. Craig (Ed.), Longman, Essex, 1986, pp. 65–110.
8. Ebinghaus, R., Turner, R.R., de Lacerda, L.D., Vasiliev, O., and Salomons, W., *Mercury Contaminated Sites: Characterization, Risk Assessment and Remediation*, Springer-Verlag, Berlin, 1999.
9. Lindqvist, O., Johannson, K., Aastrup, M., Andersson, A., Bringmark, L., Hovsenius, G., Hakanson, L., Meili, M., and Timm, B., Mercury in the Swedish environment—Recent research on causes, consequences and corrective methods, *Water Air Soil Pollut.*, Special Issue, 1991, Vol. 55.
10. Nriagu, J.O., A global assessment of the natural sources of atmospheric trace metals, *Nature*, 1989, **338**, 47.
11. Mason, R.P., Fitzgerald, W.F., and Morel, F.M.M., The biogeochemical cycling of elemental mercury: Anthropogenic influences, *Geochim. Cosmochim. Acta*, 1994, **58**, 3191.
12. Fitzgerald, W.F., Engstrom, D.R., Mason, R.P., and Nater, E.A., The case for atmospheric mercury contamination in remote areas, *Environ. Sci. Technol.*, 1998 **32**, 1.
13. Pacyna, J.M., Emission inventories of atmospheric mercury from anthropogenic sources. In *Global and Regional Mercury Cycles: Sources, Fluxes and Mass Balances*, W. Baeyens, R. Ebinghaus, and O. Vasiliev (Eds), Kluwer Academic Publishers, Dordrecht, 1996, pp. 161.
14. Porcella, D.B., Chu, P., and Allan, M.A., Inventory of North American mercury emissions to the atmosphere. In *Global and Regional Mercury Cycles: Sources, Fluxes and Mass Balances*, W. Baeyens, R. Ebinghaus, and O. Vasiliev (Eds), Kluwer Academic Publishers, Dordrecht, 1996, pp. 179–190.
15. Ebinghaus, R., Tripathi, R.M., Wallschlager, D., and Lindberg, S.E., Natural and anthropogenic mercury sources and their impact on the air-surface exchange of mercury on regional and global scales. In *Mercury Contaminated Sites: Characterization, Risk Assessment and Remediation*, Ebinghaus, R., Turner, R.R., de Lacerda, L.D., Vasiliev, O., and Salomons, W. (Eds), Springer-Verlag, Berlin, 1999, pp. 1–50.
16. Swain, E.B., Engstrom, D.R., Brigham, M.E., Henning, A.S and Brezonik, P.L., Increasing rates of atmospheric mercury deposition in midcontinental North America, *Science*, 1992, **257**, 784.
17. Engstrom, D.R., and Swain, E.B., Recent declines in atmospheric mercury deposition in the Upper Midwest, *Environ. Sci. Technol.*, 1997, **31**, 960.

18. Benoit, J.M., Fitzgerald, W.F., and Damman, A.W.H., The biogeochemistry of an ombrotrophic bog: Evaluation of use as an archive of atmospheric mercury deposition, *Environ. Res.*, 1998, **78**, 118.

19. US EPA, *Mercury Study Report to Congress*, EPA-452/R-97-004, US EPA Office of Air, Washington, DC, 1997.

20. Akagi, H., Malm, I., Branches, F.J.P., Kinjo, Y., Kashima, Y., Guimaraes, J.R.D., Oliveira, R.B., Haragughi, K., Pfeiffer, W.C., Takizawa, Y., and Kato, H., Human exposure to mercury due to goldmining in the Tapajos River basin, Amazon, Brazil: Speciation of mercury in human hair, blood and urine, *Water Air Soil Pollut.*, 1995, **80**, 85.

21. Cleary, D., *Anatomy of the Amazon Gold Rush.*, University of Iowa Press, Iowa City, 1990.

22. Seigneur, C., Lohman, K., Pai, P., Heim, K., Mitxhell, D., and Levin, L., Uncertainty analysis of regional mercury exposure, *Water Air Soil Pollut.*, 1999, **112**, 151.

23. Bullock, R.O., Benjey, W.G., and Keating, M.H., Modeling of regional scale atmospheric mercury transport and deposition using RELMAP. In *Atmospheric Deposition of Contaminants to the Great Lakes and Coastal Waters*, J.E. Baker (Ed.), SETAC Press, Florida, 1997, pp. 323–347.

24. Watras, C.J., Morrison, K.A., and Host, J.S., Concentration of mercury species in relationship to other site-specific factors in the surface waters of northern Wisconsin lakes, *Limnol. Oceanogr.*, 1995, **40**, 556.

25. Fitzgerald, W.F., Vandal, G. M., Mason, R.P., and Dulac, F., Air-water cycling of mercury in lakes. In *Mercury as a Global Pollutant: Towards Integration and Synthesis*, C.J. Watras and J.W. Huckabee (Eds), Lewis, Boca Raton, 1994, pp. 203–220

26. Compeau, G., and Bartha, R., Sulfate-reducing bacteria: Principle methylators of mercury in anoxic estuarine sediment, *Appl. Environ. Microbiol.*, 1985, **50**, 498.

27. Gilmour, C.C., Henry, E.A., and Mitchell, R., Sulfate stimulation of mercury methylation in freshwater sediments, *Environ. Sci. Technol.*, 1992, **26**, 2281.

28. See, for example, Clarkson, T.W., Human toxicology of mercury, *J. Trace Element Exp. Med.*, 1998, **11**, 303; Sigel, A. and Sigel, H. (Eds) *Metal Ions in Biological Syatems*, Vol. 34, *Mercury and Its Effects on Environment and Biology*, Marcel Dekker, New York, 1997.

29. Van Oostdam, J., Gilman, A., Dewailly, E., Usher, P., Whatley, B., Kuhnlein, H., Neve, S., Walker, J., Tracy, B., Feely, M., Jerome, V. and Kwavnick, B., Human health implications of environmental contaminants in Arctic Canada: a review, *Sci. Tot. Environ.*, 1999, **230**, 1.

30. Wheatley, B., Paradis, S., Lassonde, M., Giguere, M.F., and Tanguay, S., Exposure patterns and long term sequelae on adults and children in two Canadian indigenous communities exposed to methylmercury, *Water Air Soil Pollut.*, 1997, **97**, 63.

31. Bruhn, C.G., Rodriguez, A.A., Barrios, C.A., Jaramillo, V.H., Beccerra, J., Gras, N.T., Nunez, E., and Reyes, O.C., Total mercury and methylmercury levels in scalp hair and blood of pregnant women residents of fishing villages in the Eighth Region of Chile, *Environ. Biomon.*, 1997, **654**, 151.

32. Kehrig, H.D., Malm, O., Akagi, H., Guimaraes, J.R.D., and Torres, J.P.M., Methylmercury in fish and hair samples from the Balbina reservoir, Brazilian Amazon, *Environ. Res.*, 1998 **77**, 84.

33. Lipfert, F.W., Estimating exposure to methylmercury: Effects of uncertainties, *Water Air Soil Pollut.*, 1997, **97**, 119.

34. Stern, A.H., Re-evaluation of the reference dose for methylmercury and assessment of current exposure levels, *Risk Anal.*, 1993, **13**, 355.

35. US EPA, *National Forum on Mercury in Fish*, EPA 823-R-95-002, US EPA Office of Water, Washington, DC, 1995.

36. US EPA, *Listing of Fish and Wildlife Advisories*, CD-ROM, EPA-823-C-97-005, 1996.

37. Clarkson, T.W., Human health risks from methylmercury in fish, *Environ. Toxicol. Chem.*, 1990, **9**, 821.

38. Cox, C., Clarkson, T.W., Marsh, D.O., Amin-Zaki, L., Tikriti, S., and Myers, G.G., Dose-response analysis of infants prenatally exposed to methylmercury: An application of a single compartment model to single-strand hair analysis, *Environ. Res.*, 1989, **49**, 318.

39. Marsh, D.O., Clarkson, T.M., Cox, C., Myers, G.J., Amin-Zaki, L., and Al-Tikriti, S., Fetal methylmercury poisoning: Relationship between concentration in single strands of maternal hair and child effects, *Arch. Neurol.*, 1987, **44**, 1017.

40. Kjellstrom, T., Kennedy, P., Wallis, S., Stewart, A., Lind, B., Wutherspoon, T., and Mantell, C., Physical and mental development of children with prenatal exposure to mercury from fish. Stage 2: Interviews and psychological tests at age 6. *National Swedish Environmental Protection Board, Report 3642*, Solna, 1989.

41. McKeown-Eyssen, G.E., Ruedy, J., and Neims, A., Methylmercury exposure in northern Quebec. II. Neurologic findings in children, *Am. J. Epidemiol.*, 1983, **118**, 470.

42. Wagemann, R., Trebacz, E., Boila, G., and Lockhart, W.L., Methylmercury and total mercury in tissues of arctic marine mammals, *Sci.Tot. Environ.*, 1998, **218**, 19.

43. Myers, G.J, Davidson, P.W., and Shamlaye, C.F., A review of methylmercury and child development, *Neurotoxicol.*, 1998, **19**, 313.

44. Myers, G.J., and Davidson, P.W., Prenatal methylmercury exposure and children: Neurologic, developmental, and behavioral research, *Environ. Health Perspec.*, 1998, **106**, 841.

45. Mahaffey, K.R., Methylmercury exposure and neurtoxicity, *J. Am. Med. Assoc.*, 1998, **280**, 737.

46. Davidson, P.W., Myers, G.J., Cox, C., Axtell, C., Shamlaye, C., Sloane-Reeves, J., Cernichiari, E., Needham, L., Choi, A., Wang, Y.N., Berlin, M., and Clarkson, T.W., Effects of prenatal and postnatal methylmercury exposure from fish consumption on neurodevelopment—outcomes at 66 months of age in the Seychelles Child Development Study, *J. Am. Med. Assoc.*, 1998, **280**, 701.

47. Myers, G.J., Davidson, P.W., Shamlaye, C.F., Axtell, C.D., Cernichiari, E., Choisy, O., Choi, A., Cox, C., and Clarkson, T.W., Effects of prenatal methylmercury exposure from a high fish diet on developmental milestones in the Seychelles Child Development Study, *Neurotoxicol.*, 1997, **18**, 819.

48. Axtell, C.D., Myers, G.J., Davidson, P.W., Choi, A.L., Cernichiari, E., Sloane-Reeves, J., Shamlaye, C., Cox, C., and Clarkson, T.W., Semiparametric modeling of age at achieving developmental milestones after prenatal exposure to methylmercury in the Seychelles Child Development Study, *Environ. Health Perspec.*, 1998, **106**, 559.

49. Grandjean, P., Weihe, P., White, R.F., Debes, F., Araki, S., Yokoyama, K., Murata, K., Sorensen, N., Dahl, R., and Jorgensen, P.J., Cognitive deficit in 7-year-old children wtih prenatal exposure to methylmercury, *Neurotoxicol. Terat.*, 1997, **19**, 417.

50. Murata, K., Weihe, P., Renzoni, A., Debes, F., Vasconcelos, R., Zino, F., Araki, S., Jorgensen P.J., White, R.F., and Grandjean, P., Delayed evoked potentials in children exposed to methylmercury from seafood, *Neurotoxicol.Terat.*, 1999, **21**, 343.

51. Wolfe, M.F., Scwarzback, S., and Sulaiman, R.A., Effects of mercury on wildlife: A comprehensive review, *Environ. Toxicol. Chem.*, 1998, **17**, 146.

52. O'Connor, D.J., and Nielsen, S.W., Environmental survey of methylmercury levels in wild mink (*Mustela vison*) and otter (*Lutra canadensis*) from the northeastern United States and experimental pathology of methylmercurialism in the otter,

Proceedings: Worldwide Furbearer Conference, Frostburg, MD, USA, August 3–11, 1981, pp. 1728.

53. Woebeser, G., Nielsen, N.O., and Schiefer, B., Mercury and mink I. Use of mercury-contaminated fish as a food for ranch mink intoxication, Can. J. Comp. Med., 1976, **40**, 30.

54. Woebeser, G., Nielsen, N.O., and Schiefer, B., Mercury and mink II. Experimental methylmercury intoxication, Can. J. Comp. Med., 1976, **40**, 34.

55. Wren, C.D., A review of metal accumulation and toxicity in wild mammals I. Mercury, Environ. Res., 1986, **40**, 1737.

56. Hoffman, D.J., Ohlendorf, H.M., Marn, C.M., and Pendleton, G.W., Association of mercury and selenium with altered glutathione metabolism and oxidative stress in diving ducks from the San Francisco Bay region, USA, Environ. Toxicol. Chem., 1998, **17**, 167.

57. Evers, D.C., Kaplan, J.D., Meyer, M.W., Reaman, P.S., Braselton, W.E., Major, A., Burgess, N., and Scheuhammer, A.M., Geographic trend in mercury measured in common loon feathers and blood, Environ. Toxicol. Chem., 1998, **17**, 173.

58. Monteiro, L.R., Granadeiro, J.P., and Furness, R.W., Relationship between mercury levels and diet in Azores seabirds, Mar. Ecol. Prog. Ser., 1998, **166**, 259.

59. Scheuhammer, A.M., Atchison, C.M., Wong, A.H.K., and Evers, D.C., Mercury exposure in breeding common loons (Gavia immer) in central Ontario, Canada, Environ. Toxicol. Chem., 1998, **17**, 191.

60. Barr, J.F., Population dynamics of the common loon (Gavia immer) associated with mercury-contaminated waters in northwestern Ontario, Occasional Paper 56, Canadian Wildlife Service, Ottawa, ON, 1986.

61. Fimerite, N., Effects of methylmercury on ring-necked pheasants, with special reference to reproduction, Occasional Paper 9, Canadian Wildlife Service, Ottawa, ON, 1971.

62. Heinz, G., Effects of low dietary levels of methylmercury on mallard reproduction, Bull. Environ. Contam. Toxicol., 1974, **11**, 386.

63. Bouton, S.N., Frederick, P.C., Spalding, M.G., and McGill, H., Effects of chronic, low concentrations of dietary methylmercury on the behavior of juvenile great egrets, Environ. Toxicol. Chem., 1999, **18**, 1934.

64. Sepulveda, M.S., Frederick, P.C., Spalding, M.G., and Williams, G.E. Jr., Mercury contamination in free-ranging great egret nestlings (Ardea albus) from southern Florida, USA, Environ. Toxicol. Chem., 1999, **18**, 985.

65. Frederick, P.C., Spalding, M.G., Sepulveda, M.S., Williams, G.E., Nico, L., and Robins, R., Exposure of great egret (Ardea albus) nestlings to mercury through diet in the Everglades ecosystem, Environ. Toxicol.Chem., 1999, **18**, 1940.

66. Scheuhammer, A.M., Wong, A.H.K., and Bond, D., Mercury and selenium accumulation in common loons (Gavia immer) and common mergansers (Mergus merganser) from Eastern Canada, Environ. Toxicol.Chem., 1998, **17**, (2), 197.

67. Koeman, J.H., Peeters, W.H.M., Koudstaal-Hol, C.H.M., Tjioe, P.S., and de Goeij, J.J.M., Mercury–selenium correlations in marine mammals, Nature, 1973, **245**, 385.

68. Thompson, D.R., and Furness, R.W., The chemical form of mercury stored in South Atlantic seabirds, Environ. Pollut., 1989, **60**, 305.

69. Kari, T., and Kauranen, P., Mercury and selenium contents of seals from fresh and brackish waters in Finland, Bull. Environ. Contam. Toxicol., 1978, **19**, 273.

70. Hoffman, D.J., and Heinz, G.H., Effects of mercury and selenium on glutathione metabolism and oxidative stress in mallard ducks, Environ. Toxicol. Chem., 1998, **17**, 161.

71. Stoewsand, G.S., Bache C.A., and Lisk, D.J., Dietary selenium protection of methylmercury intoxication of Japanese quail, *Bull. Environ. Contam. Toxicol.*, 1974, **11**, 152.

72. Zillioux, E.J., Porcella, D.B., and Benoit, J.M., Mercury cycling and effects in freshwater ecosystems, *Environ. Toxicol. Chem.*, 1993, **12**, 2245.

73. Monteiro, L.R., Furness, A.J., and del Novo, A.J., Mercury levels in seabirds from the Azores, mid-North Atlantic Ocean, *Arch. Environ. Contam. Toxicol.*, 1995, **28**, 304.

74. Furness, R.W., Muirhead, S.J., and Woodburn, M., Using bird feather parts as a monitor for metal pollution, *Environ. Pollut.*, 1986, **8**, 281.

75. Braune, B.M., Comparison of total mercury levels in relation to diet and molt for nine species of marine birds, *Arch. Environ. Contam.Toxicol.*, 1987, **16**, 217.

76. Horvat, M., Mercury analysis and speciation in environmental samples. In *Global and Regional Mercury Cycles: Sources, Fluxes and Mass Balances*, W. Baeyens, R. Ebinghaus, and O. Vasiliev (Eds), Kluwer Academic Publishers, Dordrecht, 1996, pp. 1–31.

77. Weber, J.H., Analytical methods for the determination of mercury(II) and methylmercury compounds: the problem of speciation. In *Metal Ions in Biological Systems*, Volume 34. A. Sigel and H. Sigel (Eds) Marcel Dekker, New York, 1997, pp. 1–20.

78. Mason, R.P., Rolfhus, K.R., and Fitzgerald, W.F., Mercury in the North Atlantic, *Mar. Chem.*, 1998, **61**, 37.

79. Mason, R.P., Lawson, N.M., Lawrence, A.L., Leaner, J.J., Lee, J.G., and Sheu, G-R., Mercury in the Chesapeake Bay, *Mar. Chem.*, 1999, **65**, 77.

80. Bloom, N.S., Determination of picogram levels of methylmercury by aqueous phase ethylation followed by cryogenic gas chromatography with cold vapour atomic fluorescence detection, *Can. J. Fish. Aquat. Sci.*, 1989, **46**, 1131.

81. Horvat, M., Bloom, N.S., and Liang, L., Comparison of distillation with other current isolation methods for the determination of methyl mercury compounds in environmental samples, Part 1: Sediments, *Anal. Chim. Acta.*, 1993 **281**, 135.

82. Westoo, G., Determination of methylmercury salts in various kinds of biological material, *Acta Chem. Scand.*, 1968, **22**, 2277.

83. Bloom, N.S., and Fitzgerald, W.F., Determination of volatile mercury species at the picogram level by low-temperature gas chromatography with cold vapour atomic fluorescence detection, *Anal. Chim. Acta*, 1988, **208**, 151.

84. Rapsomanikis, S., Donard, O.F.X., and Weber, J.H., Speciation of lead and methyllead ions in water by chromatography/atomic absorption spectrometry after ethylation with sodium tetraethylborate, *Anal. Chem.*, 1986, **58**, 35.

85. Mason, R.P., and Fitzgerald, W.F., Alkylmercury species in the equatorial Pacific, *Nature*, 1990, **347**, 457.

86. Mason, R.P., and Fitzgerald, W.F., The distribution and biogeochemical cycling of mercury in the equatorial Pacific Ocean, *Deep-Sea Res.*, 1993, **40**, 1897.

87. Amouroux, D., Tessier, E., Pecheyran, C., and Donard, O.F.X., Sampling and probing volatile metal(loid) species in natural waters by *in-situ* purge and cryogenic trapping followed by gas chromatography and inductively coupled plasma mass spectrometry (P-CT-GC-ICP/MS), *Anal. Chim. Acta*, 1998, **377**, 241.

88. Bloom, N.S., and Von Der Geest, E.J., Matrix modification to improve recovery of methylmercury from clear water using distillation, *Water Air Soil Pollut.*, 1995, **80**, 1319.

89. Horvat, M., May, K., and Stoeppler, P., Comparative studies of methylmercury determination in biological and environmental samples, *Appl. Organomet. Chem.*, 1986, **2**, 515.

90. Horvat, M., Bloom, N.S., and Liang, L., Comparison of distillation with other current isolation methods for the determination of methyl mercury compounds in environmental samples, Part II. Water, *Anal. Chim. Acta*, 1993, **282**, 153.

91. Hintelmann, H., Falter, R., Ilgen, G., and Evans, R.D., Determination of artifactual formation of monomethylmercury (CH_3Hg^+) in environmental samples using stable Hg^{2+} isotopes with ICP-MS detection: Calculation of contents applying species specific isotope addition, *Fresenius J. Anal. Chem.*, 1997, **358**, 363.

92. Bloom, N.S., Colman, J.A., and Barber, L., Artifact formation of methylmercury during aqueous distillation and alternative techniques for the extraction of methyl mercury from environmental samples, *Fresenius J. Anal. Chem.*, 1997, **358**, 372.

93. Hintelmann, H., Comparison of different extraction techniques used for methylmercury analysis with respect to accidental formation of methylmercury during sample preparation, *Chemosphere*, 1999, **39**, 1093.

94. Falter, R., Experimental study on the unintentional abiotic methylation of inorganic mercury during analysis, Part 1. Localisation of the compounds effecting the abiotic mercury methylation, *Chemosphere*, 1999, **39**, 1051.

95. Falter, R., Experimental study on the unintentional abiotic methylation of inorganic mercury during analysis, Part 2. Controlled laboratory experiments to elucidate the mechanism and critical discussion of the species specific isotope addition correction method, *Chemosphere*, 1999, **39**, 1075.

96. Furutani, A., and Rudd, J.W.M. Measurement of mercury methylation in lakewater and sediment samples, *Appl. Environ. Microbiol.*, 1980, **40**, 770.

97. Gilmour, C.C., and Riedel, G.S., Measurement of Hg methylation in sediments using high specific-activity ^{203}Hg and ambient incubation, *Water Air Soil Pollut.*, 1995, **80**, 747.

98. Stordal, M., and Gill, G., Determination of mercury methylation rates using a ^{203}Hg radiotracer technique, *Water Air Soil Pollut.*, 1995, **80**, 725.

99. Gilmour, C.C., Riedel, G.S., Ederington, M.C., Bell, J.T., Benoit, J.M., Gill, G.A., and Stordal, M.C., Methylmercury concentrations and production rates across a trophic gradient in the northern Everglades, *Biogeochem.*, 1998, **40**, 327.

100. Ramlal, P.S., Rudd, J.W.M., and Hecky, R.E., Methods for measuring specific rates of mercury methylation and degradation and their use in determining factors controlling net rates of mercury methylation, *Appl. Environ. Microbiol.*, 1986, **51**, 110.

101. Marvin-Dipasquale, M.C., and Oremland, R.S., Bacterial methylmercury degradation in Florida Everglades peat sediment, *Environ. Sci. Technol.*, 1998, **32**, 2556.

102. Hintelmann, H., and Evans, R.D., Application of stable isotopes in environmental tracer studies—Measurement of monomethylmercury (CH_3Hg^+) by isotope dilution ICP-MS and detection of species transformation, *Fresenius J. Anal. Chem.*, 1997, **358**, 378.

103. Wilken, R.D., and Falter, R., Determination of methylmercury by the species-specific isotope addition method using a newly developed HPLC-ICP MS coupling technique with ultrasonic nebulization, *Appl. Organomet. Chem.*, 1998, **12**, 551.

104. Falter, R., and Ilgen, G., Determination of trace amounts of methylmercury in sediment and biological tissue by using water vapor distillation in combination with RP C18 preconcentration and HPLC-HPF/HHPN-ICP-MS, *Fresenius J. Anal. Chem.*, 1997, **358**, 401.

105. Wasik, A., Pereiro, I.R., Dietz, C., Szpunar, J., and Lobinski, R., Speciation of mercury by ICP-MS after on-line capillary cryofocussing and ambient temperature multicapillary gas chromatography, *Anal. Comm.*, 1998, **35**, 331.

106. Tao, H., Murakami, T., Tominaga, M., and Miyazaki, A., Mercury speciation in natural gas condensate by gas chromatography inductively coupled plasma mass spectrometry, *J. Anal. Atom. Spec.*, 1998, **13**, 1085.

107. Hirner, A.V., Krupp, E., Schulz, F., Koziol, M., and Hofmeister, W., Organometal(loid) species in geochemical exploration: preliminary qualitative results, *J. Geochem. Explor.*, 1998, **64**, 133.

108. Tseng, C.M., de Diego, A., Martin, F.M., and Donard, O.F.X., Rapid and quantitative microwave-assisted recovery of methylmercury from standard reference sediments, *J. Anal. Atom. Spec.*, 1997, **12**, 629.

109. de Diego, A., Tseng, C.M., Stoichev, T., Amouroux, D., and Donard, O.F.X., Interferences during mercury speciation determination by volatilization, cryofocusing, gas chromatography and atomic absorption spectroscopy: comparative study between hydride generation and ethylation techniques, *J. Anal. Atom. Spec.*, 1998, **13**, 623.

110. Craig, P.J., Jenkins, R.O., and Stojak, G.H., The analysis of inorganic and methylmercury by derivatisation methods; opportunities and difficulties, *Chemosphere*, 1999, **39**, 1181.

111. Puk, R., and Weber, J.H., Determination of mercury(II), monomethylmercury cation, dimethylmercury and diethylmercury by hydride generation, cryogenic trapping and atomic absorption spectrometric detection, *Anal. Chim. Acta*, 1994, **292**, 175.

112. Schmitt, V.O., de Diego, A., Cosnier, A., Tseng, C.M., Moreau, J., and Donard, O.F.X., Open focused microwave-assisted sample preparation procedures: fundamentals and application to the speciation of tin and mercury in environmental samples, *Spectroscopy*, 1996/97, **13**, 99.

113. Lee, Y.H., Mowrer, J., Determination of methylmercury in natural waters at subnanogram per liter levels by capillary gas chromatography after absorbent preconcentration, *Anal. Chim. Acta*, 1989, **221**, 259.

114. Hudson, J.M., Gherini, S.A., Fitzgerald, W.F., and Porcella, D.B., Anthropogenic influences on the global mercury cycle: A model-based analysis, *Water Air Soil Pollut.*, 1995, **80**, 265.

115. Gilmour, C.C., and Henry, E.A., Mercury methylation in aquatic systems affected by acid deposition, *Environ. Poll.*, 1991, **71**, 131.

116. Benoit, J.M., Gilmour, C.C., and Mason, R.P., Sulfide controls on mercury speciation and bioavailability in sediment pore waters, *Environ. Sci. Technol.*, 1999, **33**, 951.

117. Benoit, J.M., Mason, R.P., and Gilmour, C.C., The effect of sulfide on the octanol–water partitioning and bioavailability of Hg, *Environ. Toxicol. Chem.*, 1999, **18**, 2138.

118. St. Louis, V.L., Rudd, J.M.W., Kelly, C.A., Beaty, K.G., Bloom, N.S., and Flett, R.J., The importance of wetlands as sources of methyl mercury to boreal forest ecosystems, *Can. J. Fish. Aquat. Sci.*, 1994, **51**, 1065.

119. Gill, G.A., Bloom, N.S., Cappellino, S., Driscoll, C., Dobbs, C., McShea, L., Mason, R. and Rudd, J.W.M., Sediment-water fluxes of mercury in Lavaca Bay, Texas, *Environ. Sci. Technol.*, 1999, **33**, 663.

120. Bloom, N.S., Watras, C.J., and Hurley, J.P., Impact of acidification on the methylmercury cycle of remote seepage lakes, *Water Air Soil Pollut.*, 1991, **56**, 477.

121. Watras C.J., Bloom, N.S., Hudson, R.J.M., Gherini, S., Munson, R., Class, S.A., Morrison, K.A., Hurley, J., Wiener, J.G., Fitzgerald, W.F., Mason, R., Vandel, G., Powell, D., Rada, R., Rislov, L., Winfrey, M., Elder, J., Krabbenhoft, D., Andren, A.W., Babiarz, C., Porcella, D.B., and Huckabee, J.W., Sources and fates of mercury and methylmercury in Wisconsin Lakes. In *Mercury Pollution: Integration and Synthesis*, C.J. Watras and J.W. Huckabee (Eds), Lewis Publishers, Boca Raton, 1994, pp. 153–177.

122. Driscoll, C.T., Blette, V., Yan, C., Schofield, C.L., Munson, R., and Holsapple, J., The role of dissolved organic carbon in the chemistry and bioavailability of mercury in remote Adirondack lakes, *Water Air Soil Pollut.*, 1995, **80**, 499.

123. Mason, R.P., Rolfhus, K.R., and Fitzgerald, W.F., Methylated and elemental mercury cycling in surface and deep ocean waters of the North Atlantic, *Water Soil Air Pollut.*, 1995, **80**, 665.

124. Mason, R.P., and Sullivan, K.A., Mercury in the South and equatorial Atlantic, *Deep-Sea Res.*, 1999, **46**, 937.

125. Cossa, D., Martin, J.-M., and Sanjuan, J., Dimethylmercury formation in the Alboran Sea, *Mar. Pollut. Bull.*, 1994, **28**, 381.

126. Cossa, D., Coquery, M., Martin, J.-M., and Gobell, C., Mercury fluxes at the ocean margins. In: *Global and Regional Mercury Cycles: Sources, Fluxes and Mass Balances*, W. Baeyens, R. Ebinghaus and O. Vasiliev (Eds), Kluwer Academic Publishers, Dordrecht, 1996, pp. 292–248.

127. Watras, C.J., Morrison, K.A., and Back, R.C., Mass balance studies of mercury and methylmercury in small temperate/boreal lakes in the Northern Hemisphere. In *Global and Regional Mercury Cycles: Sources, Fluxes and Mass Balances*, W. Baeyens, R. Ebinghaus, and O. Vasiliev (Eds), Kluwer Academic Publishers, Dordrecht, 1996, pp. 329–358.

128. Meili, M., Mercury in lakes and rivers. In *Metal Ions in Biological Systems*, Volume 34, A. Siegel. and H. Siegel (Eds), Marcel Dekker, NY, 1997, pp. 21–52.

129. Boudou, A., and Ribeyre, F., Mercury in the food web: Accumulation and transfer mechanisms, In *Metal Ions in Biological Systems*, Volume 34. A. Sigel and H. Sigel (Eds), Marcel Dekker, NY, 1997, pp. 289–320.

130. Hintelmann, H., Welbourn, P.M., and Evans, R.D., Measurement of complexation of methylmercury(II) compounds by freshwater humic substances using equilibrium dialysis, *Environ. Sci. Technol.*, 1997, **31**, 489.

131. Hudson, R.J.M., Gherini, S., Watras, C., and Porcella, D., Modeling the biogeochemical cycling of mercury in lakes. In *Mercury as a Global Pollutant: Towards Integration and Synthesis*, C.J. Watras and J.W. Huckabee (Eds), Lewis Publishers, Boca Raton, 1994, pp. 473–526.

132. Dyrssen, D., and Wedborg, M., The sulphur–mercury(II) system in natural waters, *Water Air Soil Pollut.*, 1991, **56**, 507.

133. Mason, R.P., Reinfelder, J.R., and Morel, F.M.M., Uptake, toxicity, and trophic transfer of mercury in a coastal diatom, *Environ. Sci. Technol.*, 1996, **30**, 1835.

134. Dzomback, D.A., and Morel, F.M.M., *Surface Complexation Modeling*, John Wiley & Sons, NY, 1990.

135. Gobeil, C., and Cossa, D., Mercury in sediments and sediment pore water in the Laurentian Trough, *Can. J. Fish. Aquat. Sci.*, 1993, **50**, 2281.

136. Mason, R.P., and Lawrence, A.L., The concentration, distribution and bioavailability of mercury and methylmercury in sediments of Baltimore Harbor and the Chesapeake Bay, Maryland USA, *Environ. Toxicol. Chem.*, 1999, **18**, 2438.

137. Verta, M., and Matalianen, T., Methylmercury distribution and partioning in stratified Finnish forest lakes, *Water Air Soil Pollut.*, 1995, **80**, 585.

138. Iverfeldt, A., Mercury in the Norwegian fjord Framvaren, *Mar. Chem.*, 1988, **23**, 441.

139. Mason, R.P., Fitzgerald, W.F., Hurley, J., Hanson, A.K. Jr., Donaghay, P.L., and Sieburth, J.M., Mercury biogeochemical cycling in a stratified estuary, *Limnol. Oceanog.*, 1993, **38**, 1227.

140. Herrin, R.T., Lathrop, R.C., Gorski, P.R., and Andren, A.W., Hypolimnetic methylmercury and its uptake by plankton during fall destratification: a key entry point of mercury into lake food chains? *Limnol. Oceanogr.*, 1998, **43**, 1476.

141. Slotton, D.G., Reuter, J.E., and Goldman, C.R., Mercury uptake patterns of biota in a seasonally anoxic northern California reservoir, *Water Air Soil Pollut.*, 1995, **80**, 841.

142. Krabbenhoft, D.P., Hurley, J.P., Olson, M.L., and Cleckner, L.B., Diel variability of mercury phase and species distributions in the Florida Everglades, *Biogeochem.*, 1998, **40**, 311.

143. Bloom, N.S., Gill, G.A., Driscoll, C., Rudd, J., and Mason, R.P., An investigation regarding the speciation and cycling of mercury in Lavaca bay sediments, *Environ. Sci. Technol.*, 1999, **33**, 7.

144. Gagnon, C., Pelletier, E., Mucci, A., and Fitzgerald, W.F., Diagenetic behavior of methylmercury in organic-rich coastal sediments, *Limnol. Oceanogr.*, 1996, **41**, 428.

145. Mason, R.P., Lawson, N.M., and Sullivan, K.A., The concentration, speciation and sources of mercury in Chesapeake Bay precipitation, *Atmos. Environ.*, 1997, **31**, 3541.

146. Mason, R.P., and Fitzgerald, W.F., Sources, sinks and biogeochemical cycling of mercury in the ocean. In *Global and Regional Mercury Cycles: Sources, Fluxes and Mass Balances*, W. Baeyens, R. Ebinghaus, and O. Vasiliev (Eds), Kluwer Academic Publishers, Dordrecht, 1996, pp. 249–272.

147. Bloom, N.S., and Watras, C.J., Observations of methylmercury in precipitation, *Sci. Total Environ.*, 1989, **87/88**, 199.

148. Lamborg, C.H., Fitzgerald, W.F., Vandal, G.M., and Rolfhus, K.R., Atmospheric mercury in northern Wisconsin: Sources and species, *Water Air Soil Pollut.*, 1995, **80**, 189.

149. St. Louis, V.L., Rudd, J.W.M., Kelly, C.A., and Barrie, L.A., Wet deposition of methylmercury in Northwestern Ontario compared to other geographic locations, *Water Air Soil Pollut.*, 1995, **80**, 405.

150. Iverfeldt, A., Munthe, J., and Hultberg, H., Terrestrial mercury and methylmerucry budgets for Scandinavia. In *Global and Regional Mercury Cycles: Sources, Fluxes and Mass Balances*, W. Baeyens, R. Ebinghaus, and O. Vasiliev (Eds), Kluwer Academic Publishers, Dordrecht, 1996, pp. 381–402.

151. Prestbo, E.M., and Bloom, N.S., Mercury speciation adsorption (MESA) method for combustion flue gas: methodology, artifacts, intercomparison, and atmospheric implications, *Water Air Soil Pollut.*, 1995, **80**, 145.

152. Bringmark, L., Accumulation of mercury in soil and effects on the soil biota. In *Metal Ions in Biological Systems*, Volume 34, A. Sigel and H. Sigel (Eds), Marcel Dekker, New York, 1996, pp. 161–181.

153. Carpi, A., and Lindberg, S.E., Sunlight-mediated emission of elemental mercury from soil amended with municipal sewage sludge, *Environ. Sci. Technol.*, 1997, **81**, 2085.

154. Bloom, N.S., Prestbo, E.M., Tokos, J.S., Von Der Geest, E., and Kuhn, E.S., Distribution and origins of mercury species in the Pacific northwest atmosphere. Presentation made at the 4th International Conference on Mercury as a Global Pollutant, Hamburg, Germany, August 1996.

155. Nikki, H., Maker, P.D., Savage, C.M., and Brietenbach, L.P., Kinetics and mechanisms for the reaction of chloride and dimethylmercury, *J. Phys. Chem.*, 1983, **87**, 3722.

156. Nikki, H., Maker, P.D., Savage, C.M., and Brietenbach, L.P., A long-path Fourier transform infrared study of the kinetics and mechanism for the OH-radical initiated oxidation of dimethylmercury, *J. Phys. Chem.*, 1983, **87**, 4978.

157. Fitzgerald, W.F., Mason, R.P., and Vandal, G.M., Atmospheric cycling and air-water exchange of mercury over mid-continental lacustrine regions, *Water Air Soil Pollut.*, 1991, **56**, 745.

158. Mason, R.P., The chemistry of mercury in the equatorial Pacific Ocean. Ph.D. Dissertation, The University of Connecticut, 1991, 305 pages.

159. Iverfeldt, A., and Lindqvist, O., Distribution equilibrium of methyl mercury chloride between water and air, *Atmos. Environ.*, 1982, **16**, 2917.
160. Meuleman, C., Laino, C.C., Lansens, P., and Baeyens, W., A study of the behaviour of methylmercury compounds in aqueous solutions, and of gas/liquid distribution coefficients, using head space analysis, *Water Res.*, 1993, **27**, 1431.
161. Mason, R.P., Fitzgerald, W.F., and Vandal, G.M., The sources and composition of mercury in Pacific Ocean rain, *J. Atmos. Chem.*, 1992, **14**, 489.
162. Lamborg, C.H., Rolfhus, K.R., and Fitzgerald, W.F., The atmospheric cycling and air–sea exchange of mercury species in the south and equatorial Atlantic Ocean, *Deep-Sea Res.*, 1999, **46**, 957.
163. Seinfeld, J.H., and Pandis, S.N., *Atmospheric Chemistry and Physics*, John Wiley & Sons, NY, 1998.
164. Rolfhus, K.R., and Fitzgerald, W.F., Linkages between atmospheric mercury deposition and the methylmercury content of marine fish, *Water Air Soil Pollut.*, 1995, **80**, 291.
165. Jensen, S., and Jernelov, A., Biological methylation of mercury in aquatic organisms, *Nature.*, 1969, **223**, 753.
166. Nagase, H., Ose,Y., Sato, T., and Ishikawa, T., Mercury methylation by compounds in humic material, *Sci. Total Environ.*, 1984, **32**, 147.
167. Lee, Y.H., Hultberg, H., and Andersson, I., Catalytic effect of various metal ions on the methylation of mercury in the presence of humic substances, *Water Air Soil Pollut.*, 1985, **25**, 391.
168. Weber, J.H., Review of possible paths for abiotic methylation of Hg(II) in aquatic environments, *Chemosphere.*, 1993, **26**, 2063.
169. Olson, B.H., and Cooper, R.C., Comparison of aerobic and anaerobic methylation of mercuric chloride by San Francisco Bay sediments, *Water Res.*, 1976, **10**, 113.
170. Berman, M., and Bartha, R., Levels of chemical versus biological methylation of mercury in sediments, *Bull. Environ. Contam. Toxicol.*, 1986, **36**, 401.
171. Regnell, O., The effect of pH and dissolved oxygen levels on methylation and partitioning of mercury in freshwater model systems, *Environ. Pollut*, 1994, **84**, 7.
172. Robinson, J.B., and Tuovinen, O.H., Mechanisms of microbial resistance and detoxification of mercury and organomercury compounds: Physiological, biochemical and genetic analyses, *Microbiol. Rev.*, 1984, **48**, 95.
173. Wood, J.M., Kennedy, F.S., and Rosen, C.G., Synthesis of methyl-mercury compounds by extracts of a methanogenic bacterium, *Nature*, 1968, **220**, 173.
174. Compeau, G., and Bartha, R., Effect of salinity on mercury-methylating activity of sulfate reducers in estuarine sediments, *Appl. Environ. Microbiol.*, 1987, **53**, 261.
175. Chen, Y., Bonzongo, J.-C.J., Lyons, W.B., and Miller, G.C., Inhibition of mercury methylation in anoxic freshwater sediment by group VI anions, *Environ. Toxicol. Chem.*, 1997, **16**, 1568.
176. King, J.K., Saunders, F.M., Lee, R.F., and Jahnke, R.A., Coupling mercury methylation rates to sulfate reduction rates in marine sediments, *Environ. Toxicol. Chem.*, 1999, **18**, 1362.
177. Branfireun, B.A., Roulet, N.T., Kelly, C.A., and Rudd, J.W.M., *In situ* sulphate stimulation of mercury methylation in a boreal peatland: toward a link between acid rain and methylmercury contamination in remote environments, *Global Biogeochem. Cycles.*, 1999, **13**, 743.
178. Choi, S.-C., and Bartha, R., Environmental factors affecting mercury methylation in estuarine sediments, *Bull. Environ. Contam. Toxicol.*, 1994, **53**, 805.
179. Devereaux, R., Winfrey, M.R., Winfrey, J., and Stahl, D.A., Depth profile of sulfate-reducing bacterial ribosomal RNA and mercury methylation in an estuarine sediment, *FEMS Microbial. Ecol.*, 1996, **20**, 23.

180. Compeau, G., and Bartha, R., Effects of sea-salt anions on the formation and stability of methyl mercury, *Bull. Environ. Contam. Toxicol.*, 1983, **31**, 486.

181. Thayer, J.S., Effect of halide ions on the rates of reaction of methylcobalamin with heavy-metal species, *Inorg. Chem.*, 1981, **20**, 3575.

182. Bertilsson, L., and Neujahr, H.Y., Methylation of mercury compounds by methylcobalamin, *Biochem.*, 1971, **10**, 2805.

183. Imura, N., Sukegawa, E., Pan, S.-K., Nagao, K., Kim, J.-Y., Kwan, T., and Ukita, T., Chemical methylation of inorganic mercury with methylcobalamin, a vitamin B₁₂ analog, *Science*, 1971, **172**, 1248.

184. Berman, M., Chase, T., and Bartha, R., Carbon flow in mercury biomethylation by *Desulfovibrio desulfuricans, Appl. Environ. Microbiol.*, 1990, **56**, 298.

185. Choi, S.-C., and Bartha, R., Cobalamin-mediated mercury methylation by *Desulfovibrio desulfuricans* LS, *Appl. Environ. Microbiol.*, 1993, **59**, 290.

186. Stupperich, E., Eisinger, H.-J., and Schurr, S., Corrinoids in anaerobic bacteria, *FEMS Microbiol. Rev.*, 1990, **87**, 355.

187. Choi, S.-C., Chase, Jr., T., and Bartha, R., Enzymatic catalysis of mercury methylation by *Desulfovibrio desulfuricans* LS, *Appl. Environ. Microbiol.*, 1994, **60**, 1342.

188. Choi, S.-C., Chase, Jr., T., and Bartha, R., Metabolic pathways leading to mercury methylation in *Desulfovibrio desulfuricans, Appl. Environ. Microbiol.*, 1994, **60**, 4072.

189. Foster, T.J., The genetics and biochemistry of mercury resistance, *CRC Crit. Rev. Microbiol.*, 1987, **15**, 117.

190. Summers, A.O., Organization, expression, and evolution of genes for mercury resistance, *Ann. Rev. Microbiol.*, 1986, **40**, 607.

191. Silver, S., and Misra, T.K., Plasmid-mediated heavy metal resistances, *Ann. Rev. Microbiol.*, 1988, **42**, 717.

192. Moore, M.J., Distefano, M.D., Zydowsky, L.D., Cummings, R.T., and Walsh, C.T., Organomercurial lyase and mercuric ion reductase: nature's mercury detoxification catalysts, *Acc. Chem. Res.*, 1990, **23**, 301.

193. Spangler, W.J., Spegarelli, J.L., Rose, J.M., and Miller, H.M., Methylmercury: bacterial degradation in lakes sediments, *Science*, 1973, **180**, 192.

194. Winfrey, M.R., and Rudd, J.W.M. Environmental factors affecting the formation of methylmercury in low pH lakes: A review, *Environ. Toxicol. Chem.*, 1990, **9**, 853.

195. Oremland, R.S., Culbertson, C.W., and Winfrey, M.R., Methylmercury decomposition in sediments and bacterial cultures: Involvement of methanogens and sulfate reducers in oxidative demethylation, *Appl. Environ. Microbiol.*, 1991, **57**, 130.

196. Oremland, R.S., Miller, L.G., Dowdle, P., Connel, T., and Barkay, T., Methylmercury oxidative degradation potentials in contaminated and pristine sediments of the Carson River, Nevada, *Appl. Environ. Microbiol.*, 1995, **61**, 2745.

197. Pak, K.-R., and Bartha, R., Products of mercury demethylation by sulfidogens and methanogens, *Bull. Environ. Contam. Toxicol.*, 1998, **61**, 690.

198. Pak, K.-R., and Bartha, R., Mercury methylation and demethylation in anoxic lake sediments and by strictly anaerobic bacteria, *Appl. Environ. Microbiol.*, 1998, **64**, 1013.

199. Rowland, I.R., Davies, M.J., and Grasso, P., Volatilisation of methylmercury chloride by hydrogen sulphide, *Nature*, 1977, **265**, 718.

200. Craig, P.J., and Bartlett, P.D., The role of hydrogen sulphide in environmental transport of mercury, *Nature*, 1978, **275**, 635.

201. Baldi, F., Pepi, M., and Filippelli, M., Methylmercury resistance in *Desulfovibrio desulfuricans* strains in relation to methylmercury degradation, *Appl. Environ. Microbiol.*, 1993, **59**, 2479.

202. Sellers, P., Kelly, C.A., Rudd, J.W.M., and MacHutchon, A.R., Photodegradation of methylmercury in lakes, *Nature.*, 1996, **380**, 694.

203. Fitzgerald, W.F., and Mason, R.P., Biogeochemical cycling of mercury in the marine environment. In *Metal Ions in Biological Systems*, Volume 34, A. Sigel and H. Sigel (Eds), Marcel Dekker, NY, 1997, pp. 53–111.

204. Craig, P.J., and Brinckman, F.E., Occurrence and pathways of organometallic compounds in the environment—General considerations. In *Organometallic Compounds in the Environment*, P.J. Craig (Ed.), Longman, Essex, 1986, pp. 1–64.

205. Stumm, W., and Morgan, J.J., *Aquatic Chemistry*, John Wiley & Sons, NY, 1996.

206. Bodaly, R.A., Rudd, J.M.W., Fudge R.J.P., and Kelly, C.A., Mercury concentrations in fish related to the size of remote Canadian Shield Lakes, *Can. J. Fish. Aquat. Sci.*, 1993, **50**, 980.

207. Ramlal, P.S., Kelly, C.A., Rudd, J.W.M., and Furutani, A., Sites of methylmercury production in remote Canadian Shield Lakes, *Can. J. Fish. Aquat. Sci.*, 1992, **50**, 972.

208. Korthals, E.T., Winfrey, M.R., Seasonal and spatial variations in mercury methylation and demethylation in an oligotrophic lake, *Appl. Environ. Microbiol.*, 1987, **53**, 2397.

209. Callister, S.M., and Winfrey, M.R., Microbial methylation of mercury in Upper Wisconsin River Sediments, *Water Air Soil Pollut*, 1986, **29**, 453.

210. Wright, D.R., and Hamilton, R.D., Release of methyl mercury from sediments: effects of mercury concentration, low temperature, and nutrient addition, *Can. J. Fish. Aquat. Sci.*, 1982, **39**, 1459.

211. Lee, Y.-H., Bishop, K., Petterson, C., Iverfeldt, A., and Allard, B., Subcatchment output of mercury and methylmercury at Svartberget in Northern Sweden, *Water Air Soil Pollut.*, 1995, **80**, 455.

212. Branfireun, B.A., Heyes, A., and Roulet, N.T., The hydrology and methylmercury dynamics of a Precambrian Shield headwater peatland, *Water Res.*, 1996, **32**, 1785.

213. St. Louis, V.L., Rudd, J.M.W., Kelly, C.A., Beaty, K.G., Flett, R.J., and Roulet, N.T., Production and loss of methylmercury and loss of total mercury from boreal forest catchments containing different types of wetlands, *Environ. Sci. Technol.*, 1996, **30**, 2719.

214. Bodaly, R.A., Hecky, R.E., and Fudge, R.J.P., Increases in fish mercury levels in lakes flooded by the Churchill River diversion, northern Manitoba, *Can. J. Fish. Aquat. Sci.*, 1984, **41**, 682.

215. Hecky, R.E., Ramsey, D.J., Bodaly, R.A., and Strange, N.E., Increased methylmercury contamination in fish in newly formed fresh-water reservoirs. In *Advances in Mercury Toxicology*, T. Suzuki, N. Imura, and T.W. Clarkson (Eds), Plenum Press, New York, 1991, p.33–52.

216. Kelly, C.A., Rudd, J.W.M., Bodaly, R.A., Roulet, N.T., St. Louis, V.L., Heyes, A., Moore, T.R., Aravena, R., Dyck, B., Harris, R., Schiff, S., Warner, B., and Edwards, G., Increases in fluxes of greenhouse gases and methylmercury following the flooding of an experimental reservoir, *Environ. Sci. Technol.*, 1996, **31**, 1334.

217. Wren, C.D., and MacCrimmon, H.R., Mercury levels in the sunfish, *Lepomis gibbosus*, relative to pH and other environment variables of Precambrian Shield lakes, *Can. J. Fish. Aquat. Sci.*, 1983, **40**, 1737.

218. Wiener, J.G., Martini, R.E., Sheffy, T.B., and Glass, G.E., Factors influencing mercury concentrations in walleyes in northern Wisconsin lakes, *Trans. Am. Fish. Soc.*, 1990, **119**, 682.

219. Häkanson, L., Nilsson, A., and Andersson, T., Mercury in fish in Swedish lakes, *Environ. Pollut.*, 1988, **49**, 145.

220. Cope, W.G., Wiener, J.G., and Rada, R.G., Mercury accumulation in yellow perch in Wisconsin seepage lakes: relation to lake characteristics, *Environ. Toxicol. Chem.*, 1990, **9**, 931.

221. Spry, D.J., and Wiener, J.G., Metal bioavailability and toxicity to fish in low-alkalinity lakes: A critical review, *Environ. Pollut.*, 1991, **71**, 243.

222. Ramlal, P.S., Rudd, J.W.M., Furutani, A., and Xun, L., The effect of pH on methyl mercury production and decomposition in lake sediments, *Can. J. Fish. Aquat. Sci.*, 1985, **42**, 685.

223. Steffan, R.J., Korthals, E.T., and Winfrey, M.R., Effect of acidication on mercury methylation, demethylation, and volatilization in sediments from an acid-susceptible lake, *Appl. Environ. Microbiol.*, 1988, **54**, 2003.

224. Xun, L., Campbell, N.E.R., and Rudd, J.W.M., Measurement of specific rates of net methylmercury production in the water column and surface sediments of acidified and circumneutral lakes, *Can. J. Fish. Aquat. Sci.*, 1987, **44**, 750.

225. Miskimmin, B.M., Rudd, J.W.M., and Kelly, C.A., Influence of dissolved organic carbon, pH and microbial respiration rates on mercury methylation and demethylation in lake water, *Can. J. Fish. Aquat. Sci.*, 1992, **49**, 17.

226. Jackson, T.A., Kipphut, G., Hesslein, R.H., and Schindler, D.W., Experimental study of trace metal chemistry in soft-water lakes at different pH levels, *Can. J. Fish Aquat. Sci.*, 1980, **37**, 387.

227. Schindler, D.W., Hesslein, R.H., Wagemann, R., and Broecker, W.S., Effect of acidification on mobilization of heavy metals and radionuclides from the sediments of a freshwater lake, *Can. J. Fish Aquat. Sci.*, 1980, **37**, 373.

228. Mierle, G., Aqueous inputs of mercury to Precambrian shield lakes in Ontario, *Environ. Toxicol. Chem.*, 1990, **9**, 843.

229. Watras, C.J., Back, R.C., Halvorsen, S., Hudson, R.J.M., Morrison, K.A., and Wente, S.P., Bioaccumulation of mercury in pelagic freshwater food webs, *Sci. Tot. Environ.*, 1998, **219**, 183.

230. Babiarz, C.L., Hurley, J.P., Benoit, J.M., Shafer, M.M., Andren, A.W., and Webb, D.A., Seasonal influences on partitioning and transport of total and methylmercury in rivers from contrasting watersheds, *Biogeochem.*, 1998, **41**, 237.

231. Miskimmin, B.M., Effect of natural levels of dissolved organic carbon (DOC) on methyl mercury formation and sediment-water partitioning, *Bull. Environ. Contam. Toxicol.*, 1991, **47**, 743.

232. Wren, C.D., Scheider, W.A., Wales, D.L., Muncaster, B.W., and Gray, I.M., Relation between mercury concentrations in walleye (*Stizostedion vitreum vitreum*) and northern pike (*Esox lucius*) in Ontario lakes and influence of environmental factors, *Can. J. Fish. Aquat. Sci.*, 1991, **48**, 132.

233. Fjeld, E., and Rognerud, S., Use of path analysis to investigate mercury accumulation in brown trout (*Salmo trutta*) in Norway and the influence of environmental factors, *Can J. Fish. Aquat. Sci.*, 1993, **50**, 1158.

234. Grieb, T.M., Driscoll, C.T., Gloss, S.P., Schofield, C.L., Bowie, G.L., and Porcella, D.B., Factors affecting mercury accumulation in fish in the upper Michigan peninsula, *Environ. Toxicol. Chem.*, 1990, **9**, 919.

235. Lee, Y.H., and Hultberg, H., Methylmercury in some Swedish surface waters, *Environ. Toxicol. Chem.*, 1990, **9**, 833.

236. Rask, M., Metsälä, T.-R., and Salonen, K., Mercury in the food chains of a small polyhumic forest lake in southern Finland. In *Mercury Pollution: Integration and Synthesis*, C.J. Watras and J.W. Huckabee (Eds), Lewis Publishers, Boca Raton, 1994, pp. 409–416.

237. Henry, E.A., The role of sulfate-reducing bacteria in environmental mercury methylation. PhD dissertation, Harvard University, 1992, 221 pp.

238. Benoit, J.M., Gilmour, C.C., Mason, R.P., Reidel, G.S., and Reidel, G.F., Sources and cycling of mercury in the Patuxent estuary, *Biogeochem.*, 1998, **40**, 249.

239. Gutknecht, J.J., Inorganic mercury (Hg^{2+}) transport through lipid bilayer membranes, *J. Membr. Biol.*, 1981, **61**, 61.

240. Mason, R.P., Reinfelder, J.R., and Morel, F.M.M., Bioaccumulation of mercury and methylmercury, *Water Air Soil Pollut.*, 1995, **80**, 915.

241. Craig, P.J., and Moreton, P.A., Total mercury, methyl mercury and sulphide in River Carron sediments, *Mar. Poll. Bull.*, 1983, **14**, 408.

242. Blum, J.E., and Bartha, R., Effect of salinity on methylation of mercury, *Bull. Environm. Contam. Toxicol.*, 1980, **25**, 404.

243. Schwarzenbach, G., and Widmer, M., Die Löslichkeit von Metallsulfiden I. Schwarzes Quecksilbersulfid, *Helv. Chim. Acta.*, 1963, **46**, 2613.

244. Paquette, K., and Helz, G., Solubility of cinnabar (red HgS) and implications for mercury speciation in sulfidic waters, *Water Air Soil Pollut.*, 1995, **80**, 1053.

245. Gilmour, C.C., and Henry, E.A., Mercury methylation by sulfate-reducing bacteria: Biogeochemical and pure culture studies, presentation made at the American Chemical Society meeting, San Fransisco, CA, April, 1992.

246. Dyrssen, D., Sulfide complexation in surface seawater, *Mar. Chem.*, 1988, **24**, 143.

247. Paquette, K., Solubility and speciation of cinnabar (red HgS) and implications for mercury solubility in sulfidic waters, PhD Dissertation, The University of Maryland, 1994, 330 pp.

248. Wiener, J.G., and Spry, D.J., Toxicological significance of mercury in freshwater fish. In *Environmental Contaminants in Wildlife: Interpreting Tissue Concentrations*, W.N. Beyer, G.H. Heinz, and A.W. Redon-Norwood (Eds), Lewis Publishers, Boca Raton, 1996, pp. 297–339.

249. Suns, K., and Hitchin, G., Interrelationships between mercury levels in yearling yellow perch, fish condition and water quality, *Water Air Soil Pollut.*, 1990, **50**, 255.

250. Johnson, M.G., Trace element loadings to sediments of fourteen Ontario lakes and correlations with concentrations in fish, *Can J. Fish. Aquat. Sci.*, 1987, **44**, 3.

251. Lange, T.R.M., Royals, H.E., and Connor, L.L., Influence of water chemistry on mercury concentration in largemouth bass from Florida lakes, *Trans. Am. Fish. Soc.*, 1993, **122**, 74.

252. Watras, C.J., and Bloom, N.S., Mercury and methylmercury in individual zooplankton: implications for bioaccumulation, *Limnol. Oceanogr.*, 1992, **37**, 1313.

253. Back, R.C., and Watras, C.J., Mercury in zooplankton of northern Wisconsin lakes: Taxonomic and site-specific trends, *Water Air Soil Pollut.*, 1995, **80**, 1257.

254. Faust, B.C., The octano/water distribution coefficients of methylmercury species: the role of aqueous phase chemical speciation, *Environ. Toxicol. Chem.*, 1992, **11**, 1373.

255. Lawson, N.M., and Mason, R.P., Accumulation of mercury of estuarine food chains, *Biogeochem.*, 1998, **40**, 235.

256. Reinfelder, J.R., and Fisher, N.S., The assimilation of elements ingested by marine copepods, *Science*, 1991, **251**, 794.

257. Reinfelder, J.R., and Fisher, N.S., Retention of elements absorbed by juvenile fish (*Meridia memidia, Meridia berylina*) from zooplankton prey, *Limnol. Oceanogr.*, 1994, **39**, 1783.

258. Boudou, A., and Ribeyre, F., Comparative study of the trophic transfer of two mercury compounds—$HgCl_2$ and Ch_3HgCl—between *Chlorella vulgaris* and *Daphnia magna*. Influence of temperature, *Bull. Environ. Contam. Toxicol.*, 1981, **27**, 624.

259. Boudou, A., and Ribeyre, F., Experimental results of trophic contamination of *Salmo gairdneri* by two mercury compounds: Analysis at the organism and organ levels, *Water Air Soil Pollut.*, 1985, **26**, 137.

260. Mason, R.P., and Sullivan, K.A., Mercury in Lake Michigan, *Environ. Sci. Technol.*, 1997, **31**, 942.

261. Bloom, N.S., On the chemistry of mercury in edible fish and marine invertebrate tissue, *Can. J. Fish. Aquat. Sci.*, 1992, **49**, 1010.

262. Riisgård, H.U., and Famme, P., Accumulation of inorganic and organic mercury in shrimp, *Crangon crangon, Mar. Pollut. Bull.*, 1986, **17**, 255.

263. Mason, R.P., Laporte, J.-M., and Andres, S., Factors controlling the bioaccumulation of mercury, methylmercury, arsenic, selenium and cadmium in freshwater invertebrates and fish, *Arch. Environ. Contam. Toxicol.*, (in press).

264. Paterson, M.J., Rudd, J.W.M., and St. Louis, V., Increases in total and methylmercury in zooplankton following flooding of a peatland reservoir, *Environ. Sci. Technol.*, 1998, **32**, 3868.

265. Rodgers, D.W., You are what you eat and a little bit more: Bioenergetics-based models of methylmercury accumulation in fish revisited. In *Mercury Pollution: Integration and Synthesis*, C.J. Watras and J.W. Huckabee (Eds), Lewis Publishers, Boca Raton, 1994, pp. 427–439.

266. Hall, B.D., Bodaly, R.A., Fudge, R.J.P., Rudd, J.W.M., and Rosenberg, D.M., Food as the dominant pathway of methylmercury uptake by fish, *Water Air Soil Pollut.*, 1997, **100**, 13–24.

267. Rudd, J.W.M., Furutani, A., and Turner, M.A., Mercury methylation by fish intestinal contents, *Appl. Environ. Microbiol.*, 1980, **40**, 777.

268. MacCrimmon, H.R., Wren, C.D., and Gots, B.L., Mercury uptake by lake trout, *Salvelinus namaycush*, relative to age, growth, and diet in Tadenac Lake with comparative data from other Precambrian Shield lakes, *Can. J. Fish Aquat. Sci.*, 1983, **40**, 114.

269. Wren, C.D., MacCrimmon, H.R., and Loescher, B.R., Examination of bioaccumulation and biomagnificaion of metals in a Precambian Shield lake, *Water Air Soil Pollut.*, 1983, **19**, 277.

270. Francesconi, K.A., and Lenanton, R.C.J., Mercury contamination in a semi-enclosed marine embayment: Organic and inorganic mercury content of biota, and factors influencing mercury levels in fish, *Mar. Environ. Res.*, 1992, **33**, 189.

271. Kidd, K.A., Hesslein, R.H., Fudge, R.J.P., and Hallard, K.A., The influence of trophic level as measured by del ^{15}N on mercury concentrations in freshwater organisms, *Water Air Soil Pollut.*, 1995, **80**, 1011.

272. Cabana, G., Tremblay, A., Kalff, J., and Rasmussen, J.B., Pelagic food chain structure in Ontario lakes: A determinant of mercury levels in lake trout (*Salvelinus namaycush*), *Can. J. Fish. Aquat. Sci.*, 1994, **51**, 381.

273. Futter, M.N., Pelagic food web structure influences probability of mercury contamination in lake trout (*Salvelinus namaycush*), *Sci. Total Environ.*, 1994, **145**, 7.

274. Tremblay, A., Cloutier, L., and Lucotte, M., Total mercury and methylmercury fluxes via emerging insects in recently flooded hydroelectric reservoirs and a natural lake, *Sci. Total Environ.*, 1998, **219**, 209.

275. Saouter, E., Hare, L., Campbell, P.G.C., Boudou, A., and Ribeyre, F., Mercury accumulation in the burrowing mayfly *Hexagenia rigida* (Ephemeroptera) exposed to CH_3HgCl or $HgCl_2$ in water and sediment, *Wat. Res.*, 1993, **6**, 1041.

276. Wang, W.-X., Stupakoff, I., Gagnon, C., and Fisher, N.S., Bioavailability of inorganic and methylmercury to a marine deposit-feeding polychaete, *Environ. Sci. Technol.*, 1998, **32**, 2564.

277. Lawrence, A.L., and Mason, R.P., Factors controlling the bioaccumulation of mercury and methylmercury by the estuarine copepod, *Leptocheirus plumulosus, Environ. Pollut.* (under review).

278. Wiener, J.G., Fitzgerald, W.F., Watras, C.J., and Rada, R.G., Partitioning and bioavailability of mercury in an experimentally acidified Wisconsin lake, *Environ. Toxicol. Chem.*, 1990, **9**, 909.

279. Nuutinen, S., and Kukkonen, J.V.K., The effect of selenium and organic material in lake sediments on the bioaccumulation of methylmercury by *Lumbriculus variegatus* (oligochaeta), *Biogeochem.*, 1998, **40**, 267.

280. Lawrence, A.L., McAloon, K.M., Mason, R.P., and Mayer, L.M., Intestinal solubilization of particle-associated organic and inorganic mercury as a measure of bioavailability to benthic invertebrates, *Environ. Sci. Technol.*, 1999, **33**, 1871.

281. Mason, R.P., Dobbs, C., and Leaner, J.J., The bioaccumulation of mercury and methylmercury in two contrasting food chains. Presentation made at the 8th International SETAC Conference, Bordeaux, France, April, 1988.

282. Leaner, J.J., and Mason, R.P., The distribution kinetics of methylmercury in sheepshead minnows, *Cyprinodon variegatus*: An assessment of the mechanisms controlling methylmercury accumulation and redistribution in fish tissues. Presentation made at the 1999 SETAC Conference, Philadelphia, PA, November, 1999.

283. Rouleau, C., Borg-Neczak, K., Gottofrey, J., and Tjälve, H., Accumulation of waterborne mercury(II) in specific areas of fish brain, *Environ. Sci. Technol.*, 1998, **33**, 3384.

284. Laporte, J.-M., Andres, A., and Mason, R.P., Factors controlling the uptake and transfer of inorganic and methylmercury across the perfused gill and gut of the blue crab, *Callinectes sapidus*. Presentation made at the 1999 SETAC Conference, Philadelphia, PA, November, 1999.

285. Langston, W.J., The distribution of mercury in British estuarine sediments and its availability to deposit-feeding bivalves, *J. Mar. Biol. Assoc. UK*, 1982, **62**, 667.

286. Mohlenberg, F., and Rissgård, H.U., Partitioning of inorganic and organic mercury in cockles *Cardum edule* (L.) and *C. glaucum* (Bruguiére) from a chronically polluted area: Influence of size and age, *Environ. Pollut.*, 1988, **55**, 137.

287. Muhaya, B.B.M., Leermakers, M., and Baeyens, W., Total mercury and methylmercury in sediments and in the polychaete *Nereis diversicolor* at Groot Buitenschoor (Scheldt Estuary, Belgium), *Water Air Soil Pollut.*, 1997, **94**, 109.

288. Mason, R.P., and Sullivan, K.A., Mercury and methylmercury transport through an urban watershed, *Wat. Res.*, 1998, **32**, 321.

289. Gardner, W.S., Kendall, D.R., Odum, R.R., Windom, H.L., and Stephens, J.A., The distribution of methylmercury in a contaminated salt marsh ecosystem, *Environ. Pollut.*, 1978, **15**, 243.

290. Leermakers, M., Meuleman, C., and Baeyens, W., Mercury speciation in the Scheldt Estuary, *Water Air Soil Pollut.*, 1995, **80**, 641.

3 Organotin Compounds in the Environment

F. CIMA
Dipartimento di Biologia, Universita di Padova, Padova, Italy

P.J. CRAIG and C. HARRINGTON,
Department of Molecular Sciences, De Montfort University, Leicester, UK

3.1 INTRODUCTION

Both inorganic and organotin compounds have extensive industrial uses, and these uses, environmental aspects and toxicological properties have been particularly well covered in several multi-volume and monograph publications in recent years [1–6]. Within these works are detailed accounts of organotin levels in aqueous, atmospheric, surface microlayer and sediment compartments for organotin species and so indicative summary levels only will be stated in this chapter. Similarly the analytical and toxicology aspects will be covered in overview format. Total organotin production is more than 50 000 tonnes per year, with about one-quarter of this being triorganotin biocides. About 4000 tonnes per year of TBT derivatives are manufactured [7], as wood preservatives, antifoulant and disinfectant biocides. Of total tin metal prepared, 7% is for organometallic tin compounds. Useage details are given in Table 3.1.

The tin–carbon bond is stable to water, atmosphere and heat (at least to 200 °C). UV radiation, strong acid and electrophilic reagents cleave the tin–carbon bond. Solubility in water varies greatly with R and X in $RnSnX_{4-n}$ and on their relative numbers.

Toxicity is dealt with in detail in Section 3.8, but is mentioned here in view of the biocidal uses. Maximum toxic effect in $RnSnX_{4-n}$ is usually achieved for R_3SnX; however, unless X is toxic it does not have much effect on overall toxicity. Within R_3SnX the nature of R profoundly effects toxicity to a single species and relative toxicity to different species. Et_3SnX is most toxic to mammals, Bu_3SnX to aquatic life. Increase in chain length of R decreases toxicity. Detailed surveys are given in References [7–9]. Tables 3.2 and 3.3 give an overview of modes of entry and general toxicity.

The industrial uses, toxicities and modes of entry of organotin compounds are given in Tables 3.1, 3.2 and 3.3. Table 3.1 presents an overview. In environmental terms, the role of tributyltin (TBT) is the most important, and the

Organometallic Compounds in the Environment
Edited by P.J. Craig © 2003 John Wiley & Sons Ltd

Table 3.1 Industrial applications of organotin compounds

Application	Compound	Comment
	R_3SnX	
Agriculture fungicides antifeedants acaricides	$(C_6H_5)_3SnX$ (X = OH, OAc) $(C_6H_5)_3SnX$ (X = OH, OAc) $(c\text{-}C_6H_{11})_3SnX$ (X = OH, $-$N.C$=$N.C$=$N)	Total biocidal use is about 20% of total organotin production. Ph_3SnX used as an antifungal agent, $(cC_6H_{11})_3SnX$ used as acaricide. Can enter surface run off. 450 tonnes pa TBT. Used in USA in 1987 for antifouling paint useage.
	$\qquad\quad$ $\begin{array}{cc} | & | \\ H & H \end{array}$ $(C_6H_5(CH_3)_2CCH_2)_3Sn)_2O$	
Antifouling paint, biocides	$(C_6H_5)_3SnX$ (X = OH, OAc, F, Cl, SCS.N(CH$_3$)$_2$, OCOCH$_2$Cl, OCOC$_5$H$_4$N-3) $(C_6H_5)_3SnOCOCH_2CBr_2COOSn$ $(C_6H_5)_3$ $(C_4H_9)_3SnX$ (X = F, Cl, OAc) $((C_4H_9)_3Sn)_2O$ $(C_4H_9)_3SnOCOCH_2CBr_2COOSn(C_4H_9)_3$ $(C_4H_9)_3SnOCO(CH_2)_4COOSn(C_4H_9)_3$ $(-CH_2C(CH_3)(COOSn(C_4H_9)_3)-)_n$	Controlled since 1982
Wood preservative, fungicides	$((C_4H_9)_3Sn)_2O$ $(C_4H_9)_3Sn$(naphthenate) $((C_4H_9)_3Sn)_3PO_4$	Applied as 1–3 wt-% in a solvent
Stone preservation, Disinfectants	$((C_4H_9)_3Sn)_2O$ $(C_4H_9)_3SnOCOC_6H_5$ $((C_4H_9)_3Sn)_2O$	
Molluscicides (field trials)	$(C_4H_9)_3SnF$ $((C_4H_9)_3Sn)_2O$	
	R_2SnX_2	
Heat and light stabilizers for rigid PVC	$R_2Sn(SCH_2COO\text{-}i\text{-}C_8H_{17})_2$ (R = CH$_3$, C$_4$H$_9$, C$_8$H$_{17}$, $(C_4H_9)OCOCH_2CH_2$) $(R_2SnOCOCH=CHCOO)_n$ (R = C$_4$H$_9$, C$_8$H$_{17}$) $(C_4H_9)_2Sn(OCOCH=CHCOOC_8H_{17})_2$ $(C_4H_9)_2Sn(SC_{12}H_{25})_2$	70% of total organotin use. Prevents loss of HCl from polymer at 180–200°C Used at 5–20 g kg^{-1} PVC. Organotins can leach into food, beverage, waters sewage sludge.
Homogeneous catalysts for RTV silicones, polyurethane foams and transesterification reactions	$(C_4H_9)_2Sn(OCOCH_3)_2$ $(C_4H_9)_2Sn(OCOiC_8H_{17})_2$ $(C_4H_9)_2Sn(OCOC_{11}H_{23})_2$ $(C_4H_9)_2Sn(OCOC_{12}H_{25})_2$ $((C_4H_9)_2SnO)_n$	

(continues)

Table 3.1 (*continued*)

Precursor for forming SnO_2 films on glass	$(CH_3)_2SnCl_2$
Anthelmintics for poultry	$(C_4H_9)_2Sn(OCOC_{11}H_{23})_2$
	RSnX_3
Heat stabilizers for rigid PVC	$RSn(SCH_2COO\text{-}i\text{-}C_8H_{17})_3{}^a$ $(R = CH_3, C_4H_9, C_8H_{17},$ $C_4H_9OCOCH_2CH_2)$ $(C_4H_9SnS_{1.5})_4$ $(C_4H_9Sn(O)OH)_n$
Homogeneous catalysts for transesterification reactions	$C_4H_9Sn(OH)_2Cl$ $C_4H_9SnCl_3$
Precursor for SnO_2 films on glass	$C_4H_9SnCl_3$ $CH_3SnCl_3{}^a$

^{a} These compounds are used in combination with the corresponding R_2SnX_2 derivatives
Adopted with permission from this work, first edition

synthesis methods of these compounds generally may be indicated by a brief description only (Table 3.4). Essentially the tetraalkylated form may be considered to be synthesized and the desired mono di- or tri-compound produced by redistribution (e.g. Equation 3.1)

$$3Bu_4Sn + SnCl_4 \longrightarrow 4Bu_3SnCl \tag{3.1}$$

Individually, the Grignard or aluminium routes, followed by disproportionation to the compound chosen, tend to be used. Methods are discussed in detail elsewhere [5]. The required counter ion, e.g. Cl^-, is introduced by a nucleophilic substitution. The methacrylate copolymer (Figure 3.1) is synthesized by a free radical initiated attack on the $C\!=\!C$ bonds from TBT methacrylate monomer and methylmethacrylate monomer to give a copolymer. Organotin waste during manufacture is generally removed by incineration at $> 850\,^\circ C$ but UV radiation and $KMnO_4$ or ozone oxidation have been used [10].

With this in view, the question of organotin compounds in the environment can be reduced to the behaviour of TBT in aqueous and sediment matrices although organotin release from landfill sites has been researched in recent years (see below), and there is some organotin leaching from polyvinylchloride (PVC). Similarly, tripropyltin (TPT) compounds used in agriculture are also toxic to aquatic life. TBT essentially degrades sequentially to dibutyltin (DBT) and monobutyltin (MBT) so the chemistry of the latter two arising from their uses in PVC merges into the former. The use of TBT copolymer also reduces to the aqueous chemistry of the TBT cation, as the antifoulant works by sequential hydrolysis of the TBT moiety from the polymer backbone. Interestingly the

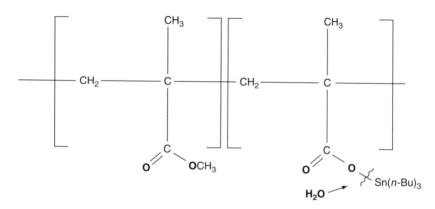

Figure 3.1 Methacrylate TBT copolymer

TBT copolymer systems also may include copper compounds and organic biocides in order to give broad spectrum effectiveness against organisms resistant to TBT. Environmental studies do not usually focus separately on this aspect of TBT-based antifouling paints. There is some evidence for direct input of MBT and DBT from PVC materials to the environment, but most comes from TBT degradation.

3.2   USE OF TBT IN ANTIFOULING PREPARATIONS

Fouling of ships means the attachment and growth of e.g. algae, shell fish and weeds on the outer hull and on large vessels hundreds of tonnes of fouling may grow within one year. Driving the non-smooth vessel through the water becomes correspondingly more difficult and expensive. Various triorganotin species were noted as being very effective as antifoulants (TBT is effective at a few ng L^{-1} as a biocide for algae, zooplankton, molluscs and the larval stages of some fish). Triphenyltin species (TPT) are also effective but less used. TBT was first used in 'free association' formulations where the biocide is physically mixed with the other paint ingredients and slowly dissolves in use to reduce TBT. These coatings last for about 2 years. By 1974 the methyl methacrylate–TBT copolymer system (TBT-MMA) had been developed [10] (Figure 3.1). These 'self polishing' systems release TBT steadily from the top few nanometers, exposing a free TBT-containing polymer surface below. This system can last for more than 5 years before exhaustion. The basic advantages are that release of TBT is controlled, constant and just sufficient for the purpose. The term 'self polishing' arises from the fact that decay of the polymer still leaves a smooth surface on the hull of the boat, reducing drag and fuel costs. When the copolymer is exhausted the boat can then be directly repainted.

Table 3.2 Possible direct modes of entry of organotin into the environment

Medium	Species	Source
Air	R_3SnX	Agricultural spraying Volatilization from biocidal treatments Antifouling paint sprays Landfill
	R_3SnX, R_2SnX_2, and $RSnX_3$	Incineration of organotin Treated or stabilized waste materials Landfill
	R_2SnX_2 and $RSnX_3$	Glass coating operations—spraying of organotins onto glass at high temperature to give SnO_2 films Landfill
Soil	R_3SnX	Agricultural applications Wood preservation
	R_3SnX, R_2SnX_2 and $RSnX_3$	Burial of organotin-containing waste materials
Water	R_3SnX	Antifouling coatings Molluscicides Overspraying from agricultural applications Land run-off from agricultural useage Industrial processes, e.g. slimicides in paper Manufacture Landfill
	R_2SnX_2 and $RSnX_3$	Leached from PVC Landfill

Taken from this work, first edition, with permission

Table 3.3 Species specific of triorganotin compounds, R_3SnX toxicity

Species	R in most active R_3SbX compound
Insects	CH_3
Mammals	C_2H_5
Gram-negative bacteria	$n\text{-}C_3H_7$
Gram-positive bacteria, fish, fungi, Molluscs, plants	$n\text{-}C_4H_9$
Fish, fungi, molluscs	C_6H_5
Fish, mites	$c\text{-}C_6H_{11}$
	$C_6H_5(CH_3)_2CCH_2$

Taken from this work, first edition, with permission

Table 3.4 Synthesis of tetrabutyl tin

(1) Grignard	$4RMgX + SnCl_4 \longrightarrow R_4Sn + 4MgClX$
(2) Wurtz	$4BuCl + 8Na + SnCl_4 \longrightarrow Bu_4Sn + 8NaCl$
(3) Aluminium	$4R_3Al + 3SnCl_4 \longrightarrow 3R_4Sn + 4AlCl_3$
(4) Direct	$BuBr + Na/Sn \longrightarrow Bu_nSnBr_{4-n}$

Treatment of the *Queen Elizabeth 2* in 1978 was reported to result in fuel cost savings of 12% annually [11]. Despite the environmental problems discussed below, TBT replacement paints (containing, for example, cuprous oxide with an organic biocide) are probably not as effective as copolymer TBT over a 5-year service life [12]. However, for small vessels, non-TBT systems are now used (see below).

Generally the TBT copolymer market has not declined, but 90% of TBT is now being used mainly by the world's commercial shipping and 10% for pleasure craft. The rate of release of TBT is given in $\mu g\ cm^{-2}\ day^{-1}$ with a design rate of 1.6 (the maximum proposed limit by the US EPA and the IMO is $4\,\mu g$). Bearing in mind the environmental decay of TBT (see below), it is assumed that open-ocean loss of TBT is not an environmental problem. However, recent work showing elevated TBT levels in higher marine animals makes this assumption less firm (see below and Section 3.8). Generally, tin concentration in the paint can be up to 3% with an initial leach rate of $6\,\mu g\ Sn\ cm^{-2}\ day^{-1}$.

Large vessels in harbour and inshore small fishing boats and pleasure craft have given rise to the main TBT environmental problem. A large vessel in port for 3 days might lose 200 g of TBT to the waters, 600 g if freshly painted. This can lead to concentrations in large marinas or dockyards of 100–200 ng Sn dm^{-3} in the water [13]. In smaller estuaries, marinas, etc., for smaller pleasure craft dissolved TBT can be in the range 10–70 ng dm^{-3} [14]. Much of this can arise from hosing of vessels prior to repainting where actual fragments of paint containing TBT may be removed from the hull during cleaning (i.e. the 'self polishing' effect is evidently not always achieved in practice!).

3.3 LEGISLATION ON TBT

The effects of TBT on non-target organisms are discussed in Section 3.8, but the actions to remove environmental problems are discussed for completeness in this section.

Essentially, legislation being a national function in a complex world, control has been imposed on a piecemeal basis. However, the degree of harmonization has been better in the case of this late-developing pollutant, with attitudes and structures being better advanced than in the case of earlier environmental pollutants.

The focus on control of TBT has been multifaceted, viz.

(i) The release rates are defined to correspond to minimal rates needed for effect, i.e. $1.6-4\,\mu g\ cm^{-2}\ d^{-1}$ maximum.

(ii) Free association (i.e. free releasing) paints are now banned in many countries.

(iii) The focus has been on near shore effects where reported environmental damage is greatest.

(iv) Prohibitions for use have therefore been focused on smaller vessels (< 25 metres in length, i.e. inshore craft).

(v) The health of local economic assets (e.g. oyster beds) has been a factor of importance.

(vi) Reduction by good management and organization of the entry of TBT into the environment from dock and boatyards, etc., has also been a tactic.

(vii) Nevertheless a total ban on TBT, ocean wide, is proposed by the IMO in 2008.

It is only reasonable to point out an environmental bonus of TBT in that its use is estimated to save 4 % of the fuel use of the worlds fleet (about 7.2 M tonnes) which is about 22 Mtonnes of emitted CO_2 and 0.6 Mtonnes of emitted SO_2 [15].

Environmental legislation in several countries has been based on a realization of the persistence of TBT in the environment (several years half-life on sediments) and damage to marine species at very low concentrations (e.g. 0.1 ppb or ng dm^{-3}). In addition, larvae of marine organisms are particularly sensitive [16]. However, there is less evidence of harm to humans (but see Section 3.8) [17].

The general pace of legislation has been rapid, e.g. EU Directive 677, 1989. The first ban (of TBT on boats of < 25 m length) was in France in 1982 (because of the damage to oyster beds). In 1987 the UK restricted the use of copolymer paints with $> 7.5\%$ TBT and effectively eliminated free-association paints [18]. Numerous states of the USA followed suit. In some countries TBT can only be supplied in containers $> 20\,dm^3$ (to discourage small scale use). Full details for actual and proposed legislation are given elsewhere [19,20]. At present only Austria, Switzerland and New Zealand have totally banned TBT. On the other hand many other countries still have little or no control over TBT.

The philosophy of control and legislation on TBT is based on a view regarding its toxicity and its environmental persistence in sediments being much greater than that in water. These properties are discussed later in this chapter (Sections 3.7 and 3.8). Continuing monitoring work gives a guide to the effectiveness of legislation and tends to demonstrate rapid clearance from water after legislation but much slower removal from sediment.

3.4 DISTRIBUTION IN THE ENVIRONMENT

Uses of methyltins are not thought to lead to the significant presence of methyltins in water and sediments. However, methyltins have been *found* in the environment probably as a result of a process of biomethylation (see Section 3.6). Similarly octyltins have not been reported away from point source use. It is also assumed that MBT and DBT found in the aquatic environment arises from decay of TBT rather than direct input of the former species.

In sea water, following hydrolysis, TBT exists as a cation complexed to chloride, carbonate and hydroxide [21]. It is rapidly absorbed onto suspended particulate material and hence, in due course, sediment [8]. This happens more rapidly than does aqueous decay, so TBT tends to be found in sediments on removal from the water column. As noted, most recent monitoring work has demonstrated clearly that TBT concentrations in water decline quite rapidly once the input of TBT is reduced or eliminated. Many surveys have illustrated this and the reader is referred to numerous detailed reference sources for quantitative information [22–26]. The initial response rate is rapid, usually less than 5 years.

With a log K_{OW} of 3.7 [27] TBT is quite lipophilic and would be expected to bioaccumulate. Bioaccumulation factors in various species range from 1000 to 30 000 (see also Section 3.8) [28]. Bioaccumulation to surface microlayer increases TBT concentrations over subsurface water two to tenfold [29]. Bioaccumulation factors to organisms are discussed in the toxicity section (Section 3.8).

Recovery from sediments is much slower and apparently variable. That the water above the sediment tends to clear rapidly suggests that, although there is a reservoir of TBT in the sediment, it is not completely available for replenishment of TBT lost from the water layer by non-use and decay. Table 3.5 gives some half-lives of TBT in sediments and other media [30–36]. In a work of this length only indicative examples can be given but these are intended to be representative.

Tables 3.6 and 3.7 show a list of organotin compounds that have been observed and measured in the natural environment (the analytical techniques developed to make these measurements are discussed below).

From Tables 3.6 and 3.7 the following deductions can be made:

(i) TBT decays sequentially to DBT and MBT.
(ii) The organotin derivatives may be reduced (i.e. converted to hydrides) in the environment.
(iii) They may also be biomethylated.
(iv) Some are volatile (TBTO, hydrides).
(v) Landfill sites make a contribution to organotin levels in the environment.

It should be noted that, as for most organometallics in the environment (except for arsenic, see Chapters 5 and 6) there is no information on the TBT or any other organotin counterion other than in the case of the saturated tetraorganotin and the volatile hydrides and TBT in seawater (where complexation with OH^-, Cl^-, and CO_3^{2-} takes place) [21]. One may speculate on the existence of organotin counter ions, but there is no evidence.

Table 3.5 Some estimated half-lives for TBT

Medium	Half-life	Reference
Freshwater	6–26 days, (light), 4 months (in dark)	8, 30, 31
Estuarine water	1–2 weeks	8, 31
Seawater	6–127 days	31
Water/sediment mixture	5 months–5 years	32, 33
Estuarine sediment	3.8 years	34
Marine sediment	1–4, 100–800 days	35, 36
	1.85–8.7 years	35, 36
Soil	140 days (TPhT)	8

Table 3.6 Organotin species detected in the natural environment. Reproduced with permission from Ref 8

Species	Compartment	References
Bu_3Sn^+	Sediment, water, surface microlayer, marine animals	Very many; see e.g. Refs 1–9
Bu_2Sn^{2+}	As above	As above
$BuSn^{3+}$	As above	As above
SnH_4	Algae, microbial cultures, sediments	37–41
Me_nSnH_{4-n}	As above	As above
Bu_3SnMe	Harbour sediments, surface water	40, 42–44
Bu_2SnMe_2	As above, sediment, surface water	40, 42–44
$BuSnMe_3$	Sediment, surface water	40, 44, 45
Me_4Sn	Atmosphere, surface water, landfill gas, sewage gas	40, 41, 46–49
Me_3Sn^+, Me_2Sn^{2+}, $MeSn^{3+}$	Water, domestic waste deposit water	38, 43, 47, 50, 51
$BuSnH_3$	Sediment, landfill gas, sewage gas, waste water	40, 41, 48, 49, 51
Bu_2SnH_2	As above	40, 48, 49, 51
$EtMe_3Sn$	Landfill gas	48
Et_2Me_2Sn	As above	48
Et_3MeSn	As above	48
Et_4Sn	As above	48

Transfers of organotins from sediment to water have been estimated at between 50 and 790 nmol m^{-2} y^{-1}. Water-to-air fluxes have been calculated at between 20 and 510 nmol m^{-2} y^{-1}. In conjunction with this, methylation and reduction have lead to the observation of volatile organotin species suggesting a considerable mobilization of tin, including to atmosphere [52].

Prior to legislation, or to its effects becoming noticed, water concentrations of TBT in polluted zones were recorded at up to 1–2000 ng Sn L^{-1}. Surface microlayer levels were up to 25 000–36 000 ng Sn L^{-1}. Sediment levels of up to 5500 ng g^{-1} were noted although levels of hundreds of ng Sn g^{-1} were more common. Sewage sludge had similar levels, with sewage treatment plant effluent up to 55 200 ng SnL^{-1} Bioconcentration factors in organisms are discussed below.

Table 3.7   Some butyltin levels measured in the natural environment. Adapted with permission from Hoch, Ref. 8, pp. 732 and 735 and refs therein

Butyltin concentrations in water (ng Sn L^{-1})

Location	MBT	DBT	TBT
Sado Estuary, Portugal	18–60	52–160	39–870
San Pedro River, Spain	6.9–14.4	5.5–14.3	9.3–16.3
Cadiz Bay, Spain	8.90–41	8.3–68	8.3–488
Guadalete River, Spain	9.8–25	5.7–39.5	9.9–116
Antwerp harbour, The Netherlands	51–76	217–283	765–1000
Ganga Plain, India	2–70	2–101	3–20
Marinas in The Netherlands	3–310	0.1–810	0.1–3620

Concentrations of butyltin compounds in municipal wastewater (ng Sn L^{-1}) and sewage sludge (μg Sn kg^{-1} dry wt)

	MBT	DBT	TBT
Zürich, Switzerland	245	523	157
Goslar, Ost	< 10	60	15
Bomlitz	35	40	15
Hildesheim (1994)	135	215	140
Hildesheim (1996)	136	269	83
Harsefeld	64	78	12
Sarnia, Canada	16–31	11–61	5–175
Toronto, Canada	440	210–305	245–277

Comparison of butyltin concentrations in river, lake, marine and harbour sediments (ng Sn g^{-1} dry wt)

Location	Depth (cm)	MBT	DBT	TBT
San Pedro River, Spain	0–10	1.9–6.1	1.5–8.7	0.7–11.1
Guadalete River, Spain	0–10	16.1–129	20.5–510	26.5–601
Cadiz Bay, Spain	0–10	1.2–31	1.1–52	1.6–225
Bay of Arcachon, France	0–51	6–156	5–141	16–161
San Diego Bay, California	0–6	2–185	2–265	2–242
Pearl Habor, Hawaii	0–6	5–533	4–367	4–2830
Lake System Westeinder, The Netherlands	*a*	6–100	6–96	6–520
Leman Lake, Switzerland	*a*	186	295	627
Hamburg, Germany	*a*	430	610	5200
Toronto, Canada	*a*	580	248	539
Vancouver, Canada	*a*	3360	8510	10780

Table 3.7 *continued*
Comparison of the concentration of butyltin compounds in body and tissues of various organisms (ng Sn g^{-1} wet wt)

Animal	Tissue	MBT	DBT	TBT
Bottlenose dolphins, Italian coastal waters	Blubber	55	16	41
	Liver	150	800	250
Bluefin tunas, Italian coastal waters	Liver	38	125	46
	Muscle	15	8.6	39
Blue sharks, Italian coastal waters	Liver	6.6	5.1	19
	Kidney	8.7	26	105
Harbour porpoises, Black Sea;	Liver	8–35	50–164	15–42
Sea otters, Californian	Liver	<7–360	21–5820	19–3020
coastal waters	Kidney	<7–61	3.7–200	4–210
	Brain	<2.4–24	1.8–105	2.7–140
Zebra mussels, lake system Westeinder, The Netherlands	Body	21–120	20–160	180–2500
Eel	Body	13–63	9–40	50–390
Roach	Body	7–34	20–210	160–2500

a = not stated

3.5 BIOMETHYLATION

As discussed above, it may be taken that methyl tin species found in the environment, other than at the site of manufacture, have been formed there from other tin substrates. There are relatively few examples of this (Table 3.6). *In vivo* and *in vitro* laboratory work has demonstrated that the main mechanisms for methylation (methyl carbanion, methyl carbonium ion) could be successful for tin. Both CH_3CoC_{12} and yeast (source of CH_3^+) have been demonstrated to methylate tin [53–55]. There has been little environmental methylation mechanistic work. The normal redox range of environmental sediments is sufficient to hydrogenate tin and other metals, for example, in the electron carrier nicotinamide adenine dinucleolide (NAD^+), the $NAD^+/$ NADH has a very negative redox couple (-0.32 volts) and can transfer hydrogen to many metals [56]. Although a reasonable biogeochemical cycle may be constructed (Figure 3.2) there is little quantitative information about fluxes, despite levels in water and sediments having been well studied. CH_3CoB_{12}, CH_3I and certain pseudomonads can produce methyltin compounds.

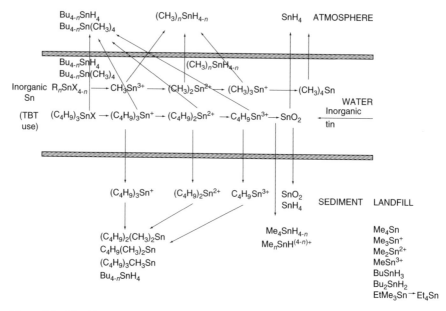

Figure 3.2   The biogeochemical cycle for tin

3.6   OVERVIEW OF DECAY RATES OF TBT AND OTHER ORGANOTINS IN WATER AND SEDIMENTS

Field studies demonstrate that decay of TBT takes place, quickly in water but slowly in sediments. A number of half-lives have been estimated and are listed in Table 3.5. Decay occurs by stepwise debutylation, with microbial or chemical attack on the butyl group leading to hydroxylation and loss of the butyl group as butanol or butene. There has been no observation of isomeric butyl groups. Evidently, tin(IV) as inorganic tin oxide is formed [57, 58].

As noted, half-lives have been estimated at from 4 days to 4 weeks in water, up to several months in freshwater sediments and years in anoxic sediments. Tin–carbon bond cleavage occurs by UV irradiation, biological and chemical cleavage (Section 3.8).

Use of triphenyl (TPhT) and tricyclohexyltins (Cy_3Sn) is mainly in agriculture. Again they degrade sequentially, with periods of only a few days being reported for TPhT. Half-lives of between 60–140 days are estimates for sediments [55]. Cy_3Sn degradation half-lives are estimated at between 50 days and 3 years. Ph_3Sn_n has been found in sediments at levels up to $20\,\mu g\ kg^{-1}$.

Photolysis by sunlight (where feasible) is the fastest route to degradation in water. The mean bond dissociation energies for tin-carbon bonds are in the range 190–$220\,kJ\ mol^{-1}$. UV radiation at $290\,nm$ corresponds to an energy of about 300 kJ mol^{-1}, i.e. sufficient to break these bonds [8]. However, this route is probably not important in water at depth or on sediments. Here degradation is caused by various bacteria or macroalgae, e.g. pseudomonads, chlorella. Chemical reagent decay is not important in most natural environments.

3.7 METHODS FOR THE ANALYSIS OF ORGANOTIN SPECIATION IN THE ENVIRONMENT

The state-of-the-art method for the analysis of organotin compounds in the environment, described in the previous edition [59], was based on the separation of the analytes by liquid (HPLC) or gas chromatography (GC), followed by detection using spectroscopic methods such as graphite furnace atomic absorption spectroscopy (GF-AAS), hydride generation quartz furnace atomic absorption spectroscopy (HG-QF-AAS), flame photometric detection (FPD), or mass spectrometry (electron impact-MS). The lowest detection limits achievable in 1986 were at the nanogram $(10^{-6}\ \text{ng}\ \text{g}^{-1})$ level and analysis of a full range of environmental samples (sediment, biomaterials, water, and materials) was reported. In the 16 years since that publication, the use of chromatography coupled to an elemental specific detector is now widely used in speciation analysis. The lowest detection limit reported in the current treatise is at the femtogram $(10^{-9}\ \text{fg}\ \text{g}^{-1})$ level and methods for a wide range of sample types are reported.

The greatest change in analytical methodology since the publication of the first edition in 1986 [59] is that the analysis of organotin compounds is routinely used to determine the speciation of organotin compounds in environmental matrices. This change has required the development of user-friendly instrumentation and acquisition software, certified reference materials (CRMs), robust extraction methods, high purity calibration standards and an understanding of the major analytical pitfalls. The speed of analysis gives a good indication of how far organotin speciation has come since the original publication. Complete analytical methods, from preparation to final determination using multicapillary gas chromatography coupled with atomic emission detection (GC-AED), can now be performed in less than 15 minutes with no loss in accuracy or precision reported [60]. The most recent approach to the analysis of TBT does away with the need for a separation step completely and relies on the differences in vapour pressure of the three buytltin chlorides. The greater volatility of TBT-Cl then allows it to be sampled from the gas phase via solid phase microextraction (SPME) prior to detection by ICP-MS [61].

A number of reviews dealing with the analysis of organometallic speciation by chromatography coupled to spectroscopic detectors, describe techniques for organotin compounds in environmental samples. Some of these are focused on different types of chromatography [62–66], others deal specifically with high performance liquid chromatography (HPLC) [67–70] or gas chromatography

(GC) [71]. In general, the methods used for the determination of organotin compounds in environmental samples are based on four common steps: (i) extraction (air, water, sediment/soil and biota); (ii) derivatization (reserved for analysis using gas chromatography (GC) and applicable to the non-volatile organotin compounds); (iii) separation (usually some form of chromatography); and (iv) detection (usually specifically for tin and dependent on the type of chromatography used). In some cases steps (i) and (ii) are combined. The following discussion will focus on each of these steps and deal with the major developments in the area since 1986.

3.7.1 SAMPLE PREPARATION METHODS

Extraction of the organotin compounds prior to analysis is required for air, water, sediment/soil and biota samples. The extraction of organotin compounds from environmental matrices is prone to possible problems, including: (i) under-extraction or analyte loss/degradation; (ii) contamination; (iii) species interconversion; and (iv) the requirement for clean-up steps if numerous interferences remain in the extract. However, by use of a suitable experimental design that incorporates the effective use of blanks and a sufficient degree of replication, and the adequate validation of the complete analytical procedure using spike recovery experiments, these problems can be assessed and corrected for if necessary. The use of analytical procedures tailored to a specific sample matrix has also been shown to overcome a number of the aforementioned problems. The availability of a suitable range of certified reference materials (CRMs) for different environmental matrices, has made evaluation of the analytical procedure more reliable and helped in the assessment of each analytical step, particularly the extraction of the organotin species in the presence of other interfering compounds.

The requirements for extraction depend on the organotin species of interest and the separation method to be used. Most environmental studies have focused on three distinct groups of organotin compounds, differing in the type of organic substituent bonded to the tin atom. Thus butyltin compounds, phenyltin compounds and methyltin compounds have been the focus of the majority of these studies. The differences in the chemical nature of the different degree of alkyl substitution (e.g. mono, di, tri and tetra), particularly between the ionic and neutral species, mean that it is difficult to find an extraction method suitable to all species in one step, so compromise conditions are used to achieve this.

The analysis of organotin compounds in water samples, both freshwater and marine, is facilitated by extraction into a non-polar solvent, which is then removed from the water layer. Complete extraction of all the variously substituted organotin compounds (mono, di, tri and tetra) is accomplished either by complexing the more ionic species (mono and di) with tropolone, or by *in-situ* conversion to the ethylated species using sodium tetraethylborate. This converts the organotin species to the correspondingly peralkylated form which is more soluble in the organic extraction solvent. Acids are often added to the water sample to aid dissolution of any organotin species from suspended particulate matter, and addition of salt facilitates phase transfer of the analytes into the organic layer. The organic layer is removed and the solvent is evaporated prior to being redissolved in a solvent suitable for gas chromatographic analysis. If the organotin compounds have been removed by tropolone, derivatization with a Grignard reagent to a peralkylated form is required before GC analysis. In the case of liquid chromatography, alkylation is unnecessary and the tropolone complexes can be analysed either directly or after some pretreatment to liberate the organotin compounds.

Analysis of surface seawater from the East China Sea for the mono-, di- and tri-substituted forms of butyltin and phenytin compounds involved the evaluation of two different extraction methods [72]. One method involved adding hydrochloric acid and sodium chloride to a 1-L water sample, followed by an internal standard. Tropolone, hexane and dichloromethane were then added and the sample stirred for 30 minutes. The two phases were separated and the organic phase dried by passing it down an anhydrous sodium sulphate column. A Grignard reagent was added to propylate or pentylate the organotin species and the mixture reacted for 30 minutes. Acid was then added to decompose any residual Grignard reagent and the aqueous layer extracted a number of times with hexane. The organic layer was again dried using a column and the volume reduced, prior to GC analysis with ICP/MS detection. The second method used a 1-L sample adjusted to pH 5 with an ammonium acetate/acetic acid buffer, followed by addition of an internal standard, sodium tetraethylborate and hexane. The mixture was then stirred for 20 minutes, which converted the organotin species to their ethylated equivalent (TBT to TBTEt, etc.). The hexane layer containing the derivatized organotin compounds was then removed, dried with anhydrous sodium sulphate and preconcentrated to a smaller volume, prior to GC-ICP-MS analysis. Evaluation of the two procedures showed that derivatization by aqueous phase ethylation provided the greatest yields and was used for the environmental samples collected.

The analysis of sediment samples for organotin pollution has received considerable interest recently because of the residual organotin contamination present after restrictions in the use of tributyltin containing antifouling preparations. Environmental studies have shown that marine, estuarine and river sediments act as the ultimate sink for non-volatile organotin compounds and can be a continuing source of contamination in the water column due to resuspension of sediment particles. A number of papers have been published showing comparisons of extraction efficiency for the analysis of butyltin compounds from sewage sludge [73] and sediment reference materials [74, 75]. The most useful of these for those interested in developing methods for the analysis of organotin species in sediment, is the comprehensive report and comparison of the methods used during the certification of a harbour sediment (RM 424,

BCR), a fresh water sediment (CRM 646, BCR) and a coastal sediment (CRM 462, BCR) [74, 75]. However it should be borne in mind that the samples encountered in environmental analysis are considerably different in terms of homogeneity, physical nature and water content as compared with CRMs. The differences in the nature of the CRM and the natural environmental sample may well affect the extraction procedure, so the analyst engaged in environmental studies should consider carefully the comparability of the results from the analysis of a CRM compared with their natural sample. A comparison of ten methods for butyltin compounds from sediment showed that only three methods produced extraction efficiencies of greater than 90 % for TBT, whereas none of them gave reproducible results for MBT [76].

Similar methods are available for the extraction of organotin species from solid matrices such as sediment, soil, sewage sludge and biota. Combinations of several chemical reagents have been used to extract organotin compounds from such samples. Acids are used to help leach out the organotin species (e.g. acetic acid, hydrobromic acid, hydrochloric acid, etc.), a complexing or chelating agent is used to facilitate phase transfer of the more lipophobic mono- and di-substituted organotin compounds (e.g. tropolone), and an organic solvent (e.g. hexane, pentane, etc.) is used to extract the lipophilic tetra- and tri-substituted compounds along with the complexed/chelated organotin species. For GC analysis derivatization via hydride generation, a Grignard reaction or aqueous phase ethylation is used, whereas for HPLC the organic phase is evaporated to dryness and redissolved in the mobile phase prior to the separation and detection.

The analysis of sediments from harbours in southern India for butyltin and methylbutyltin compounds exemplifies the approach to sediment extraction often used [77]. The sediment sample was air dried prior to addition of the internal standard tripropyltin. A solution of tropolone in toluene was added along with sodium chloride and a solution of hydrochloric acid in methanol, the mixture was shaken for 1 hour and then water was added. After being centrifuged, the toluene layer was removed and reduced in volume and the toluene was then buffered at pH 5 using an acetate buffer. Water and a solution of sodium tetraethylborate were added and after 10 minutes the upper organic layer containing the ethylated organotin compounds was removed; this was then dried with anhydrous sodium sulfate prior to analysis by GC-ICP/MS. The difference between this method and some other methods based on aqueous-phase ethylation is that the organotin compounds were removed from the matrix prior to derivatization. In this way ethylation of interfering compounds, which might then be transferred into the organic layer, is eliminated.

Two further approaches to the extraction of biomaterials, which are not applicable to sediment analysis, include the use of enzymatic digestion with lipase and protease followed by extraction of the liberated organotin compounds with a suitable solvent. An alternative method that has been widely used is digestion with tetramethylammonium hydroxide and in some cases further extraction into an organic solvent. The extraction efficiencies of 12 selected methods for the analysis of butyltin and phenyltin in mussel tissue were compared by analysis of a certified reference material (CRM 477) [78]. All of the methods involved solvent extraction, some in combination with tropolone and/or an acid, and various traditional techniques such as shaking and soxhlet extraction were used to aid the process. The use of GC-MS analysis for all the extraction methods allowed for direct comparison of each method. The results showed a wide variation in efficiency, but some general trends, such as the strong influence of tropolone on the extraction of the mono- and di-substituted organotin compounds, were highlighted.

Traditionally, shaking, sonication, soxhlet extraction or refluxing facilitates the extraction process. However, recent studies have concentrated on methods for reducing the overall time taken for sample preparation and also on attempts to automate the analytical procedure fully (although complete automation from sample extraction to analysis has yet to be realized). The use of microwave heating, particularly the use of low-power focused microwave technology, has been used to reduce sample preparation time to less than the time taken for a GC run [79]. This approach also reduces the use of solvents and in some cases improves the extraction efficiency. A number of reports on the use of supercritical fluid extraction (SFE) for a variety of organotin compounds from environmental matrices have been summarized [80]. The increased use of this technology will depend to a large extent on the availability of the expensive hardware associated with the technique. The use of accelerated solvent extraction (ASE) where the extraction medium is forced through the sample at high pressure, is a technique originally designed for extraction of organic compounds from environmental samples. However, it is also applicable to organotin analysis [81] and should help with the automation and reproducibility of sample extraction.

The extraction step for the analysis of air samples is considerably different to that for the other sample matrices mentioned and involves trapping the orga-notin compounds present in the gas sample on either a packed or unpacked column at low temperature. This can be accomplished on-site and the traps stored in liquid nitrogen, prior to desorption into the detector, which is usually ICP-MS. An alternative method is to sample the gas into a suitable bag, prior to transporting the sample back to the laboratory, where the gaseous analytes are recovered using a cooled U-trap as described above [82]. This approach has the advantage for on-site sample collections of simplicity and does not require the use of complicated sampling apparatus and a power supply.

3.7.2 ANALYTICAL METHODS BASED ON GAS CHROMATOGRAPHY

The type of gas chromatographic approach used in organotin analysis is deter-mined to some extent by the group of compounds that are of interest. The majority of studies have focused on the series of butyltin compounds (mono-, di-, tri- and tetrabutyltin), but there is a smaller number of methods for the methyl-, phenyl-, cyclohexyl- and octyltin compounds (again mono-, di-, tri- and tetra-substituted tin). Although some organotin compounds, particularly the tetra-substituted compounds, are readily amenable to analysis by GC methods, the more environmentally important tri-, di- and mono-substituted compounds require derivatization to thermally stable, volatile derivatives before analysis. Both packed and capillary column methods have been reported, but the advantages of using capillary columns means that these methods are becoming increasingly important. A distinct advantage of GC compared with HPLC separations is it's general applicability to a wide range of different alkyl-substituted organotin compounds in one run. Thus it is possible to analyse for a wide range of compounds that appear in the environment more quickly and at a lower cost using GC rather than HPLC. A summary of some analytical speciation methods using gas chromatography that have been used for the analysis of organotin compounds in gases, liquids, soil and biota samples is given in Table 3.8.

The main criteria for the GC detector is that it gives a response specifically for tin, rather than the compound. This is because of the detection limits required for environmental analysis, not attainable with conventional GC detectors, which also simplify sample preparation. With tin-specific detectors, sample clean-up is less critical than with conventional GC detectors. Obviously, both in terms of the time required and because of the possibility of losses, it is preferable to keep sample clean-up steps to a minimum. The most popular detection methods available for organotin analysis are flame photometry (FPD), using a tin selective filter, mass spectrometry (EI-MS), microwave-induced plasma atomic emission spectroscopy (MIP-AES), and inductively coupled plasma mass spectrometry (ICP-MS). Not all of these separation–detection configurations are available commercially. Methods such as GC coupled to ICP-MS require in-house fabrication of a transfer line from the column to the plasma. However, this is readily accomplished and a number of reliable and simple coupling methods have been published [83]. Three different derivatization approaches have been used with methods based on gas chromatography and these are described below. A comparison of the efficiency of these methods for the derivatization of MBT, DBT, TBT, DPhT and TPhT in mussel tissue has been reported [84].

Hydride generation methods

The longest established derivatization method available involves the formation of an organotin hydride, which confers the volatility required for chromatography to take place. This can be carried out off- or on-line by reducing the organotin compound with sodium borohydride in the presence of acid (Equation 3.2).

Table 3.8 A summary of the methods used for organotin speciation by gas chromatography (GC)

Species sample matrix	Column or U-tube packing	Detector used	Detection limit	Derivatization reagent	Derivatization on/off-line	Date	Reference
MMT, MBT, TMT, TBT River sediment	U-tube, no packing −100–200°C	HG-QF/AAS Et-QF/AAS	33 ng/g	NaBH$_4$	On	1993	95
Seven volatile tin species Land fill gas	U-trap 10 % SP-2100 60/80 Supelcosport	HG-ICP/MS			All gases	1994	96
TMT, DMT, TPRT, TEBT, MMT, TBT, MBT, DBT, TEPT Water, sediment	Capillary HP-1	GC-QF/AAS CGC-MIP/AED	0.05 pg 50 pg ml^{-1}		Off	1992	97
MBT, MPhT, DPhT, TrPeT, DBT, TBT, TPhT, TPrT, TeBT	Capillary DB-1	CGC-ICP/MS	0.01 pg L	Grignard NaBET$_4$	Off	1999	72
Open ocean seawater MBT, DBT, TBT, TPhT	Capillary DB-210	CGC-FPD GC-AED	23–40 ng g^{-1} 3–10 ng g^{-1}		Off	1996	98
MBT, TBT Sediment		CGC-ICP/MS			Off	1997	99
MMT, DMT, TMT Marine sediments	Capillary SPB-1	CGC-pulsed-FPD	21 ng Sn g^{-1}	NaBEt$_4$	Off	2001	100

continues

Table 3.8 (*continued*)

Species sample matrix	Column or U-tube packing	Detector used	Detection limit	Derivatization reagent	Derivatization on/off-line	Date	Reference
DBT, TBT, DPhT, TPhT Marine biota and sediment	Capillary HP-5	CGC-FPD	$3-4\,ng\,g^{-1}$	$NaBEt_4$	Off	2000	77
MBT, DBT, TBT, MPhT, DPhT, TPhT, MOcT, DOcT, TOcT, TBMeT Marine water and sediment	nr	CGC-ICP/MS	nr	$NaBEt_4$	Off	2001	101
TET, trimethyl-phenyltin, TeBT	Capillary DB5 $30\,m \times 0.32\,mm \times 0.25\,\mu m$	CGC-(LP/RP-He)-ICP/MS	$0.12-0.56\,pg$	nr		1998	102
MBT, DBT, TBT Sediment	Capillary HP-5 $30\,m \times 0.32\,mm \times 0.25\,\mu m$	CGC-ICP/MS ID-MS calibration	nr	$NaBEt_4$	Off	2000	93
TeMT, TeEtT Standards	Capillary DB5 $30\,m \times 0.25\,mm$	CGC-ICP/ TOFMS	$11-13\,fg$	NR, volatile	—	2000	103
MBT, DBT, TBT Sediment	Capillary DB5 $25\,m \times 0.5\,mm$	CGC-GD/OES	$21-75\,ng\,L^{-1}$	$NaBEt_4$	Off	2001	104

LP/RP-He: low power reduced pressure helium plasma. TOFMS: time-of-flight MS. GD/OES: glow discharge optical emission spectroscopy nr: not reported

$$R_xSn_{(4-x)} + NaBH_4 \longrightarrow R_xSnH_{(4-x)} + H_2 \qquad (3.2)$$

$$x = 1,\ 2,\ 3 \quad R = Me,\ n\text{-}Bu,\ Ph$$

Typically the sodium borohydride solution (1–20 % m/v) is stored in a solution of sodium hydroxide (0.02–0.06 M) for increased stability. The presence of the acid during hydride generation helps to neutralize the sodium hydroxide and to produce hydrogen from $NaBH_4$, which reacts with the organotin compound to produce an organotin hydride. The gaseous organotin hydrides are purged from the reaction vessel using helium and trapped on the chromatographic material, which is contained in a U-tube immersed in liquid nitrogen. The use of an optional water trap immersed in dry ice/acetone prior to the chromatographic trap in the original instrumental set-up, has now generally been replaced with a Nafion type drying tube. This helps to protect the chromatographic material from damage by water, which can significantly reduce its separation properties. The organotin species is then separated by slowly heating the U-tube electrothermally to a temperature above the boiling point of the least volatile organotin hydride, which causes the analytes to elute in order of boiling point. The organotin hydrides are then detected using some form of tin-specific detector. The most widely used type of detector is the quartz furnace atomic absorption spectrometer (HG-QT/AAS), which utilizes electrothermal heating (950 °C) to produce ground-state atoms, which are then detected by atomic absorption. A more up-to-date detector for the hydride generation approach is the use of inductively coupled plasma mass spectrometry (ICP-MS) in place of the QT-AAS. A comprehensive description of these types of analytical set-up has been described elsewhere [85].

The use of hydride generation derivatization for the analysis of organotin compounds in water samples works well. It is easy to use, provides high preconcentration factors (a large sample can be used with a long trapping period) and hydride generation acts as an effective clean-up step, because the organotin compounds are purged from the sample as gases. However, the use of this method for the analysis of more complex matrices, such as sediment or biota, is less successful because of the interfering compounds present in the sample. Compounds such as other metals and organic compounds interfere with the reaction between $NaBH_4$ and the organotin compound, either by breaking down the hydrides formed or by preferentially reacting with the borohydride reagent. Another problem that often arises with environmental samples is excessive foaming when the hydrides are purged into the detector, which leads to inconsistent and non-reproducible results.

A few reports of the generation of a hydride on-line prior to separation by GC have been published [86]. Derivatization of the organotin compounds is achieved by packing the top of the GC column with $NaBH_4$, through which the compounds of interest must pass. This approach has not been widely adopted. On-line hydride generation has also been used for HPLC detection and this will be dealt with below.

Alkylation methods

The most widely used method to convert the organotin analytes to forms suitable for gas chromatography is the formation of tetra-substituted alkyltin compounds. This is usually carried out off-line, although a few on-line reports are available. Two different approaches are possible, the first being carried out in a non-polar solvent by using a Grignard reagent, $R'MgX$ (see Equation 3.3) to methylate, ethylate, propylate, butylate, pentylate or hexylate the analyte. The Grignard derivatization reaction is carried out in a suitable organic solvent and parameters such as temperature and nature of agitation differ between methods. However, studies have shown that only the concentration of Grignard reagent used significantly affected the extraction efficiency [87]. The reaction requires only 5 to 10 minutes before the excess Grignard reagent is neutralized with an acid (Equation 3.3):

$$R_xSn_{(4-x)} + R'MgX \longrightarrow R_xSnR'_{(4-x)} + H_2 \qquad (3.3)$$

$$x = 1, 2, 3 \quad R' = Me,\ n\text{-Bu},\ Pe,\ Et,\ Pr \quad R = Me,\ n\text{-Bu},\ Ph$$

Problems with the Grignard derivatization include: losses when used for methyl derivatives, due to the volatility of the derivative, interconversion of MPhT and TPhT, and interference of sulfur derivatives with GC-MS and GC-FPD analysis of sediment.

The second method is carried out in the aqueous phase and involves ethylation by sodium tetraethylborate (Equation 3.4). The use of sodium tetraethylborate for derivatization of organotin compounds in environmental samples has been extensively reviewed [88]; in summary the pH is adjusted to between 4–5 and then the derivatization reagent, dissolved in ethanol, is added. The organotin compounds are either derivatized *in situ* and then extracted into an organic solvent, or else extracted from the sample and then derivatized. The latter method produces an extract with fewer interferences in most cases. In general the Grignard derivatization gives a higher yield than does aqueous-phase ethylation for the same samples. However the Grignard reaction is sensitive to the presence of water so the derivatization step is not as robust.

$$R_xSn_{(4-x)} + NaB(C_2H_5)_4 \longrightarrow (C_2H_5)_xSnR_{(4-x)} \qquad (3.4)$$

$$x = 1, 2, 3 \quad R = Me,\ n\text{-Bu},\ Ph$$

A number of detectors have been used for the analysis of organotin compounds with gas chromatography. Some conventional detectors such as flame photometry and mass spectrometry are specific enough to deliver the detection limits required for environmental analysis, but the small sample volumes used in GC analysis limit most other detectors. However, plasma-based techniques such as inductively coupled plasma (ICP) and microwave-induced plasma (MIP) (either in the atomic emission or mass spectrometric configuration) have been successfully used as GC detectors in speciation analysis. Until recently the only commercially available instrumentation of this type was the MIP-AES system, but a GC system coupled to ICP-MS is currently being developed. This will bring the speciation of organotin compounds by GC-ICP/MS out of the research laboratory and into more routine use, because it removes the need for building in-house interfaces for coupling the GC to the ICP-MS detector.

The use of ICP-MS detection has significant advantages over other detection systems. The high power of ICP compared to other plasmas reduces problems associated with other organic compounds eluting from the system, and the long dynamic range of the detector means calibration is limited by the volume of sample that can be handled by the chromatographic step. The multi-elemental nature of detection means that determination of a range of volatile organometallic compounds is limited only by their reactivity to the derivatizing agent and solubility in the solvent used. Introduction of a gaseous sample can reduce the formation of polyatomic interferences, a major problem in solution ICP-MS analysis, and mass spectrometric detection allows the measurement of isotopic ratios, which can aid in the evaluation of isotopic interferences and allows for the use of isotope dilution analysis (ID-MS).

3.7.3 ANALYTICAL METHODS BASED ON LIQUID CHROMATOGRAPHY

The use of high performance liquid chromatography (HPLC) for the separation of organotin compounds prior to detection, has a number of differences compared to gas chromatography. These include the ease of interfacing the liquid flow from the column to the detection system (no need for a heated transfer line) and no derivatization step is required to confer thermal stability, so the analytical methodology is simplified and contamination from the derivatization reagents is eliminated. However, it is not possible to determine such a broad range of organotin species differing in alkyl substituents (methyl, ethyl, etc.) and degree of substitution (mono-, di-, etc.) in one chromatographic run because of the difference in solubility. The time for a particular separation is generally longer than with GC separations, particularly when a solvent gradient is used.

Ion exchange, size exclusion, reversed and normal phase separations have all been developed for organotin speciation [67] and some recent methods are shown in Table 3.9. Ion exchange chromatography is a popular separation methodology for a wide range of organotin species (not necessarily all by the same method) including methyl, ethyl, butyl, propyl, phenyl and cyclohexyl substituted species. A wide range of environmental matrices has also been analysed, including sediment, biota and water. Reversed phase separations and similar types of chromatography (micellar, ion-pair, etc.) are also widely reported for separation of a similarly wide variety of organotin species in different environmental matrices.

Table 3.9 A summary of the methods used for organotin speciation by high performance liquid chromatography (HPLC)

Species Sample matrix	Column packing	Detector used	Detection limit	Date	Ref.
DBT, TBT Harbour sediment	Partisil SCX Silica-based ion exchange	HPLC-ICP/MS	0.02 ng as Sn	1990	105
MBT, DBT, TBT Harbour sediment and marine biota	Spherisorb SCX Silica-based ion exchange	HPLC-GF/AAS	0.5–1.1 ng as Sn	1993	106
TBT, TPhT Fish	PRP-1 Resin-based ion exchange column	HPLC-ICP/MS	1.5–2.3 pg as Sn	1993	107
MBT, DBT, TBT, TPhT Mussel tissue	Partisil SCX Silica-based ion exchange	HPLC-HG/ICP/ AES	7 ng as Sn	1995	108
TBT, TPhT Sediment	Reversed phase column	HPLC-APCI/MS		1997	109
TBT Lake water and marine sediment	Supelcosil LC-18	HPLC-ESI/MS In-tube SPME	0.05 ng mL^{-1}	2001	110
MBT, DBT, TBT Sediment	Whatman SCX Silica-based ion exchange	HPLC-ICP/MS	0.03–0.08 μg g^{-1}	2001	111
MBT, DBT, TBT Sediment, mussel tissue	Whatman SCX Silica-based ion exchange	HPLC-HG-ETAAS On-line hydride generation	135–942 pg	2001	112

Methods based on conventional, non-specific detection have been developed with detection limits able to determine the concentration of butyltin compounds in sediments. UV and fluorescence spectroscopy have been used by reacting the organotin analyte with a suitable chromophore either pre- or post-separation. The most widely used and most successful HPLC detection methods for organotin speciation analysis are based on atomic spectroscopy. Flame, on-line hydride generation and off-line graphite furnace AAS, ICP-AES and ICP-MS have all been used for detection of organotin speciation in a wide range of matrices. The methods with the lowest detection limits are based on detection by ICP-MS and this still remain an easy and sensitive method to use.

Coupling the separation to the detector is simple for aqueous phase eluents with < 5% methanol, higher volumes of organic substituent, or less polar solvents, require the use of a cooled spray chamber and addition of oxygen to the plasma. By cooling the spray chamber or using a more sophisticated nebulizer, the loading of volatile solvent to the plasma is reduced and by adding 1–2% oxygen, excess carbon is burned off the mass spectrometer interface. Eluent systems based on gradient elution have been developed but are difficult and time consuming to use because the original chromatographic conditions have to be established for every run and the change in eluent composition can affect some detection systems, particularly ICP-MS.

Recently, detection based on atmospheric ionization methods such as electrospray, ion spray or chemical ionization have been used to detect the organotin species. These methods have the advantage that molecular information concerning the analyte is available, rather than just atomic information. In this way greater certainty of identification is achieved because it is not based solely on comparison of retention time with a standard.

3.7.4   ANALYTICAL METHODS BASED ON OTHER TECHNIQUES

Only a limited number of other techniques have been reported for organotin analysis, specifically supercritical fluid chromatography (SFC), tandem mass spectrometry (MS/MS), and selective extraction using solid phase microextraction. Table 3.10 shows an assortment of such methods. Supercritical fluid chromatography involves using a liquified gas as an eluent at high pressure. For the analysis of organotin compounds the use of carbon dioxide as the eluent with detection by ICP-MS has been reported. The SFC work that has been reported, deals with a more limited number of organotin compounds in environmental samples compared with HPLC or GC separations. The technique has been used with a limited number of eluents so its general applicability to the speciation of organotin compounds has not yet been shown. Atmospheric pressure ionization tandem mass spectrometry has been used for the quantitation of tributyltin in a sediment reference material. The method relied on molecular fragmentation and tandem mass spectrometry for the specificity necessary in environmental speciation analysis [89]. A recent method for the analysis of TBT does away with the need for a separation step completely and relies on the differences in vapour pressure of the three buytltin chlorides. In this approach, the greater volatility of TBT-Cl allows it to be sampled from the gas phase via solid phase microextraction (SPME) prior to detection by ICP-MS [61].

3.7.5   CALIBRATION TECHNIQUES

The calibration step in analytical procedures for organotin speciation analysis in environmental analysis is another area of possible systematic error. The recommendations by bodies involved in certification of reference materials are that standard additions should be performed if possible. Internal standards have been widely used in environmental analysis: with GC-based methods, the

Table 3.10   A summary of the methods used for organotin speciation by other analytical methods

Species Sample matrix	Column packing	Detector used	Detection limit	Date	Ref.
TBT, TeBT Standards	SB-octyl-50 capillary column 2.5 m long	ICP-MS	0.04 pg as Sn	1991	113
TBT, TeBT, TPhT, TePhT Standards	SB-octyl-30 capillary column 2.5 and 10 m long	ICP-MS	0.2–0.8 pg as Sn	1992	114
TBT, TeBT Standards	SB-octyl-30 capillary column 2 m long	ICP-MS	0.03 pg as Sn	1994	115
TBT Sediment	No column Separation by fragmentation potential and MS/MS detection	API-MS/MS	64 fg as TBT	1997	89
TBT Sea water, leached sediment	Selective extraction of TBT-Cl by SPME	ICP-MS	5.8 pg mL^{-1}	2001	61

tripropyltin is often used because it is not found in environmental samples. By using tri- and tetra-substituted tin compounds as internal standards in the same method, the former standard can give an indication of derivatization efficiency, whilst the later indicates the efficiency of the solvent extraction. With HPLC separations the choice of internal standard is limited because of the differences in polarity and solubility of different alkyl substituted organotin compounds.

The use of calibration by isotope dilution mass spectrometry (ID-MS) has been used for many years in organic analysis by LC or GC-MS because of the availability of spike materials containing ^2H or ^{13}C isotopes, which can be used as the internal standard. The analysis of TBT using GC-MS analysis with ID-MS calibration has been reported [81] using a spike containing ^{13}C. However, an alternative approach for organometallic analysis is the use of a spike material containing an enriched isotope of the metal. In this case one of the tin isotopes that occurs at low abundance naturally is present in the spike material at a higher abundance, and can be used as the internal standard.

Two possible approaches based on ICP-MS detection can be used for ID-MS analysis of organotin compounds. Their applicability will depend on the organotin compound of interest and the availability of an isotopically altered organotin compound as spike. These approaches have been referred to as *species-specific* and *species non-specific* spiking [90]. With species-specific spiking an isotopically enriched analogue of the organotin compound of interest is added to the sample prior to extraction. This approach is only possible if the structure and composition of the organotin species of interest is known and an appropriate spike is available. For this calibration method to work, it is important that no isotopic exchange occurs between the analyte and spike (to date this has not been reported) and that the analyte and spike reach equilibrium in the extract. If these criteria are met, any loss of analyte, such as during clean-up and derivatization steps, will not affect the final result as the measurement process is based on the isotopic ratio between the analyte and spike, rather than the absolute amount of analyte. The advantage of this approach to speciation analysis is that the calibrant and analyte are measured simultaneously rather than in a time-resolved manner such as with the use of an internal standard. This means that any changes in plasma conditions during elution of the calibrant and spike, which would affect the generation of ions, has no effect on the final result. Also, any loss or change of organotin species during the extraction/clean-up steps is readily apparent but, more importantly, will not affect the final result. Species non-specific spiking is also applicable to organotin analysis, but has not been reported to date. In this case the isotopically enriched spike is added after the complete separation of the different organotin species. The spike here does not exist in the same chemical form as the analyte and in most cases will be present as the inorganic ion. This type of ID-MS technique is necessary if the structure and composition of the different species to be analysed are not exactly known, or if a species-specific spike is not available.

The use of on-line isotope dilution calibration using species-specific spiking and analysis by HPLC-ID-ICP/MS has been reported for the analysis of TBT [91, 92] and for the analysis of DBT by GC-ICP/MS [93]. A small-scale preparative method for butyltin and phenyltin compounds enriched with ^{117}Sn has also been reported [94].

3.7.6 CERTIFIED REFERENCE MATERIALS

A major advance since publication of the first edition of *Organometallic Compounds in the Environment* is the possibility of establishing a quality control system for the analysis of organotin compounds based on the use of certified reference materials. The advent of reference materials containing certified levels of, mainly, butyltin but some phenyltin compounds, in a number of matrices (see Table 3.11) has allowed the validation of analytical procedures for organotin analysis. A number of international agencies have developed CRMs mainly in environmental matrices such as freshwater, marine and estuarine sediments, marine biota and fish.

3.8 GENERAL TOXICOLOGY OF ORGANOTIN COMPOUNDS

On the basis of toxic effects organotin derivatives can be divided into four classes:

Table 3.11 A summary of the certified reference materials available for the validation of analytical methods to determine organotin species in environmental samples

Reference material	Sample matrix	Organotin species	Certified concentration	Originator
CRM 462	Coastal sediment	MBT		BCR
		DBT	65 ± 8 ng g^{-1} Sn	
		TBT	70.5 ± 13.2 ng g^{-1}	
RM 646	Fresh water sediment	MBT		BCR
		DBT		
		TBT	20 ± 5 µg kg^{-1}	
		DPhT		
		TPhT		
CRM 477	Mussel tissue	MBT	1.50 ± 0.28 mg kg^{-1}	BCR
		DBT	1.54 ± 0.12 mg kg^{-1}	
		TBT	2.20 ± 0.19 mg kg^{-1}	
PACS-1	Marine sediment	MBT		NRCC
		DBT	1160 ng g^{-1}Sn	
		TBT	1.22 ± 0.22 mg kg^{-1}	
PACS-2	Freshwater sediment	MBT	0.45 ± 0.05 µg g^{-1}Sn	NRCC
		DBT	1.09 ± 0.15 µg g^{-1}Sn	
		TBT	0.98 ± 0.13 µg g^{-1}Sn	
NIES 11	Tuna tissue	Total Sn	As compound	NIES
		TBT	1.3 ± 0.1 µg g^{-1}	
		TPhT	6.3 (indicative) µg g^{-1}	

BCR: Bureau of Community Reference (European Union).
NRCC: National Research Council Canada.
NIES: National Institute of Environmental Standards (Japan).

(i) *low-molecular-weight trialkyltin compounds*, like trimethyltin (TMT) and triethyltin (TET);

(ii) *butyltin compounds*, like monobutyltin (MBT), dibutyltin (DBT), and tributyltin (TBT);

(iii) *triphenyltin compounds* (TPT), like triphenyltin acetate or Fentin acetate (TPTA), triphenyltin chloride or Fentin chloride (TPTC), triphenyltin hydroxide or Fentin hydroxide (TPTH);

(iv) *phenyl-alkyltin compounds*, like fenbutatin oxide or Lexitin or Torque (FBTO)

(v) *cycloalkyltin compounds*, like tricyclohexyltin or Cyhexatin or Plictran (TCHT), and azocyclotin or Peropal (ACT).

All these compounds are toxic, but the effect varies according to the number and type of organic moiety present. They can easily move along the trophic chains and have an impact on natural aquatic environments, particularly in estuarine and coastal ecosystems, leaching out from various sources, as a result of chemical industry, agriculture and maritime activities. Other sources of environmental exposure include discard and sanitary landfill disposal of plastic, PVC food wrappings, bottles and rigid potable water pipes. Their stable presence in both freshwater and sea water ecosystems represents a high risk in provoking deleterious effects on biocenoses already at low concentrations $(0.1–1.0\,\mu g\,L^{-1})$. They are rapidly bioaccumulated in tissues of organisms living in the water–sediment interface causing severe, long-term, toxic effects on local epifauna with repercussions on biodiversity and human health. The latter may be exposed to a risk mainly through the diet, although a number of studies of human illnesses, some of them serious, after occupational exposure or suicide attempts have been reported [116].

An outbreak of almost epidemic nature took place in humans in France in 1954 due to the oral ingestion of an antibacterial medication (Stalinon) containing diethyltin diiodide for the treatment of skin disorders: 102 fatalities occurred out of 217 contaminated people, and symptoms (headache, cerebral oedema) persisted for years in survivors [117–119].

Another important and historical impact of organotins has been represented by non-predicted, adverse effects of TBT on non-target organisms, particularly molluscs, leading to their reproductive failure and population decline. From 1976 to 1981 the oyster industry in the Bassin d'Arcachon (France) was disturbed by a drastic reduction of reproduction so that the number of oyster farmers declined by 50%. As a consequence, France has been the first country in the world (1982) to ban the use of TBT in antifouling paints in neighbourhood of mollusc farmers [120].

Acute toxicity of organotin compounds (LD_{50}) to experimental mammals and of TBT to some aquatic invertebrates and vertebrates are reported in Tables 3.12 and 3.13, respectively. Except for the three acaricides (ACT, TCHT, FBTO), mice appear to be generally the most sensitive species relative to the other mammals, and the oral route seems less dangerous because organotins undergo limited gastrointestinal absorption. The tissue distribution of tin from organotin compounds shows the highest concentrations in the blood and in the liver, with smaller amounts in the muscle, spleen, heart or brain. Among aquatic invertebrates, the most sensitive species to TBT are the filter-feeding, benthonic ones, like bivalve molluscs, bryozoans and ascidians. Among aquatic vertebrates, marine fish appear more sensitive to TBT than freshwater fish.

3.8.1 HUMAN TOXICOLOGY

Absorption of organotins, either through the skin or the gastrointestinal tract, is generally limited and few poisoning and deaths have been reported as a result of occupational exposure (manufacturing, formulating, painting, spraying).

Table 3.12 Acute toxicity (LD_{50} mg kg^{-1} body weight) of organotin compounds to laboratory mammals

		MBTC	DBTC	TBTC	TBTO	TPTA	TPTC	TPTH	ACT	TCHT	FBTO
Mus musculus	oral	1400	35	117	85	81		209		710	1450
	i.p.				16	7.9					
Rattus	oral	2140	150	129	122	136	80	171	209	540	4400
norvegicus	i.p.				19	8.5				13	
Cavia	oral		190			24		27	261	780	
porcellus	i.p.					3.7				9	
Oryctolagus	oral					40				800	
cuniculus	i.p.									>126	

The manifestations of poisoning are local or systemic [121,122]. All organotins are local irritants producing skin lesions (erythematous eruption, contact dermatitis, ulceration), eye irritation (profuse lacrimation, eye inflammation, photophobia, blurred vision), and respiratory damage (shortness of breath, coughing, wheezing) after a few minutes contact.

The following systemic effects are reported for some widespread organotins but not for the acaricides (ACT, TCHT, FBTO).

(i)   Neurological effects: disturbance of the sense of smell (TBT), neural necrosis (TMT, TET), encephalopathy (TPT), seizures (TMT), amnesia (TMT, TPTA), headache (TET, TBT), sensorimotor polyneuropathy (TPTA).

(ii) Gastrointestinal effects: pharyngitis, nausea, vomiting, loss of appetite, abdominal pain (TBT, TPTA).

(iii) Hepatic effects: elevated serum transaminase activities, mucoprotein and C-reactive protein, generalized hepatomegaly, biopsy evidence of hepatic inflammation (TPTA).

(iv) Genitourinary effects: acute renal failure (TPTA).

(v)   Haematological effects: mild anaemia, leukopenia (TPTA).

There are no specific antidotes and treatment involves decontamination and supportive therapies. Moreover, the long-term human health effects due to low-level exposure to organotins are unknown.

3.8.2 BIOACCUMULATION

The lipophilic properties of organotin compounds contribute to bioaccumulation in living organisms, and their liposolubility gives them easy access to the various links of trophic chains.

Table 3.13 Acute toxicity (LD_{50} mg kg^{-1} body weight) of TBT to some aquatic organisms

Annelids	
Nereis diversicolor	0.002
Tubifex tubifex	0.005
Crustaceans	
Acartia tonsa	0.001
Nitocra spinipes	0.002
Gammarus sp.	0.005
Daphnia magna	0.006
Crangon crangon	0.073
Carcinus maenas	0.110
Freshwater molluscs	
Bulinus tropicus	0.010
Limnaea stagnalis	0.060
Biomphalaria glabrata	0.070
Marine molluscs	
Crassostrea virginica	0.0018
Ostrea edulis	0.0032
Mytilus edulis	0.3000
Crassostrea gigas	1.8000
Bryozoans	
Schizoporella errata	0.0001
Ascidians	
Botrylloides sp.	<0.0001
Fish	
Agonus cataphractus	0.026
Salmo gairdneri	0.028
Leuciscus idus	0.050
Cyprinus carpio	0.075
Solea solea	0.088
Lepornis macrochirus	0.240

The potential risk of bioaccumulating-organotin compounds is high for aquatic organisms, because these compounds are easily and rapidly accumulated in algae, thus interfering with their life-cycle [123], and in animals by dietary and branchial uptake from water into the lipophilic compartments [124], whereas their elimination from the body is slow, as reported in fish for butyltins [125] and TPT [126]. These are therefore considered to be slow-acting toxic compounds with long-term effects. Surprisingly, the three acaricides ACT, TCHT and FBTO are generally much more toxic to aquatic species than to terrestrial ones and even more toxic than either TBT or TPT, but their potential for environmental contamination is low. TET and monosubstituted organotins appear to be more effective in growth inhibition of methanogenic and sulphate reducing bacteria than trisubstituted alkyltins [127,128] but also TBT tolerant marine bacteria, like *Alteromonas* spp. have been recently found

[129]. Some *Pseudomonas* spp. have been reported to accumulate TBT up to 2% of their dry weight [130].

In vertebrates, organotins accumulate in specific target organs, i.e., brain, liver, kidney and lymphatic tissues. Metabolites can also be found in tissues, since the hepatic microsomal system is able to metabolize TBT in DBT by a dealkylation process in mammals that has been widely described by Aldridge [131], in fish by Tsuda *et al.* [132], and in tadpoles by Semlitsch *et al.* [133], and cycloalkyltin compounds undergo a progressive scission of the cycloalkyl groups from the tin atom [134]. Debutylation of TBT is known to take place also in bacteria, fungi and algae and this provides one route for detoxification. Bacteria tolerate high levels of TBT due to their ability to dealkylate it, forming less toxic compounds, which may also utilized as carbon sources, and/or to exclude the toxicants outside the cell [129].

Marine vertebrates of the higher trophic chains, like cetaceans (*Neophocaena phocaenoides, Tursiops truncatus*), tunas (*Thunnus thynnus*) and sharks (*Prionace glauca*) bioaccumulate great amounts of butyltin compounds: TBT is the predominant species in the blubber of dolphins (0.2 mg kg^{-1}), muscle of tunas (0.039 mg kg^{-1}) and kidney of sharks (0.1 mg kg^{-1}), whereas DBT is preferentially concentrated in the liver of dolphins (1.6 mg kg^{-1}) and tunas (0.12 mg kg^{-1}), and in all tissues of sharks (0.021 mg kg^{-1}) [135, 136].

Tin levels of 0.2 to 20 mg kg^{-1} have been demonstrated in various marine invertebrates [137, 138], but organotin levels have only recently been documented, e.g. TBT (from 3.3 to 45.6 µg L^{-1} dry weight) in mussels from the Adriatic Sea [139]. In molluscs, organotin bioaccumulation appears to be related to seasons and ship activities, and is more rapid when animals feed microalgae, supporting the hypothesis of the contaminant transfer through the trophic chain [140]. Molluscs show the highest concentrations of organotin compounds in their tissues relative to vertebrates, probably because of the presence, in the latter, of enzyme systems able to degrade TBT allowing the excretion of its metabolites [141]. Bivalves such as *Mytilus edulis* and *Crassostrea virginica* accumulate organotin compounds to the greatest extent in gill tissue [142] and metabolize these pollutants slowly in comparison with gastropods, crustaceans, echinoderms and fish [143–145]. Organotin degradation appears to be relatively inefficient in *Mercenaria mercenaria* and *Mya arenaria* [146], hence bioaccumulation and bioavailability represent a potential risk to human health by their intake from contaminated food. High levels of consumption of fish and molluscs from contaminated areas may be of some toxicological concern. Organotin residues in fish and shellfish for human consumption occur at typical levels of 0.01–1.0 mg L^{-1}, primarily as TBT and TPT, but levels in shellfish may be higher.

3.8.3 NEUROTOXICITY

Lower trialkyltin compounds are known to provoke central nervous system (CNS) injury in mammals. TMT and TET are highly neurotoxic, each producing distinct effects [147,148]. TMT is primarily a CNS neurotoxin inducing gliosis [149] and destroying neurones within the hippocampal pyramidal band and the fascia dentata [150]. Additional sites include the amygdala, pyriform cortex, and the neocortex. TET is a neurotoxin that produces a pathological picture dominated by depression, brain and spinal cord oedema, and resulting hyperglycaemia may be related to the centrally mediated depletion of catecholamines from the adrenals. Moreover, TET causes demyelination and affects GABAergic neurones [151].

In particular, these organotins cause neuronal and immune cell apoptosis *in vitro*, and this is mediated in part by stannin, a target protein which was isolated from organotin-sensitive tissues [152].

3.8.4 HEPATOTOXICITY

In experimental mammals, TBT and DBT have been shown to induce hepato-toxicity resulting in the inflammation of the bile duct associated with hepatic lesions, hepatomegaly and necrosis, while MBT is relatively inert, probably due to low absorption through the digestive tract. In mice, which are the most sensitive mammals, butyltin compounds negatively affect the activity of ornithine carbamyl transferase in serum and hepatotoxicity induced by DBT is more rapid than that induced by TBT [153]. Moreover, most of the total tin compounds in the mice liver after TBT administration is in the form of DBT, which thus represents the main active metabolite in the hepatocytes responsible for induction of a widespread liver damage. In rats, a single dose of DBT of $50 \, mg \, kg^{-1}$ body weight produces congestion and inflammation, especially in the lower part of the bile duct. It is suggested that damage to enzymes and lipids of plasma membranes of hepatocytes occurs due to the direct interaction of DBT with membrane components, causing alteration in their structural and functional organization [154]. Since rabbits with dietary exposure to DBT die with dystrophic changes in the liver, without bile duct or liver necrosis, and guinea pigs tolerate high doses without any signs of adverse effects, it has been suggested that bile duct injury occurs only in those species where the bile duct and the pancreatic duct have a common course [155].

On the other hand, TPTA administration produces fatty degeneration of the liver in guinea pigs and TCHT administration triggers cholangitis in rats [156]. Moreover, TCHT causes a significant and prolonged induction of haem oxy-genase and a sustained decrease in haemoprotein content (cytochrome P_{450} and cytochrome b_5) in the liver in a dose-dependent manner [157].

In fish, depletion of glycogen in liver cells of DBT, TBT and TPT exposed fish has been reported, whereas no signs of toxicity were observed in the case of TMT [158]. The interaction of TBT and TPT with microsomal cytochrome P_{450} has been demonstrated in *Stenotomus chrysops* [159], *Oncorhynchus mykiss, Anguilla anguilla* and *Cottus gobio* [160], in which TPT led to a greater inactivation of P_{450} enzyme than TBT. Moreover, cytochrome b_5 content is never affected, whereas TBT and TPT inhibit NADH cytochrome c reductase and NADPH cytochrome c reductase, respectively, with species-related significant and selective effects on different components of the microsomal monooxygenase system.

3.8.5 HAEMOLYSIS

As organotins are lipophilic and amphipathic, they are regarded as being membrane active like detergents, with general membrane-disruptant properties at 0.3–3.5 mg L^{-1} concentrations. In human erythrocytes, they exhibit the following order of potency for induction of *in vitro* haemolysis: TBT > TPT > TET > DBT > TMT = MBT [161]. In fish, plasma membrane perturbations induced by TBT have been described for trout erythrocytes [162]. In invertebrates, the unique data reported concern cytolysis of haemerythrocytes of the marine worm, *Sipunculus nudus*, exposed to 3.2 mg L^{-1} TBT [163].

The differences in potency are explained on the basis of the different locations of the above compounds within the bilayer membrane.

3.8.6 IMMUNOTOXICITY

In vertebrates, organotin compounds have been shown to interfere selectively with the immune system, causing atrophy of the thymic cortex and lymphoid tissues by apoptosis with consequent depletion of T lymphocytes in peripheral blood. TBT is significantly more potent than TPT in the thymolytic action [164], whereas DBT and MBT do not produce atrophy of lymphoid organs [165]. Organotins also have deleterious effects on human B lymphocytes derived from tonsil tissue at concentrations of 35 μg L^{-1} [166]. TPT and TCHT are more active in decreasing spleen weight in mice dosed orally, while TBT and FBTO are less active [167]. Dietary TBT is suspected of increasing susceptibility to infections and causing deaths in several mammalian wildlife species [168, 169]. TBT and TPT have been shown to inhibit the tumour killing activity of human NK lymphocytes at submicromolar concentrations *in vitro* [170, 171]. This effect can be related to a B2 class carcinogen property ('probable human carcinogen') of TPT, despite chronic dietary studies at comparable dose rates in mice and rats having proved negative for carcinogenicity [172]. So TPT is the exception to the rule that organotins are generally not genotoxic [173], although it has been shown to be clastogenic or co-clastogenic *in vitro* and *in vivo* [174]. However, the immunosuppression caused by TPT (lymphopenia and lymphocyte depletion in the spleen and thymus) seems to be transient and tends to diminish on longer exposure [170]. Inhibition of phagocytosis and cytolysis of polymorphonuclear leukocytes (PMNs) after inhibition of chemotaxis and respiratory burst, with resultant depression of cell-mediated immune responses, have been also observed. Even low concentrations are immunotoxic as reported for TBT in rats [174] and rabbits [175], for DBT in rats [176], for TPT in guinea pigs [177], mice[178], rats[179, 180] and chickens[181]. In mammalian leukocytes, organotin compounds cause disaggregation of actin filaments and inhibition of tubulin polymerization to which cytoskeletal modifications have been ascribed [182], hindering chemotactic movements and particle adhesion [183].

In fish, decreased resistance to infections has been found even at the lowest-effect concentrations of TPT as well as TBT and DBT after continued exposure [158].

Butyltin derivatives and TPT both impair cell-mediated immunity, thus compromising many functions in marine invertebrates. They have inhibitory effects on yeast phagocytosis by cultured haemocytes of bivalves [184–186], sipunculids [163] and benthonic tunicates [187–189]. The potency of inhibition is TBT > DBT > MBT and TPTC > TPTA > TPTH. TBT and TPTC inhibit phagocytosis both at $0.5\,mg\,L^{-1}$ in the solitary ascidian *Ciona intestinalis*, at $32.5\,\mu g\,L^{-1}$ and $38.5\,\mu g\,L^{-1}$, respectively, in the colonial ascidian *Botryllus schlosseri*, and at $3.25\,\mu g\,L^{-1}$ and $3.85\,\mu g\,L^{-1}$ in the bivalve mollusc *Tapes philippinarum*. No cytolytic response was observed in invertebrate leukocytes, as opposed to the PMNs of mammals. However, it has been hypothesized that the enhancement of the progression of infection by the pathogenic protozoan *Perkinsus marinus* in bivalves exposed to environmentally relevant TBT levels was not attributed to effects on cell-mediated responses, but to the inhibition of the production of reactive oxygen intermediates [190]. Other toxic effects consist of (i) inhibition of respiratory burst associated with phagocytosis [188]; (ii) inhibition of the activity of hydrolytic, oxidative, and active transport enzymes [187,188,191–193]; (iii) decrease in GSH content [194]; (iv) increase in cytosolic Ca^{2+} [187]; (v) alterations in cell morphology closely related to severe interference with cytoskeletal proteins [194] and (vi) apoptosis [195].

3.8.7 MUTAGENICITY/GENOTOXICITY

In a standard battery of *in vitro* tests for mutagenicity, pesticidal organotins are generally negative although some have been shown to be clastogenic or co-clastogenic *in vitro* and *in vivo* [124, 173, 196] and butyl and methyl compounds were positive in the Ames mutagenicity test [197]. In chronic dietary studies, chromosomal aberrations have been reported in human peripheral blood lymphocytes: chromatid and chromosome breaks and gaps, dicentrics, increased sister-chromatid exchanges, altered cell cycle kinetics [198] and induction of aneuploidy [199]. Also in bone marrow cells of mice, chromosomal aberrations were observed following i.p. injection of TMT [200] as well as increased frequency of micronuclei in mitomycin C-induced peripheral reticulocytes in mice treated orally [124, 201].

TPT gave a positive result for chromosomal aberrations after metabolic activation in human lymphocytes [202] and positive results were also obtained in two mouse lymphoma mutation assays [172], tentatively interpreted in terms of the toxicity of TPT to the lymphocytes rather than as a specific clastogenic action.

Recent studies on fish nucleated erythrocytes by 'comet' assay show that TBT presents a marked genotoxic effect, less pronounced for DBT and completely absent for MBT, in *Salmo irideus* [203], but present for the latter in *Sparus aurata* [204].

Among marine invertebrates, only the bivalve *Mytilus edulis*, has been reported to have a TBT capability of inducing cytogenetic damage [205], and so very recently has the polychaete worm, *Platynereis dumerilii* [206].

3.8.8 CARCINOGENICITY

No evidence of carcinogenicity was observed from long-term dietary studies in mice and rats [207]. However, TBT can suppress NK cells in mice and this inhibition may predispose animals to malignancy [208]. It is also possible that butyltins can initiate the multistage process of carcinogenesis by influencing the steroid hormonal metabolism. FBTO is classified as a group E carcinogen ('evidence of non-carcinogenicity in humans') by the US EPA. TPT is the exception to this rule and classified as B2 carcinogen ('probable human carcinogen') by the US EPA: the value calculated for the unit of cancer potency factor from a cancer risk model (Q_1^*) is $18.3\,mg\,kg^{-1}day^{-1}$.

3.8.9 STERILITY

The first known effect of organotin compounds in marine ecosystems was the development of male sexual characteristics in female molluscs in populations subjected to high TBT pollution by antifouling paints. This malformation, namely 'imposex', causes sterilization and reproduction failure in these invertebrates leading to their local extinction. This effect has been reported in both gastropods, like *Ilyanassa obsoleta* [209], *Thais clavigera* and *T. bronni* [210], *Nucella lapillus* [211], *Hexaples trunculus* [212], and bivalves, like *Tapes decussatus* (= *Ruditapes decussata*) [212], *Ostrea edulis* [214] and *Crassostrea virginica* [215].

At concentrations of less than $10\,ng\,L^{-1}$ TBT, populations of *O. edulis* and *C. virginica* have shown a high incidence of imposex. Histological observations indicate that butyltins may retard the hormonally determined change in sex from male to female that normally occurs during the gametogenic cycle [214, 215].

Possible explanation of imposex include: (i) inhibition of aromatase activity *in vivo* (estradiol level decreased and testosterone levels increased specifically) [213]; (ii) effect on the enzymes that degrade sex hormones and excretory transport systems [216]; (iii) direct stimulatory effect on androgen receptor. The latter mechanism has been proved for TBT and mainly TPT (at concentrations of 0.4 μg L^{-1}) to be androgen-dependent transcription and cell proliferation in human prostate cancer cells, but it is not known whether such direct effects occur in molluscs [217].

In male rats, diminution in the size of the testes with less advanced maturation of the germinal epithelium, and reduced fertility were observed after 64 days of dietary exposure to 100 mg L^{-1} TPTH [156, 218].

Sperm toxicity resulting in significantly reduced mobility and viability has been reported in some marine invertebrates. Inhibition of egg fertilization after organotin exposure occurs in the bivalve *Isognomon californicum* with 5 μg L^{-1} TBT [219], in the sea-urchins *Echinometra mathaei* with 2.5 μg L^{-1} [219] and *Paracentrotus lividus* with 0.1 μg L^{-1} TPT [210], and in the ascidians *Ciona intestinalis*, *Ascidia malaca* and *Ascidiella aspersa* with 30.4 mg L^{-1} DBT [211].

The depression of sperm fertility could be attributed (i) to reduction of the sperm ability to move and/or to the blockage of important enzymatic activities, (ii) to modifications of the acrosomal structures and/or components, (iii) to interference of the compounds with the cellular membrane that regulates some important biochemical processes, and (iv) to alteration of genetic paternal material. Until now, no evidence of any adverse effects on reproduction from TCHT has been found in either vertebrates or invertebrates.

3.8.10   EMBRYOTOXICITY

In vertebrates, evidence of teratogenic effects from TBT, TPT and TCHT, but not from TMT, TET, ACT and FBTO, has been reported, with differences among species. In laboratory mammals, TBT-induced teratogenesis has a very low potential and appears to be controversial [220] and TPT yielded variable results [218]. However, it has been reported recently that butyltins produce foetal malformations when administered at different stages of embryogenesis. In particular, DBT administered during early embryogenesis and TBT administered during late embryogenesis cause foetal malformations such as cleft jaw, ankyloglossia, omphalocele, open eyelids, club foot and tail anomaly [221]. Teratogenic effects by TCHT are still of doubtful significance.

In fish, these compounds are known to cause embryonic and larval mortality or malformations that appear to be higher at warmer water temperatures. Among organotins, TCHT appears to be the most toxic. Embryonic-larval toxicity [222] has been reported for TCHT in *Oncorhynchus mykiss*

$(1.2\,\mu g\ L^{-1})$ [158], for TBT in *Solea solea* $(2\,\mu g\ L^{-1})$ [220], *Pimephales promelas* $(2.6\,\mu g\ L^{-1})$ [223], *Oncorhynchus mykiss* $(4.8\,\mu g\ L^{-1})$ [159] and *Phoxinus phoxinus* $(> 4.3\,\mu g\ L^{-1})$ [224,225], for TPT in *Pimephales promelas* $(7.1\,\mu g\ L^{-1})$ [226], *Oncorhynchus mykiss* $(6.1\,\mu g\ L^{-1})$ [158] and *Phoxinus phoxinus* $(> 4\,\mu g\ L^{-1})$ [227], and for DBT in *Oncorhynchus mykiss* $(> 40\,\mu g\ L^{-1})$ [159]. TBT causes skeletal malformations, eye opacity, degenerative hydropic vacuolation of cytoplasm in skeletal muscles, skin, kidneys and CNS, including spinal cord, with resulting vertebral malformation and tail paralysis. TPT is more toxic to fish in larval stages and acts more selectively on the lens and CNS.

In amphibians, few data on embryo-larval toxicity are generally reported. TBT affects embryonic survival of *Rana temporaria* at high concentrations, i.e., from $30\,\mu g\ L^{-1}$ [228]. Tadpoles of *Rana esculenta* exposed to $20\,\mu g\ L^{-1}$ TPTC undergo reduced swimming activity and food intake which results in decreased growth and development with consequent negative influence on size at maturity, survival to first reproduction and fecundity in adults [133].

Among marine invertebrates bivalves, echinoderms and tunicates are the most sensitive during their embryonic and larval stages, and teratogenic properties producing deformities in limbs have been reported in crustaceans [229]. Molluscs exposed to TBT and DBT undergo embryonic mortality and block in the early developmental stages [230–232], reduction of veliger larvae and post-larvae length and absence of metamorphosis [228, 233, 234]. In the sea-urchin *Paracentrotus lividus*, interruption at the morula stage is observed beginning from exposition to $5\,\mu g\ L^{-1}$ TBT and $10\,\mu g\ L^{-1}$ TPT. At $0.5\,\mu g\ L^{-1}$ TBT and $1.5\,\mu g\ L^{-1}$ TPT, significant growth reduction of plutei occurs, the latter exhibiting clear disturbance of skeletal symmetry. The embryotoxicity level of TBT is at least 100-times higher than that of DBT, and DBT is about 50-times more toxic than MBT [210, 235]. Solitary ascidians suffer from a blocking or delay in embryonic and larval development, as reported for TBT in *Ciona intestinalis* [236–238] and *Phallusia mammillata* [239], and for butyltins (MBT, DBT, TBT), triphenyltins (TPTA, TPTC, TPTH), and TCHT in *Styela plicata* [240]. Organotins strongly affect all stages of development in a dose-dependent manner, but the most sensitive and critical stage is gastrula, in which developmental block and morphological anomalies take place. In ascidians, the effective concentrations are higher than those having effects in vertebrates and other invertebrates, probably because in tunicate embryos the egg envelopes act as protective barriers against contaminants in seawater. The order of inhibition of the ascidian embryonic development appears to depend strongly on organotin liposolubility: TBT > DBT > MBT and TPTA > TPTC > TPTH ≥ TCHT. TBT, TPTA and TPTC are the most embryotoxic, because at $0.3\,mg\ L^{-1}$ they block development to the larval stage, and at $3\,mg\ L^{-1}$ prevent the embryonic cleavage. At $3\,mg\ L^{-1}$ DBT induces larval mortality, while MBT, TPTH and TCHT allow up to 50% of embryos to complete normal metamorphosis.

3.8.11   MECHANISMS OF TOXICITY

Organotin lipophilicity plays a key role in both bioaccumulation and mechanism of action, since their potential toxicity depends on their high affinity for the cell membranes. Their ability to cross the biological membranes permits them to penetrate easily into cells and their organelles and to interact with several intracellular targets. The biocidal properties are influenced strongly by the $R_{(1-3)}$ Sn$^+$ cation and are relatively independent of the anionic radical associated anions [241]. MBT, TPTH, and TCHT appear less active, probably owing to their less lipophilic property. However, although butyltin toxicity shows a pattern increasing with the number of butyl groups binding from one to three to the tin atom, it decreases remarkably in the presence of the fourth one, i.e. in tetrabutyltin, the latter compound representing a potential risk only in the case of dealkylation by the biochemical systems of the organisms [242]. Therefore, the general sequence of toxicity is trisubstituted organotins > disubstituted organotins > monosubstituted organotins $\geq$ tetrasubstituted organotins.

Most of their toxic effects are irreversible, dose- and time-dependent, and can be divided in Ca^{2+}-independent and Ca^{2+}-dependent mechanisms, but they are strictly connected.

3.8.11.1   Ca^{2+}-dependent mechanisms

Organotin compounds induce alterations in the calcium homeostasis of the cells, increasing concentrations of internal calcium ions [243, 244]. This effect appears to be a multifactorial process, involving a release of Ca^{2+} from intracellular stores due to an increase of the membrane Ca^{2+} permeability of cell compartments, and a decrease of Ca^{2+}-ATPase activity which represents the main Ca^{2+} extrusion system [245]. Recently, a direct interaction between butyltins and calmodulin, probably hydrophobic in nature (between the aliphatic chains of butyltins and hydrophobic pouches of Ca^{2+}-activated calmodulin), has been proposed. It follows a saturation pattern which leads to the formation of a complex preventing the regulative activity of calmodulin on the membrane-bound calmodulin-dependent Ca^{2+}-ATPase [246]. Moreover, addition of calmodulin in cultured cells exposed to TPT is able to reverse the inhibition of Ca^{2+}-ATPase [186]. Altered Ca^{2+} homeostasis leads to apoptosis in the case of prolonged exposure to TBT through cytosol calcification and endonuclease activation [195, 247].

Other irreversible damage related to alteration in cytosolic Ca^{2+} content is represented by depolymerization of cytoskeletal components. It has been demonstrated *in vitro* that TBT and TPT both hinder the polymerization of G-actin in F-actin, followed by disaggregation [248], and inhibit polymerization of tubulin with a mechanism of action similar to that of heavy metals [249, 250]. In cell cultures, microfilaments assemble in clusters around the peripheral cytoplasm, indicating massive disassembly, with the exception of unaltered adhesion plaques. Microtubules reveal extensive disaggregation being scattered in the cytoplasm and are not recognizable as single filaments, whereas the microtubule organizing centre is still visible [251]. In embryos of the ascidian *Styela plicata*, it had already been observed that organotin compounds cause cleavage block, ascribed to inhibition of microtubule polymerization and deep changes in blastomere morphology [240]. Interaction of butyltin compounds with calcium homeostasis also occurs in sea-urchin development by altering cleaving stages [252] and deposition of the larval skeleton [235] and in appreciable thickening of bivalve shell by inhibition of the H^+ transporting pump, coupled with inhibition of bicarbonate transport into the haemolymph, and enhancement of the diffusion of Ca^{2+} across the outer mantle epithelium [253].

3.8.11.2 Ca^{2+}-independent mechanisms

The main mechanism of action of all organotin compounds is their ability to inhibit mitochondrial oxidative phosphorylation by inhibiting the fundamental energy processes of the cell system [131, 254, 255]. In particular, triorganotin compounds derange mitochondrial function in different ways:

  (i) an accumulation in the inner membrane of mitochondria forming stable electron-dense precipitates recognizable through ultrastructural observations; they vary in size, ranging from 45 to 170 nm for butyltins and 55 nm for TPT [236, 240];
 (ii) an oligomycin-like inhibition of coupled phosphorylation, more by TBT than TPT [256];
(iii) an alteration of Cl^-/OH^- exchange across lipid membranes, producing a reduction of intramitochondrial substrate and phosphate concentrations followed by structural damage [257];
 (iv) an opening effect of the mitochondrial permeability transition pore leading to rapid and severe mitochondrial swelling [258];
  (v) an inhibition of ATP synthesis owing to a reaction of the phenyl groups of TPT with the thiols of the lipoic acid, followed by enzymatic inhibition of lipoic acid acetyltransferase and lipoamide dehydrogenase [259].

Until now it has not been clear which (if any) of these mitochondrial actions predominates as the ultimate cause of toxic effects observed *in vivo*. Organotin compounds inhibit protein synthesis by the formation of coordination bonds with amino acids [260] and can also interact directly with proteins to cause conformational changes. Activities of many enzymes are thus deranged as reported for cytosolic enzymes in some bacteria [261] and for digestive enzyme activity such as amylase and protease in insects [262, 263].

Several important thiol-containing and GSH-dependent enzymes are inhibited by organotins. These enzymes include aromatase, Na^+, K^+-ATPase

[264], caspases [265], glutathione S-transferase (GST) and glutathione peroxidase (GPX) [266, 267]. Moreover, haem proteins, such as haemoglobin [268, 269] and cytochromes [270] are inhibited.

Organotin compounds, by interacting with vicinal thiol proteins, may be able both to activate the caspases, and therefore apoptosis, at low concentrations and to inhibit caspases, with consequent necrosis activation, at higher concentrations. Caspase activation occurs for the cytochrome c export from its mitochondrial intermembrane space to the cytosol due to a direct interaction, i.e. oxidation, of the organotins with the vicinal thiols present on the permeability transitional pore [265].

Inhibition of the detoxicant system constituted of GST and GPX antioxidant enzymes makes exposed organisms more vulnerable to oxidative stress and aggravates the toxicity of other chemical pollutants present in the environment.

TBT, TCHT and TPT led to a time- and concentration-dependent decrease in spectral total microsomal P_{450} content, formation of cytochrome P_{420} through a direct interaction, and corresponding losses in oxidative activity [271]. Cytochrome P_{450} monooxygenases are also key enzymes in the biotransformations (both activative and inactivative) of xenobiotics, and self-synergism of organotins seems possible, since their degradation occurs primarily through P_{459} catalysed oxidation.

3.8.11.3   Combined mechanisms

Cytoskeletal protein depolymerization is due to both cytosolic Ca^{2+} increase and extensive oxidation of their thiol groups.

The increase in cytosolic Ca^{2+} concentration is not only due to a direct interaction of TBT with calmodulin and therefore to the inhibition of Ca^{2+}-ATPase activity, but also to interaction with GSH, since the redox state of GSH represents a control system for the activity of the InsP$_3$ receptor. The cytosolic presence of high amounts of GSSG causes stimulation of Ca^{2+}-releasing property of the InsP$_3$ receptor. Both the resulting calcium mobilization and inhibition of calmodulin-dependent Ca^{2+}-ATPase increase the intracellular Ca^{2+} content.

As a consequence, the two proposed Ca^{2+}-dependent and Ca^{2+}-independent mechanisms of action of TBT are linked and synergistic in triggering the cascade of secondary events that lead to toxic activities and cell death.

3.9   REFERENCES

1. Abel, E.W., Stone, F.G.A., and Wilkinson, G. (Eds), *Comprehensive Organometallic Chemistry*, Pergamon Press, Oxford, First edn, 1982; Second edn, 1995.
2. Champ, M., and Selgman, P.F. (Eds), *Organotin: Environmental Fate and Effects*, Chapman and Hall, London, 1996.

3. Davies, A.G., *Organotin Chemistry*, VCH, Weinheim, 1997.
4. Patai, S. (Ed.), *The Chemistry of Organic Germanium, Tin and Lead Compounds*, Wiley, Chichester, UK, 1995.
5. De Mora, S.J. (Ed.), *Tributyltin: Case Study of an Environmental Contaminant*, Cambridge University Press, 1996.
6. Smith, P.J. (Ed.), *Chemistry of Tin*, Second edn, Blackie Academic and Professional, London, 1998.
7. Maguire, R.A., *Appl.Organomet. Chem.*, 1987, **10**, 475.
8. Hoch, M., *Appl. Geochem.*, 2001, **16**, 719.
9. Donard, O.F.X., Lespes, G., Amouroux, D., and Morabito, R., in *Trace Element Speciation for Environment Food and Health*, Ebdon, L.E., Pitts, L., Cornelis, R., Crews, H., Donard, O.F.X., and Quevauviller, Ph. (Eds), Royal Society of Chemistry, London, 2001, pp. 142–175.
10. Bennett, R.F., in Reference 5, p. 43.
11. Bennett, R.F., in Reference 5, p. 45.
12. Bennett, R.F., in Reference 5, p. 47.
13. Batley, G.E., in Reference 5, p. 141.
14. Batley, G.E., Mann, K.J., Brockbank, C.I., and Maltz, A., *Aust. J. Mar. Freshwater Res.*, 1989, **40**, 39.
15. Bennett, R.F., in Reference 5, p. 56.
16. See Section 3.8.10.
17. CEFIC Submission to IMO, *TBT Copolymer Anti-fouling Paints: The Facts*, 33rd Session of the MEPC, 1992, p. 9.
18. Champ, M.A., *Sci. Total Environ.*, 2000, **258**, 21.
19. Bosselman, K., in Reference 5, pp. 237–263.
20. Stewart, C., in Reference 5, pp. 264–297.
21. Laughlin, R.B., Guard, H.E., and Coleman, W.M., *Environ. Sci. Technol.*, 1986, **20**, 201.
22. Maguire, R.J., *Water Qual. Res. J. Canada*, 2000, **35**, 633.
23. Maguire, R.J. in Reference 5, pp. 96–138.
24. Batley, G., in Reference 5, pp. 141–161.
25. See Reference 9, pp. 155–157.
26. See, e.g., Reference 2, Chapters 21 and 27.
27. Laughlan, R.B., Johannsen, R.B., French, W., Guard, H., and Brickmann, F.E., *Environ. Sci. Technol.*, 1985, **4**, 343.
28. See Ref. 2, pp. 331–357.
29. Cleary, J.J., *Mar. Environ. Res.*, 1991, **32**, 213.
30. Maguire, R.J., *Water Qual. Res. J. Can.*, 2000, **35**, 633.
31. Maguire, R.J., *Water Qual. Res. J. Can.*, 2000, **35**, 641.
32. Dawson, P.H., Bubb, J.M., and Lester, J.N., *Mar. Pollut. Bull.*, 1992, **24**, 492.
33. Dawson, P.H., Bubb, J.M., and Lester, J.N., *Mar. Pollut. Bull.*, 1993, **26**, 487.
34. Kilby, G.W., and Batley, G.E., *Aust. J. Mar. Freshwater Res.*, 1993, **44**, 635.
35. Stang, P.M., Lee, R.F., and Seligwain, P.F., *Environ. Sci. Technol.*, 1992, **26**, 1382.
36. Watanabe, N., Sakai, S., and Takatsuki, H., *Chemosphere*, 1995, **31**, 2809.
37. Weber, J.H., *Mar. Chem.*, 1999, **65**, 67–75.
38. Jackson, J.-A., Blair, W.R., Brickmann, F.E., and Iverson, W.P., *Environ. Sci. Technol.*, 1982, **16**, 110–119.
39. Donard, O.F.X., and Weber, J.H., *Nature*, 1988, **332**, 339.
40. Amouroux, D., Tessier, E., and Donard, O.F.X., *Environ. Sci. Technol.*, 2000, **34**, 988.
41. Haas, K., Feldman, J., Wennrich, R., and Stark, H.-J., *Fresenius J. Anal. Chem.*, 2001, **370**, 587.

42. Maguire, R.J., Tkacz, R.J., Chau, Y.K., Bergert, G.A., and Wong, P.T.S., *Chemosphere*, 1986 **15**, 253.
43. Yozenawa, Y., Fukui, M., Yoshida, T., Ochi, A., Tanaka, T., Noguti, Y., Kowata, T., Sato, Y., Masunuga, S., and Urushigawa, Y., *Chemosphere*, 1994, **29**, 1349.
44. Vella, A.J., and Vassallo, R., *Appl. Organomet. Chem.*, 2002, **16**, 1.
45. Vella, A.J., Mintoff, B., Axiak, V., Agius, D., and Casson, P.R., *Toxicol. Environ. Chem.*, 1998, **67**, 491.
46. Pecheyran, C., Quetel, C.R., Martin, F.F., and Donard, O.F.X., *Anal. Chem.*, 1998 **70**, 2639.
47. Weber, J.H., *Mar. Chem.*, 1999, **65**, 67.
48. Feldmann, J., Koch, I., and Cullen, W.R., *Analyst*, 1998, **123**, 815.
49. Feldmann, J., and Hirner, A.V., *Int. J. Environ. Anal. Chem.*, 1995, **60**, 339.
50. Craig, P.J., and Rapsomanikis, S., *Environ. Technol. Lett.*, 1984, **5**, 407.
51. Feldmann, J., Grumping, R., and Hirner, A.V., *Fresenius J. Anal. Chem.*, 1994, **350**, 228.
52. See Reference 9, pp. 148–49.
53. Ashley, J., and Craig, P.J., *Appl. Organomet. Chem.*, 1987, **1**, 275.
54. Craig, P.J., and Rapsomanikis, S., *Environ. Sci. Technol.*, 1985, **19**, 726.
55. Fanchiang, Y.-T., and Wood, J.M., *J. Am. Chem. Soc.*, 1981, **103**, 5100.
56. See for example, Prescott, L.M., Harley, J.P., and Klein, D.A., in *Microbiology*, Wm. C. Brown Publishers, 3rd edn, 1996, pp. 156–7.
57. See Ref. 7, pp. 149–150
58. Chau, Y.I.C., Maguire, R.J., Brown, M., Yang, F., and Batchelor, S.P., *Water Qual. Res. J. Can.*, 1997, **32**, 453.
59. Blunden, S.J., and Chapman, A. In *Organometallic Compounds in the Environment* P.J. Craig (Ed.), Longman, London.
60. Lobinski, R., Pereiro, I.R., Chassaigne, H., Wasik, A., and Szpunar, J. *J. Anal. Atm. Spectrom.*, 1998, **13**, 859.
61. Mester, Z., Sturgeon, R.E., Lam, J.W., Maxwell, P.S., and Peter, L. *J. Anal. Atm. Spectrom.*, 2001, **16**, 1313.
62. Ebdon, L., Hill, S., and Ward, R.W. *Analyst*, 1987, **112**, 1.
63. Chau, Y.K., and Wong, P.T.S., *Fresenius J. Anal. Chem.*, 1991, **339**, 640.
64. Hill, S.J., Bloxham, M.J., and Worsfold, P.J., *J. Anal. Atm. Spectrom.*, 1993, **8**, 499.
65. Byrdy, F.A., and Caruso, J.A., *Environ. Sci. Technol.*, 1994, **28**, 528A.
66. Lobinski, R., *App. Spectroscopy*, 1997, **51**, 260A.
67. Harrington, C.F., Eigendorf, G.K., and Cullen, W.R., *Appl. Organomet. Chem.*, 1996, **10**, 339.
68. Vela, N.P., and Caruso, J.A., *J. Anal. Atm. Spectrom.*, 1993, **8**, 787.
69. Sarzanini, C., and Mentasti, E., *J. Chromatogr.*, 1997, **789**, 301.
70. Sutton, K.L., and Caruso, J.A., *J. Chromatogr.*, 1999, **856**, 243.
71. Ebdon, L., Hill, S., and Ward, R.W., *Analyst*, 1986, **111**, 1113.
72. Tao, H., Babu Rajendran, R.B., Quetal, C.R., Nakazato, T., Tominaga, M., and Miyazaki, A., *Anal. Chem.*, 1999, **71**, 4208.
73. Chau, Y.K., Zhang, S., and Maguire, R.J., *Analyst*, 1992, **117**, 1161.
74. Quevauviller, P., Astruc, M., Morabito, R., Ariese, F. and Ebdon, L., *Trends Anal. Chem.*, 2000, **19**, 180.
75. Quevauviller, P., and Morabito, R., *Trends Anal. Chem.* 2000, **19**, 86.
76. Zhang, S., Chau, Y.K., and Chau, A.S.Y. *Appl. Organomet. Chem.*, 1991, **5**, 431.
77. Babu Rajendran, R.B., Tao, H., Miyazaki, A., Ramesh, R., and Ramachandran *J. Environ. Mon.*, 2000, **3**, 627.
78. Pelegrino, C., Massanisso, P., and Morabito, R., *Trends Anal. Chem.*, 2000, **19**, 97.

79. Szpunar, J., Schmitt, V.O., Donard, O.F.X., and Lobinski, R. *Trends Anal. Chem.*, 1996, **15**, 181.
80. Bayona, J.M., and Cai, Y., *Trends Anal. Chem.*, 1994, **13**, 327.
81. Arnold, C.G., Berg, M., Muller, S.R., Dommann, U., and Schwarzebach, R.P., *Anal. Chem.*, 1998, **70**, 3094.
82. Haas, K., and Feldmann, J., *Anal. Chem.*, **72**, 4205.
83. De Smaele, T., Vercauteren, J., Moens, L., and Dams, R. *HP Peak*, 1999, **2**, 10.
84. Morabito, R., Massanisso, P., and Quevauviller, P., *Trends Anal. Chem.*, 2000, **19**, 113.
85. Donard, O.F.X., and Pinel, R. In *Environmental Analysis Using Chromatography Interfaced with Atomic Spectroscopy*, R.M. Harrison and S. Rapsomanikis (Eds) Wiley, New York, 1989.
86. Clark, S., and Craig, P.J., *Appl. Organomet. Chem.*, 1998, **2**, 33.
87. Calle-Guntinas, M.B. de la, Scerbo, R., Chiavarini, S., Quevauviller, P., and Morabito, R., *Appl. Organomet. Chem.*, 1997, **5**, 431.
88. Rapsomanikis, S., *Analyst*, 1994, **119**, 1429.
89. Corr, J.J., *J. Anal. Atm. Spectrom.*, 1997, **12**, 537.
90. Heumann, K.G., Rottmann, L., and Vogl, J., *J. Anal. Atm. Spectrom.*, 1994, **9**, 1351.
91. Harrington, C.F., *LC:GC Europe*, 2000, **6**, 420.
92. Hill, S.J., Pitts, L.J., and Fisher, A.S., *Trends Anal. Chem.*, 2000, **19**, 120.
93. Sutton, P.G., Harrington, C.F., Fairman, B., Evans, E.H., Ebdon, L., and Catterick, T., *Appl. Organomet. Chem.* 2000, **14**, 268.
94. Cai, Y., Rapsomanikis, S., Andreae, M.O., *Talanta*, 1994, **41**, 589.
95. Feldman, J., Frumping, R., and Hirner, A.V., *Fresenius J. Anal. Chem.*, 1994, **350**, 228.
96. Lobinski, R., Dirkx, W.M.R., Ceulemans, M., and Adams, F.C., *Anal. Chem.*, 1992, **64**, 159.
97. Pereiro, I.R., Schmitt, V.O., Szpunar, J., Donard, O.P.X., and Lobinski, R., *Anal. Chem.*, 1996, **68**, 4135.
98. Moens, L., De Smaele, T., Dams, R., Van Den Broeck, P., and Sandra, P., *Anal. Chem.*, 1997, **69**, 1604.
99. Gomez-Ariza, J.L., Mingorance, F., Velasco-Arjona, A., Giraldez, I., Sanchez-Rodas, D., and Morales, E., *Appl. Organomet. Chem.*, 2002, **16**, 210.
100. Babu Rajendran, R.B., Tao, H., Miyazaki, A., Ramesh, R., and Ramachandran, *J. Environ. Mon.*, 2001, **3**, 627.
101. Waggoner, J.W., Belkin, M., Sutton, K.L., Caruso, J.A., and Fannin, H.B., *J. Anal. Atm. Spectrom.*, 1998, **13**, 879.
102. Encinar, J.R., Alonso, J.I.G., and Sanz-Medel, A., *J. Anal. Atm. Spectr.*, 2000, **15**, 1233.
103. Leach, A.M., Heisterkamp, M., Adams, F.C., and Hieftie, G.M., *J. Anal. Atm. Spectrom.*, 2000, **15**, 151.
104. Orellana-Vellado, N.G., Pereiro, R., and Sanz-Medel, A., *J. Anal. Atm. Spectrom.* 2001, **16**, 376.
105. McLaren, J.W., Siu, K.W.M., Lam, J.W., Willie, S.N., Maxwell, P.S., Palepu, A., Koether, M., and Berman, S.S., *Fresenius J. Anal. Chem.*, 1990, **337**, 721.
106. Pannier, F., Dauchy, X., Poutin-Gautier, M., Astruc, A., and Astruc, M.J., *Appl. Organomet. Chem.*, 1993, **7**, 213.
107. Kumar, U.T., Dorsey, J.G., Caruso, J.A., and Evans, E.H., *J. Chromatog.*, 1993, **654**, 261.
108. Rivaro, P., Zaratin, L., Frache, R., and Massucotelli, A., *Analyst*, 1995, **120**, 1937.
109. Fairman, B., White, S., and Catterick, T., *J. Chromatog.* 1998, **794**, 211.

110. Wu, J., Mester, Z., and Pawliszyn, J., *J. Anal. Atm. Spectrom.*, 2001, **16**, 159.
111. Yang, L., and Lam, J.W.H., *J. Anal. Atm. Spectrom.*, 2001, **16**, 724.
112. Grotti, M., Rivaro, P. and Frache, R., *J. Anal. Atm. Spectrom.* 2001, **16**, 270.
113. Shen, W.-L., Vela, N.P., Sheppard, B.S. and Caruso, J.A., *Anal. Chem.*, 1991, **63**, 1491.
114. Vela, N.P., and Caruso, J.A., *J. Anal. Atm. Spectrom*, 1992, **7**, 971.
115. Blake, E., Raynor, M.W. and Cornell, D.J., *J. Chromatog.*, 1994, **683**, 223.
116. Sluis-Cremer, G.K., Thomas, R.G., Goldstein, B. and Solomon, A., *S. Afr. Med. J.*, 1989, **75**, 124.
117. Anon, *Br. Med. J.*, 1958, **1**, 515.
118. Alajouanine, T., Derobert, L., and Thieffry, S., *Rev. Neurol.*, 1958, **98**, 85.
119. Barnes, J.M., and Stoner, H.B., *Pharmacol. Rev.*, 1959, **11**, 211.
120. His, E. In *Organotin*, M.A. Champ and P.F. Seligman (Eds), Chapman & Hall, London, 1996, pp. 239–8
121. Boyer, I.J., *Toxicology*, 1989, **55**, 253.
122. Reigart, J.R., and Roberts, J.R., *Recognition and Management of Pesticide Poisonings*, 5th edn, EPA Document 735-R-98-003, US EPA, Washington, DC, 1999.
123. Huang, G., Bai, Z., Dai, S., and Xie, Q., *Appl. Organometal. Chem.*, 1993, **7**, 373.
124. Yamada, H., Tateishi, M., and Takayanagi, K., *Environ. Toxicol. Chem.*, 1994, **13**, 1415.
125. Wester, P.W., Canton, J.H., Van Iersel, A.A., Krajnc, E.I., and Vaessen, M.G., *Aquat. Toxicol.*, 1990, **16**, 53.
126. Tas, J.W., Opperhuizen, A., and Seinen, W., *Toxicol. Environ. Chem.*, 1990, **28**, 129.
127. Belay, N., Rajagopal, B.S., and Daniels, L., *Curr. Microbiol.*, 1990, **20**, 329.
128. Boopathy, R., and Daniels, L., *Appl. Environ. Microbiol.*, 1991, **57**, 1189.
129. Fugakawa, T., Kanno, S., Takama, K., and Suzuki, S., *J. Mar. Biotechnol.*, 1994, **1**, 211.
130. Gadd, G.M., *Sci. Total Environ.*, 2000, **258**, 119.
131. Aldridge, W.N., *The Toxicology and Biological Properties of Organotin Compounds*, CRC Press, Orlando, Florida, USA, 1986, pp. 1–373.
132. Tsuda, T., Nakanishi, H., Aoki, S., and Takebayashi, J., *Water Res.*, 1988, **22**, 647.
133. Semlitsch, R.D., Foglia, M., and Mueller, A., *Environ. Toxicol. Chem.*, 1995, **14**, 1419.
134. Blair, E.H., *Environ. Qual. Saf.*, 1975, **3**, 406.
135. Iwata, H., Tanabe, S., Mizuno, T., and Tatsukawa, R., *Environ. Sci. Technol.*, 1995, **29**, 2959.
136. Kannan, K., Corsolini, S., Focardi, S., Tanabe, S., and Tatsukawa, R., *Arch. Environ. Contam. Toxicol.*, 1996, **31**, 19.
137. Bowen, H.J.M. In *Trace Elements in Biochemistry*, Academic Press, New York, 1996, pp. 203–204.
138. Monniot, F., Martoja, R., and Monniot, C., *C.R. Acad. Sci. Paris*, 1993, **316**, 588.
139. Bressa, G., Cima, F., Fonti, P., and Sisti, E., *Fresenius Environ. Bull.*, 1997, **6**, 16.
140. Laughlin, R.B., French, W., and Guard, H.E., *Environ. Sci. Technol.*, 1986, **20**, 884.
141. Laughlin, R.B. In *Organotin*, M.A. Champ and P.F. Seligman (Eds), Chapman and Hall, London, 1996, pp. 331–355.
142. Page, D.S., Dassanayake, T.M., and Gilfillan, E.S., *Mar. Environ. Res.*, 1995, **40**, 409.
143. Lee, R.F., *Mar. Environ. Res.*, 1985, **17**, 145.
144. Bryan, G.W., Gibbs, P.E., Hummerstone, L.G., and Burt, G.R., *Mar. Environ. Res.*, 1989, **28**, 241.
145. Mercier, A., Pelletier, E., and Hamel, J.F., *Aquat. Toxicol.*, 1994, **28**, 259.

146. Butler, P.A., *Proc. R. Soc. London B*, 1971, **177**, 321.
147. McMillan, D.E., and Wenger, G.R., *Pharmacol. Rev.*, 1985, **37**, 365.
148. Aschner, M., and Aschner, J.L., *Neurosci. Biobehav. Rev.*, 1992, **16**, 427.
149. Karpiak, V.C., and Eyer, C.L., *Cell Biol. Toxicol.*, 1999, **15**, 261.
150. Chang, L.W., and Dyer, R.S., *Neurobehav. Toxicol. Teratol.*, 1983, **5**, 673.
151. Mehta, P.S., Bruccoleri, A., Brown, H.W., and Harry, G.J., *J. Neuroimmunol.*, 1998, **88**, 154.
152. Viviani, B., Galli, C.L., and Marinovich, M., *Neurosci. Res. Commun.*, 1998, **23**, 139.
153. Ueno, S., Susa, N., Furukawa, Y., and Sugiyama, M., *Arch. Toxicol.*, 1994, **69**, 30.
154. Srivastava, S.C., *Toxicol. Lett.*, 1990, **52**, 287.
155. Kimbrough, R.D., *Environ. Health Perspect.*, 1976, **14**, 51.
156. FAO/WHO, *1970 Evaluations of Some Pesticide Residues in Food*, FAO and WHO, Rome, 1971.
157. Rosenberg, D.W., Drummond, G.S., Cornish, H.C., and Kappas, A., *Biochem. J.*, 1980, **190**, 465.
158. Devries, H., Pennincks, H., Snoeji, A.H., and Seinen, W., *Sci. Total Environ.*, 1991, **103**, 229.
159. Fent, K., and Stegeman, J.J., *Aquat. Toxicol.*, 1993, **24**, 219.
160. Fent, K., and Bucheli, T.D., *Aquat. Toxicol.*, 1994, **28**, 107.
161. Gray, B.H., Porvaznik, M., Flemming, C., and Lee, L.H. *Toxicology*, 1987, **47**, 35.
162. Santroni, A.M., Fedeli, D., Zolese, G., Gabbianelli, R., and Falcioni, G., *Appl. Organometal. Chem.*, 1999, **13**, 777.
163. Matozzo, V., Ballarin, L., and Cima, F., *Fresenius Environ. Bull.*, 2002, **11**, 568.
164. Aw, T.Y., Nicotera, P., Manzo, L., and Orrenius, S., *Arch. Biochem. Biophys.*, 1990, **283**, 46.
165. Seinen, W., Vos, J.G., Van Spanje, I., Snoek, M., Brands, R., and Hooykaas, H., *Toxicol. Appl. Pharmacol.*, 1977, **42**, 197.
166. De Santiago, A., and Aguilar-Santelises, M., *Hum. Exp. Toxicol.*, 1999, **18**, 619.
167. Ishaaya, I., Engel, J.L., and Casida, J.E., *Pestic. Biochem. Physiol.*, 1976, **6**, 270.
168. Kannan, K., Senthilkumar, K., Loganathan, B.G., Takahashi, S., Odell, D.K., and Tanabe, S., *Environ. Sci. Technol.*, 1997, **31**, 296.
169. Kannan, K., Guruge, K.S., Thomas, N.J., Tanabe, S., and Giesy, J.P., *Environ. Sci. Technol.*, 1998, **32**, 1169.
170. Whalen, M.M., Longanathan, B.G., and Kannan, K., *Environ. Res.*, 1999, **81**, 108.
171. Whalen, M.M., Hariharan, S., and Longanathan, B.G., *Environ. Res.*, 2000, **84**, 162.
172. FAO/WHO, In *Pesticide Residues in Food – 1991. Evaluations 1991*, Part II – *Toxicology*, WHO/PCS/92.52. WHO, Geneva, 1992, pp. 73–208.
173. US EPA, *Reregistration Eligibility Decision (RED): Triphenyltin Hydroxide (TPTH)*, EPA 738-R-99-010, US EPA, Washington, DC, 1999.
174. Funahashi, N., Iwasaki, I., and Ide, G., *Acta Pathol. Jpn*, 1980, **30**, 955.
175. Elferink, J.G.R., Deierkauf, M., and Van Steveninck, J., *Biochem. Pharmacol.*, 1986, **35**, 3737.
176. Snoeji, N.J., Penninks, A.H., and Seinen, W., *Int. J. Immunopharmacol.*, 1988, **10**, 891.
177. Verschuuren, H.G., Ruitenberg, E.J., Peetoom, F., Helleman, P.W., and Van Esch, G.J., *Toxicol. Appl. Pharmacol.*, 1970, **16**, 400.
178. Dacasto, M., Nebbia, C., and Bollo, E., *Pharmacol. Res.*, 1994, **29**, 179.
179. Vos, J.G., Van Logten, M.J., Kreeftenberg, J.G., and Kruizinga, W., *Toxicology*, 1984, **29**, 325.

180. Snoeji, N.J., Van Jersel, A.A.J., and Pennincks, A.H., *Toxicol. Appl. Pharmacol.*, 1985, **81**, 274.
181. Guta-Socaciu, C., Giurgea, R., and Rosioru, C., *Arch. Exper. Vet. Med. Leipzig*, 1986, **40**, 307.
182. Marinovich, M., Sanghvi, A., Colli, S., Tremoli, E., and Galli, C.L., *Toxicol. In Vitro*, 1990, **4**, 109.
183. Arakawa, Y., and Wada, O., *Biochem. Biophys. Res. Commun.*, 1984, **123**, 543.
184. Beckmann, N., Morse, M.P., and Moore, C.M., *J. Invertebr. Pathol.*, 1992, **59**, 124.
185. Cajaraville, M.P., Olabarrieta, I., and Marigomez, I., *Ecotoxicol. Environ. Saf.*, 1996, **35**, 253.
186. Cima, F., Marin, M.G., Matozzo, V., Da Ros, L., and Ballarin, L., *Chemosphere* 1998, **37**, 3035.
187. Cima, F., Ballarin, L., Bressa, G., and Sabbadin, A., *Appl. Organometal. Chem.*, 1995, **9**, 567.
188. Cima, F., Ballanin, L., Bressa, G., Sabbadin, A., and Burighel, P., *Mar. Chem.*, 1997, **58**, 267.
189. Cooper, E.L., Arizza, V., Cammarata, M., Pellerito, L., and Parrinello, N., *Comp. Biochem. Physiol. C*, 1995, **112**, 285.
190. Anderson, R.S., Unger, M.A., and Burreson, E.M., *Mar. Environ. Res.* 1996, **42**, 177.
191. Cima, F., Marin, M.G., Matozzo, V., Da Ros, L., and Ballarin, L., *Mar. Poll. Bull.*, 1999, **39**, 112.
192. Tujula, N., Radford, J., Nair, S.V., and Raftos, D.A., *Aquat. Toxicol.*, 2001, **55**, 191.
193. Cima, F., Dominici, D., Ballarin, L., and Burighel, P., *Fesenius Environ. Bull.*, 2002, **11**, 573.
194. Cima, F., Ballarin, L., Bressa, G., and Burighel, P., *Ecotoxicol. Environ. Saf.*, 1998, **40**, 160.
195. Cima, F., and Ballarin, L., *Appl. Organometal. Chem.*, 1999, **13**, 697.
196. Sasaki, Y.F., Yamada, H., Sugiyama, C., and Kinae, N., *Mutat. Res.*, 1993, **300**, 5.
197. Hamasaki, T., Sato, T., Nagase, H., and Kito, H., *Mutat. Res.*, 1993, **300**, 265.
198. Ghosh, B.B., Talukder, G., and Sharma, A., *Mech. Ageing Dev.*, 1991, **57**, 125.
199. Jensen, K.G., Andersen, O., and Ronne, M., *Mutat. Res.*, 1991, **246**, 109.
200. Ganguli, B.B., *Mutat. Res.*, 1994, **312**, 9.
201. Yamada, H., and Sasaki, Y.F., *Mutat. Res.*, 1993, **301**, 195.
202. Moriya, M., Ohta, T., Watanabe, K., Miyazawa, T., Kato, K., and Shirasu, Y., *Mutat. Res.*, 1983, **116**, 185.
203. Tiano, L., Fedeli, D., Moretti, M., and Falcioni, G., *Appl. Organometal. Chem.*, 2001, **15**, 575.
204. Gabbianelli, R., Villarini, M., Falcioni, G., and Lupidi, G., *Appl. Organometal. Chem.*, 2002, **16**, 163.
205. Jha, A.N., Hagger, J.A., and Hill, S.J., *Environ. Mol. Mutagen.*, 2000, **35**, 343.
206. Haggar, J.A., Fisher, A.S., Hill, S.J., Depledge, M.H., and Jha, A.N., *Aquat. Toxicol.*, 2002, **57**, 243.
207. Whinship, K.A., *Adv. Drug React. Acute Poison Rev.*, 1988, **7**, 19.
208. Ghoneum, M., Hussein, E., Gill, G., and Alfred, L.J., *Environ. Res.*, 1990, **52**, 178.
209. Smith, B.S., *J. Appl. Toxicol.*, 1981, **1**, 22.
210. Horiguchi, T., Shiraishi, H., Shimizu, M., and Morita, M., *J. Mar. Biol. Ass. UK*, 1994, **74**, 651.
211. Gibbs, P.E., and Bryan, G.W., In *Organotin*, M.A. Champ and P.F. Seligman PF (Eds), Chapman & Hall, London, 1996, pp. 259–280.

212. Axiak, V., Vella, A.J., Micallef, D., Chircop, P., and Mintoff, B., *Mar. Biol.*, 1995, **121**, 423.
213. Morcillo, Y., Ronis, M.J.J., and Porte, C., *Aquat. Toxicol.*, 1998, **42**, 1.
214. Thain, J.E., *Oceans '86 Proceedings of the International Organotin Symposium*, Vol. 4, Institute of Electrical and Electronic Engineers, New York, 1986, pp. 1306–1313.
215. Roberts, M.H., Bender, M.E., DeLisle, P.F., Sutton, P.F., and Williams, R.L., *Oceans '87 Proceedings of the International Organotin Symposium*, Vol. 4, Institute of Electrical and Electronic Engineers, New York, 1987, pp. 1471–1476.
216. Ronis, M.J.J., and Mason, A.Z., *Mar. Environ. Res.*, 1996, **42**, 161.
217. Yamabe, Y., Hoshino, A., Imura, N., Suzuki, T., and Himeno, S., *Toxicol. Appl. Pharmacol.*, 2000, **169**, 177.
218. Gaines, T.B., and Kimbrough, R.D., *Toxicol. Appl. Pharmacol.*, 1968, **12**, 397.
219. Ringwood, A.H., *Arch. Environ. Contam. Toxicol.*, 1992, **22**, 288.
220. Nemec, M.D., *A Teratology Study in Rabbits with TBTO: Final Report*, Wil Research Laboratories, Inc., 1987, pp. 1–210.
221. Ema, M., Kurosaka, R., Amano, H., and Ogawa, Y., *J. Appl. Toxicol.*, 1996, **16**, 71.
222. Thain, J.E., *The Acute Toxicity of Bis(tributyltin) Oxide to the Adults and Larvae of Some Marine Organisms.* International Council for the Exploration of the Sea, Copenhagen, paper CM1983/E:13, pp 1–5
223. Brooke, L.T., Call, D.J., Poirier, S.H., Markee, T.P., Lindberg, C.A., McCauley, D.J., and Simmonson, P.G. *Acute Toxicity and Chronic Effects of Bis(tri-n-butyltin)oxide to Several Species of Freshwater Organisms.* Report to Battelle Memorial Research Institute, Columbus, Ohio, 1986, pp. 1–20.
224. Fent, K., *Aquat. Toxicol.*, 1992, **20**, 147.
225. Fent, K., and Meier, W., *Arch. Environ. Contam. Toxicol.*, 1992, **22**, 428.
226. Jarvinen, A.W., Tanner, D.K., Kline, E.R., and Knuth, M.L., *Environ. Pollut.*, 1988, **52**, 289.
227. Fent, K., and Meier, W., *Arch. Environ. Contam. Toxicol.*, 1994, **27**, 224.
228. Laughlin, R., and Linden, O., *Bull. Environ. Contam. Toxicol.*, 1982, **28**, 494.
229. Weis, J.S., and Kim, K., *Arch. Environ. Contam. Toxicol.*, 1988, **17**, 583.
230. Roberts, M.H., *Bull. Environ. Contam. Toxicol.*, 1987, **39**, 1012.
231. May, L., Whalen, D., and Eng, G., *Appl. Organometal. Chem.*, 1993, **7**, 437.
232. Ruiz, J.M., Bryan, G.W., Wigham, G.D., and Gibbs, P.E., *Mar. Environ. Res.*, 1995, **40**, 363.
233. Laughlin, R.B., Gustafson, R., and Pendoley, P., *Mar. Ecol. Prog. Ser.*, 1988, **48**, 29.
234. Ruiz, J.M., Bryan, G.W., and Gibbs, P.E., *J. Exp. Mar. Biol. Ecol.*, 1995, **186**, 53.
235. Marin, M.G., Moschino, V., Cima, F., and Celli, C., *Mar. Environ. Res.*, 2000, **50**, 231.
236. Mansueto, C., Gianguzza, M., Dolcemascolo, G., and Pellerito, L., *Appl. Organometal. Chem.*, 1993, **7**, 391.
237. Mansueto, C., Lo Valvo, M., Pellerito, L., and Girasolo, M.A., *Appl. Organometal. Chem.*, 1993, **7**, 95.
238. Gianguzza, M., Dolcemascolo, G., Mansueto, C., and Pellerito, L., *Appl. Organometal. Chem.*, 1996, **10**, 405.
239. Franchet, C., Godeau, M., and Godeau, H., *Aquat. Toxicol.*, 1999, **44**, 213.
240. Cima, F., Ballarin, L., Bressa, G., Martinucci, G.B., and Burighel, P., *Ecotoxicol. Environ. Saf.*, 1996, **35**, 174.
241. Champ, M.A., and Seligman, P.F. In *Organotin*, M.A. Champ and P.F. Seligman (Eds), Chapman and Hall, London, 1996, pp. 1–25.

242. Laughlin, R.B., Johannesen, R.B., French, W., Guard, H.E., and Brinkman, F.E., *Environ. Toxicol. Chem.*, 1985, **4**, 343.
243. Chikahisa, L., and Oyama, Y., *Pharmacol. Toxicol.*, 1992, **71**, 190.
244. Oyama, Y., Ueha, T., Hayashi, A., and Chikahisa, L., *Eur. J. Pharmacol.*, 1994, **270**, 137.
245. Orrenius, S., McCabe, M.J. Jr, and Nicotera, P., *Toxicol. Lett.*, 1992, **64/65**, 357.
246. Cima, F., Dominici, D., Mammi, S., and Ballarin, L., *Appl. Organometal. Chem.*, 2002, **16**, 182.
247. Raffray, M., and Cohen, G.M., *Arch. Toxicol.*, 1993, **67**, 231.
248. Galli, C.L., Viviani, B., and Marinovich, M., *Toxicol. In Vitro*, 1993, **7**, 559.
249. Tan, L.P., Ng, M., and Kumar Das, V.G., *J. Neurochem.*, 1978, **31**, 1035.
250. Jensen, K.G., Önfelt, A., Wallin, M., Lidums, V., and Andersen, O., *Mutagenesis*, 1991, **6**, 409.
251. Chow, S.C., and Orrenius, S., *Toxicol. Appl. Pharmacol.*, 1994, **127**, 19.
252. Girard, J.-P., Graillet, C., Pesando, D., and Payan, P., *Comp. Biochem. Physiol. C*, 1996, **113**, 169.
253. Machado, J., Coimbra, J., and Sa, C., *Comp. Biochem. Physiol. C*, 1989, **29**, 77.
254. Aldridge, W.N. and Cremer, J.E., *Biochem. J.*, 1955, **61**, 406.
255. Aldridge, W.N., Casida, J.E., Fish, R.H., Kimmel, E.C., and Street, B.W., *Biochem. Pharmacol.*, 1977, **26**, 1997.
256. Stockdale, M., Dawson, A.P., and Selwin, M.J., *Eur. J. Biochem.*, 1970, **15**, 342.
257. WHO, *Environmental Health Criteria 15: Tin and Organotin Compounds*, WHO, Geneva, 1980.
258. Bragadin, M., Marton, D., Toninello, A., and Viola, E.R., *Inorg. Chem. Commun.*, 2000, **3**, 255.
259. Ascher, K.R.S., and Nissim, S., *World Rev. Pestic. Control*, 1964, **3**, 188.
260. Costa, L.G., and Sulaiman, R. *Toxicol. Appl. Pharmacol.*, 1986, **86**, 321.
261. White, J.S., Tobin, J.M., and Cooney, J.J., *Can. J. Microbiol.*, 1999, **45**, 541.
262. Ascher, K.R.S., and Ishaaya, I., *Pestic. Biochem. Physiol.*, 1973, **3**, 326.
263. Khan, K.A., *Hamd Med.*, 1996, **39**, 79.
264. Rao, K.S., Chetty, S.C., and Desaiah, D., *J. Biochem. Toxicol.*, 1987, **2**, 125.
265. Stridh, H., Orrenius, S., and Hampton, M.B., *Toxicol. Appl. Pharmacol.*, 1999, **156**, 141.
266. Al-Ghais, S.M., and Ali, B., *Environ. Contam. Toxicol.*, 1999, **62**, 207.
267. Di Simplicio, P., Dacasto, M., Giannerini, F., and Nebbia, C., *Vet. Human Toxicol.*, 2000, **42**, 159.
268. Taketa, F., Siebenlist, K., Kasten-Jolly, J., and Palosaari, N., *Arch. Biochem. Biophys.*, 1980, **203**, 466.
269. Santroni, A.M., Fedeli, D., Gabbianelli, R., Zolese, G., and Falcioni, G., *Biochem. Biophys. Res. Commun.*, 1997, **238**, 301.
270. Nebbia, C., Ceppa, L., Dacasto, M., and Carletti, M., *J. Toxicol. Environ. Health A.*, 1999, **56**, 433.
271. Rosenberg, D.W., Anderson, K., and Kappas, A., *Biochem. Biophys. Res. Commun.*, 1984, **119**, 1022.

4 Organolead Compounds in the Environment

JUN YOSHINAGA

Institute of Environmental Studies, The University of Tokyo, Japan

4.1 INTRODUCTION

Lead is a common metal and its use by mankind dates back 6000 years [1]. Until recently its use had been limited to those as a metal and inorganic compound. The use of lead metal accompanies smelting, in which a high-temperature process is inevitably involved. Since fine lead oxide particles are emitted into the atmosphere from such processes, the history of environmental pollution with anthropogenic lead could also date back to ancient times. Figure 4.1 shows chronology of lead pollution recorded as total lead in ice cores from central Greenland [2–6]. There was a small peak in the core corresponding to 2000 BP, which was attributed to mining of silver and lead by the Roman and Greek civilization [4, 7] (Figure 4.1a). The gradual increase thereafter reached a peak in the 1960s and 70s, which was followed by a sharp decrease toward the end of the Twentieth century (Figure 4.1b). It was this peak in the 1960s and 70s that had been attributed to a massive usage of organolead compounds in that period.

Organolead compounds having a carbon–lead bond in the molecular structure were first synthesized in 1853 by Löwig and since then a number of compounds have been in commercial and industrial use. In the early 1920s, one such compound was discovered as a very efficient antiknock agent for gasoline engines: tetraethyllead (TEL). An industry of tetraalkyllead compounds (TALs), including TEL and other methyl and methyl–ethyl mixed compounds introduced later, quickly developed thereafter and supported the rapid development of worldwide motorization, despite the concern among health specialists over the public health consequences of this usage [8]. The consumption of lead in this industry occupied as much as 15% of world lead consumption in the early 1970s (2.5 million tons) and TALs were known to be one of the largest volumes of organic compounds being produced at that time [9].

As a result of this worldwide massive usage of TALs added to gasoline, lead contamination can now be found everywhere in the world, even in the remotest environments such as the polar regions, as can be seen in Figure 4.1. It was

Organometallic Compounds in the Environment
Edited by P.J. Craig © 2003 John Wiley & Sons Ltd

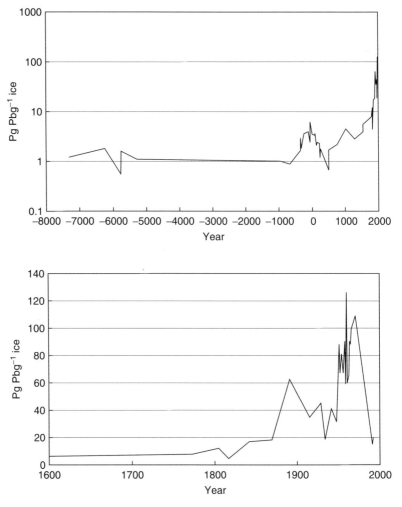

Figure 4.1 (a) Chronological trend of lead concentration in the ice core of central Greenland covering the last 10 000 years. [2–6]. Vertical axis denotes log-transformed value. (b) Chronological trend of lead concentration in the ice core of central Greenland. Expansion of Figure 1(a) covering the last 400 years. Note that the vertical axis denotes crude concentration in this figure unlike that in part (a)

estimated in 1983 that mobile sources accounted for 66–86 % of the worldwide atmospheric lead burden [10]. Although global contamination by lead has been generally recognized as that by inorganic lead (the combustion product of TALs in gasoline) recently analytical science has revealed that organolead contamination, obviously derived from leaded gasoline usage, was also global [11, 12]. However, our knowledge of the occurrence and dynamics of organolead compounds, including TALs and their degradation products, trialkyllead

(TriAL) and dialkyllead (DiAL), in the environment has still been sporadic and incomplete as compared with that of inorganic lead. It is the aim of this chapter to add newer information to earlier extensive reviews on this subject [13–16].

4.2 TETRAALKYLLEAD IN GASOLINE

Organolead compounds in the environment can be almost exclusively traced back to TALs added to gasoline, though controversy has been continuing regarding biological alkylation of lead in the environment, which will be discussed in a later section of this chapter.

The internal combustion engine works by pushing a piston and turning a driving shaft using the energy generated by the combustion of a compressed air–fuel mixture. By increasing the compression ratio (the degree of compression of the air–fuel mixture) the fuel economy and power output increase. However, when the ratio is too high, detonation or explosion takes place inside the engine. This is known as 'knock' and it results in loss of power and damage to the engine.

In 1922, a research team in General Motors Research Laboratory, USA, discovered the antiknock property of TEL. TEL prevents the spontaneous premature combustion of the air–fuel mixture and increases the octane rating of the fuel, thus enhancing engine performance. TEL is also known to work as a valve lubricant by preventing abrasive and adhesive wear on the soft (cast iron or soft steel) valve seat, which can lead to engine damage.

Soon after the introduction of the TEL additive, it was found that lead oxide deposition on the engine parts shortened engine lifetime. Addition of 1,2-dichloroethane and 1,2-dibromoethane to the leaded gasoline was found to be effective in solving this problem. It works by converting lead oxide into more-volatile lead halogenides, such as lead bromochloride (PbBrCl) and mixed salts (PbBrCl.2NH$_4$Cl, α-2PbClBr.NH$_4$Cl, etc.). These lead halides are emitted into environment via the exhaust [17].

As an organolead additive, TEL had been the major compound since the 1920s but tetramethyllead (TML) and mixed methyl–ethyl compounds [trimethylethyllead (TMEL), dimethyldiethyllead (DMDEL) and methyltriethyllead (MTEL)] have also been used since the 1960s. Table 4.1 shows the major physical properties of these five TALs [18–22] and Figure 4.2 shows a typical example of the measured composition of the five TALs in a major commercial gasoline sold in Belgium at that time [23]. The composition varied to a considerable extent from one gasoline brand to another. The differences in the thermal stability and volatility of the five TALs (Table 4.1) were considered to be factors in their effectiveness as antiknock agents, however, it is now thought that the effectiveness is almost identical across all five TALs [22]. In Figure 4.2 TAL concentrations in gasoline, as expressed as ppm Pb, are also indicated. Gasoline

Table 4.1 Physical properties of TAL [18–22]

	B.p. (°C)	M.p. (°C)	Vapor pressure (Torr /20°C)	Density (d_4^{20})	Solubility in water (mg L^{-1})
TEL	200	−130 ∼ −136	0.26	1.65–1.66	0.2–0.3
TML	108–110	− 30	22.5	2.00	15
DEDML	159		2.2	1.79	
TMEL	137		7.3	1.88	
MTEL	179		0.75	1.71	

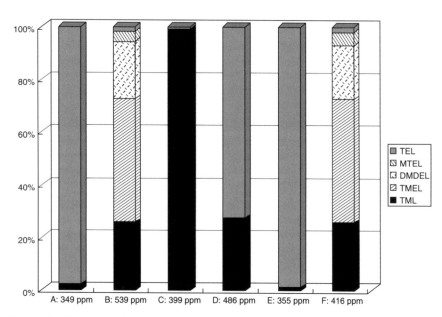

Figure 4.2 Examples of TALs composition of commercial gasoline [23]

containing more than 1000 ppm Pb ($>$ 1 g Pb L^{-1}) was used in some countries in the 1970s.

4.3 LEADED GASOLINE PHASEOUT [24]

In the 1970s public health specialists addressed serious concern over the adverse health effects, particularly neurotoxic effects, of inorganic lead emitted into the environment via the combustion of leaded gasoline. From a technical viewpoint, leaded gasoline was found not to be suitable for catalytic converters, a new technology introduced in the 1970s to reduce hydrocarbon, nitrogen dioxide and carbon monoxide emission from the exhaust of cars. Introduction of sophisticated gasoline refinery technology and alternative additives, such as methyl tertiary-butyl ether (MTBE), made TALs an unnecessary option as an octane enhancer. Further, TALs used as valve lubricants could be replaced with other additives and, even when leaded gasoline was a better option, lower lead content gasoline (0.02–0.05 g L^{-1}) was found to be tolerable.

All these situations combined to launch a leaded-gasoline phaseout in the USA and Japan in 1975. Leaded gasoline phaseout was also launched in many developed countries in the 1980s and 90s. Figure 4.3 shows the share trend of unleaded gasoline consumed in selected OECD countries during 1985–1992 [25]. It clearly shows the increasing share of unleaded gasoline, with varying pace according to the country.

It was evident that local phaseout dramatically decreased atmospheric (inorganic) lead concentrations in that area [26–30], and, consequently in, human blood lead level [31–34]. Table 4.2 shows recent lead content and unleaded gasoline share by country [24, 35]. Although leaded gasoline phaseout is proceeding among OECD countries, as shown in Figure 4.3, leaded gasoline is still used in many developing countries, particularly in Africa and Asia. Lead emission from usage in developing countries, though the amount from each of the countries may be less than that used formerly in Western countries, must be a continuing source of lead compounds in the environment. Figure 4.4 shows motor gasoline consumption and total added lead in the world by area in 1993 [36]. It can be seen that lead consumption (world total: 70 000 tons/year) was almost evenly shared by six areas, namely Western Europe, former USSR,

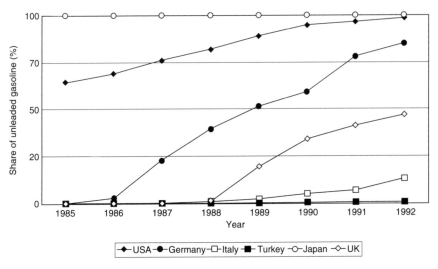

Figure 4.3 Trend of market share (%) of unleaded gasoline (1985–1992) in selected OECD countries [25]

Table 4.2 Lead content and share of unleaded gasoline by country (as of 1996 unless otherwise indicated) [24, 35]

Countries	Maximum Pb content ($g\ L^{-1}$)	Market share of unleaded gasoline (%)	Countries	Maximum Pb content ($g\ L^{-1}$)	Market share of unleaded gasoline (%)
Africa			Sri Lanka	0.2	0 (1995)
			Taiwan	0.15	59 (1994)
Angola	0.77	0	Thailand	0.15	100
Benin	0.84	0	Vietnam	0.40	0
Botswana	0.44	0			
Burkina Faso	0.84	0	*America and Caribbean*		
Burundi	0.84	0			
Cameroon	0.84	0	Argentina	0	100
Chad	0.84	0	Bahamas	0.84	58 (1993)
Côte d'Ivoire	0.26	0	Barbados	0.79	40 (1995)
Ethiopia	0.76	0	Bermuda	0	100 ($\sim$ 1990)
Gabon	0.8	0	Bolivia	0	100 ($\sim$ 1995)
Ghana	0.63	0	Brazil	0	100 ($\sim$ 1991)
Kenya	0.4	0	Canada	0	100 ($\sim$ 1990)
Liberia	0.77	0	Chile	0.6	28
Madagascar	0.8	0	Colombia	0	100 ($\sim$ 1990)
Malawi	0.53	0	Costa Rica	0	100
Mauritania	0.25	0	Cuba	0.84	0 (1993)
Mauritius	0.4	0	Ecuador	0.50	24
Mozambique	0.65	0	Guatemala	0	100 ($\sim$ 1991)
Namibia	0.4	0	Jamaica	0.77	30
Niger	0.65	0	Mexico	0.26	56
Nigeria	0.66	0	Nicaragua	0	100
Senegal	0.6	0	Panama	0.82	93
South Africa	0.4	12	Paraguay	0.2	1
Uganda	0.84	0	Peru	0.75	25
Zimbabwe	0.84	0	United States	0	100
			Uruguay	No limit	6
Asia-Pacific			Venezuela	0.85	0
Australia	0.25–0.45	45 (1994)	*North Africa and Middle East*		
Bangladesh	0.8	0			
Brunei	0.11	43 (1993)	Algeria	0.63	0
China	0.33	60	Bahrain	0.4	0
Hong Kong	0.15	80 (1995)	Egypt	0.8	introduced recently
India	0.42	in 4 largest cities			
Indonesia	0.45	<1, Jakarta only (1995)	Iran	0.19	0
			Iraq	0.6	0
Japan	0	100	Israel	0.15	10 (1994)
Korea	0.3	83 (1992)	Jordan	0.3	0
Lao PDR	0.4	0 (1993)	Kuwait	0.53	0
Malaysia	0.15	54 (1994)	Lebanon	0.84	0
New Zealand	NA	44 (1995)	Libya	0.8	0
Pakistan	0.42	0	Morocco	0.5	0.4 (1992)
Philippines	0.15	10	Oman	0.62	0
Singapore	0.15	60 (1994)	Qatar	0.4	0

Table 4.2 Lead content and share of unleaded gasoline by country (as of 1996 unless otherwise indicated) [24, 35]

Countries	Maximum Pb content (g L^{-1})	Market share of unleaded gasoline (%)	Countries	Maximum Pb content (g L^{-1})	Market share of unleaded gasoline (%)
Saudi Arabia	0.4	0	Netherlands	0.15	86
Syria	0.4	0	Norway	0.15	98
Tunisia	0.5	0	Portugal	0.4	39
UAE	0.4	0	Spain	0.4	23
Yemen	0.45	0	Sweden	0	100 ($\sim$ 1994)
			Switzerland	0.15	87
Western Europe			Turkey	0.4	18
Austria	0	100 ($\sim$ 1991)	United Kingdom	0.15	67
Belgium	0.15	74	*Central and Eastern Europe*		
Denmark	0	100			
Finland	0	100	Bulgaria	0.15	5
France	0.15	62	Croatia	0.6	30
Germany	0.15	97	Czech Republic	0.15	55
Greece	0.4 (0.15 in Athens)	33 (1995)	Hungary	0.15	64
			Moldova	0.4	0
			Poland	0.15	70
Iceland	0.15	85	Romania	0.6	6
Ireland	0.15	65	Russian Federation	0.6	50
Italy	0.15	44	Slovak Republic	0	100 ($\sim$ 1995)
Luxembourg	0.15	82			

NA, not available.

Middle East, Central and South America, Africa and Asia (Figure 4.4b), though motor gasoline consumption in these areas amounts only half of that of the world (total: 10^{12} liter per year) (Figure 4.4a). It is noteworthy that North America, including the USA, Canada and Mexico, accounts for approximately 50% of world gasoline consumption (Figure 4.4a) but only 2% of gasoline lead (Figure 4.4b). This is obviously attributable to leaded gasoline phaseout in the USA (where >90% of gasoline in this area is consumed) and Canada.

4.4 ANALYTICAL METHODS FOR ORGANOLEAD COMPOUNDS

Earlier studies for the analysis of TALs in environmental matrices were based on traditional wet chemistry. For instance Henderson and Snyder [37] used a colorimetric method for the determination of TriEL, DiEL and inorganic lead after converting them into their respective dithizonates. However, these classical analytical methods had been replaced by the species-specific and lead-specific analytical methods by the 1980s.

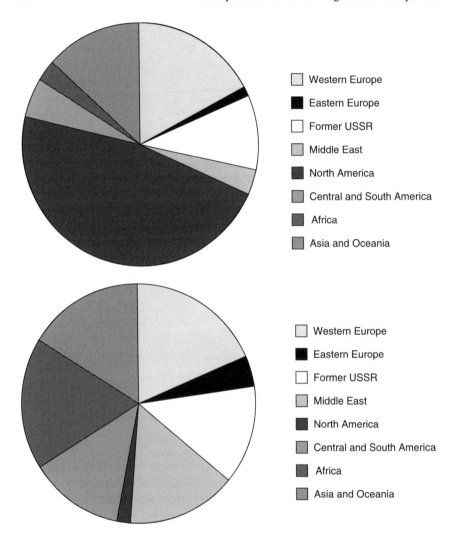

Figure 4.4   (a) Motor gasoline consumption (10^9] liter/year) by area (1993) [36] (b) Total lead used in gasoline (tons of Pb/year) by area (1993) [36]

Species-specific analytical methods for organolead compounds involve separation and detection. Gas chromatography (GC) has been predominantly used for the separation of organolead compounds in the environment. Analysis by GC is advantageous because of the high sensitivity and resolution capability, particularly when capillary column is used, and it allows simultaneous determination of TALs and ionic species when appropriate pretreatment is done. Pretreatment for GC analysis is, however, often complicated involving chelation, extraction, dehydration, and derivatization to convert ionic TriALs and DiALs into volatile compounds suitable for GC analysis. Another method of choice is liquid chromatography (LC), with reversed phase, ion-pair separation modes. It is easier than GC because no derivatization is needed for ionic species and it allows simultaneous analysis of inorganic lead if necessary. However, sensitivity and resolution of the species is generally poorer than for GC analysis.

Detection systems used for GC and LC include atomic absorption spectrometry (AAS), atomic emission spectrometry (AES) and mass spectrometry (MS). These detection systems are superior to conventional GC and LC detection systems, such as UV and ECD, in that these are highly specific to lead atoms and ions. Flame AAS was the first lead specific detector applicable both for GC and LC but the sensitivity is low: the detection limit is typically in the 10–100 ng Pb range [38, 39]. Use of electrically heated quartz (or ceramic) furnaces for lead atomization dramatically increased the sensitivity by three orders of magnitude, reaching 17–100 pg [40–42] and GC-AAS (particularly quartz furnace AAS (QAAS)) has long been the predominant analytical method for organolead compounds in the environment. Figure 4.5 shows an example of a GC-AAS chromatogram of the five TALs [38].

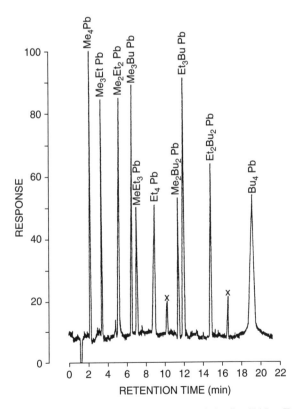

Figure 4.5 A GC-AAS chromatogram of *ca* 5 ng each of the five TALs. From Chau *et al.* (1976) [40] reproduced with permission

Inductively coupled argon plasma (ICP) and microwave induced helium plasma (MIP) have been introduced as an emission source of lead for organolead analysis. ICP atomic emission spectrometry (ICP-AES) is suitable for both LC and GC but MIP-AES is only suitable for GC. The main advantage in the use of AES as GC and LC detector is that it allows multi-element simultaneous (or high-speed sequential) analysis in one chromatographic sample run. Detection of carbon and sulfur as well as analyte metal helps in analyzing the molecular structure of the organometallic compounds. However, ICP-AES is not sensitive enough to obtain a lower detection limit than that of QAAS for organolead analysis [43, 44]. MIP-AES was first used as a GC detector in 1965 and it is characterized by its high sensitivity to lead (and other metals), better than QAAS by one order of magnitude [45]. Currently GC-MIP-AES (called GC-atomic emission detection (AED) in commercially available instruments) is gaining wider acceptance for trace organolead and other organometallic analysis.

ICP mass spectrometry (ICP-MS) has emerged as a newer detector for both GC and LC. It is a highly sensitive multi-element detector [46–48] that also allows accurate analysis by isotope dilution technique. GC- and LC-ICP-MS is one other method of choice for ultratrace organolead analysis.

Hydride generation-AAS has also been used for the speciation of organolead compounds [49, 50] but there is little advantage over the GC-based analytical method in terms of sensitivity and specificity.

Excellent reviews of organometal analysis with GC- and LC-atomic spectrometry (AAS, AES, and MS) are available [51–54].

4.4.1 ATMOSPHERIC SAMPLES

The classical analytical method for atmospheric TALs was as follows: collection and conversion into DiALs in iodine monochloride solution, extraction of the DiALs after chelation, back-extraction and colorimetric [55] or AAS detection [56, 57]. This method was not only non-species-specific (only 'total' TAL concentration measured) but also liable to interference from other constituents, particularly inorganic lead in particulates during the collection process. Vapor phase inorganic lead compounds, such as lead halides, may also be included in the 'total' organolead determined by the monochloride method [58].

Most of the more recent studies employ species-specific GC analysis. TALs in air were sampled by passing air through a column with a variety of solid adsorbents at room temperature, such as activated carbon [37], silicon rubber on Chromosorb P [59], Porapack-Q [60], and Porapak:Tenax (65:35) [61]. Cold trapping with solid adsorbent, such as 3 % OV-1 or OV-101 on Chromosorb W (chromatographic column packing), in a dry ice–ethanol bath [38, 62] and glass beads immersed in a liquid nitrogen bath [63] were also used. The size of the air sampler varied from 110 mm to 500 mm in length and maximum sampling volume accordingly varied from 70 L to 2200 L. The adsorbed TALs were liberated either by thermal desorption or by solvent extraction and introduced to GC coupled with a lead specific detector. Detection limits varied from *ca* 0.1 μg m^{-3} to 0.1 ng m^{-3} depending on the detector sensitivity and sampling volume. Cold-trap followed by GC-QAAS allowed the most sensitive detection because of its ability to sample large volumes of air (up to 50 m^3) [64]. The lowest detection limit reported so far is 0.004 ng m^{-3} for 15 L air, obtained by GC-ICP-MS [65]. Recovery of organolead compounds was reported to be satisfactory (>80%) in all of the studies.

The TALs cryogenically adsorbed on the sampling column were reported to be stable when stored at the temperature of dry ice for 4 days [38]. Stability of TALs sampled on solid adsorbent at room temperature was reported for up to 45 days when stored in a refrigerator [61].

It was the mid 1980s when the first analytical methods for gaseous TriALs and DiALs in the atmosphere were reported [66]. It was this method that allowed us to obtain the detailed knowledge on the chemistry and environmental dynamics of organolead compounds that we now have. In this method, ionic species were trapped by bubbling air in water, and dissolved compounds were complexed with a chelate (sodium diethyldithiocarbamate (NaDDTC)), and extraction and derivatization followed (described in detail later). Recovery of ionic organoleads ranged from 46–79% with minimum interference from coexisting TALs.

Organolead compounds have also been analyzed in particulate matter and dust. Particulate matter collected on filter paper was first extracted with hexane for TAL analysis, then with water for ionic species [67]. Ionic species in the aqueous layer were complexed with sodium diethyldithiocarbamate (NaDDTC) at pH 9 in the presence of ethylenediaminetetraacetic acid (EDTA) and the chelate was extracted into pentane, followed by butylation and GC-AAS analysis. The first solvent extraction may be omitted [68]. TriML was extracted from a dust sample by shaking 1 g of sample in 100 mL of water with NaCl (10 g), but without a chelate, for 30 min [69]. TriML concentration in the extract was measured by *in situ* derivatization using sodium tetraethylborate (NaBEt$_4$) as described in Section 4.2 below. Ultrasonic agitation of the sample–water mixture resulted in lower recovery of TriALs due to excess inorganic lead leaching, which may affect NaBEt$_4$ derivatization and detection, so shaking was recommended for extraction. Supercritical fluid extraction of TriML from dust and sediment was also tested with good recovery [70].

Analytical methods for organolead in precipitation (e.g., rainwater and snow) are described in Section 4.2 below.

4.4.2   WATER SAMPLES

Separation of organolead compounds from environmental waters has been carried out either by GC and LC. In GC analysis, organolead compounds in sample water have to be extracted into a suitable solvent. The commonly used method, as already given in Section 4.1 above for particulate matter, is as follows: extraction of organolead compounds into solvent after complexation with a chelate, followed by dehydration, alkylation with Grignard reagent and re-extraction into solvent which can be injected into GC [71–74]. EDTA is commonly used to mask inorganic lead, which may interfere with the formation of organolead–chelate complex in water samples. The solvent (hexane, hexane/benzene, octane, pentane, etc.), chelate (NaDDTC or dithizone), and Grignard reagent (*n*-butylmagnesium halide, *n*-propylmagnesium halide, *n*-phenylmagnesium halide, etc.) used in this extraction procedure may be different from one study to another. Figure 4.6 shows a typical GC-AAS chromatogram of organolead compounds obtained after butylation [71]. Detection limits achieved by this extraction-GC method variys from $\mu g\ L^{-1}$ to sub-ng L^{-1} levels, primarily depending on the concentration factor (sample:solvent ratio) and detector used (AAS, MIP-AES or ICP-MS). An analysis of organolead in polar ice/snow with GC-MIP-AES, in which the detection limit reaching 0.01 pg g^{-1} ice (corresponds to 0.01 ng L^{-1}) was reported [75], and the study also used this extraction method.

When TALs are the only analytes of interest, simpler extraction methods can be used: e.g. shaking the sample water with hexane and injecting the hexane

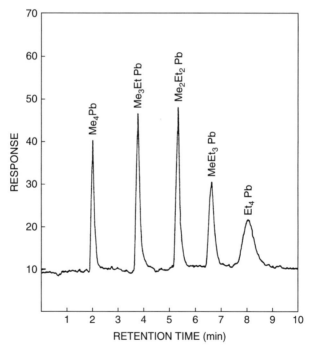

Figure 4.6 A GC-AAS chromatogram of organolead compounds including five TALs (10 ng each), four butylated DiALs and TriALs (8 ng each), and Pb^{2+} (15 ng). From Chau *et al.* (1983) [71] reproduced with permission

phase directly to GC [76]. An *in situ* purge and trap sampling of TALs in natural waters followed by GC-ICP-MS analysis reported very low detection limit of 0.08 pg L^{-1} [77].

Since alkylation of extracted compounds with Grignard reagents involves a complicated process, use of NaBEt₄ for ethylation has been the method of choice [78]. This requires only pH adjustment and NaBEt₄ addition, while evolved volatile ethylated species may be purged directly to GC [79] or they may be extracted into a solvent [80]. Detection limit for TriML and DiML by NaBEt₄ ethylation purge and trap GC-AAS is sub-ng L^{-1} [79]. A major drawback is that it only allows the speciation of methylated lead compounds because it cannot speciate ethylated compounds.

Pretreatment is not necessary for LC separation of organolead compounds in water samples. A reversed phase ion-pair separation mode has been employed, however, chromatographic conditions for organolead separation were reported to be difficult to optimize. Using a methanol–acetate buffer (60:40), a mobile phase containing sodium 1-pentanesulfonic acid as ion-pairing agent, and ICP-MS detection using the isotope dilution technique, a detection limit of 1.5 ng g^{-1} ($\cong \mu g\ L^{-1}$) was reported [81].

The stability of organolead compounds in water varies depending on the compounds and storage conditions. It is known that TALs readily decompose in water under sunlight. Significant decrease of TML concentration was observed when a sample was stored both at room temperature and at 4 °C. This might be due to intact TAL adsorption onto the glass vessel and not necessarily to decomposition. Addition of 5 mL hexane to 200 mL of sample water followed by shaking for 5 min was effective to maintain TML concentration for 1 week [76]. TriEL in a simulated rainwater was reported to be stable for at least 12 months when stored in a Nalgene bottle wrapped in aluminum foil at 4 °C but an approximately 10 % decrease was noticed after 37 months [82]. A significant decrease in TriEL concentration took place after the same sample was stored at 37 °C for 15 days.

4.4.3 SEDIMENT AND SOIL SAMPLES

Since organolead compounds are not involved in geological processes, their presence is assumed to be limited to the surface of sediment or soil particles. Therefore simple extraction with solvent (hexane, benzene, etc.), similar to that described for dust samples, can be employed for TALs [68, 76, 83]. Use of chelates, such as NaDDTC, and EDTA for masking inorganic lead is essential for quantitative extraction of TriAL. However, the extraction yield of DiALs is lower probably because of strong affinity to the particles. A typical procedure is as follows [68]: 10 g of sample is dispersed in 100 mL of purified water with 5 g of NaCl and NaDDTC solution (0.5 M) followed by shaking with 5 mL of hexane for 30 min. The complexed and extracted TriALs and DiALs in the hexane phase were alkylated with Grignard reagent.

4.4.4 BIOLOGICAL SAMPLES

Organolead compounds are assumed to be present in the cells and therefore biological tissues should be physically or chemically destroyed prior to extraction. Use of Polytron homogenizer [76], decomposition of the cells with enzymatic hydrolysis (lipase and protease) [84] or tetramethylammonium hydroxide (TMAH) [83, 85–87] were tested. Temperature and heating time should be optimized in the TMAH dissolution of tissues otherwise loss or disproportionation can occur [87].

The total TALs in seafoods were directly extracted from homogenized tissues into benzene by shaking with aqueous EDTA. A portion of the benzene layer was digested and total lead content was measured by AAS to obtain 'total' TALs [88]. Recovery of TML (100 and 500 ng Pb) was 103–130% and that of TEL was 70–119%. Chau et al. [89] used a similar procedure but used hexane as an extracting solvent and GC-AAS for species-specific analysis and found 72–76% recovery for the five TAL compounds. Detection limits in these studies were 10 and 25 ng Pb g^{-1}.

TriML recovery was satisfactory (94–103%) when a 13–95 ng Pb g^{-1} level of it was spiked to cod, shrimp and scallop samples. This was also the case for avian tissues (82–103%) [85]. TriEL recovery was somewhat lower for seafood (81–90%) and avian tissues (71–95%). Recoveries of DiALs were variable depending on the sample and were generally low (DiML: 51–91%; DiEL: 76–91%) for seafood [90]. This was also the case for avian tissues (DiML: 23–35%; DiEL: 60–79%) [85]. TriALs and DiALs recoveries from glass and leaves were 84–109% [87].

For TML analysis of human blood, cellular components were released by hemolysis and this was followed by extraction, derivatization and GC-AAS analysis [91]. Analytical methods for organolead concentrations in human urine, based on extraction without a chelate and using GC-ECD detection (TALs and TriALs) [92], and on hydride generation AAS (TriALs and DiALs) [49] were reported with satisfactory recoveries. The latter method utilized selective generation of TriEL and DiEL hydrides with different reagents (0.5 M DL-malic acid for TriEL and 0.75 M H_2O_2–0.004 M $HClO_4$ for DiEL) added to urine. A standard analytical method for organolead compounds in water samples (extraction–alkylation–GC) has also been applied to urine analysis [93]. Hexane containing NaCl (1 g for 5 mL hexane) was the best solvent for extraction of TriALs from urine. The formation of a thick foam or gel prevented quantitative extraction of analytes from a urine matrix, when another solvent (pentane with a variety of added substances) was used.

4.4.5   ANALYTICAL QUALITY ASSURANCE

A certified reference material (CRM), Urban Dust (CRM 605) certified for TriML content, was prepared by the Standards, Measurement and Testing programme (formerly BCR) of the European Union [69]. The certified value of TriML in CRM 605 is $7.9 \pm 1.2 \, \mu g \, kg^{-1}$ (as TriML). This CRM is the only one available at present for quality assurance of organolead analysis. An attempt was made to certify TriML content in another candidate CRM, Rainwater. However, certification wasn't completed because its long term stability had turned out to be doubtful [82]. Rainwater will be distributed as a (non-certified) reference material (RM 604).

4.5   OCCURRENCE IN NATURE

4.5.1   AIR

Atmospheric organolead concentrations have been measured since the mid-1970s. In those early days, 'total' organic lead was measured but as better analytical methods developed, speciation data became available. Table 4.3 lists atmospheric gaseous TAL concentrations ($ng \, m^{-3}$) in urban, rural and indoor environments [61–65, 94–98]. Particulate lead concentrations ($ng \, m^{-3}$) as well as gaseous TALs/particulate Pb (%) are included in this table where available. Reported gaseous TAL composition and their concentrations vary depending on the sampling sites and time. Traffic density and TAL composition in the gasoline used in the location of sampling must be the reasons for the observed variations.

The ionic TriALs and DiALs had been found in the gaseous phase of the atmosphere. Concentrations of these compounds in the atmosphere have been reported in a limited number of studies as compared with TALs. Table 4.4 shows TriAL and DiAL concentrations in urban and rural air samples collected in the UK and Ireland [64, 96, 97, 99]. It was noted that some rural air, particularly maritime air, contained a high TriALs + DiALs concentration relative to particulate inorganic lead (up to 33 %). This should be compared with 5 % in urban air (Tables 4.3 and 4.4). This could be explained by the different lifetimes between inorganic lead and ionic alkyllead species in air because the maritime air of this region had been long-range transported [100], and also, in part, by an indication of natural alkylation of lead in marine environment [100–102].

Organolead compounds, including TALs, TriALs and DiALs, were detected in the particulate phase (Table 4.5) [67, 96, 97]. They constituted only a small fraction of the atmospheric concentrations of the compounds and therefore are of limited geochemical significance.

In urban air samples, a highly positive correlation was observed between gaseous total alkyllead and particulate inorganic lead concentrations. It was inferred that the emission source of the components was common, i.e., vehicle

Table 4.3 Concentrations of TALs in atmosphere

Columns under "Concentration (ng Pb m^{-3})": TML, TMEL, DMDEL, MTEL, TEL, Total TAL, Particulate Pb, TAL/Pb(%)

Country	Location	Year	TML	TMEL	DMDEL	MTEL	TEL	Total TAL	Particulate Pb	TAL/Pb(%)	Ref.
USA	Baltimore, tunnel	76 Jul	—	—	—	—	—	57–130	12200–21600	0.37–0.83	94
USA	Baltimore, tunnel	77 May	21–66	<0.5–9	11–16	1–6	11–42	74–100	12600–23200	0.40–0.69	94
USA	Baltimore, tunnel	77 Mar	36–64	1–4	0.5–2	1	4–9	42–75	1200–1700	3.5–5.2	94
USA	Baltimore, tunnel	77 May	12–41	2–3	2–3	2–4	7–23	26–74	1800–3100	1.5–2.4	94
Canada	Toronto, urban street	—	—	1	2	1	8	14	650	—	62
Denmark	Copenhagen, busy street, in bus	79 Sept	140–150	—	—	—	45	—	—	—	95
Sweden	Stockholm, busy street	80 Mar	47–77	—	—	—	—	—	—	—	95
Sweden	Stockholm, quiet street	80 Mar	11–22	—	—	—	—	—	—	—	95
Belgium	shopping center	80 Feb	21.7	4.4	1.5	0.8	2.9	31.2	—	—	63
UK	Lancaster, urban busy street	—	19.5–38.5	1.1–3.0	1.7–3.0	0.9–2.5	25.3–42.6	50.3–89.6	1500–2700	—	96
UK	Colchester, urban	86 Jan–Feb	0.7–10.9	<0.14–2.6	<0.28–1.6	<0.49–3.3	<0.83–6.6	—	45–768	—	97
Spain	Zaragoza, heavy traffic	95 Sep–96 May	—	—	3.4	—	—	—	—	—	61
France	Bordeaux, urban	95 Sep–96 May	—	—	—	—	—	15.5	—	—	98
France	Bordeaux, underground car park	—	—	—	—	—	—	210.7	—	—	98
Denmark	suburban, nine cities	79 Jan–Feb	2–54	—	—	—	<1–36	—	—	—	95
Denmark	suburban, nine cities	79 May–Jun	0.9–58	—	—	—	<1–36	—	—	—	95
Belgium	residential area	80 Feb	4.7	2.6	1	0.4	<0.3	8.6	—	—	63
UK	Essex University Campus, semirral	86 Jan–Feb	<0.14–11	<0.14–1.7	<0.28	<0.49–1.2	<0.83–5.0	—	<13–397	—	97
France	Bordeaux, industrial and suburb	95 Sept	—	—	—	—	—	1.6	—	—	98
Denmark	rural	80 Apr	0.5–2.5	—	—	—	—	0.2–1.3	—	—	95
UK	Hazelrigg, rural	—	0.2–1.3	<0.225	<0.25	<0.275	<0.3	—	32–82	—	96
UK	NW Scotland, coast	88 Jun–Jul	0.17–0.71	—	<0.006	<0.004	<0.004	—	6.1–10.5	—	64
Ireland	Eire, coast	88 Aug	0.32–1.46	—	<0.006	<0.004	<0.004	—	8.4–14	—	64
France	Bordeaux, on board ship	96 Jun	0.02–0.23	<0.004–0.02	<0.004–0.04	<0.004–0.01	<0.004–0.42	—	—	—	65
USA	laboratory	77 May	7	4	4	2	9	26	—	—	94
Belgium	laboratory	80 Feb	3.8	2.2	1.4	0.3	7.4	11.5	—	—	63
Belgium	office	80 Feb	5.6	3.5	0.4	<0.2	<0.3	9.5	—	—	63
Belgium	office	80 Feb	2	0.5	<0.2	<0.2	<0.3	2.4	—	—	63
France	Bordeaux, day nursery	96 Apr	—	—	—	—	—	5.2	—	—	98

Table 4.4 Concentrations of ionic alkyl lead in UK and Ireland (ng Pb m^{-3})

Location		Time	Ionic alkyl Pb	Total TAL	Total gaseous organolead	Particulate Pb	Alkyl Pb/ particulate Pb (%)
Lancaster	Urban	—	15.7–24.7	50.3–89.6	75.0		2.8–4.2
Lancaster	Urban	—	18	178	218	4767	4.6
Colchester	Urban	86 Feb	TML <0.11–19.8; DMEL <0.16–1.2; MDEL <0.2; TEL <0.25–1.1; DML <0.18–2.6; DEL <0.65–5.0 ML <0.65–1.0		0.7–35.9	45–768	1.3–27
Essex	Semirural	86 Jan–Feb	TML <0.11–2.1; DMEL <0.16; MDEL <0.2; TEL <0.25–2.6; DML <0.18–0.6; DEL <0.65–1.0 ML <0.65		0.4–11.0	<13–397	0.6–20<
Hazelrigg	Rural	—	0.06–1.57	0.2–1.3	0.35–2.4		0.4–7.5
Hazelrigg	Rural, air over UK	84 Apr	1.7	0.7	2.4	115	2.7
Cockerham	Rural, air over UK	84 Apr	2.5	0.8	3.3	101	3.5
Hazelrigg	Rural, maritime air	84 Jun	2.2	0.7	2.9	56	5.3
Cockerham	Rural, maritime air	84 Jun	2	0.8	2.8	24	13.1
Harris		83	—	—	3.2	16.4	22
Harris		84	7.2	0.1	7.3		32
Hazelrigg	Rural	—	0.5	2.6	3.5	114	3.1
Cockerham	Rural	—	0.6	3.3	4.6	128	3.6
Harris	Rural	88 Jun	TriML <0.005 TriEL <0.006 DML <0.01–0.68 DEL <0.01		0.17–1.39	6.1–10.5	1.8–15
Eire	Rural, maritime air	88 Aug	TriML <0.00–0.19 TriEL <0.006 DML <0.01 DEL		0.32–1.65	8.4–14	2.7–20

Table 4.5 Concentrations of organolead compounds in atmospheric particulate (ng Pb m^{-3})

Country	Location	Year	TML	TMEL	DMDEL	MTEL	TEL	TriML	DMEL	MDEL	TriEL	DML	DEL	Inorganic Pb	Ref.
UK	Lancaster	—	<0.009–0.23	<0.0013–0.15	<0.0016–0.129	0.005–0.118	<0.0019–0.172	<0.002–0.01	—	—	0.023–0.09	<0.015	<0.035	1500–2700	96
UK	Hazelrigg, rural	—	<0.009	<0.0013	<0.0016	<0.0015 (0.0003)	<0.0019–0.0088	(0.0004)–0.0026	—	—	(0.0011)–0.006	<0.015	<0.035	32–83	96
UK	Colchester, urban	86 Feb	<0.00006	<0.00004	<0.00007	<0.0001	<0.0001	<0.00009–0.0023	<0.00012–0.0002	<0.00015–0.0003	<0.00018–0.0026	<0.00011–0.0011	<0.00029–0.0095	45–768	97
UK	Essex, semirural	86 Jan–Feb	<0.00006–0.0013	<0.00004	<0.00007	<0.0001	<0.0001–0.0004	<0.00009–0.002	<0.00012–0.0005	<0.00015–0.0002	<0.00018–0.0004	<0.00011–0.0005	<0.00029–0.0156	<13–397	97
Belgium	Antwerp, residential	86 Feb–Mar	—	—	—	—	—	<0.0006–0.0104	<0.0006–0.0055	<0.0006–0.0011	0.008–0.094	<0.001–0.01	0.06–0.295	—	67

Table 4.6 Concentrations of organolead compounds in rainwater (ng Pb L^{-1})

Country	Location	Year	TML	TMEL	DMDEL	MTEL	TEL	TriML	DMEL	MDEL	TriEL	DML	DEL	RPb3+	Inorganic Pb[a]	Ref.
Belgium	Antwerp, city center	Mar 84	—	—	—	—	—	35–61	—	—	12–86	7–25	9–195	—	104–510	103
Belgium	Antwerp, city center	Jan 85 (snow)	—	—	—	—	—	48–91	—	—	37–93	5–14	21–92	—	272–315	103
Belgium	Antwerp, near	Mar 84	—	—	—	—	—	29–30	—	—	6–23	11–15	7–33	—	138–281	103
Belgium	Antwerp, near	Jan 85 (snow)	—	—	—	—	—	42	—	—	11	7	16	—	183	103
UK	Colchester, urban	Nov 85–Feb 86	<0.2	<0.1	<0.2	<0.3	<0.3	8.6–56.5	<0.4–5	<0.5–15	22.2–68.7	<0.4–22.6	14.3–165	<1–165	18–51	104
UK	Colchester, urban	Feb 86 (snow)	<0.2	<0.1	<0.2	<0.3	<0.3	111.1	<0.4	<0.5	55.5	<0.4	119	39.7	116	104
Canada	urban	—	—	—	—	—	—	<15–74	—	—	<20–107	—	—	—	0.29–0.92	105
Croatia	Zagreb,	May 87–Sep 89	—	—	—	—	—	—	26–208[a]	—	<1–17[a]	—	—	1.6–67	—	106
Croatia	Sibenik, urban	Mar 88–Mar 89	—	—	—	—	—	—	24–50[a]	—	<1[a]	—	—	12–57	—	106
UK	Birmingham, urban	Nov 92–Jun 93	<2–3	<2–6	<2–14	<2	<2–12	3–49	<2–30	<2–36	8–207	<2–25	7–141	<2–61	3.1–39.8	107
Belgium	Antwerp, city suburb	Jan 85 (snow)	—	—	—	—	—	38–67	—	—	23–54	6–7	22–30	—	194	103
UK	Lancaster, suburb	Apr–May 84	<20	<2.8	<3.5	<3.3	<4.2	17	<3.1	<3.7	<4.4	<21	<50	<50	119	104
UK	Sudbury, surburb	Aug 85	<0.2	<0.1	<0.2	<0.3	<0.3	15	12	<0.5	<0.6	4.0	<1.0	<1.0	4	104
Canada	residential	—	—	—	—	—	—	18–62	—	—	<20–147	—	—	—	—	105
Belgium	Antwerp, residential	Mar 84	—	—	—	—	—	19–33	—	—	3–22	5–11	7–29	—	87–156	103
Belgium	Antwerp, residential	Jan 85 (snow)	—	—	—	—	—	11	—	—	10	1	2	—	81	103
Belgium	Antwerp, residential	86–87	—	—	—	—	—	2.3–90.6	—	—	<0.59–81.3	<0.59–23.3	<0.74–73.4	—	2.8–107	108
UK	Bailrigg, semirural	Nov 83–Apr 84	<20	<2.8–37	<3.5–25	<3.3–27	<4.2–15	<2.5–23	<3.1	<0.5–2.4	<4.4–344	<21	<50	<50–71	5–70	104
UK	Essex Univ., semirural	Mar 85–Mar 86	<2–78	<3	<3	<5	<5–18	5–97	<4–23	<4–11	<5–57	<3–68	<8–732	<8–62	<1.3–80	104
UK	Essex Univ., semirural	Feb 86 (snow)	<0.2	<0.1	<0.2	<0.3	<0.3	20	<0.4	<0.5	10	<0.4	20	30	22	104
UK	Essex Univ., semirural	May 93–Jun 93	<2	<2–5	<2	<2	<2	9–19	<2–8	<2	4–11	6–10	17–29	<2–9	1.9–2.8	107
Belgium	Kalmthout, rural	Mar 84	—	—	—	—	—	15–18	—	—	<3–14	8–11	<5–11	—	40–67	103
UK	Burton in Kendal,	Apr–May 84	<20	<2.8	<3.5	<3.3	4.2–72	<2.5–18	<3.1	<3.7	<4.4	<21	<50	<50	48	104
UK	Hazelrigg, rural	May–Jun 85	<20	<2.8	<3.5	<3.3	4.2–16	9–49	<3.1	<3.7	<4.4–23	<21	<50	<50	11–66	104
Ireland	Bantry bay, rural	Mar 86	<0.2	<0.1	<0.2	<0.3	<0.3	0.23–2.04	0.4–0.84	<0.5–0.84	<0.12–0.28	<0.4–4.96	0.6–3.41	2.03–6.7	—	104
UK	Suffolk, rural	Nov 92–May 93	<2	<2–5	<2	<2	<2	<2–13	<2–9	<2–8	4.43	<2–9	3–32	<2–3	1.7–5.7	107
Denmark	Greenland	87–89	—	—	—	—	—	<0.01	—	—	0.027–0.61	<0.01	0.15–2.3	—	0.012–0.19	11
Denmark	Greenland	Jan–Jul 89	—	—	—	—	—	<0.01–0.037	—	—	<0.005–0.208	<0.01–0.26	0.011–0.703	—	0.013–0.27	11

[a] Dissolved inorganic Pb (μg L^{-1}) Total TriAL and DiAL

exhaust. However, this correlation became insignificant in semirural and rural air samples [97].

Concentrations of organolead compounds in rain and snow are listed in Table 4.6. Ionic TriALs are found most frequently, and at higher concentration than other species. TALs have not been detected because of their instability in atmosphere and their insolubility. Seasonal variation has been found in the concentrations of ionic species in rainwater: higher TriALs in winter and lower in summer in rainwater [108]. This trend was also observed in snow samples from Greenland [11], although a strong seasonal variation was observed for methyllead species in the latter study in contrast with the Belgian study, in which this was observed for ethyllead species. The seasonal variation was explained by the elevated photochemical decomposition of atmospheric TriALs in summer. On the other hand, no seasonal variation was observed for the concentrations and speciation of organolead compounds in rainwater samples in a UK study [104].

A declining trend was observed in rainwater TriALs and DiALs concentrations between 1984–1993 in the UK [107]. However, the magnitude of the decline was somewhat smaller (50 %) when compared with the decline in the concentration of inorganic lead (90 %) or with the decline in alkyllead usage (72 %) during the same period [107]. No clear reason for this observation could be found. In this study a significant correlation was also found between the concentrations of organolead and inorganic lead in the rainwater that was sampled at urban sites but not in those sampled in rural sites. A similar finding was obtained in the correlation between the concentrations of gaseous organolead and particulate lead in the atmosphere [97].

Considerably elevated TriEL in rainwater (< 2000–$62\,000$ ng L^{-1}) found in Germany was attributed to Black Forest decay [109]. If the values were valid, they would be toxic to plant cells. However, the analytical method used was a less specific bioassay and the reliability of the method has been questioned [16, 110].

The detection of monoalkyllead in rainwater is noteworthy. Monoalkyllead is considered to be an intermediate in the dealkylation process of organoleads; however, its presence cannot theoretically be detected because of the considerable instability of the compound. So far, the detection of monoalkyllead is attributed to an analytical artifact rather than its actual presence [111].

4.5.2 ENVIRONMENTAL WATERS

Concentrations of organolead compounds in environmental waters have been reported for surface waters and potable waters (Table 4.7) [77, 105,112–115]. TALs have been hardly detected in surface water. Only a recent study using highly sensitive GC-ICP-MS could detect TALs at pg L^{-1} level in estuarine waters [77].

TriEL and DiEL were the main species generally detected at orders of ng L^{-1} or less; occasional elevated values were found in rainwater runoff sampled near gasoline stations and streets with heavy traffic density. Ionic species

Table 4.7 Concentrations of organolead compounds in environmental waters (ng Pb L^{-1})

Sample	Location	Sampling	TML	TMEL	DMDEL	MTEL	TEL	TriML	DMEL	MDEL	TriEL	DML	DEL	MEL	Ref.
Lake and river water	Canada	—	<500	<500	<500	<500	<500	—	—	—	—	—	—	—	112
Lake water	UK	—	<2	<2	<3	<5	<5	<2	<3	<4	<5	<3	<8	—	113
River water	Germany	—	—	—	—	—	—	<15	—	—	<20	—	—	—	105
Creek water	Germany	—	—	—	—	—	—	<15	—	—	<20	—	—	—	105
Estuarine water	UK	—	<0.2	<0.1	<0.2	<0.3	<0.3	<0.3–0.8	<0.4	<0.5	<0.6	<0.4	<1.0	—	113
Estuarine water	UK, near port	—	<2	<2	<3	<5	<5	97	<3	<4	50	219	<11	—	113
Estuarine water	Germany	1996	<0.00008–0.0018	<0.00008–0.00042	<0.00008–0.0006	<0.00008–0.00072	<0.00008–0.0015	—	—	—	—	—	—	—	
Harbor water	Germany	—	<0.2	<0.1	<0.2	<0.3	<0.3	123	<0.4	<0.5	40	<0.4	<1.0	—	105
Seawater	UK	—	<0.2	<0.1	<0.2	<0.3	<0.3	<0.3	<0.4	<0.5	<0.6	<0.4	<1.0	—	113
Groundwater	UK	—	—	—	—	—	—	<0.3	<0.4	<0.5	<0.6	<0.4	<1.0	—	113
Potable water	Belgium	1986	<0.04	<0.02	<0.04	<0.06	<0.06	2.17	0.18	<0.09	0.31	0.47	0.28	<0.13	114
Potable water	UK	—	—	—	—	—	—	<0.06–3.6	<0.08–0.1	<0.10–0.2	<0.12–0.2	<0.08–1.2	<0.2–0.4	—	113
Potable water	UK	1986	—	—	—	—	—	1.35	0.01	<0.10	0.02	0.16	0.10	<0.16	114
Potable water	Germany	—	—	—	—	—	—	<15	—	—	<20	—	—	—	105
Potable water	Belgium	—	—	—	—	—	—	<0.08	—	—	<0.13	1.91	1.76	—	113
Road surface	UK, urban	—	<2	<2	<3	<5	<5	499	57	<4	129	108	1132	—	113
Road surface	UK, suburban	—	<2	<2	<3	<5	<5	66	<3	<4	<5	<3	<8	—	113
Road surface	UK, rural	—	<0.2	<0.1	<0.2	<0.3	<0.3	10	<0.4	<0.5	5	<0.4	6	—	113
Runoff	UK, near gas station	—	<0.2	<0.1	<0.2	<0.3	<0.3	15	<0.4	<0.5	3	5	5	—	113
Rain water runoff	Canada, near gas station	—	—	—	—	—	—	—	—	—	160	—	1800	—	115
Rain water runoff	Canada, urban	—	—	—	—	—	—	—	—	—	280–400	—	360–470	—	115

have been detected in potable waters. Their presence might be a health concern since there is a direct human exposure source, even though the level was not high.

4.5.3 SEDIMENT AND SOIL

Information on organolead concentrations in sediment and soil is much more limited than that on air and water (Table 4.8) [68, 112, 114–116]. Ionic species, particularly ethylated ones, have been the species sporadically detected at $< 1\,ng\,g^{-1}$ level.

4.5.4 ORGANISMS

Reported organolead concentrations in biological materials are listed in Table 4.9 [85–88, 93, 112, 116–120].

TALs were detected in 17 out of 107 fish samples collected in Canada. The concentrations ranged from 0.2–9.3 ng g^{-1} wet weight. TEL was the most frequently (12 samples) detected followed by DMDEL ($n=9$), TML ($n=6$), TMEL ($n=5$) and MTEL ($n=4$) [112]. Considerably elevated TAL concentrations, up to 62 000 ng g^{-1}, and alkyllead/total lead ratios (50–75%) were found in the fish taken from contaminated river systems in Canada [86]. Closure of an alkyllead manufacturer resulted in lower alkyllead burden of fish.

The concentrations of ionic alkyllead compounds in liver, kidney and whole egg of herring gulls from Great Lakes were measured [118]. The concentrations varied depending on the alkyllead compound, sampling localities and tissues, but not on the maturity of the birds. The concentrations of methylated species were highest in the samples from Lake Ontario, which was followed by Lake Huron and Lake Erie, and Lake Superior, while ethylated species were highest in Lake Superior and lowest in Lake Ontario and Lake Erie.

Organolead compounds were analyzed in fresh and canned fish [117]. Many of the samples contained < 0.09–0.7 ng g^{-1} of TriML and DiML and some contained TriEL at similar levels.

Concentrations of ionic alkyllead compounds were determined in the brains of 22 subjects in Denmark [119]. The concentration ranged from < 0.008–0.050 µg g^{-1} wet tissue. A negative correlation with age was found for ionic alkyllead concentrations with no sex difference.

4.6 ENVIRONMENTAL DYNAMICS

4.6.1 SOURCE OF TALS IN THE ENVIRONMENT

Organolead compounds in the environment are almost exclusively derived from TALs added to gasoline. TALs are emitted from the exhaust of automobile, even

Table 4.8 Organolead compounds concentrations in soil and sediment (ng Pb g⁻¹)

Sample	Location	Sampling	TML	TMEL	DMDEL	MTEL	TEL	TriML	DMEL	MDEL	TriEL	DML	DEL	MEL	Ref.
Soil	UK	—	<4	<6	<8	<7	<9	—	—	—	—	—	—	—	68
Soil	UK	1986	—	—	—	—	—	15	7	12	13	25	18	—	68
Soil	UK	1986	—	—	—	—	—	<0.02	<0.03	<0.03	<0.06	<0.03	<0.10	—	68
Soil	Canada	—	—	—	—	—	—	—	—	—	0.7–1.2	—	4–10	—	115
Soil	Canada	1986	—	—	—	—	—	<0.1	—	—	0.95	<0.1	7.0	—	114
Soil	Belgium	1987	—	—	—	—	—	0.22	0.04	0.02	0.08	0.02	1.02	<0.02	114
Sediment	Canada	—	<10	<10	<10	<10	<10	—	—	—	—	—	—	—	112
Sediment	Canada	—	<0.5	<0.5	<0.5	<0.5	<0.5	—	—	—	—	—	—	—	116
Sediment	UK, Essex, intertidal	<0.02	<0.02	<0.02	<0.04	<0.06	<0.06–0.4	<0.02–0.1	<0.03	<0.03	<0.06–0.2	<0.04–1.1	<0.1	—	68
Sediment	Ireland, intertidal	—	—	—	—	—	—	0.2	<0.03	<0.03	<0.06	<0.04	<0.1	—	68
Sediment	UK, underground channel near gas station	—	1.3	<0.6	<0.8	<0.7	<0.9	<0.8	<1.1	<1.5	<1.8	<4	<10	—	68

Table 4.9 Organolead compounds concentrations in biological and clinical samples (ng g^{-1} wet)

Sample	Tissue type	Total TAL	TriML	DMEL	MDEL	TriEL	DML	DEL	MEL	Ref.
Cod[a]	Frozen liver	10–37	—	—	—	—	—	—	—	88
Cod[a]	Liver	125	—	—	—	—	—	—	—	88
Cod[a]	Lobe	28–44	—	—	—	—	—	—	—	88
Lobster[a]	Digestive gland	162	—	—	—	—	—	—	—	88
Mackerel[a]	Muscle	54	—	—	—	—	—	—	—	88
Flounder[a]	Muscle	4790	—	—	—	—	—	—	—	88
Fish										116
Various lake fish										112
Contaminated river fish	Whole-intestine	<8–61713				40				86
Contaminated river fish	Intestine	<8–30608				<0.13	1.91	1.76		86
River fish	Muscle	<8–3585								86
River fish	Carcass	<8–1716								86
River fish	Whole	<8–2524								86
Clam	Fresh		<0.09			<0.1	<0.1			117
Mussel	Fresh		<0.09			0.3	<0.1			117
Lobster	Claw+tail		<0.09			0.3	<0.1			117
Lobster	Hepatopancreas		<0.09			0.1	0.9			117
Lobster	Abdominal meat		0.2			<0.1	<0.1			117
Oyster	Fresh		<0.09			0.5	<0.1			117
Scallop	Fresh and frozen		<0.09			<0.1	<0.1			117
Baby clam	Canned		<0.09			<0.1	<0.1			117
Mussel	Canned		<0.09–0.7			0.2–0.3	<0.1			117
Cockles	Canned		<0.09–0.6			<0.1–0.2	<0.1–1.0			117
Crab	Canned meat		<0.09–0.2			<0.1	<0.1			117
Lobster	Pate		<0.09			<0.1	<0.1			117
Oyster	Canned		<0.09			0.4	<0.1			117
Cod	Canned liver		0.7	—	—	0.2	<0.1			117
Herring gull	liver		1.3–7.9	ND[b]–0.3	ND[b]	0.8–7.3	ND[b]–4.2	ND[b]–2.2	ND	118
Herring gull	kidney		2.3–18.7	ND[b]–1.2	ND[b]	0.8–8.8	1.2–5.8	ND[b]–1.6	ND	118

Sample	Tissue/site								Reference
Pigeon	muscle	ND*	—	—	ND^b–1.5	—	—	—	85
Pigeon	liver	ND*	—	—	ND^b–10.7	—	—	—	85
Pigeon	kidney	—	—	—	1.3–29.9	—	—	—	85
Mallard	liver (male)	0.5–4.3	ND^b–1.2	ND^b–1.3	0.5–2.0	ND^b–2.2	0.7–1.5	ND^b–0.6	85
Mallard	liver	0.4–1.3	ND^b	ND^b	ND^b–0.5	ND^b–1.7	ND^b–5.5	ND^b–0.6	85
Mallard	kidney	4.0–4.3	1.7–1.8	1.7–1.8	3.3–3.5	1.9–2.0	1.3–3.0	ND^b–0.7	85
Mallard	kidney (female)	1.4	0.3–0.4	0.3–0.4	0.6–0.7	ND^b	0.4–1.0	ND^b	85
Duck	brain (male)	1.1	0.5	0.8	1.2	1.4	0.5	ND^b	85
Mallard	brain	0.6	ND^b	ND^b	0.3	ND^b	0.7	ND^b	85
Mallard	muscle	1.6	0.6	0.9	1.4	1.7	0.8	ND^b	85
Mallard	muscle (female)	0.4	ND^b	0.4	ND^b	ND^b	0.5	ND^b	85
Duck	brain (female)								
Human[a]	brain	—	<0.01–0.05^c	—	—	—	—	—	119
Grass[a]	(near gas station)	1.8–15.8	<0.7–5.5	<0.8–1.5	3.4–20.2	<1.0–3.5	<1.3–64.8	<1.2	87
Grass[a]	(residential)	0.9–1.1	<0.7–1.2	<0.8	1.2–2.3	<1.0	1.3–2.6	<1.2	87
Tree leaves[a]	(residential)	<0.6–3.2	<0.7–1.1	<0.8	1.0–4.3	<1.0	1.8–3.6	<1.2	87

[a] ng Pb/g wet, others are ng compound/g wet
[b] Not detected.
[c] Total TriAL

TML, TMEL, DMDEL, MTEL, TEL, were not measured except for fish (0.1–1.7, 1.3–4.2, 0.4–3.5, <0.1–0.1, <0.1–0.2 respectively [116] and Lake fish (<0.1–2, <0.1–5.2, <0.1–7.1, <0.1–4.4 <0.1–9.3) respectively [112].

though the compounds in gasoline have gone through high temperatures inside the fuel chamber and exhaust system. About 0.023 % of the TALs in gasoline survive and is emitted to atmosphere via the exhaust [121]. However, under the same experimental conditions, it was estimated that 1.7 % of the gasoline was emitted from exhaust unburned, therefore TALs were much more thermally unstable than other gasoline components. Another estimate of TAL emission via the exhaust was 0.7–1.1 % [13]. Other routes for release of TALs into the environment include evaporation and spillage at gasoline stations, which is evident from the elevated atmospheric TAL concentrations near gasoline stations, and distribution and emission from TAL manufacturers [13]. Total loss of TALs in gasoline by exhaust and via the above routes has been estimated to be 2 % [14] and 2–3 % [16]. If world total lead consumption in gasoline in 1993 was 70 000 tons [36] and loss of lead from gasoline was 2 %, then 1400 tons of lead were emitted into the environment as TALs and this estimate may be compared with 5000 tons in a 1982 estimation [16].

Improper disposal of gasoline tank sludge was suggested as another source of TAL contamination [122]. Accidental massive spillage of TALs or leaded gasoline can occur [123, 124].

4.6.2   DECOMPOSITION IN THE ATMOSPHERIC ENVIRONMENT

Harrison and coworkers experimentally investigated TALs decomposition in the atmosphere [125]. Decomposition of TALs is caused mainly by the attack of OH radicals in the atmosphere, while ozone and $O(^3P)$ attack and direct photochemical decomposition takes place to a much lesser extent. Decomposition rate in a typical polluted atmosphere in sunny summer conditions was estimated to be 21 % h^{-1} for TML and 88 % h^{-1} for TEL, the latter being four-times faster than TML. Under coexistence of ozone, the rates increased a little to 29 % h^{-1} and 93 % h^{-1}, respectively. On the other hand, under dark conditions no appreciable decomposition of atmospheric TALs was found. The total decomposition rates were estimated to be 4 and 23 % h^{-1} for TML and TEL, respectively, in winter. Heterogeneous reactions involving atmospheric particulate matter, such as direct adsorption and surface reactions, was found not to contribute to TALs decomposition. The rate constants of OH radical attack on TEL and TML obtained in this study were confirmed in a later study (Table 4.10) [126].

The decomposition products of TALs were identified as TriALs and DiALs in both gaseous and particulate phases, and as inorganic lead in particulate phase [127]. The majority of TriALs was present in the gaseous phase and only a minor fraction was found in the particulate phase, and also on the reaction chamber wall. DiALs distributed much more to chamber wall than did TriALs. It is also OH radicals that decompose TriALs in the atmosphere. The decomposition took place at a slower rate than for TALs: half-lives were estimated to be 126 h for TriML and 34 h for TriEL, as compared with 41 h for TML and 8 h for TEL.

Table 4.10   Rate constants for the reaction of OH radicals with TALs [126]

	Constant ($\times 10^{12}$ cm^3 molecule^{-1} s^{-1})	Method^a	Ref.
TML	9.0±1.6	RR	125
	5.9±1.2	PR-KS	126
	3.9±0.2	RR	126
TEL	80±12	RR	125
	68±16	PR-KS	126
	61±5	RR	126

^{a} RR, relative method, PR-KS, pulse radiolysis–kinetic spectroscopy.

TriALs and DiALs thus formed will be dissolved in rainwater because of their high solubility. Wet and dry deposition is the main pathway to the hydrosphere and troposphere. On the other hand, deposition of TALs is unlikely, or at least insignificant, because of its instability in the atmosphere. Radojevic and Harrison [104] estimated wet deposition rates of organolead compounds to be 5.6 ng Pb cm^{-2} year^{-1} at semirural UK sites, and an almost equal dry deposition rate was assumed. Estimated deposition rates in urban and rural UK sites was 1.1 and 0.7 ng Pb cm^{-2} month^{-1}, (13 and 8.4 ng Pb cm^{-2} year^{-1}) respectively [97]. In a similar Belgian estimate, 32 μg m^{-2} year^{-1} (3.2 ng Pb cm^{-2} year^{-1}) of ionic alkyllead wet deposition was calculated in rainwater samples obtained near Antwerp [108].

4.6.3   DECOMPOSITION IN THE AQUATIC ENVIRONMENT

Abiotic organolead decomposition in an aqueous medium has been experimentally investigated [128]. TALs at 30–50 mg L^{-1} levels were fairly stable in pure water under dark conditions: TEL was more stable (2 % decomposition in 77 days) than TML (16 % in 22 days). The order of stability was reversed to that in the atmosphere. Decomposition was accelerated in the presence of metal ions (Cu and Fe) and silica, as well as by exposure to sunlight, indicating a higher decomposition rate in natural waters. The decomposition products of TALs were mainly TriALs, and traces of DiALs were also detected. Evaporation into the headspace was another route of removal of TALs in water [129].

TriALs were more stable than TALs in pure water under dark conditions: no TriEL decomposition took place in 12 months and only 1 % in 220 days for TriML. Unlike for TALs, metal ions did not accelerate TriAL decomposition. Some anions promoted disproportionation of TriALs: sulfide ion disproportionates TriML to produce TML and DiML via bis(trimethyllead) sulfide [130]. Like TALs, sunlight accelerated decomposition of TriALs. Decomposition took place to the extent of 99 % for TriEL and 4 % for TriML in 15 days under sunlight.

Reaction of DiALs was slower. Some 10 and 6% of DiML and DiEL, respectively, reacted in one month under darkness and the figures were 5 and 25% under light conditions. The products were TriALs and inorganic Pb, indicating that disproportionation had taken place.

Similar findings were observed in natural rainwater samples spiked with TALs, TriALs and DiALs at 1–20µg Pb L^{-1} levels [104].

When a similar experiment was done with lower TAL loading (low µg L^{-1}), a different result was obtained: 100% of TALs decomposed within 5 days even under dark conditions [131]. Moreover, inorganic lead, but not DiALs, was identified as a major decomposition product of TriALs (Figure 4.7) [131]. The latter result indicated that TriAL decomposition did not follow the traditionally postulated pathway, i.e. TALs → TriALs → DiALs → inorganic lead, and that direct decomposition (or disproportionation) of TriALs to inorganic lead (and also TALs) took place in the water media. Therefore, the decomposition pathway in water may be written as

$$\text{TALs} \longleftrightarrow \text{TriALs} \longrightarrow \text{inorganic lead}$$

If this is the case, DiALs found in various environmental surface waters might be originate predominantly from DiALs produced in the atmosphere but not from decomposition of TALs or TriALs in the medium [131].

Microbial transformation of TALs in environmental water has also been found. This was deduced from the accelerated disappearance of added TALs

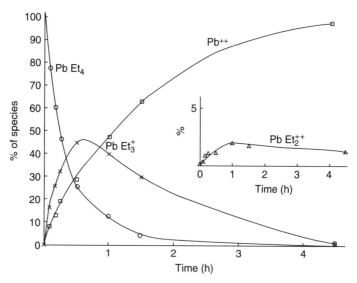

Figure 4.7 Degradation of TEL in water under UV light (254 nm) and production of TriEL, DiEL and inorganic lead. From Van Cleuvenbergen *et al.* (1992) [131] reproduced with permission.

in groundwater samples by nutrient supplementation [132], however, the organism(s) responsible for the decomposition have not been identified. The decomposition took place at low rates ($< 66\,\mu g$ compound L^{-1} day^{-1} for $10\,mg\,L^{-1}$ addition) and thus is of limited significance when compared with abiotic decomposition.

Information so far is missing on the ultimate fate of dissolved ionic species in the water column. Some will decompose to form inorganic lead as described; however, taking their affinity for suspended particulate matter and their persistence into consideration, some of the ionic species, particularly DiALs, must precipitate with particulates. Sediment, or suspended particulate, has some effect on the rate of aquatic organolead transformation. It accelerated decomposition of TALs [128] and reduced TriAL disproportionation in aquatic media [130].

4.6.4   DECOMPOSITION IN THE SOIL

An experiment using ^{14}C labelled TEL ($1\,mg$ TEL $100\,g^{-1}$ soil) revealed that TEL decomposed rapidly either in sterile (half life: 7–$14\,h$) and non-sterile (12–$17\,h$) forest soil to form ionic organolead compounds [133]. $^{14}CO_2$ was present over the non-sterile soil but not over sterile soil, thus confirming that the TEL decomposition in non-sterile soil was a biologically mediated process. Mass calculations of ^{14}C after complete TEL decomposition indicated that the majority of the applied ^{14}C-TEL was converted to unknown products (non-hexane extractable and non-aqueous EDTA extractable).

Aerobic microbial decomposition of TEL in soil was confirmed in another experiment [134]. At a lower dose ($2\,g$ Pb kg^{-1} dry soil), the rate of microbial TEL decomposition was $780\,\mu mol$ day^{-1} kg^{-1} while abiotic decomposition was estimated to be $50\,\mu mol$ day^{-1} kg^{-1}. At a higher dose ($10\,g$ Pb kg^{-1} dry soil), abiotic decomposition increased almost proportionally but microbial decomposition decreased slightly, probably due to TEL toxicity. Decomposition products were largely inorganic lead with traces of TriEL. These TEL decomposing soil microorganism(s) have not been identified.

Decomposition of TriML and DiML (up to $10\,mg\,kg^{-1}$ addition) has been assessed in agricultural soil [135]. Recovery was examined for TriML and DiML added to both intact and sterilized soil suspensions and supernatant in the dark. A higher recovery of TriML was found after 10 days' incubation when the compound was added to the supernatant of the soil suspension than to the suspension itself, indicating a higher decomposition rate or higher irreversible adsorption of TriML in the presence of soil particles. Recovery of 26 and 11.8 % was observed for intact soil suspension and supernatant, respectively, and 35.5 and 19.1 % recovery for sterilized soil. DiML was also found as 8.6–26.9 % of the added TriML, indicating sequential dealkylation. Recovery profile of added DiML was similar to those of TriML, but with a lower recovery rate: only 0.6

and 1.8% of added DiML was recovered from intact soil and sterilized soil suspension, respectively. Detection of TriML in all of the incubation with DiML (2–5% of added DiML) indicated that a disproportionation reaction had taken place. Detection of TML in TriML added intact soil suspension, but not in sterilized soil, also indicated the same reaction. This might indicate that the disproportionation reaction was biologically mediated but the sterilization procedure might have changed the chemical composition of the soil. The ratio of total alkyllead recovered from intact soil suspension assays to total alkyllead recovered from corresponding sterilized assays was constant at around 0.6:1 and the authors concluded that the major contributor to the decomposition was chemical, perhaps via the well known sulfide route [135].

The fate of ionic organolead species in soil may also be deduced from the result of a sample storage experiment [114]. When a soil sample (<60 mesh) containing ionic species was stored in the laboratory in daylight (but not in sunlight), DiEL concentration decreased most steeply and TriEL the next, while virtually no decrease in concentration took place for TriML in 100 days. There was no information on the decomposition products of the ethylated species.

4.6.5   METABOLISM IN ORGANISMS

TEL was not metabolized by the chrysophycean flagellate *Poterioochromonas malhamensis* [136]. A freshwater green alga, *Ankistrodesmus falcatus*, bioaccumulates and metabolizes TriALs and DiALs [137]. Bioconcentration factors (concentration in organism/concentration in water) for TriALs, DiALs and inorganic lead was 100, 2000 and 20 000, respectively. The organolead decomposition pathway in this organism was assumed to be sequential dealkylation, i.e. TriALs → DiALs → inorganic lead, but organism-mediated chemical disproportionation of DiML into TriEL and inorganic lead was also indicated. An *Arthrobacter* sp. and a wood decay macrofungus, *Phaeolus schweinitzii*, had a metabolic TriALs decomposing activity [138].

An experiment with Japanese quail indicated a complicated metabolism of administered TriEL and DiEL, i.e. a combination of *in vivo* methylation or disproportionation and dealkylation [139]. When DiEL was administered via drinking water for 8 weeks, TriEL, DiML and TriML, as well as DiEL, were detected in eggs from the birds but no methylated species could be found in the soft tissues. It was therefore suggested that methylation took place either in the reproductive system or in the egg. It can be assumed that detection of TriEL was due to chemical disproportionation. When TriEL was administered, TriEL and DiEL, but not methylated species, were detected indicating that dealkylation was the dominant metabolism of TriEL.

TALs are known to be readily absorbed via the gastrointestinal tract, respiratory tract, and skin of mammals and humans. TALs administered orally are thought to be absorbed after decomposition to TriALs in the presence of gastric juice [13]. An inhalation experiment using ^{203}Pb-labeled TALs (around 1 μg m^{-3}) revealed 30–40 % absorption of the inhaled TALs vapor [140]. The absorbed ^{203}Pb-TALs disappeared from the body by exhalation and by conversion to TriALs, the latter reappearing in the erythrocytes after 8 (TML) and 20 (TEL) hours of exposure. The rate of dermal absorption is not clear but it was estimated to be negligible in humans at the concentration levels found in gasoline.

The metabolism of absorbed TALs in mammals is basically a dealkylating process. Enzymatic TEL dealkylation takes place predominantly in the microsomes of hepatic cells of experimental animals at a high rate [141, 142]. The main product was identified as TriEL, which was distributed mainly to blood, kidney and brain as well as to liver. The sequential dealkylation pathway, i.e. TEL → TriEL → DiEL → inorganic lead, has been postulated in the organisms; however, failure to detect DiEL in TEL administered rats, despite the persistent nature of the compound, indicated *in vivo* direct dealkylation of TriEL to inorganic lead [143] or disproportionation. Traces of DiML were detected in the liver, but not in any other organs and tissues, of TML administered rabbits [144] and at least a part of the TML should be dealkylated via the sequential pathway.

On the other hand, organolead compounds excreted in urine were predominantly DiALs and inorganic lead. DiEL was the major organolead in the urine of TEL-administered rabbits one day after injection (69 % DiEL, 27 % inorganic and 4 % TriEL) [145]. This was also the case for an intoxicated human (50 % DiEL, 48 % inorganic and only 2 % TriEL) even after 15 days from exposure [146] and for exposed workers [147]. DiML was also the major compound found in the urine of TML administered rabbits [144]. The fact that DiALs are hardly detected in TALs-exposed animals, but were the major species excreted in urine, indicates that DiALs are quite efficiently excreted through the kidneys.

Although urine had been believed to be the main excretion pathway for organolead compounds [13], experiments with rats and rabbits revealed that fecal excretion was 6 to 20 times urinary excretion [143–145]. In addition, lead in the feces of TAL-administered rabbit was almost 100 % inorganic. An experiment showed that the DiAL excreted in bile was dealkylated to inorganic lead in the cecum, colon and rectum [148]. High concentrations of total lead were also found in the feces of TAL-poisoned patients [149].

Excretion of organolead compounds is faster and more efficient than that of inorganic lead. In rabbits administered TEL, 58 % of the administered dose was excreted in the 4 days after injection, while this was only 4 % in lead-acetate-treated rabbits (Table 4.11) [150]. In this table comparison is given for lead distribution in the organs of rabbits at 24 hours and 30 days after TEL and inorganic-lead intravenous injection. It is clear from this table that administered lead disappears rather quickly in TEL-treated rabbits than inorganic lead treated rabbits and feces constitute the dominant route of excretion.

Table 4.11 Distribution of lead in organs of rabbits administered TEL and lead acetate (7.7 mg Pb kg⁻¹ body weight) [150]

	TEL administered				Lead acetate administered		Control
	24 h after injection		30 days after injection		24 h	30 days	
	Total Pb^a	TriEL^a	Total Pb^a	TriEL^a	Total Pb^a	Total Pb^a	Total Pb
Brain	3.15	2.67	0.08	0.03	0.11	0.34	—
Spine	2.24	1.43	0.06	<0.001	0.16	0.23	—
Liver	29.91	25.62	1.42	1.36	73.3	65.1	—
Kidney	13.57	10.40	0.28	0.17	14.61	4.37	—
Skeletal muscle	2.84	2.24	0.07	0.06	0.10	0.05	—
Cardiac muscle	2.62	2.05	0.02	<0.001	0.27	0.08	—
Blood	0.70	0.26	0.09	0.002	1.53	0.70	—
Femur	1.7	—	2.6	—	17.7	29.7	1.6
4-days cumulative excretion (μg)	Urine		814		401		
	Feces		12812		561		

^{a} Concentration is expressed as μg Pb g⁻¹ wet tissue except for femur where μg Pb g⁻¹ dry bone is given.

4.6.6   NATURAL ALKYLATION OF LEAD COMPOUNDS

t was in 1975 when the first indication was proposed on the biological alkyla-
ion of lead compounds to yield TML [151]. In this model environmental
xperiment, lake sediments were anaerobically incubated in a glass vessel and
TML was detected in the headspace, but no TML was found over autoclaved
ediment. Addition of inorganic lead nitrate or TriML acetate enhanced the
volution of TALs in some sediments. A maximum of 6% of added TriML was
converted to TML in a week. Similar, if not identical, findings that supported
viological methylation of inorganic lead and/or TriALs were obtained for
vacteria-seeded water [152], marine sediment [153], intertidal sediment [99],
estuary sediment [154] and agricultural soil [135]. On the other hand, Reisinger
et al. [155] demonstrated absence of biological methylation of inorganic lead in
vacterial cultures, sediment and sludge using a double tracer (^{14}C and ^{210}Pb)
tudy. Extensive surveys involving a number of sediment and sludge types
and cultures of microorganism with a variety of lead compounds (inorganic
alts, monoalkyllead compounds) failed to detect biological methylation
156].

Production of TEL, but not MTEL, by addition of TriEL to sulfide-contain-
ng distilled water, as well as lake sediment, was taken as the evidence of sulfide-
nediated chemical disproportionation in such systems(see Chapter 1) [157].
This experimental result that showed no TML production with autoclaved
ediment [151] could be explained by the loss of free sulfide in such sediment
via the autoclaving. Later studies supported the supposition that sulfide-mediated chemical reaction of TriALs could account for the experimental observa-
ions that had been attributed to biological methylation in the environment
155, 156, 158].

However, sulfide-mediated disproportionation cannot explain the observed
TML production from inorganic lead salts in sediment systems [99, 152] and
evidence of chemical methylation of inorganic lead salts was sought. Methylcobalamin is a well known methylating agent for inorganic mercury salts, and
t is the only agent known to be capable of transferring carbanion methyl
groups (CH_3^-) [159]. However, methylation of lead by methylcobalamin has
not been successful [160–162]. Another type of methylating agent is the
methyl carbonium (CH_3^+) donor, i.e. S-adenosylmethionine (SAM) and N^5-
methyltetrahydrofolate [159]. Craig and Rapsomanikis [163] examined lead
methylation by environmentally realistic CH_3^+ donors, such as methyl iodide,
SAM, and betaine, and found that only methyl iodide could methylate inorganic and metallic lead to TML (total organolead yield: 0.037–1.18% of applied
inorganic lead). Methylation of metallic lead and inorganic lead compounds by
methyl iodide has also been reported in several other works [164–166]. This
finding was important because methyl iodide is abundant in marine environment, particularly in the vicinity of kelp. Methylation of inorganic lead in kelp-
rich marine environments will explain the excess atmospheric organolead

observed in maritime air mass by Harrison and coworkers [99–102] and the significance of kelp in natural organolead production has been stressed [167].

In a culture experiment in line of this hypothesis, Harrison and Allen detected TriML and traces of TML in laboratory cultures of marine phytoplankton (*Emiliana huxleyi* and *Gyrodinium auroleum*, but not of *Thalassiora eccentrica* and *Amphidinium carterae*) and marine macrophyte (*Chaetomorpha aerea* and *Fucus spiralis*) added with inorganic lead(II) [100] but it was not clear from this study whether the production of the organolead was biological or chemical.

Apart from the findings from model environments, there are also several studies in which the biological methylation of lead was suggested, based on the occurrence of organolead compounds in the organisms, although the evidence is circumstantial in nature. Detection of methylated species in the tissues of birds collected from Canada, where TEL was exclusively used as a gasoline additive, is one example [85, 115]. Craig and Van Elteren [168] ascribed these observations to long-range transport of methylated species into the ecosystem. However, detection of methylated species in the tissues of experimental animals exclusively administered ethylated species [139] cannot be explained. It is necessary to study further whether this result is an indication of biological methylation in higher animals or is due to other reason, e.g. analytical artifact.

In summary, natural alkylation of lead could occur, though it may be inefficient, and most of the experimental facts, which had been attributed to biological alkylation, have turned out to be explained by environmental chemical reactions. However, those explanations do not necessarily exclude the possibility of the existence of a biological alkylating process.

Regardless of whether alkylation is chemically or biologically mediated environmental organolead burdens derived from natural alkylation have not been estimated. Concentrations of organolead in Greenland ice core corresponding to 1980–1985, when the world total TAL emission from gasoline was estimated to be 5000 tonnes [16], were around $500 \, fg \, g^{-1}$ [11]. Concentrations of organolead in 1990–1992 snow are estimated to be $150 \, fg \, g^{-1}$ from the average total lead concentration in the snow ($15 \, pg \, g^{-1}$) [169] and average organolead/total lead ratio (1.13 %) in the snow of the period [170]. The total world TAL emission via gasoline, is estimated in Section 4.6.1 to be 1400 tons. The two sets of data suggest that 10 tonnes of world TAL emission from gasoline corresponds to $1 \, fg \, g^{-1}$ organolead in central Greenland snow/ice. Organolead concentrations in the ice core formed before 1922, when no anthropogenic TALs had been produced, were $< 10 \, f \, g^{-1}$ [11]. It can then be roughly estimated that natural TALs production amounts to $< 100 \, tons \, year^{-1}$. It must be noted, however, that Lobinski denied the presence of natural biomethylation of lead from the fact that it was not possible to detect organolead in the ice core formed before 1922 [171] in his experiments.

4.7 TOXICITY TO ORGANISMS

4.7.1 TOXICITY TO AQUATIC ORGANISMS

TALs are toxic to freshwater green algae, *Scenedesmus quadricauda*, *Ankistrodesmus falcatus*, and *Chlorella pyrenoidosa* [172]. Although exposure levels were not quantitatively determined, TML at an estimated concentration of 0.5 mg Pb in 1.5 L culture medium, 7 days) inhibited productivity and cell growth of the three algae by 32–85%. *Chlorella pyrenoidosa* was most affected. Lead was detected in the cytoplasm of the exposed cells. Relative toxicity was reported to be TML(20)>TriML(10)>Pb^{2+}(1) (relative toxicity in parentheses).

Acute toxicity of TALs has been tested in marine organisms [173]. A 48-h ED$_{50}$ of TML for mixed coastal marine bacteria, algae (*Dunaliella tertiolecta*), crustaceans (*Artemia salina*) and larval fish (*Morone labrax*) was 1900, 1650, 250, 100 ppb Pb, respectively, and the figures for TEL were 200, 150, 85, and 65 ppb Pb, when oxygen consumption (bacteria), photosynthetic activity (algae) and mortality (crustacean and fish) were used for the toxicity indicators.

In a series of detailed toxicity assessments for TEL, TriEL, DiEL and inorganic lead salts for a freshwater chrysophyte *Poterioochromonas malhamensis*, Roderer made the following observations [174–176]:

(i) Relative toxicity under dark conditions was TriEL>DiEL>TEL=Pb^{2+} and essentially there was no toxicity of TEL in darkness.

(ii) Under light conditions, TEL toxicity increases by the production of TriEL.

(iii) Major toxicity signs include formation of giant cells, polyploid nuclei and lorica.

Several thousands of mortalities of waders, wildfowl and gulls were observed in 1979–1981 in the Mersey Estuary, UK [177]. A number of sick birds were also found. Elevated ionic alkyllead burdens were found in dead, sick and even in apparently healthy birds: the concentrations (mg Pb kg^{-1} wet weight) in the liver were 10.1–11.4 and 0.7–1.9 for dead and apparently healthy birds, respectively, which is to be compared to 0.03 for non-Mersey birds. The values for kidney were 13.8–17.7, 1.2–3.3 and 0.14, and those for muscle were 3.0–4.1, 0.33–1.0 and 0.05, respectively [177]. Contamination of the estuary by effluent from an alkyllead manufacturer was suspected to be the cause. The symptoms that the sick birds exhibited included tremor, uncoordinated behavior and inability to fly. These symptoms were similar to those in starlings orally administered with low doses of TriML (200 µg TriML chloride day^{-1} bird^{-1} or 2.8 mg kg^{-1} day^{-1}) rather than to those with TriEL at the same administration levels [178]. At high doses (10-times those of the low-dose birds) TriML and TriEL killed the birds within 6 days of administration, but the pre-death symptoms were less severe in TriEL dosed birds.

Oral exposure to TriEL (0.25 or 2.5 mg L^{-1} as chloride), DiEL (25 mg L^{-1} as chloride) and inorganic lead (250 mg L^{-1} as nitrate) via drinking water for 8 weeks did not result in any overt signs of toxicity in Japanese quail (*Coturnix coturnix japonica*) except for reduced drinking water consumption and slightly elevated kidney weight/body weight sporadically observed [139]. In this experiment the doses received by the bird (mg compound $bird^{-1}$ day^{-1}) were 0.061 and 0.425 for TriEL, 5.66 for DiEL and 61.2 for lead nitrate.

4.7.2   TOXICITY TO TERRESTRIAL ORGANISMS

TriALs and DiALs inhibited nitrogen transformation by agricultural soil microflora [135]. DiALs affected more than TriALs and no difference was found between the toxicities of ethylated and methylated species.

4.7.3   TOXICITY TO EXPERIMENTAL ANIMALS AND HUMANS

Table 4.12 lists LD_{50} values, as expressed in mg Pb kg^{-1} body weight basis, of organolead compounds [13, 141, 179–185]. The reported LD_{50} values (10–20 mg Pb kg^{-1} body weight) of TEL were fairly consistent. Those for TML were higher (toxicity lower) than TEL by one order of magnitude except in the case of mice. TriALs are most toxic, and TALs the next, while DiALs and inorganic were the least toxic. It is recognized that toxic actions of organolead compounds are exclusively due to TriALs, the metabolic products of TALs.

Toxic symptoms by exposure to TALs in rats included irritability, convulsions, coma and death. Similar symptoms (tremors, hyperactivity, convulsions, coma and death) were noticed in dogs exposed to TALs via inhalation [186]. These symptoms clearly indicate damage to the central nervous system. TriEL is known to inhibit energy production of brain cells, and neurotransmitter and amino acid metabolism, and it is known to induce hypomyelination [187, 188].

The symptoms, observed in a number of accident victims and abusers exposed to TALs and/or leaded gasoline, were similar to those observed in exposed experimental animals and birds: tremors, anorexia, insomnia, ocular disturbances, irritability, hallucination, hypothermia, convulsion, incoordination, ataxia, paralysis and death [189, 190]. Although quantitative data is not available, TML is less toxic to humans than TEL as seen for rats [149].

4.7.4   ENVIRONMENTAL HEALTH ASPECTS

Apart from the acute toxicity observed in accident victims, health effects arising from chronic exposure to organolead, particularly carcinogenic effects, are of concern because genotoxic activity of ionic alkyllead compounds has been suggested [191, 192].

Table 4.12 LD_{50} of organolead compounds in experimental animals

Compound	Animal	Administration route	LD_{50} mg Pb kg^{-1} bw	Ref.
TEL	Rat	i.v.	9.9	141
	Rat	i.g.	11[a]	179
	Rat	i.g.	14	180
	Mouse	i.p.	19	181
TML	Rat	i.g.	109	182
	Rat	i.g.	83[a]	179
	Mouse	i.p.	11	181
TriEL	Rat	i.p.	7.0	141
	Mouse	i.p.	8.1	181
TriML	Rat	i.p.	25.5	182
	Mouse	i.p.	6.8	181
DiEL chloride	Rat	—	120[a]	183
DiML acetate	Rat	—	120	184
Inorganic	Rat	i.v.	70	185
	Rat	i.v.	120[a]	13

[a] Approximate value

Occupational epidemiologic studies of TEL workers failed to detect any excess death from cancer [193, 194]. On the other hand, recent case-control study of TEL manufacturing workers revealed a strong association between exposure to the TEL manufacturing process and colorectal cancer [195]. The causality of this result has to be carefully evaluated because no sound biological mechanism(s) had been proposed for the association between organolead and colorectal cancer.

Subclinical neurotoxicity and other actions of TALs have also been indicated among exposed workers. Workers at gasoline depots in China, who were severely exposed to TEL (maximum TEL: 84.8 μg Pb m^{-3}), had more tremor and sinus bradycardia than did non-exposed office workers [120]. Neurobehavioral abnormalities (deficits in attention and concentration, verbal and visual memory, and psychomotor speed) and sensorimotor polyneuropathies were seen among TEL manufacturing workers [196]. Blood lead level is associated with variation in the result of the simple visual reaction time test (SVRT), an indicator of psychomotor speed, among TEL manufacturing workers, suggesting that TEL exposure deteriorates central and peripheral nervous system integrity [197].

Information, however, is scarce on the possible health effects due to environmental organolead exposure, despite the earlier warning by Grandjean and Andersen [192] that the genotoxic effects of organolead compounds might occur at concentrations that might not be unrealistic in environmental settings. There is no estimation of average daily intake of organolead compounds in the general population. Organolead levels in occupationally non-exposed human organs and body fluids have been only sporadically measured (Table 4.9) [119, 120]. Figure 4.8 shows the relationship between TALs concentration

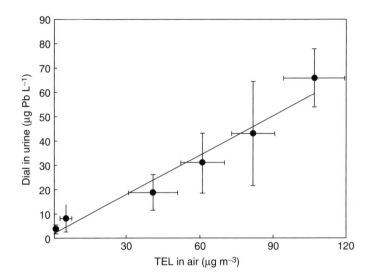

Figure 4.8   The relationship between ambient air TEL and urinary DiEL concentrations. Bars attached to the dot denote standard deviation [urinary DiEL(μg Pb L^{-1})] = 0.54× [air TEL(μg m^{-3})] + 1.58, r = 0.980 [120]

in ambient air and DiALs concentrations in urine [120]. This figure shows that urine DiALs levels are a suitable indicator of TALs exposure levels and future environmental health research may use urine for biological monitoring of organolead compounds.

4.8   CONCLUSIONS

Owing to the extensive studies carried out in the 1980s, our knowledge on the occurrence and fate of organolead compounds in the environment has considerably expanded. This was mainly due to the advances in speciation analytical chemistry and health concerns over organolead compounds. However, the extension of our knowledge seems to be slowing in the last decade. It is not because we do not have ample knowledge of organolead compounds in the environment but probably because leaded gasoline phaseout is proceeding in developed countries. It must be noted, however, that leaded gasoline is still used heavily in developing countries and this usage may pose a health risk to the local population and ecosystem. Application of cost–benefit analysis for the use/ban of leaded gasoline [24, 198] may suggest that these countries should phase out lead.

A future research area of particular importance includes organolead metabolism in biological systems (including biological alkylation). It is one of the research areas suffering from a paucity of reliable data. The dearth of knowledge over the health impacts of low-level and chronic exposure to organolead compounds also warrants further epidemiologic research on non-exposed populations as well as occupationally exposed populations.

Fortunately, analytical sensitivity for organolead speciation has developed considerably in the 1990s with the introduction of GC-MIP-AES and GC-ICP-MS. These new analytical methods may play an important role in the future research, as GC-QAAS did in the 1980s.

4.9 REFERENCES

1. Nriagu, J.O., *Lead and Lead Poisoning in Antiquity*, Wiley, New York, 1983.
2. Boutron, C.F., Görlach, U., Candelone, J.-P., Bolshov, M.A., and Delmas, R.J., *Nature*, 1991, **353**, 153.
3. Rosman, K.J.R., Chisholm, W., Boutron, C.F., Candelone, J.-P., and Görlach, U., *Nature*, 1993, **362**, 333.
4. Hong, S., Candelone, C.F., Patterson, C.C., and Boutron, C.F., *Science*, 1994, **265**, 1841.
5. Savarino, J., Bouron, C.F., and Jaffrezo, J.-L., *Atmos. Environ.*, 1994, **28**, 1731.
6. Candelone, J.-P., Hong, S., Pellone, C., and Boutron, C.F., *J. Geophys. Res.*, 1995, **100**, 16605.
7. Rosman, K.J.R., Chisholm, W., Hong, S., Candelone, J.-P., and Boutron, C.F., *Environ. Sci. Technol.*, 1997, **31**, 3413.
8. Rosner, D., and Markowitz, G., *Am. J. Pub. Health*, 1985, **75**, 344.
9. Nriagu, J.O., *Sci. Total Environ.*, 1990, **92**, 13.
10. Nriagu, J.O., and Pacyna, J.M., *Nature*, 1988, **333**, 134.
11. Lobinski, R., Boutron, C.F., Candelone, J.-P., Hong, S., Szpunar-Lobinska, J., and Adams, F.C., *Environ. Sci. Technol.*, 1994, **28**, 1459.
12. Lobinski, R., Boutron, C.F., Candelone, J.-P., Hong, S., Szpunar-Lobinska, J., and Adams F.C., *Environ. Sci. Technol.*, 1994, **28**, 1467.
13. Grandjean, P., and Nielsen, T., *Res. Rev.*, 1979, **92**, 97.
14. Hewitt, C.N., and Harrison, R.M., in *Organometallic Compounds in the Environment*, P.J. Craig (Eds.), Longman, Harlow, 1986, p. 160.
15. Radjevic, M., and Harrison, R.M., *Sci. Total Environ.*, 1987, **59**, 157.
16. Van Cleuvenbergen, R.J.A., and Adams, F.C., in *The Handbook of Environmental Chemistry*, O. Hutzinger (Ed.), Springer-Verlag, Berlin-Heidelberg, 1990, pp. 97–153.
17. Biggins, P.D.E., and Harrison, R.M., *Environ. Sci. Technol.*, 1979, **13**, 558.
18. Feldhake, C.J., and Stevens, C.D., *J. Chem. Eng.*, 1963, **8**, 196.
19. *Gmelin Handbook of Inorganic Chemistry*, Springer-Verlag, Berlin, New York, 1987, pp. 116–119.
20. *Gmelin Handbook of Inorganic Chemistry*, Springer-Verlag, Berlin, New York, 1990, pp. 105–109.
21. *Gmelin Handbook of Inorganic Chemistry*, Springer-Verlag, Berlin, New York, 1995, pp. 39–40.
22. *Gmelin Handbook of Inorganic Chemistry*, Springer-Verlag, Berlin, New York, 1996, pp. 14–24.
23. De Jonghe, W.R.A., Chakraborti, D., and Adams, F.C., *Environ. Sci. Technol.*, 1981, **10**, 1217.

24. Lovei, M., Phasing out Lead from Gasoline, *World Bank Technical Paper No. 397.* World Bank, Washington DC, 1997.
25. *CONCAWE Report No. 5/95 Motor Vehicle Emission Regulations and Fuel Specification in Europe and the United States 1995 Update*, cited in Nishikawa, T., *Petrotech*, 1995, **20**, 463–469 (in Japanese).
26. Eisenreich, S.J., Metzer, N.A., and Urban, N.R., *Environ. Sci. Technol.*, 1986, **20**, 171.
27. Jensen, R.A., and Laxen, D.P.H., *Sci. Total Environ.*, 1987, **59**, 1.
28. Page, R.A., Cawse, P.A., and Baker, S.J., *Sci. Total Environ.*, 1988, **68**, 71.
29. Simmons, J.A.K., and Knap, A.H., *Atmos. Environ.*, 1993, **27A**, 1729.
30. Loranger, S., and Zayed, J., *Atmos. Environ.*, 1994, **28**, 1645.
31. Hayes, E.B., McElvaine, M.D., Orbach, H.G., Fernandez, A.M., Lyne, S., and Matte, T.D., *Pediatr.*, 1994, **93**, 195.
32. Wietlisbach, V., Rickenbach, M., Berode, M., and Guillemin, M. *Environ. Res.*, 1995, **68**, 82.
33. Bono, R., Pignata, C., Scursatone, E., Rovere, R., Natale, P., and Gilli, G., *Environ. Res.*, 1995, **70**, 30.
34. Thomas, V.M., Socolow, R.H., Fanelli, J.J., and Spiro, T.G., *Environ. Sci. Technol.*, 1999, **33**, 3942.
35. *Research Report on the Use of Liquefied Coal as Automotive Fuel*, Petroleum Energy Center, Tokyo, 1999, pp. 71–85 (in Japanese).
36. Thomas, V.M., *Annu. Rev. Energy Environ.*, 1995, **20**, 301.
37. Henderson, S.R., and Snyder, L.J., *Anal. Chem.*, 1961, **33**, 1172.
38. Chau, Y.K., Wong, P.T.S., and Saitoh, H., *J. Chromatogr. Sci.*, 1976, **14**, 162.
39. Messman, J.D., and Rains, T.C., *Anal. Chem.*, 1981, **53**, 1632.
40. Chau, Y.K., Wong, P.T.S., and Goulden, P.D., *Anal. Chim. Acta*, 1976, **85**, 421.
41. De Jonghe, W., Chakraborti, D., and Adams, F.C., *Anal. Chim. Acta*, 1980, **115**, 89.
42. Ebdon, L., Ward, R.W., and Leathard, D.A., *Analyst*, 1982, **107**, 129.
43. Ibrahim, M., Gilbert, T.W., and Caruso, J.A., *J. Chromatogr. Sci.*, 1984, **22**, 111.
44. Ibrahim, M., Nisamaneepong, W., Haas, D.L., and Caruso, J.A., *Spectrochim. Acta*, 1900, **40B**, 367.
45. Greenway, G.M., and Barnett, N.W., *J. Anal. At. Spectrom.*, 1989, **4**, 783.
46. Al-Rashdan, A., Heitkemper, D., and Caruso, J.A., *J. Chromatogr. Sci.*, 1991, **29**, 98.
47. Prange, A., and Jantzen, E., *J. Anal. At. Spectrom.*, 1995, **10**, 105.
48. De Smaele, T., Moens, L., Dams, R., and Sandra, P., *Fresenius J. Anal. Chem.*, 1996, **355**, 778.
49. Yamauchi, H., Arai, F., and Yamamura, Y., *Ind. Health*, 1981, **19**, 115.
50. Bettmer, J., and Cammann, K. *Appl. Organomet. Chem.*, 1994, **8**, 615.
51. Ebdon, L., Hill, S., and Ward, R.W., *Analyst*, 1986, **111**, 1113.
52. Ebdon, L., Hill, S., and Ward, R.W., *Analyst*, 1987, **112**, 1.
53. Lobinski, R., Dirkx, W.M.R., Szpunar-Lobinska, J., and Adams, F.C., *Anal. Chim. Acta*, 1994, **286**, 381.
54. Lobinski, R., and Adams, F.C., *Spectrochim. Acta*, 1997, **52B**, 1865.
55. Moss, R., and Browett, E.V., *Analyst*, 1966, **91**, 428.
56. Purdue, L.J., Enrione, R.E., Thompson, R.J., and Bonfield, B.A., *Anal. Chem.*, **45**, 527.
57. Hancock, S., and Slater, A., *Analyst*, 1975, **100**, 422.
58. Robinson, J.W., Rhodes, L., and Wolcott, D.K., *Anal. Chim. Acta*, 1975, **78**, 474.
59. Cantuti, V., and Cartoni, G.P., *J. Chromatogr.*, 1968, **32**, 641.
60. Hewitt, C.N., and Harrison, R.M., *Anal. Chim. Acta*, 1985, **167**, 277.
61. Nerín, C., and Pons, B., *Appl. Organomet. Chem.*, 1994, **8**, 607.

62. Radziuk, B., Thomassen, Y., and Van Loon, J.C., *Anal. Chim. Acta*, 1979, **105**, 255.
63. De Jonghe, W.R.A., Chakraborti, D., and Adams, F.C., *Anal. Chem.*, 1980, **52**, 1974.
64. Hewitt, C.N., and Metcalfe, P.J., *Sci. Total Environ.*, 1989, **84**, 211.
65. Pécheyran, C., Quetel, C.R., Lecuyer, F.M.M., and Donard, O.F.X., *Anal. Chem.*, 1998, **70**, 2639.
66. Hewitt, C.N., Harrison, R.M., and Radojevic, M., *Anal. Chim. Acta*, 1986, **188**, 229.
67. Chakraborti, D., Van Cleuvenbergen, R.J.A, and Adams, F.C., *Intern. J. Environ. Anal. Chem.*, 1987, **30**, 233.
68. Radojevic, M., and Harrison, R.M., *Environ. Technol. Lett.*, 1986, **7**, 525.
69. Quevauviller, Ph., Ebdon, L., Harrison, R.M., and Wang, Y., *Appl. Organomet. Chem.*, 1999, **13**, 1.
70. Johansson, M., Berglöf, T., Baxter, D.C., and Frech, W., *Analyst*, 1995, **120**, 755.
71. Chau, Y.K., Wong, P.T.S., and Kramar, O., *Anal. Chim. Acta*, 1983, **146**, 211.
72. Forsyth, D.S., and Marshall, W.D., *Anal. Chem.*, 1985, **57**, 1299.
73. Chakraborti, D., De Jonghe, W.R.A., Van Mol, W.E., Van Cleuvenbergen, R.J.A., and Adams, F.C., *Anal. Chem.*, 1984, **56**, 2692.
74. Lobinski, R., and Adams, F.C., *Anal. Chim. Acta*, 1992, **262**, 285.
75. Lobinski, R., Boutron, C.F., Candelone, J.-P., Hong, S., Szpunar-Lobinska, J., and Adams, F.C., *Anal. Chem.*, 1993, **65**, 2510.
76. Chau, Y.K., Wong, P.T.S., Bengert, G.A., and Kramar, O., *Anal. Chem.*, 1979, **51**, 186.
77. Amouroux, D., Tessier, E., Pécheyran, C., and Donard, O.F.X., *Anal. Chim. Acta*, 1998, **377**, 241.
78. Rapsomanikis, S., *Analyst*, 1994, **119**, 1429.
79. Rapsomanikis, S., Donard, O.F.X., and Weber, J.H., *Anal. Chem.*, 1986, **58**, 35.
80. Witte, C., Szpunar-Lobinska, J., Lobinski, R., and Adams, F.C., *Appl. Organomet. Chem.*, 1994, **8**, 621.
81. Ebdon, L., Hill, S., and Rivas, C., *Spectrochim. Acta*, 1998, **53B**, 289.
82. Quevauviller, Ph., Ebdon, L., Harrison, R.M., and Wang, Y., *Analyst*, 1998, **123**, 971.
83. Chau, Y.K., Wong, P.T.S., Bengert, G.A., and Dunn, J.L., *Anal. Chem.*, 1984, **56**, 271.
84. Forsyth, D.S., and Marshall, W.D., *Anal. Chem.*, 1983, **55**, 2132.
85. Forsyth, D.S., Marshall, W.D., and Collette, M.C., *Appl. Organomet. Chem.*, 1988, **2**, 233.
86. Wong, P.T.S., Chau, Y.K., Yaromich, J., Hodson, P., and Whittle, M., *Appl. Organomet. Chem.*, 1989, **3**, 59.
87. Van Cleuvenbergen, R., Chakraborti, D., and Adams, F.C., *Anal. Chim. Acta*, 1990, **228**, 77.
88. Sirota, G.R., and Uthe, J.F., *Anal. Chem.*, 1977, **49**, 823.
89. Chau, Y.K., Wong, P.T.S., Bengert, G.A., and Kramar, O., *Anal. Chem.*, 1979, **51**, 186.
90. Forsyth, D.S., and Iyengar, J.R., *Appl. Organomet. Chem.*, 1989, **3**, 211.
91. Nygren, O., and Nilsson, C.-A., *J. Anal. At. Spectrom.*, 1987, **2**, 805.
92. Hayakawa, K., *Jpn. J. Hyg.*, 1971, **26**, 377.
93. Pons, B., Carrera, A., and Nerín, C., *J. Chromatogr. B*, 1998, **716**, 139.
94. Reamer, D.C., Zoller, W.H., and O'Haver, T.C., *Anal. Chem.*, 1978, **50**, 1449.
95. Nielsen, T., Egsgaard, H., and Larsen, E., *Anal. Chim. Acta*, 1981, **124**, 1.
96. Harrison, R.M., and Radojevic, M., *Sci. Total Environ.*, 1985, **44**, 235.
97. Allen, A.G., Radojevic, M., and Harrison, R.M., *Environ. Sci. Technol.*, 1988, **22**, 517.

98. Pécheyran, C., Lalère, B., and Donard, O.F.X., *Environ. Sci. Technol.*, 2000, **34**, 27.
99. Hewitt, C.N., and Harrison, R.M., *Environ. Sci. Technol.*, 1987, **21**, 260.
100. Harrison, R.M., and Allen, A.G., *Appl. Organomet. Chem.*, 1989, **3**, 49.
101. Harrison, R.M., and Laxen, D.P.H., *Nature*, 1978, **275**, 738.
102. Hewitt, C.N., Harrison, R.M., and De Mora, S.J., *Atmos. Environ.*, 1984, **15**, 189.
103. Van Cleuvenbergen, R.J.A., Chakraborti, D., and Adams, F.C., *Environ. Sci. Technol.*, 1986, **20**, 589.
104. Radojevic, M., and Harrison, R.M., *Atmos. Environ.*, 1987, **21**, 2403.
105. Blaszkewicz, M., Baumhoer, G., and Neidhart, B., *Intern. J. Environ. Anal. Chem.*, 1987, **28**, 207.
106. Mikac, N., and Branica, M., *Atmos. Environ.*, 1994, **28**, 3171.
107. Turnbull, A.B., Wang, Y., and Harrison, R.M., *Appl. Organomet. Chem.*, 1993, **7**, 567.
108. Van Cleuvenbergen, R.J.A., and Adams, F.C., *Environ. Sci. Technol.*, 1992, **26**, 1354.
109. Faulstich, H., and Stournaras, C., *Nature*, 1985, **317**, 714.
110. Unsworth, M.H., and Harrison, R.M., *Nature*, 1985, **317**, 674.
111. Van Cleuvenbergen, R.J.A., Chakraborti, D., and Adams, F.C., *Anal. Chim. Acta*, 1986, **182**, 239.
112. Chau, Y.K., Wong, P.T.S., Kramar, O., Bengert, G.A., Cruz, R.B., Kinrade, J.O., Lye, J., and Van Loon, J.C., *Bull. Environ. Contam. Toxicol.*, 1980, **24**, 265.
113. Radojevic, M., and Harrison, R.M., *Environ. Technol. Lett.*, 1986, **7**, 519.
114. Chakraborti, D., Dirkx, W., Van Cleuvenbergen, R.J.A., and Adams, F.C., *Sci. Total Environ.*, 1989, **84**, 249.
115. Blais, J.S., and Marshall, W.D., *J. Environ. Qual.*, 1986, **15**, 255.
116. Cruz, R.B., Lorouso, C., George, S., Thomassen, Y., Kinrade, J.D., Butler, L.R.P., Lye, J., and Van Loon, J.C., *Spectrochim. Acta*, 1980, **35B**, 775.
117. Forsyth, D.S., Dabeka, R.W., and Cléroux, C., *Food Addit. Contam.* (1991) **8**, 477.
118. Forsyth, D.S., and Marshall, W.D., *Environ. Sci. Technol.*, (1986) **20**, 1033.
119. Nielsen, T., Jensen, K.A., and Grandjean, P., *Nature*, 1978, **274**, 602.
120. Zhang, W., Zhang, G.-G., He, H.-Z., and Bolt, H.M., *Int. Arch. Occup. Environ. Health*, 1994, **65**, 395.
121. Hirschler, D.A., and Gilbert, L.F., *Arch. Environ. Health*, 1964, **8**, 297.
122. Rhue, R.D., Mansell, R.S., Ou, L.-T., Cox, R., Tang, S.R., and Ouyang, Y., *Crit. Rev. Environ. Cont.*, 1992, **22**, 169.
123. Horowitz, A., and Atlas, R.M., *Appl. Environ. Microbiol.*, 1977, **33**, 1252.
124. Tiravanti, G., and Boari, G., *Environ. Sci. Technol.*, 1979, **13**, 849.
125. Harrison, R.M., and Laxen, D.P.H., *Environ. Sci. Technol.*, 1978, **12**, 1384.
126. Nielsen, O.J., O'Farrell, D.J., Treacy, J.J., and Sidebottom, H.W., *Environ. Sci. Technol.*, 1991, **25**, 1098.
127. Hewitt, C.N., and Harrison, R.M., *Environ. Sci. Technol.*, 1986, **20**, 797.
128. Jarvie, A.W.P., Markall, R.N., and Potter, H.R., *Environ. Res.*, 1981, **25**, 241.
129. Robinson, J.W., Kiesel, E.L., and Rhodes, I.A.L., *J. Environ. Sci. Health*, 1979, **A14**, 65.
130. Jarvie, A.W.P., Whitmore, A.P., Markall, R.N., and Potter, H.R., *Environ. Poll. Ser. B*, 1983, **6**, 69.
131. Van Cleuvenbergen, R.J.A., Dirkx, W., Quevauviller, Ph., and Adams, F.C., *Intern. J. Environ. Anal. Chem.*, 1992, **47**, 21.
132. DuPont, E.I., *Bull. Environ. Contam. Toxicol.*, 1994, **53**, 603.
133. Ou, L.-T., Thomas, J.E., and Jing, W., *Bull. Environ. Contam. Toxicol.*, 1994, **52**, 238.
134. Teeling, H., and Cypionka, H., *Appl. Microbiol. Biotechnol.*, 1997, **48**, 275.

135. Blais, J.S., Doige, C.A., Marshall, W.D., and Knowles, R., *Arch. Environ. Contam. Toxicol.*, 1990, **19**, 227.
136. Röderer, G., *Environ. Res.*, 1980, **23**, 371.
137. Wong, P.T.S., Chau, Y.K., Yaromich, J.L., and Kramar, O., *Can. J. Fish Aquat. Sci.*, 1987, **44**, 1257.
138. Macaskie, L.E., and Dean, A.C.R., *Appl. Microbiol. Biotechnol.*, 1990, **33**, 81.
139. Krishnan, K., and Marshall, W.D., *Environ. Sci. Technol.*, 1988, **22**, 1038.
140. Heard, M.J., Wells, A.C., Newton, D., and Chamberlain, A.C., in *Proc. Int. Conf. Heavy Metals in the Environment, London*, CEP Ltd, Edinburgh, 1979, pp. 103–108 (cited in Refs 14 and 16).
141. Cremer, J.E., *Br. J. Ind. Med.*, 1959, **16**, 191.
142. Bolanowska, W., and Wisniewska-Knypl, J.M., *Biochem. Pharmacol.*, 1971, **20**, 2108.
143. Bolanowska, W., *Br. J. Ind. Med.*, 1968, **25**, 203.
144. Arai, F., and Yamamura, Y., *Ind. Health*, 1990, **28**, 63.
145. Arai, F., Yamamura, Y., and Yoshida, M., *Jpn. J. Ind. Health*, 1981, **23**, 496 (in Japnese with English abstract).
146. Yamamura, Y., Arai, F., and Yamauchi, H., *Ind. Health*, 1981, **19**, 125.
147. Turlakiewicz, Z., and Chmielnicka, J., *Br. J. Ind. Med.*, 1985, **42**, 682.
148. Arai, F., Yamamura, Y., Yamauchi, H., and Yoshida, M., *Jpn. J. Ind. Health*, 1983, **25**, 175 (in Japanese with English abstract).
149. Gething, J., *Br. J. Ind. Med.*, 1975, **32**, 329.
150. Yamamura, Y., Arai, F., Yoshida, M., and Shimada, E., *St. Marianna Med J.*, 1979, **7**, 10 (in Japanese with English abstract).
151. Wong, P.T.S., Chau, Y.K., and Luxon, P.L., *Nature*, 1975, **253**, 263.
152. Schmidt, U., and Huber, F., *Nature*, 1976, **259**, 157.
153. Thompson, I.A.J., and Crerar, J.A., *Mar. Poll. Bull.*, 1980, **11**, 251.
154. Walton, A.P., Ebdon, L., and Millward, G.E., *Appl. Organomet. Chem.*, 1988, **2**, 87.
155. Reisinger, K., Stoeppler, M., and Nürnberg, H.W., *Nature*, 1981, **291**, 228.
156. Jarvie, A.W.P., Whitmore, A.P., Markall, R.N., and Potter, H.R., *Environ. Poll. Ser. B*, 1983, **6**, 81.
157. Jarvie, A.W.P., Markall, R.N., and Potter, H.R., *Nature*, 1975, **255**, 217.
158. Craig, P.J., *Environ. Technol. Lett.*, 1980, **1**, 17.
159. Ridley, W.P., Dizikes, L.J., and Wood, J.M., *Science*, 1977, **197**, 329.
160. Wood, J.M., *Naturwissenschaften*, 1975, **62**, 357.
161. Taylor, R.T., and Hanna, M.L., *J. Environ. Sci. Health*, 1976, **3**, 201.
162. Rapsomankis, S., Donard, O.F.X., and Weber, J.H., *Appl. Organomet. Chem.*, 1987, **1**, 115.
163. Craig, P.J., and Rapsomanikis, S., *Environ. Sci. Technol.*, 1985, **19**, 726.
164. Ahmad, I., Chau, Y.K., Wong, P.T.S., Carty, A.J., and Taylor, L., *Nature*, 1980, **287**, 716.
165. Jarvie, A.W.P., and Whitmore, A.P., *Environ. Technol. Lett.*, 1981, **2**, 197.
166. Snyder, L.J., and Bentz, J.M., *Nature*, 1982, **296**, 228.
167. Jarvie, A.W.P., *Sci. Total Environ.*, 1988, **73**, 121.
168. Craig, P.J., and Van Elteren, J.T., in *The Chemistry of Organic Germanium, Tin and Lead Compounds*, S. Patai (Ed.) Wiley, Chichester, 1995, pp. 843–855.
169. Candelone, J.-P., Jaffrezo, J.-L., Hong, S., Davidson, C.I., and Boutron, C.F., *Sci. Total Environ.*, 1996, **193**, 101.
170. Adams, F.C., Heisterkamp, M., Candelone, J.-P., Laturnus, F., van de Velde K., Boutron, C.F., *Analyst*, 1998, **123**, 767.
171. Lobinski, R., *Analyst*, 1995, **120**, 615.

172. Silverberg, B.A., Wong, P.T.S., and Chau, Y.K., *Arch. Environ. Contam. Toxicol.*, 1977, **5**, 305.

173. Marchetti, R., *Mar. Poll. Bull.*, 1978, **9**, 206.

174. Röderer, G., *Environ. Res.*, 1981, **25**, 361.

175. Röderer, G., *Environ. Exp. Bot.*, 1984, **24**, 17.

176. Röderer, G., *Environ. Res.*, 1986, **39**, 205.

177. Bull, K.R., Every, W.J., Freestone, P., Hall, J.R., and Osborn, D., *Environ. Poll. Ser. B*, 1983, **31**, 239.

178. Osborn, D., Every, W.J., and Bull, K.R., *Environ. Poll. Ser. B*, 1983, **31**, 261.

179. Schepers, G.W.H., *Arch. Environ. Health*, 1964, **8**, 89.

180. Schroeder, T., Avery, D.D., and Cross, H.A., *Experientia*, 1972, **28**, 425

181. Hayakawa, K., *Jpn. J. Hyg.*, 1972, **26**, 526.

182. Cremer, J.E., and Callaway, S., *Br. J. Ind. Med.*, 1961, **18**, 277.

183. Springman, F., Bingham, E., and Stemmer, K.L., *Arch. Environ. Health*, 1963, **6**, 469.

184. Gras, G., and Boucard, M., *C. R. Soc. Biol.*, 1968, **162**, 1456 (cited in [13]).

185. Fried, J.F., Rosenthal, N.W., and Schubert, J., *Proc. Soc. Exp. Biol. Med.*, 1956, **92**, 331.

186. Davis, R.K., Horton, A.W., Larson, E.E., and Stemmer, K.L., *Arch. Environ. Health*, 1963, **6**, 476.

187. Konat, G., *Neurotoxicol.*, 1984, **5**, 87.

188. Chang, L.W., *J. Toxicol. Sci.*, 1990, **15**, suppl. 4, 125.

189. Sanders, L.W., *Arch. Environ. Health*, 1964, **8**, 270.

190. Tenenbein, M., *Hum. Exp. Toxicol.*, 1997, **16**, 217.

191. Ahlberg, J., Ramel, C., and Wachtmeister, C.A., *Ambio*, 1972, **1**, 29.

192. Grandjean, P., and Andersen, O., *Lancet*, 1982, 333.

193. Robinson, T.R., *J. Occup. Med.*, 1974, **16**, 601.

194. Sweeney, M.H., Beaumont, J.J., Waxweiler, R.J., and Halperin, W.E., *Arch. Environ. Health*, 1986, **41**, 23.

195. Fayerweather, W.E., Karns, M.E., Nuwayhid, I.A., and Nelson, T.J., *Am. J. Ind. Med.*, 1997, **31**, 28.

196. Mitchell, C.S., Shear, M.S., Bolla, K.I., and Schwartz, B.S., *J. Occup. Environ. Med.*, 1996, **38**, 372.

197. Balbus, J.M., Stewart, W., Bolla, K.I., and Schwartz, B.S., *Am. J. Ind. Med.*, 1997, **32**, 544.

198. Schwartz, J. *Environ. Res.*, 1994, **66**, 105.

5 Organoarsenic Compounds in the Marine Environment

J. S. EDMONDS

National Institute for Environmental Studies, Tsukuba, Japan

and

K. A. FRANCESCONI

Institute of Chemistry, Analytical Chemistry, Karl-Franzens University, Graz, Austria

5.1 ARSENIC CONCENTRATIONS IN SEA WATER, MARINE ALGAE AND MARINE ANIMALS

Sea water contains arsenic at a concentration of 1 to 2 μg per litre, and marine plants and animals naturally contain arsenic at 1 to 100 mg per kg on a 'wet' or live weight basis [1]. Tables 5.1 and 5.2 show arsenic concentrations in marine animals and algae from a range of geographical locations. Whereas arsenic in sea water is almost entirely present in inorganic forms, both marine plants and animals contain arsenic almost entirely in organic forms [1]. Thus marine organisms bioaccumulate arsenic by a factor of from 1000 to 100 000 as compared with sea water, and transform the arsenic they accumulate by the formation of arsenic–carbon bonds. Whereas marine plants, specifically marine algae, must accumulate the arsenic that they contain directly from sea water, animals could both absorb arsenic from ambient water and also retain arsenic contained in food organisms.

Surface sea water that receives sufficient light to support photosynthetic organisms also contains small concentrations of methylarsonic acid and dimethylarsinic acid [2]. These are presumably there as a result of phytoplankton activity, but whether they result from the breakdown of larger molecules (see below) or represent the end-products of phytoplankton arsenical biosynthesis is unclear. Both are possible and certainly dimethylarsinic acid has been shown to be an algal product [3]. Seasonal fluctuations in their concentrations reinforce the notion that they are products of phytoplankton [4–6].

Organometallic Compounds in the Environment
Edited by P.J. Craig © 2003 John Wiley & Sons Ltd

Table 5.1 Arsenic concentrations in marine animals

Type of animal (No. of species)	Location	Arsenic concentration (mg kg^{-1}) (wet or dry weight basis as shown)			Reference
		Range	Mean		
Teleost fish (three spp.)	N.E. Atlantic	2.5–5.4	3.5	Dry	74
Teleost fish (several spp.)	UK (inshore waters)	1.7–8.7	5	Dry	75
Teleost fish (five spp.)	Greenland (West)	7.6–307	60	Dry	76
Teleost fish (four spp.)	Canada	0.2–11.7	—	Wet	77
Teleost fish (nine spp.)	Goa	0.3–12.6	—	Wet	78
Teleost fish (nine spp.)	Australia (S.E.)	0.1–4.4	1.0	Wet	79
Teleost fish (eight spp.)	Norway	0.6–7.8	2.7	Wet	80
Teleost fish (many spp.)	USA (All areas)	Range of means 1–7		Wet	81
Teleost fish (plaice, – *Pleuronectes platessa*)	Netherlands (North Sea)	3–166	23	Wet	82
Teleost fish (fifteen spp.)	Scotland	0.2–89.9	—	Wet	83
Teleost fish (four spp.)	Australia	0.82–13.8	6.5	Dry	84
Elasmobranch fish (*Carcharias sorrakowah*)	Goa	6.3–10.8	8.5	Wet	78
Elasmobranch fish (several spp.)	USA (all areas)	Range of means 7–20		Wet	81
Elasmobranch fish (two spp.)	Australia (S.E.)	5–30	15	Wet	85
Elasmobranch fish (three spp.)	Papua New Guinea	0.2–7.5	2.5	Wet	86
Crustaceans (prawn, *Pandalus borealis*)	Greenland, west	53–80	68	Dry	76
Crustaceans (three spp.)	Goa	9.3–25.2	14	Dry	78
Crustaceans (sixteen spp.)	USA (all areas)	Range of means 3–50		Wet	81
Crustaceans (krill, whole animals)	Antarctic, Scotia Sea	1.2–3.5	2.0	Wet	87

Crustaceans (three spp.)	Scotland	3.3–38.2	11.5	Wet	83
Crustaceans (five spp.)	Australia	7.1–91	26.7	Dry	84
Bivalve molluscs (five spp.)	UK (inshore waters)	2.6–15	7.8	Dry	75
Bivalve molluscs (mussels, *Mytilus edulis*)	Greenland, West	14–17	15	Dry	76
Bivalve molluscs (four spp.)	Goa	2.3–10.9	5	Dry	78
Bivalve molluscs (twelve spp.)	USA (all areas)	Range of means 2–20		Wet	81
Bivalve molluscs (mussels, *M. edulis*)	Adriatic Sea	10–15.3	13	Dry	88
Bivalve molluscs (five spp.)	Japan	1.0–10.4	3.6	Wet	89
Bivalve molluscs (two spp., scallops)	Canada	2.3–9.4	4.2	Dry	90
Gastropod molluscs (three spp.)	Thailand	112–339	170	Dry	11
Gastropod molluscs (two spp.)	Australia	74–233	154	Dry	91
Jellyfish (two spp.)	Japan	0.039–0.135	0.067	Wet	92
Mammals (whales, three spp.)	Alaska	0.17–1.27	0.57	Wet	93
Mammals (seals, two spp.)	Alaska	0.48–2.4	1.44	Wet	93

Table 5.2 Concentrations of arsenic in some marine algae

Type of alga: Phaeophyta, Rhodophyta or Chlorophyta (No. of species)	Location	Arsenic concentration (mg kg^{-1}, dry weight basis)		Reference
		Range	Mean	
Algae: Phaeophyta (eight spp.)	India	8–68	30	94
Rhodophyta (five spp.)		0–5	1.5	
Chlorophyta (five spp.)		0.1–6.3	2.2	
Algae: Phaeophyta (seven spp.)	Norway	15–109	44	95
Rhodophyta (two spp.)		10–13	12	
Algae: Phaeophyta (three spp.)	UK (inshore waters)	26–47	39	75
Rhodophyta (two spp.)		11–39	25	
Algae: Phaeophyta (fifteen spp.)	Japan	<1–230	46	96
Rhodophyta (eight spp.)		<1–12	4.4	
Chlorophyta (three spp.)		<1–8	3.8	
Algae: Phaeophyta (twenty-four spp.)	USA	1.1–31.6	10.3	97
Rhodophyta (fifteen spp.)		0.4–3.2	1.4	
Chlorophyta (sixteen spp.)		0.2–23.3	1.5	
Algae: Phaeophyta (fourteen spp)	British Columbia	40.8–92.4	57	98
Algae: Phaeophyta (fourteen spp)	Australia	21.3–179	62	99
Rhodophyta (ten spp)		12.5–31.3	19.2	
Chlorophyta (nine spp)		6.3–16.3	10.7	
Algae: Phaeophyta (seven spp)	Goa	4.5–20.9	8.1	100
Algae: Phaeophyta (two spp)	Denmark	21.3–43	31.9	69,101
Algae: Phaeophyta (two spp)	Canada	18.4–39	28.7	102

5.2 CHEMICAL FORMS OF ARSENIC IN MARINE ALGAE AND MARINE ANIMALS

5.2.1 INTRODUCTION

Arsenic is chiefly present in marine algae as relatively complex carbohydrate derivatives (arsenosugars, dimethylarsinoylribosides and trimethylarsonioribosides) [1, 7, 8] (Figure 5.1, structures **1** to **15**) and in marine animals as comparatively simple forms (Figure 5.2) The most abundant and commonly encountered arsenic compound in the majority of the classes of marine animals that have been examined is arsenobetaine (Figure 5.2; Table 5.3) [8]. Although arsenobetaine has never been found in marine algae, arsenic-containing carbohydrates, which were at first thought to be specific to marine algae or to certain organs of molluscs that contain symbiotic algae [1, 7], have now been identified in marine animals [1, 9–11]. This may reflect the direction of the food-chain; marine plants do not eat animals whereas some marine animals do eat plants or exist with them in a symbiotic relationship that supplies plant products to the

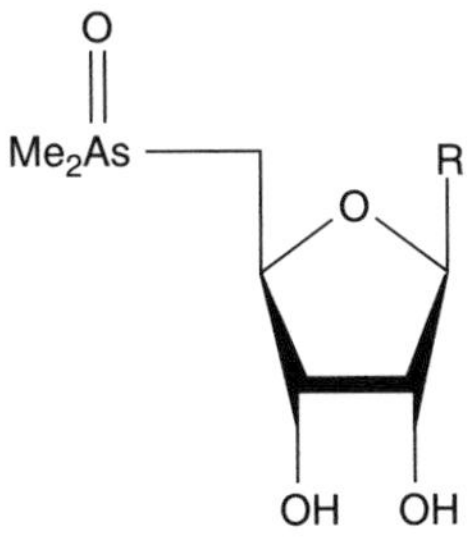

Dimethylarsinoylribosides (**1** to **11**)

(**1**) R =

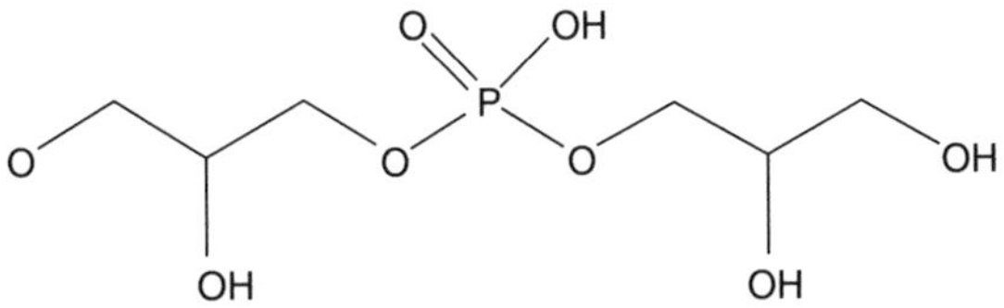

(**2**) R =

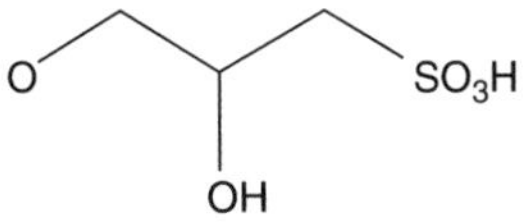

(**3**) R =

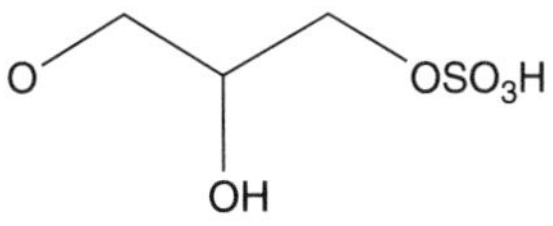

(**4**) R =

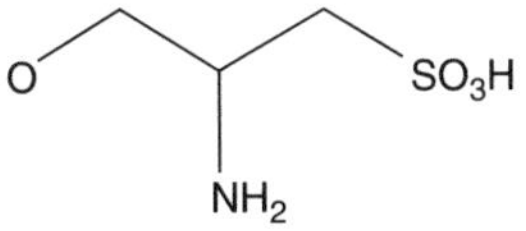

(**5**) R =

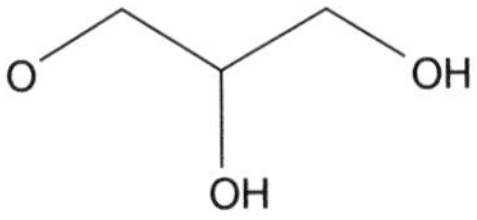

(**6**) R =

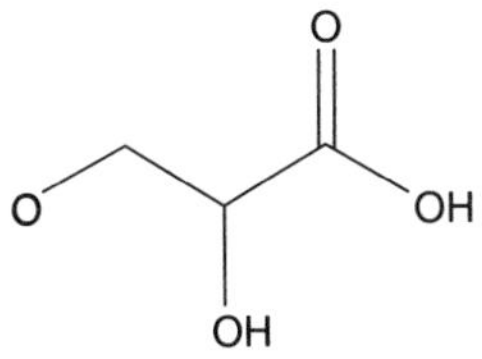

(7) R =

(8) R = OMe

(9) R =

(10) R =

(11) R =

Trimethylarsonioribosides (**12**, **13**)

Figure 5.1 Organic arsenic compounds from marine algae and from the kidney of the giant clam *Tridacna maxima*. Although the β-ribosyl group is shown as the D form, consistent with a biosynthetic pathway involving *S*-adenosylmethionine and with X-ray data on compounds **3** and **4**, the stereochemistry of the aglycones is not displayed. Those interested in the stereochemistry of the aglycones are referred to References [1] and [7]. Stereochemical aspects of compounds **14** and **15** are shown because it is possibly relevant to their biosynthesis (see Reference 38). The stereochemistry of **17** is also shown because it is a consequence of the opening of the D ribose ring. Some of the dimethylarsinoylribosides and trimethylarsonioribosides have also been found in marine animals

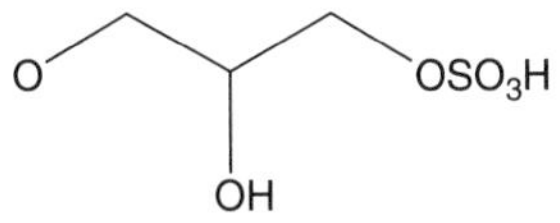

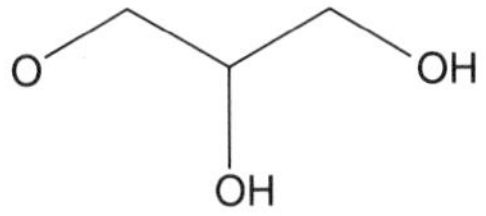

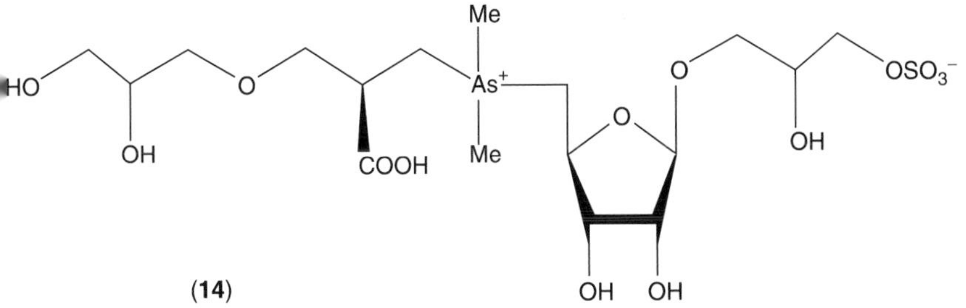

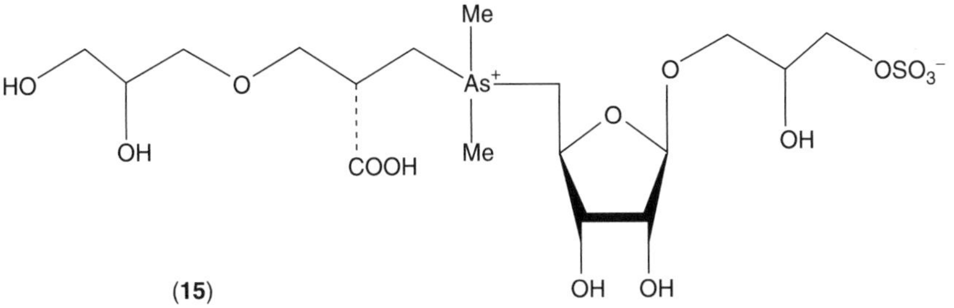

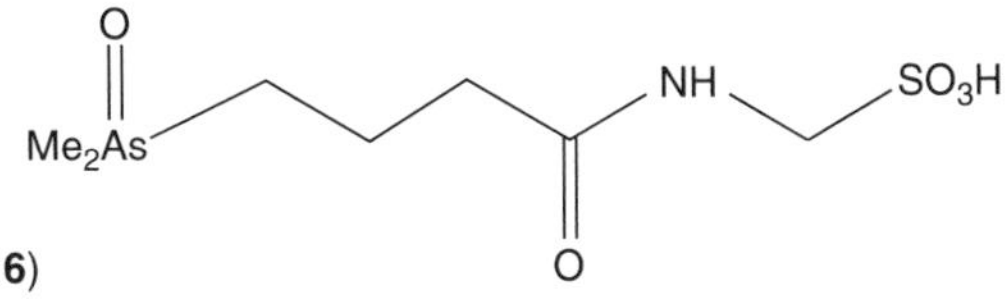

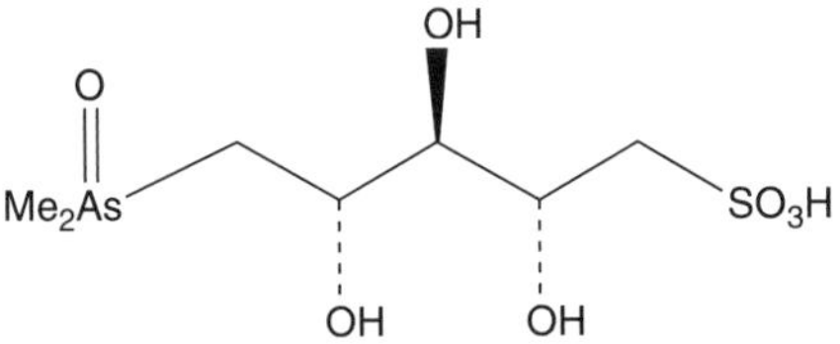

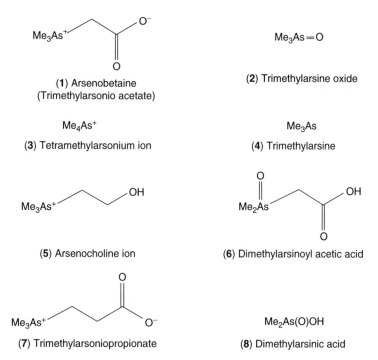

Figure 5.2   Organic arsenic compounds and ions from marine animals

Table 5.3   Arsenobetaine in marine animals

Animal (No. of species)	Arsenic concentration (mg kg^{-1}, wet weight)	Arsenic present as arsenobetaine (%)
FISH		
Elasmobranchs (seven spp)	3.1–44.3	94–>95
Teleosts (seventeen spp)	0.1–166	48–>95
CRUSTACEANS		
Lobsters (four spp)	4.7–26	77–>95
Prawns/shrimps (five spp)	5.5–20.8	55–>95
Crabs (six spp)	3.5–8.6	79–>95
MOLLUSCS		
Bivalves (four spp)	0.7–2.8	44–88
Bivalves (seven spp)	1.0–2.3	12–50
Gastropods (six spp)	3.1–116.5	58–>95
Cephalopods (three spp)	49	72–>95
ECHINODERMS (one sp)	12.4	60
COELENTERATES (one sp)	7.5	15
PORIFERA (two spp)	3.2–6.8	13–15

Data from [8].

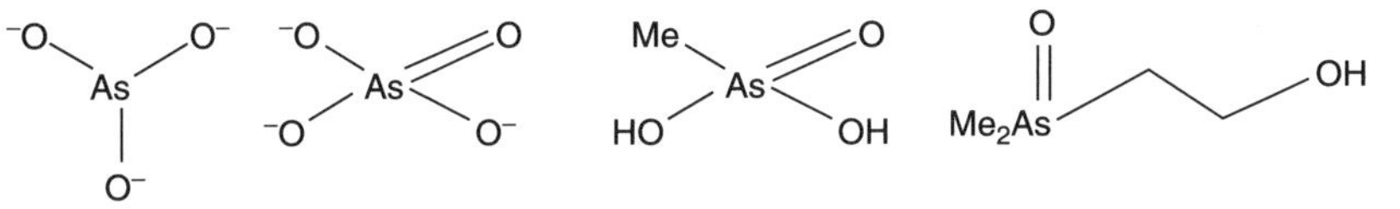

(1) Arsenite   **(2)** Arsenate   **(3)** Methylarsonic acid   **(4)** Dimethylarsinoylethanol

Figure 5.3   Additional arsenic compounds and ions referred to in the text

animal. There is also the possibility that animals are, in some cases, making arsenosugars *de novo* (see below). For other arsenic compounds referred to in the text, see Figure 5.3.

5.2.2   BIOSYNTHESIS OF ARSENOSUGARS

It is not entirely clear why marine algae absorb arsenate from sea water and convert it into arsenosugars. Possibly arsenate is absorbed because it is so similar to the essential phosphate that it cannot be excluded from the phosphate uptake mechanism [12]. The absorbed arsenate must then be detoxified before it can cause cellular damage by uncoupling oxidative phosphorylation [13]. If this is the case, and if algal arsenosugars are detoxification products, the evolution of their biosynthetic pathway has presumably replaced the development of a more specific phosphate uptake mechanism. Whatever the reason for the formation of these compounds in algae (and we shall return to the topic again), it is probable that they are biosynthesized by a variation of the pathway proposed by Challenger [4, 15] for the methylation of arsenic by microorganisms (Figure 5.4) (see also Chapter 1, Section 1.13). This process involves the sequential reduction and oxidative methylation of arsenic to produce first methylarsonic acid, second dimethylarsinic acid, and third trimethylarsine oxide. Reduction of trimethylarsine oxide yields the relatively volatile trimethylarsine which was the end product of the microbial pathway investigated by Challenger.

Challenger considered [16] that the methylating agent was *S*-adenosylmethionine (AdoMet), which had been shown by Cantoni [17, 18] to be an active methyl donor in some enzymatic systems (see Chapter 1, Figure 1.5). AdoMet is now known to be the main biological donor of methyl groups. It was postulated [19] that algae, instead of attaching a third methyl group, transferred the adenosyl group of AdoMet to the arsenic atom. Enzymatic removal of adenine and glycosidation of suitable algal metabolites (e.g. glycerol, cysteinoleic acid, etc.) would, it was suggested [20], produce the range of compounds found in algae and in the kidney of the giant clam (where their presence was likely to result from the presence of symbiotic unicellular algae in the mantle) (Figure 5.5). This pathway, first suggested in 1987, received support when the proposed intermediate dimethylarsinoyladenosine

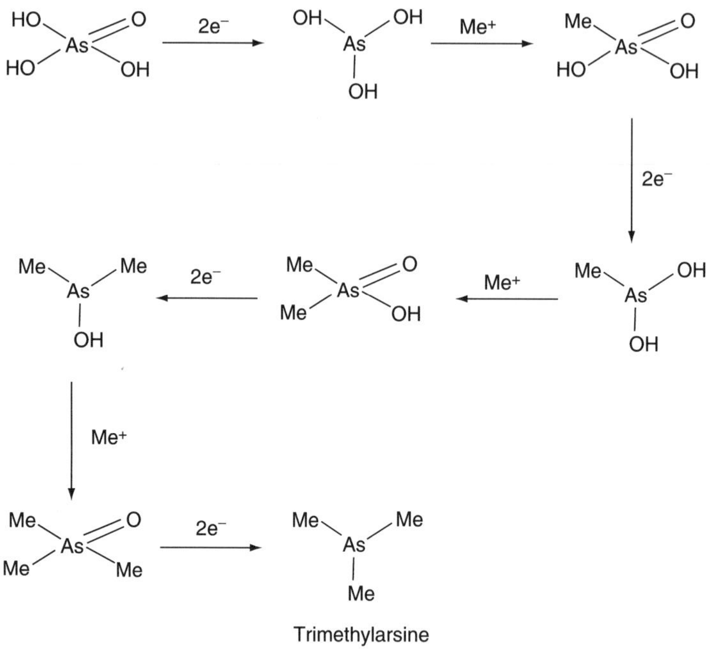

Figure 5.4 The 'Challenger' pathway for the methylation of arsenic

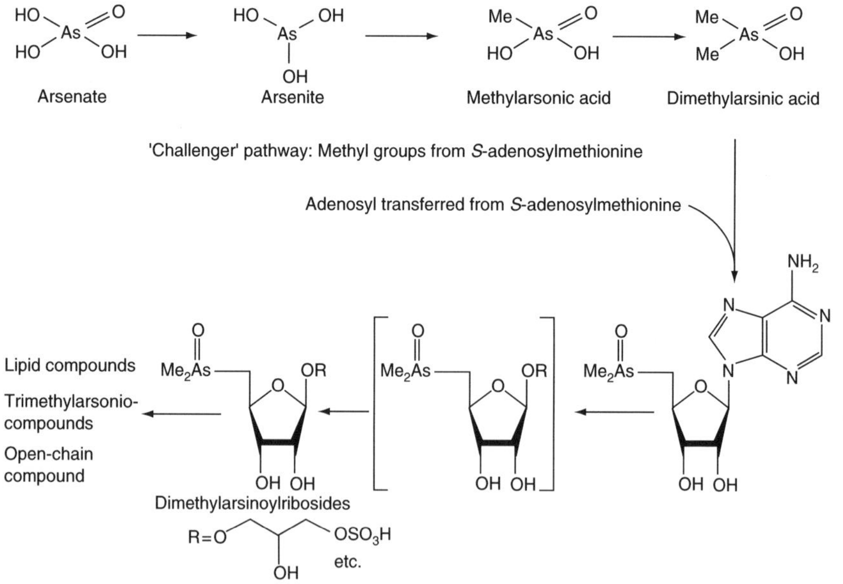

Figure 5.5 Proposed pathway for the methylation and adenosylation of arsenic to yield dimethylarsinoylriboses

Figure 5.1, **11**) was later discovered [21] as a natural product in the giant clam kidney. The acyclic compound (Figure 5.1, **17**), isolated from *Chondria crassicaulis* was an apparent exception to the algal dimethylarsinoylribosides, but it was shown that its stereochemistry followed that of ribitol (adonitol) and hence it was probably formed by opening of the ribose ring followed by reduction and sulfate ester formation at what was the anomeric centre [22].

Of more possible biochemical interest is the phosphate diester (Figure 5.1, **1**). This compound, which has been found in all organisms that contain arsenosugars with the exception of clam kidney, where its absence may reflect a need of the clam to conserve phosphorus. It has also been found [23] in terrestrial animals (earthworms, see Chapter 6), has been isolated, acylated with palmitic acid (**10**, Figure 5.1) from the brown alga *Undaria pinnatifida* [24], and its presence demonstrated [25] in the mussel, *Mytilis edulis*, and the digestive gland of the western rock lobster, *Panulirus cygnus* [9]. In the latter case the compound was also present in lipid-soluble acylated forms. The observation that the phosphate diester compound was the *only* arsenosugar detected in the rock lobster digestive gland and in mussels, despite other arsenosugars predominating in probable diet algae, suggests that the compound might well have been, at least partly, synthesized *de novo* by the animal in question. This is particularly so because, of all the arsenosugars, the phosphate diester is both the most obvious analogue of central metabolites (directly analogous to phosphatidylcholine with the dimethylarsinoylribosyl moiety serving as the polar head portion instead of choline) and possibly the most susceptible to enzymatic breakdown. All this suggests that the compound might have some cellular role or, at least, be central to detoxification processes. Such processes might bind arsenic in membranes where it can do no cellular damage, or facilitate its passage across membranes and thereby out of cells.

5.2.3 BIOSYNTHESIS OF ARSENOBETAINE: THE CONVERSION OF ARSENOSUGARS

The biosynthesis of arsenobetaine, the main arsenic compound in marine animals, is not so easily explained. Although the attachment of methyl groups bound to arsenic presents no biosynthetic problems, there is no ready way to explain the origin of the carboxymethyl group. It is reasonable, because of food-chain considerations, to seek a biogenetic pathway in which algal arsenosugars are transformed into the arsenobetaine found in animals. There is, though, no obvious structural connection between arsenosugars and arsenobetaine and it is possible that they are generated independently.

However, the first experimental attempts to link the compounds in a biogenetic scheme produced a surprising result. Arsenosugars, either as pure compounds or naturally contained in algal fragments, when subjected to the anaerobic microbial activity found in sediments, were converted rapidly and virtually quantitatively into dimethylarsinoylethanol (**4**, Figure 5.3) which could be easily envisaged as a precursor of arsenobetaine [26]. Importantly, the ribose residue attached to arsenic had been degraded to, or possibly replaced by, a hydroxyethyl group that could be converted to the required carboxymethyl group by a facile oxidation. Reduction and additional methylation at arsenic was also necessary to produce arsenobetaine (Figure 5.6). Although these last reactions also appeared to offer no biosynthetic impediment between dimethylarsinoylethanol and arsenobetaine, no marine animal supplied with the former, through the diet or otherwise, has converted it into the latter in measurable quantity [27]. Furthermore, arsenobetaine has been shown to occur in concentrations of 20 to 30 mg As kg^{-1} (wet weight) in the pelagic and *non-feeding* puerulus larvae of the western rock lobster, accounting for more than 99% of the arsenic present [28]. Concentrations of 20 to 30 mg kg^{-1} (wet weight) are normal for adult western rock lobsters also; indeed, although there is some variation (a total range of about 10 to 60 mg kg^{-1} across the size/age range) there is a remarkable consistency in arsenic (arsenobetaine) concentrations from larvae to ageing adults [28]. This consistency also applies to other organisms; for example, the mussel, *Mytilus edulis*, appears to contain 1–2 mg kg^{-1} of arsenic from whatever geographic location they are collected.

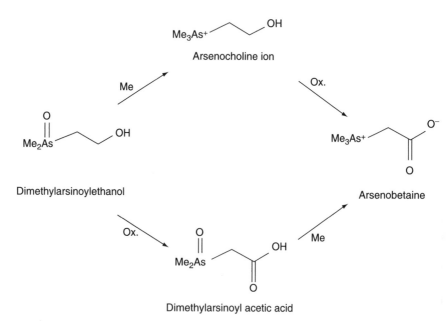

Figure 5.6 The transformations required for the conversion of dimethylarsinoylethanol to arsenobetaine

An extension of the work on the anaerobic microbial degradation of dimethylarsinoyl ribosides to produce dimethyarsinoyl ethanol, showed that similar anaerobic degradation of trimethylarsonioribosides gave the expected arsenocholine in high yield [29]. This, of course, only required oxidation to produce arsenobetaine and thus a more direct route for its biogenesis was possible. However, although trimethylarsonioribosides do occur naturally, the concentrations discovered to date have been very low and the contribution to the overall arsenic burden in the organisms that contain them has been similarly low. It is doubtful if there are enough trimethylarsonioribosides available at one end of the food chain to account for all the arsenobetaine at the other.

There is reason then to suppose that not only is the biogenetic pathway involving an anaerobic microbial stage unnecessary for the production of arsenobetaine but also that animals might be making it themselves, presumably from arsenate ingested from sea water. If the latter is the case, then it must be a strictly controlled process because subjecting rock lobsters [25] or mussels [30] to elevated arsenate in their ambient water does not result in any measurable increase in arsenobetaine concentration in animal tissues. In addition, providing catfish or whiting with food dosed with arsenate did not increase the concentration of arsenobetaine in the fish tissues although there was a detectable increase in trimethylarsine oxide concentrations [31]. The accepted view is that arsenic (arsenobetaine) is passed along food chains and retained, but the actual point of production of arsenobetaine, apart from the suggested anaerobic stage in sediments for the significant transformation of arsenosugar into a putative arsenobetaine precursor, has remained vague. There is, though, at least one piece of evidence that supports the biogenesis of arsenobetaine from arsenosugars [32], namely that when trimethylarsonioribosides were fed to shrimps, *Crangon crangon*, there was a measurable rise in arsenobetaine concentration in shrimp tissues. About 4% only of arsenic fed to the shrimps in this form was retained as arsenobetaine. When dimethylarsinoylribosides were fed to the shrimps there was no measurable increase in arsenobetaine concentration. This experiment might be thought to have demonstrated the breakdown of trimethylarsonioribosides to arsenocholine in the anaerobic environment of the shrimp gut with subsequent oxidation of the arsenocholine to arsenobetaine.

If marine animals are indeed absorbing arsenate, especially in a controlled way, the question arises as to why? Could this be another entanglement with phosphate metabolism? Is there a more specific reason? Is arsenobetaine the non-toxic fully methylated, fully oxidized end product of a metabolic pathway that starts with arsenate and involves as-yet-unidentified and relatively transient intermediates of which the phosphate diester (1 in Figure 5.1) may be one of the more enduring?

It is of course possible that transformation of arsenosugars to arsenobetaine, whether or not via dimethylarsinoylethanol, is a normal part of organismic metabolism and is not restricted to anaerobic microbial activity. If this is the case a situation could be envisaged whereby marine algae absorbed arsenate from sea water and converted it to arsenosugars, and marine animals absorbed arsenate also and converted it to arsenobetaine, possibly also via arsenosugars. The kinetics of such pathways would need to ensure that arsenosugars were rarely accumulated in animals in measurable quantity, but that arsenobetaine, once produced, was retained. Arsenobetaine passing from animal to animal along food chains would also be largely retained in the predator organism. Radiolabelled methylarsonic acid and dimethylarsinic acid were converted to arsenobetaine and accumulated in mussels in simple aquarium experiments [33, 34] but the transformation could have involved a number of agencies (micro-algae, bacteria, etc.) as well as the mussels themselves, and any intermediate compounds were not revealed. Presumably they could have included arsenosugars.

The possibility that arsenosugars are converted to arsenobetaine within single organisms received support recently from an LC-MS study [35] which identified several, obviously related, arsenic metabolites in kidney of the clam, *Tridacna derasa*. It appeared that arsenosugars contained therein were degraded by successive oxidation and decarboxylation, first of the aglycone and then of the opened ribose ring, to yield dimethylarsinoylacetic acid (**6** Figure 5.2) which only requires methylation to be converted to arsenobetaine. Thus, arsenosugars, several successive degradation products of arsenosugars (including dimethylarsinoylacetic acid) and arsenobetaine were all present in the same organism.

Much of the above discussion linking algal and animal arsenic compounds is speculation. Other observations that appear to deny the possibility of the involvement of algal compounds in the biogenesis of arsenobetaine also need to be accommodated. Arsenobetaine (and, indeed, arsenosugars) has been identified [36] in deep-sea hydrothermal vent animals (i.e. in organisms that do not have a plant stage in their food chain); and in some terrestrial organisms, namely mushrooms [37] and earthworms [23] (see Chapter 6). Although in both these instances the intermediacy of arsenosugars cannot be ruled out (the phosphate diester sugar coexisted with arsenobetaine in the earthworm as noted above), the role of marine algae in providing those sugars can certainly be excluded. It is necessary therefore to consider biogenetic schemes for arseno-betaine that do not involve algal compounds. One recently proposed [38] biogenetic scheme is considered below.

5.2.4   AN ALTERNATIVE SCHEME FOR THE FORMATION OF ARSENOBETAINE

The likely involvement of *S*-adenosylmethionine (AdoMet) in both the methyl-ation and adenosylation of arsenic, and the possibility that the carboxymethyl group in arsenobetaine is ultimately derived from an adenosyl group, suggest that all arsenic–carbon bonds in arsenic-containing marine natural products might be formed through the agency of the one compound (AdoMet). However two very minor arsenical constituents of the brown alga, *Sargassum lacerifolium* [39], (**14** and **15**, Figure 5.1) and one from the kidney of the giant clam *Tridacna maxima* [40] (**16**, Figure 5.1) contain subsituents on arsenic that seem unrelated to any that could be provided by AdoMet. It has been argued [38] that these groups could be transferred to arsenic by processes that, in some ways, parallel those involved in the biosynthesis of amino acids. Thus, whereas a core reaction in the biosynthesis of amino acids is the amination of glutarate to yield glutamate, a parallel process of 'arsenylation' with, it has been suggested, a reduced dimethylated arsenic species (dimethylarsinous acid, an intermediate in the Challenger pathway) would lead, after decarboxylation, to the arsenic-containing moiety in compound **16** (Figure 5.1). Conjugation with taurine would then yield compound (**16**).

Similarly, the four-carbon arsenic-containing portion of **14** and **15** (Figure 5.1) could be derived from from 'arsenomethionine' with, again, dimethylarsinous acid as the 'arsenylating' species. In general, then, there does appear to be the possibility that some naturally occurring arsenic compounds are derived from 'arsenylation' of 2-oxo acids. Figure 5.7 shows a possible, although speculative, mechanism. If arsenobetaine has its origin in such a route, a similar 'arsenylation' of either glyoxylate or pyruvate might be involved. If the former, further reduction and methylation at arsenic would be required to produce arsenobetaine; and if the latter, loss of the methyl group from the 2-position would also be necessary; or, possibly, transfer of this methyl to the arsenic atom.

If arsenic–carbon bonds are indeed formed in natural systems by the arsenylation of 2-oxo acids we might expect that other compounds will be discovered that have their origins in such a route. These compounds might involve other commonly occurring 2-oxo acids. Interestingly, just such a compound (**7**, Figure 5.2) was recently identified [41] in a species of fish, *Abudefduf vaigiensis*, from Thailand. It is simple to envisage this novel compound (directly analogous to arsenobetaine with a carboxyethyl group replacing the carboxymethyl group) as resulting from the arsenylation (as discussed above) of oxaloacetate, perhaps by the mechanism outlined in Figure 5.7. Again, only reduction and methylation at arsenic would be required to convert the arsenylated oxaloacetate into compound **7**.

Compound **7** was identified by electrospray mass spectrometry and such MS techniques represent the most important recent advance in analytical methods for the rapid identification of organic arsenic compounds in biological and environmental matrices. In many instances they can, much more rapidly, provide similar structural information to that provided by spectroscopic methods, e.g. NMR, after the full isolation of the compounds. Recent advances in the application of these methods to the identification of naturally occurring marine organoarsenic compounds are discussed in detail below.

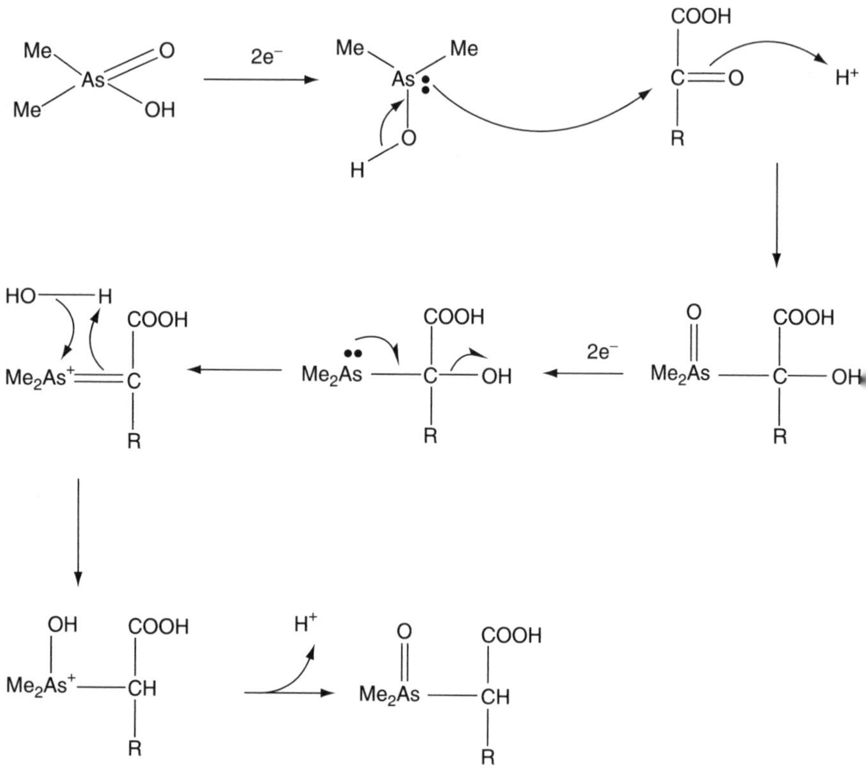

Figure 5.7 Possible mechanism for 'arsenylation' of 2-oxo acids

Most work on the metabolism of arsenic by marine organisms has relied on isolation and identification, usually by NMR spectroscopy but single crystal X-ray analyses have also provided useful data, for pure, relatively stable metabolites. Looking at metabolites in their 'native' state in biological fluids and other media has been, through lack of suitable techniques, a relatively neglected area. In a recent publication [42] EXAFS was used to identify an unusual and unexpected arsenic–selenium species in rabbit bile after the animal was administered salts of these elements and it was suggested that the technique may have some value in certain cases.

5.3   OTHER ORGANOARSENIC COMPOUNDS FROM MARINE ANIMALS

A number of other simple methylated arsenic compounds have been isolated from, or their presence demonstrated in, marine animals (Figure 5.2). Some, trimethylarsine oxide [31, 43, 44] tetramethylarsonium ion [10, 45–49] and dimethylarsinic acid [25] for example, can be envisaged as being derived from the Challenger pathway, but some (again trimethylarsine oxide and dimethylarsinic acid are examples) could also be produced by the degradation of arsenobetaine. Their relationship to arsenobetaine and its biosynthesis remains uncertain. Some, such as arsenocholine [44] and dimethylarsinoyl acetic acid [44], are probably related to arsenobetaine either biosynthetically or through its degradation. Trimethylarsine has also been identified by headspace gas chromatography at very low concentrations (μg kg^{-1}) in some species of prawns and in two species of lobster [50]. Presumably this could have arisen through reduction of trimethylarsine oxide which could in turn have been formed as discussed above.

The status of arsenocholine as a natural product in marine animals requires some discussion. It has been implied on several occasions that the reported occurrence of substantial amounts of arsenocholine in shrimps (probably *Pandalus borealis*) and other species is probably mistaken [1]. Nevertheless, arsenocholine has been reliably reported to occur in some marine animals [44] at low concentrations (less than 1 % of total arsenic) and it has been discovered bound into phospholipids, i.e. with arsenocholine replacing choline, in low concentrations in the digestive gland of the western rock lobster, *Panulirus cygnus* [9]. It is noteworthy in regard to these observations that, when synthetic arsenocholine was administered to fish in an experimental aquarium, most was speedily oxidized in the fish tissues to arsenobetaine, but some 4% was bound into glycerophosphoryl and phosphatidyl compounds. No detectable free arsenocholine remained [51]. The highest concentration of free arsenocholine that has been reliably reported [52] was in the leatherback turtle, *Dermochelys coriacea*, where it contributed up to 15% of the total arsenic in some tissues.

This review is concerned with organoarsenic compounds and much of the work that led to the identification of the compounds discussed here was undertaken to reassure consumers of seafoods that the natural presence of arsenic compounds in their diet did not pose a toxic hazard. In this regard it is worth noting that most seafoods also contain very small amounts of inorganic arsenic [53]. Recent reports [54] on the ability of inorganic arsenic to disrupt some endocrine functions at very low (10 ppb) concentrations might be taken to suggest that even the low levels of inorganic arsenic in seafoods might give cause for concern in certain cases.

As just noted, most work on the identification of arsenic compounds in marine animals has, naturally enough, concentrated on animals that contribute to the human diet. However, there have been other studies and one of these, on arsenic in the marine worm, *Tharyx marioni*, has produced results [55] so strange and potentially illuminating that they are worth relating in some detail, even though no specific arsenic compound contained therein has yet been identified. *T. marioni* accumulates the highest concentrations of arsenic ever recorded in an animal (more than 2000 mg kg^{-1} dry weight). This value is remarkably consistent from individual to individual and, most importantly, is irrespective of the external (ambient) concentration of arsenic. It is difficult to imagine that this consistently high concentration of arsenic, accumulated whether the environment is polluted with arsenic or not, has no cellular or metabolic function. The arsenic is mostly concentrated in the palps (which contain 6000 to 13000 mg kg^{-1} of arsenic on a dry weight basis) and, intriguingly, is lipid soluble rather than water soluble. If the arsenic indeed has some function it is very specific to *Tharyx marioni*, at least in elevated concentrations, because the authors point out that other marine and estuarine worms, including another species of *Tharyx*, do not accumulate arsenic to anything like the same degree. Advance on this front presumably awaits the identification of the arsenic compound(s) in *Tharyx marioni*.

5.4 THE MICROBIAL DEGRADATION OF MARINE ORGANOARSENIC COMPOUNDS

The anaerobic microbial degradation of dimethylarsinoylribosides and tri-methylarsonioribosides to yield dimethylarsinoylethanol and arsenocholine respectively and the possible important roles of these processes have been discussed above. Preliminary observations on the aerobic degradation of dimethylarsinoylribosides, e.g. in sea water, indicated that the breakdown product was dimethylarsinic acid [25]. There has been no work yet carried out to identify the microorganisms involved in these transformations.

The microbial degradation of arsenobetaine has been studied in more detail. The breakdown of arsenobetaine by unspecified microorganisms contained in a number of different media including sea water has been described [56]. The result was essentially the same in each case, namely, the production of trimethy-larsine oxide and dimethylarsinic acid. In some cases further breakdown resulted in the production of inorganic arsenic. In unenriched sea water the breakdown of arsenobetaine was slow. Essentially the same results were obtained in another study [57] that monitored the course of the reaction by ^{1}H NMR spectroscopy, namely, slow breakdown of arsenobetaine to trimethylar-sine oxide in unenriched sea water but a very rapid breakdown to the same end-product in sea water that had contained living mussels, *Mytilis edulis*, for a few hours. That microbial activity in sea water enriched in this way would bring about the speedy transformation of arsenobetaine to trimethylarsine oxide was also shown in experiments on the accumulation of a range of arsenic compounds from sea water by *Mytilis edulis* [30]. In the ^{1}H NMR spectroscopic study [57], the methyl region of the spectrum was monitored to observe the degradation of arsenobetaine in sea water (methyl groups attached to pentavalent arsenic resonate between δ 1.4 and 2.2) [1]. In a very clean series of spectra no arsenic-containing intermediates were seen; the arsenobetaine peak decreased as that resulting from trimethylarsine oxide increased by a corresponding extent. The transformation was shown to result from microbial activity because addition of sodium azide prevented any reaction from occurring.

A recent study [58] has produced interestingly different results. It was clearly demonstrated that when arsenobetaine was subjected to the microbial activity in sea water collected from Odense Fjord, Denmark (salinity about 20 parts per thousand) it yielded dimethylarsinoyl acetic acid (6, Figure 5.6), which was, in turn, degraded to dimethylarsinic acid. In other words, the loss of one methyl group preceded the loss of the carboxymethyl group. This was in marked contrast to the earlier work (in Japan and Australia) where only the carboxymethyl group was lost. The work in Denmark was particularly elegant because precise quantification of arsenic compounds in the microbial medium at all stages, through the use of LC-MS, and the absorption by the microbial cells of arsenic in one form and their subsequent release in another was clearly demonstrated. It is evident that more work is needed on this topic, particularly to identify the organisms involved in the transformations.

5.5 ELECTROSPRAY IONIZATION MASS SPECTROMETRY FOR DETERMINING ARSENIC COMPOUNDS IN MARINE ENVIRONMENTAL SAMPLES

The identification of the major naturally occurring arsenic compounds, mostly carried out in the late 70s and early 80s, spurred the development of analytical methods for their individual quantification. These methods were initially based on atomic optical spectroscopy (e.g. atomic absorption, emission and fluorescence). Later, atomic mass spectrometric techniques were developed, chiefly inductively coupled plasma mass spectrometry (ICPMS). When used in combination with liquid chromatography (LC) it provided a sensitive and robust method (LC-ICPMS) for determining arsenic species.

Application of LC-ICPMS to the analysis of marine samples has provided a wealth of data on the distribution of the various arsenicals in the marine environment. The work has also revealed the presence of many unknown arsenic compounds for which LC-ICPMS could provide no structural information whatsoever. This fundamental limitation of ICPMS has fueled interest in developing molecular mass spectrometric methods for determining arsenic compounds. The most promising results have come from LC electrospray ionization MS, and the following discussion will focus on this technique.

Electrospray ionization for the mass spectrometric determination of organoarsenic compounds was first reported in 1988 [59]. The same group (Sui and coworkers) then used an electrospray interface to couple a liquid chromatograph with a triple quadrupole mass spectrometer to determine the major arsenic compound (arsenobetaine) in fish [60]. An electrospray interface is particularly suitable for LC-MS coupling because it can transfer ions in solution directly to the gas phase at low energy, and hence fragmentation is minimal.

Despite these successful applications, LC electrospray MS was not widely used for the analysis of arsenic compounds in the period immediately following the work by Sui and coworkers [59, 60]. Possibly, this was a consequence of the growing popularity of LC-ICPMS as an arsenic speciation method at that time. Additionally, although electrospray MS provided data on specific arsenic compounds, it gave no direct measurement of arsenic *per se* and hence could not give a complete picture of the arsenic species present.

In 1996, however, the applications of electrospray (ionspray) ionization to arsenic compounds were extended by the use of a triple quadrupole MS, coupled to an LC, for sequential elemental arsenic and molecular analyses [61]. By modifying the mass spectrometer to allow higher orifice (declustering) potentials (250 V) good intensities were obtained for m/z 75 (bare As$^+$ ions) from organoarsenic compounds. In some cases, the As$^+$ signal was > 80% of the intensity for the molecular species obtained at low declustering potential (35 V). Both sets of data were obtained with the triple quadrupole MS being operated essentially as a single quadrupole instrument. The triple quadrupole was also operated in the product ion scanning mode, and characteristic fragments or product ions were reported for common arsenosugars following collision-induced dissociation in the positive ion mode. A product ion at m/z 237 was of particular diagnostic value because it was common to all arsenosugars, and could thereby provide evidence for the presence or absence of arsenosugars at the generic level. Finally, the technique was applied to the quantification of organoarsenic compounds in biological samples [61, 62].

Subsequently, electrospray tandem mass spectrometry (MS/MS) has been increasingly used to analyse arsenic compounds. Mass spectra, in both positive ion and negative ion mode, have been provided for standard organoarsenic compounds and the two inorganic species arsenite and arsenate [63]. In general, positive ion spectra were superior for the organoarsenic compounds whereas the negative ion spectra were most suitable for the inorganic arsenicals. The technique was then applied to estuarine water samples, and comparison of the data with those for standard compounds enabled some tentative structural assignments [64]. LC electrospray tandem mass spectrometry was also used in positive ion mode to study the separation and detection of ten organoarsenic compounds, including six species commonly found in marine samples [65]. In this case the triple quadrupole was operated in the selected reaction monitoring mode.

LC electrospray MS/MS has also been used to good effect to identify arsenosugars in algae [66–68, 103, 104] and arsenosugars in addition to other arsenic species in oyster [105] and clam [106] [65–67]. This work has shown that such analyses can be greatly affected by other components in the sample matrix, and considerable sample clean-up may be necessary. In view of the limitations imposed by the matrix, tandem MS was considered mandatory to confirm the identity of arsenosugars in biological samples [67].

All of the above studies were carried out with tandem MS systems, in most cases with triple quadrupole instruments. However, simpler single quadrupole mass spectrometers can also be used for determining organoarsenic compounds. LC electrospray MS with in-tube solid phase microextraction was used to obtain good quantitative data for arsenobetaine in the fish reference material DORM-2 [69]. LC electrospray MS, without sample clean-up, was used to determine arsenosugars in two species of brown algae [70]. In that study, the principle earlier described [61] was applied whereby the potential difference at the interface, between the capillary exit and the first skimmer, was varied to effect different degrees of fragmentation. The fragmentation processes are similar to the collision-induced dissociation processes occurring in the collision chamber of a triple quadrupole instrument [71]. At low potential difference (70 V) the protonated molecule ion (MH^+) was obtained whereas at high potential difference (240 V) the organoarsenic compounds fragmented to bare As^+ ions. Intermediate potential differences produced the characteristic product ion (m/z 237) common to all arsenosugars.

The single quadrupole MS used [70] by Pedersen and Francesconi had one other useful feature—the potential difference between the capillary exit and the first skimmer could be changed from high to low voltages continually and rapidly. This feature enabled, within a single chromatographic run, simultaneous recording of the protonated molecule ion (MH^+) and characteristic product ions (including As^+) from organoarsenic compounds. The application of the technique to the identification of compounds 1 and 2 (Figure 5.1) in the brown alga, *Laminaria digitata*, is depicted in Figure 5.8. The method was also shown to provide good quantitative data—comparable with those produced by LC-ICPMS [72].

5.6 ADVANTAGES OF ELECTROSPRAY IONIZATION MS OVER ICPMS

Liquid chromatography with ICPMS detection is a robust and sensitive analytical method, and is likely to remain, for some years at least, the most commonly used method for determining arsenic compounds. However, in some areas, electrospray MS offers analytical capabilities that ICPMS cannot match. Four such areas will now be briefly discussed, and illustrated with examples from recent work.

First, electrospray MS can provide qualitative information about organoarsenic compounds without the need for corresponding standards. In ICPMS, the arsenic compounds in samples can only be identified by comparing chromatographic retention times with those of standard compounds. Many of the environmental arsenic compounds are difficult to obtain in a pure form, and are not readily available as standards. Consequently, a sample may contain many described arsenic compounds that cannot be identified because suitable

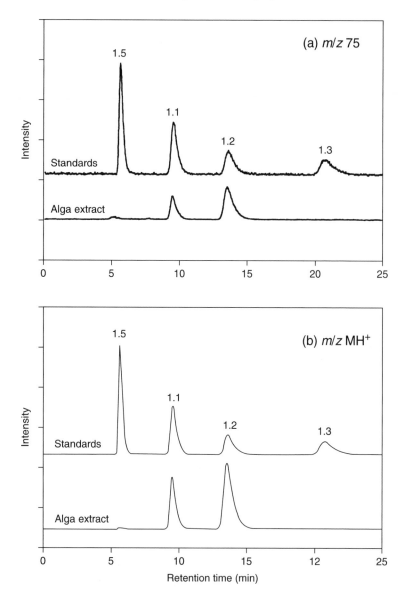

Figure 5.8 LC electrospray MS chromatograms of arsenic compounds. Four arsenosugar standards **1.1** to **1.4** (each $\equiv$ 2 ng As) and an aqueous extract of a brown alga, *Laminaria digitata*, measured at *m/z* 75 (a) and *m/z* [MH]$^+$ (b). The compounds **1.1** and **1.2** were identified in *Laminaria* by matching retention times with standards for both As (*m/z* 75) and the protonated molecule ion (MH$^+$). Chromatography was performed using a PRP-X100 anion-exchange column (250 × 4.1 mm) equilibrated with 10 % methanol in 20 mM NH$_4$HCO$_3$, pH 10.3 (flow rate 0.4 ml min^{-1}); and compounds were detected with a Hewlett Packard G1946A single quadrupole mass spectrometer using electrospray ionization in positive ion mode at variable fragmentor voltages (70 volts for *m/z* [MH]$^+$ and 240 volts for *m/z* 75) [70]

tandards are not available. Electrospray tandem MS, however, can provide useful data in the absence of standards. For example, the instrument can be set to allow transmission of molecule ions [(MH)$^+$) or (M—H)$^-$] of arsenic species of interest to a collision chamber where product ions from collision induced dissociation can provide a fingerprint of the species [61]. The data thus produced for arsenic compounds in samples can then be compared with literature data for standards to provide the necessary structural information. Such an approach was used to identify arsenosugars in ribbon kelp, *Alaria marginata* [66]. Other studies [67, 68] have similarly identified arsenosugars in algae, although on those occasions suitable arsenic standards were available in-house, allowing direct comparison with sample data.

A second advantage of electrospray MS over ICPMS is that compounds can be identified and quantified even when they are not resolved from other arsenic compounds. In ICPMS, poor signal resolution compromises the quantification of the compounds, and, perhaps more importantly, may result in some minor compounds being undetected because they coelute with the major arsenicals. Analytical problems with ICPMS associated with coelution of arsenic compounds in algae have been clearly demonstrated [67]. Electrospray MS, however, can select the molecule ion of interest to provide quantitative data even when arsenic compounds are not chromatographically resolved. A recent example of this was shown in a laboratory study on the degradation of arseno-betaine in seawater [58]. Two arsenic metabolites were produced but, because they had similar retention times, they could not be individually quantified based on arsenic detection. By detecting the relevant protonated molecule ions (MH$^+$), however, the two metabolites were effectively 'separated' and their formation throughout the experiment could be easily followed (Figure 5.9).

A third and major advantage of electrospray MS is that it can provide structural information about unknown compounds. Although this has typically been the domain of tandem MS systems, structural information can also be obtained with single quadrupole instruments as demonstrated recently by the identification of an unknown arsenic compound in a tropical fish [41]. The fish contained arsenobetaine and one unknown cationic arsenic compound as determined by LC electrospray MS in high energy (bare As$^+$ ion) mode. When the chromatography was repeated with the source set for gentler ionization and the quadrupole scanned, a signal with *m/z* 193 was evident at a retention time identical with that of the arsenic signal. Fragmentation of the molecule gave product ions characteristic of an arsenic betaine which, together with the molecular mass, indicated that the unknown compound was trimethyl-arsoniopropionate. The structural assignment was then confirmed by comparing these results with LC electrospray MS data for a chemically synthesized sample.

A similar approach (with a single quadrupole mass analyser) was taken in the identification of a new compound, 5-dimethylarsinoyl-2,3,4-trihydroxy penta-noic acid, as the major arsenic species in kidney of the clam *Tridacna derasa*

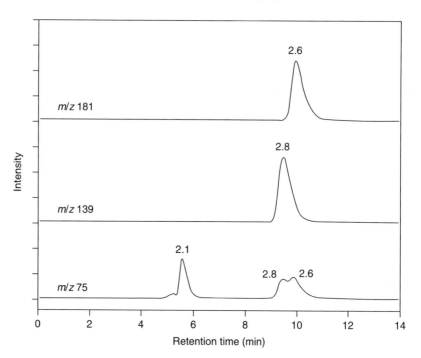

Figure 5.9 Signal resolution of two arsenic metabolites (compound **2.6** at m/z 181, retention time 9.93 min; and compound **2.8** at m/z 139, retention time 9.48 min) by selected ion monitoring with LC electrospray MS. The compounds, which are produced from microbial degradation of arsenobetaine (compound **2.1**, retention time = 5.56 min), can be individually measured despite poor chromatographic resolution. Chromatographic and mass spectrometric conditions were as described in Figure 5.8

[106]. On that occasion the structure was not confirmed by comparison with a synthetic sample. However, a separate study [35] using electrospray ionization with tandem mass spectrometry (quadrupole/time-of-flight) has supported the proposed structure. In addition, McSheehy *et al.* [35] were able to propose structures for four novel arsenic species in the clam, thereby clearly demonstrating the power of tandem mass spectrometric techniques in structural elucidation of arsenic compounds.

The fourth advantage of electrospray MS applies only to tandem mass spectrometers. Because such systems can separate the ion of interest before it is fragmented, they can provide valuable data without chromatography. This application was recently demonstrated by the use of a hybrid quadrupole time-of-flight mass spectrometer to obtain mass spectra in the negative ion mode for standard arsenosugars [72]. An algal extract (partially purified) was then introduced directly into the mass spectrometer and the ions corresponding to the arsenosugars were selected as precursors and subjected to collision-induced dissociation. In this manner, an arsenosugar was positively identified in the algal extract. Although this approach holds much promise for qualitative analyses, the suppression of ionization by the matrix is likely to preclude quantitative work.

Identification of the many unknown arsenic compounds in marine samples will help to elucidate biogeochemical cycles for arsenic. These compounds are often present at low concentrations, so they cannot readily be identified using natural products chemistry techniques. Although atomic spectroscopic methods such as LC-ICPMS will continue to provide the bulk of data for known arsenic compounds, molecular mass spectrometric techniques appear most suited to providing the necessary structural information for new compounds. It is likely that electrospray ionization mass spectrometry will play a future leading role in this area.

5.7 REFERENCES

1. Francesconi, K.A., and Edmonds, J.S., *Adv. Inorg. Chem.*, 1997, **44**, 147.
2. Andreae, M.O., *Deep-Sea Res.*, 1978, **25**, 391.
3. Jin, K., Hayashi, T., Shibata, Y., and Morita, M. *Appl. Organomet. Chem.*, 1988, **2**, 365.
4. Santosa, S.J., Wada, S., and Tanaka, S., *Appl. Organomet. Chem.*, 1994, **8**, 273.
5. Howard, A.G., Arbab-Zavar, M.H., and Apte, S., *Mar. Chem.*, 1982, **11**, 493.
6. Howard, A.G., and Apte, S.C., *Appl. Organomet. Chem.*, 1989, **3**, 499.
7. Edmonds, J.S., Francesconi, K.A., and Stick, R.V., *Nat. Prods Reports*, 1993, **10**, 421.
8. Francesconi, K.A., and Edmonds, J.S. In *Oceanography and Marine Biology: An Annual Review* A.D. Ansell, R.N. Gibson and M. Barnes (Eds), Vol. 31 UCL Press, London, 1993, pp. 111–151.
9. Edmonds, J.S., Shibata, Y., Francesconi, K.A., Yoshinaga, J., and Morita, M., *Sci. Total Environ.*, 1992, **122**, 321.
10. Morita, M., and Shibata. Y., *Analyt. Sci.*, 1987, **3**, 575.
11. Francesconi, K.A., Goessler, W., Panutrakul, S., and Irgolic, K., *Sci. Tot. Environ.*, 1998, **221**, 139.
12. Maugh, T.H. *Science*, 1979, **203**, 637.
13. Stryer, L., *Biochemistry*, 4th ed, W.H. Freeman, New York, 1995, p. 503.
14. Challenger, F., *Chem. Rev.*, 1945, **36**, 315.
15. Challenger, F., *Adv. Enzymology*, 1951, **12**, 429.
16. Challenger, F., Lisle, D.B., and Dransfield, P.B., *J. Chem. Soc.*, 1954, 1760.
17. Cantoni, G.L., *J. Am. Chem. Soc.*, 1952, **74**, 2942.
18. Cantoni, G.L., *J. Biol. Chem.*, 1953, **204**, 403.
19. Edmonds, J.S., and Francesconi, K.A. *J. Chem. Soc. Perkin. Trans.*, 1983, **1**, 2375.
20. Edmonds, J.S., and Francesconi, K.A. *Experientia*, 1987, **43**, 553.
21. Francesconi, K.A., Edmonds, J.S., and Stick, R.V., *J. Chem. Soc. Perkin Trans. 1*, 1992, 1349.
22. Edmonds, J.S., Shibata, Y., Yang, F., and Morita, M., *Tetrahedron Lett.*, 1997, **38**, 5819.
23. Geiszinger, A., Goessler, W., Kuehnelt, D., Francesconi, K., and Kosnus, W., *Environ. Sci. Technol.*, 1998, **32**, 2238.

24. Morita, M., and Shibata, Y., *Chemosphere*, 1988, **17**, 1147.
25. Edmonds, J.S., and Francesconi, K.A., unpublished results.
26. Edmonds, J.S., Francesconi, K.A., Hansen, J.A., *Experientia*, 1982, **38**, 643.
27. Francesconi, K.A., Edmonds, J.S., and Stick, R.V. *Sci. Total Environ.*, 1989, **79**, 59.
28. Francesconi, K.A., and Edmonds, J.S., unpublished results.
29. Francesconi, K.A., Edmonds, J.S., and Stick, R.V., *Appl. Organomet. Chem.*, 1992, **6**, 247.
30. Gailer, J., Francesconi, K.A., Edmonds, J.S., and Irgolic, K.J., *Appl. Organomet. Chem.*, 1995, **9**, 341.
31. Edmonds, J.S., and Francesconi, K.A., *Sci. Total Environ.*, 1987, **64**, 317.
32. Francesconi, K.A., Hunter, D.A., Bachmann, B., Raber, G., and Goessler, W., *Appl. Organomet. Chem.*, 1999, **13**, 669.
33. Cullen, W.R., and Nelson, J., *Appl. Organomet. Chem.*, 1993, **7**, 319.
34. Cullen, W.R., and Pergantis, S.A., *Appl. Organomet. Chem.* 1993, **7**, 329.
35. McSheehy, S., Szpunar, J., Lobinski, R., Haldys, V., Tortajada, J., and Edmonds, J.S. *Anal. Chem.*, 2002, **74**, 2370.
36. Larsen, E.H., Quetel, C.R., Munoz, R. Aline, F.-M., and Donard, O.F.X., *Mar. Chem.*, 1997, **57**, 341.
37. Kuehnelt, D., Goessler, W., and Irgolic, K.J., *Appl. Organomet. Chem.*, 1997, **11**, 289.
38. Edmonds, J.S., *Biorg. Med. Chem. Lett.*, 2000, **10**, 1105.
39. Francesconi, K.A., Edmonds, J.S., Stick, R.V., Skelton, B.W., and White, A.H., *J. Chem. Soc. Perkin Trans. 1*, 1991, **2707**.
40. Francesconi, K.A., Edmonds, J.S., and Stick, R.V. *J. Chem. Soc. Perkin Trans. 1*, 1992, 1349.
41. Francesconi, K.A., Khokiattiwong, S., Goessler, W., Pedersen, S.N., and Pavkov, M., *Chem. Comm.*, 2000, 1083.
42. Gailer, J., George, G.N., Pickering, I.J., Prince, R.C., Ringwald, S.C., Pemberton, J.E., Glass, R.S., Younis, H.S., DeYoung, D.W., and Aposhian, H.V., *J. Amer. Chem. Soc.*, 2000, **122**, 4637.
43. Norin, H., Christakopoulos, A., Sandström, M., and Ryhage, R., *Chemosphere*, 1985, **14**, 313.
44. Larsen, E.H., Pritzl, G., and Hansen, S.H., *J. Analyt. Atom. Spectrom.*, 1993, **8**, 1075.
45. Shiomi, K., Kakehashi, Y., Yamanaka, H., Kikuchi, T., *Appl. Organomet. Chem.*, 1987, **1**, 177.
46. Francesconi, K.A., Edmonds, J.S., and Hatcher, B.G., *Comp. Biochem. Physiol.*, 1988, **90C**, 313.
47. Shiomi, K., Aoyama, M., Yamanaka, H., Kikuchi, T., *Comp. Biochem. Physiol.*, 1988, **90C**, 361.
48. Shiomi, K., Sakamoto, Y., Yamanaka, H., and Kikuchi, T., *Nippon Suisan Gakkaishi*, 1988, **54**, 539.
49. Cullen, W.R., and Dodd, M., *Appl. Organomet. Chem.*, 1989, **3**, 79.
50. Whitfield, F.B., Freeman, D.J., and Shaw, K.J., *Chem. Ind.*, 1983, **20**, 786.
51. Francesconi, K.A., Stick, R.V., and Edmonds, J.S., *Experientia.*, 1990, **46**, 464.
52. Edmonds, J.S., Shibata, Y., Prince, R.I.T., Francesconi, K.A., and Morita, M., *J. Mar. Biol. Assoc. UK*, 1994, **74**, 463.
53. Edmonds, J.S., and Francesconi, K.A., *Mar. Pollut. Bull.*, 1993, **26**, 665.
54. Kaltreider, R.C., Davis, A.M., Lariviere, J.P., and Hamilton, J.W., *Environ. Health Perspect.*, 2001, **109**, 245.
55. Gibbs, P.E., Langston, W.J., Burt, G.R., and Pascoe, P.L., *J. Mar. Biol. Assoc. UK*, 1983, **63**, 313.
56. Hanaoka, A.G., Tagawa, S., and Kaise, T., *Appl. Organomet. Chem.*, 1992, **6**, 139.
57. Edmonds, J.S., and Byrne, L.J., unpublished results.

58. Khokiattiwong, S., Goessler, W., Pedersen, S.N., Cox, R., Francesconi, K.A., *Appl. Organometal. Chem.*, 2001, **15**, 481.
59. Siu, K.W.M., Gardner, G.J., and Berman, S.S., *Rapid Commun. Mass Spectrom.*, 1988, **2**, 69.
60. Siu, K.M., Guevremont, R., Le Blanc, J.Y., Gardner, G.J., and Berman, S.S. *J. Chromatog.*, 1991, **554**, 27.
61. Corr, J.J., and Larsen, E.H., *J. Anal. At. Spectrom.*, 1996, **11**, 1215.
62. Corr, J.J., *J. Anal. At. Spectrom.*, 1997, **12**, 537.
63. Florêncio, M.H., Duarte, M.F., de Bettencourt, A.M., Gomes, M.L., and Boas, L.F. *Rapid Commun. Mass Spectrom.*, 1997, **11**, 469.
64. Florêncio, M.H., Duarte, M.F., Facchetti, S., Gomes, M.L., Goessler, W., Irgolic, K.J., Vantklooster, H.A., Montanarella, L., Ritsema, R., Boas, L.F., de Bettencourt, A.M., *Analysis*, 1997, **25**, 226.
65. Pergantis, S.A., Winnik, W., and Betowski, D., *J. Anal. At. Spectrom.*, 1997, **12**, 531.
66. Gallagher, P.A., Wei, X.Y., Shoemaker, J.A., Brockhoff, C.A., Creed, J.T., *J. Anal. At. Spectrom.*, 1997, **14**, 1829.
67. McSheehy, S., and Szpunar, J., *J. Anal. At. Spectrom.*, 2000, **15**, 79.
68. McSheehy, S., Marcinek, M., Chassaigne, H., and Szpunar, J., *Analyt. Chim. Acta*, 2000, **410**, 71.
69. Wu, J.C., Mester, Z., and Pawliszyn, J., *Analyt. Chim. Acta*, 2000, **424**, 21.
70. Pedersen, S.N., and Francesconi, K.A., *Rapid Commun. Mass Spectrom.*, 2000, **14**, 641.
71. Voyksner, R.D., and Pack, T., *Rapid Commun. Mass Spectrom.*, 1996, **5**, 263.
72. Madsen, A.D., Goessler, W., Pedersen, S.N., and Francesconi, K.A., *J. Anal. At. Spectrom.*, 2000, **15**, 657.
73. Pergantis, S.A., Wangkarn, S., Francesconi, K.A., and Thomas-Oates, J.E., *Anal. Chem.*, 2000, **72**, 357.
74. Leatherland, T.M., Burton, J.D., Culkin, F., McCartney, M.J., and Morris, R.J., *Deep-Sea Res.*, 1973, **20**, 679.
75. Leatherland, T.M., and Burton, J.D., *J. Mar. Biol. Assoc. UK*, 1974, **54**, 457.
76. Bohn, A., *Mar. Pollut. Bull.*, 1995, **6**, 87.
77. Kennedy, V.S., *J. Fish Res. Bd Canada*, 1976, **33**, 1388.
78. Zingde, M.D., Singbal, S.Y.S., Moraes, C.F., and Reddy, C.V.G., *Ind. J. Mar. Sci.*, 1976, **5**, 212.
79. Bebbington, G.N., Mackay, N.J., Chvojka, R., Williams, R.J., Dunn, A., and Auty, E.H., *Austral. J. Mar. Freshw. Res.*, 1977, **28**, 277.
80. Egaas, E., and Braekkan, O.R., *Fiskeridirektoratets skrifter Serie Ernaering*, 1977, **1**, 93.
81. Hall, R.A., Zook, E.G., and Meaburn, G.M., *NOAA Technical Report*, NMFS SSRF-721, 1978, 313 pp.
82. Luten, J.B., Riekwel-Booy, and Rauchbaar, A., *Environ. Health Perspect.*, 1982, **45**, 165.
83. Falconer, C.R., Shepherd, R.J., Pirie, J.M., and Topping, G. *J. Exp. Mar. Biol. Physiol.*, 1983, **71**, 193.
84. Maher, W.A., *Mar. Pollut. Bull.*, 1983, **14**, 308.
85. Glover, J.W., *Austral. J. Mar. Freshw. Res.*, 1979, **30**, 505.
86. Powell, J.H., Powell, R.E., and Fielder, D.R., *Water Air Soil Pollut.*, 1981, **16**, 143.
87. Stoeppler, M., and Brandt, K., *Z. Lebensmittelunters. Forsch.*, 1979, **169**, 95.
88. Kosta, L., Ravnik, V., Byrne, A.R., Stirn, J., Dermelj, M., and Stegnar, P., *J. Radioanalyt. Chem.*, 1978, **44**, 317.

89. Shiomi, K., Shinagawa, A., Igarashi, T., Hirota, K., Yamanaka, H., and Kikuchi, T., *Bull. Jap. Soc. Sci. Fish.*, 1984, **50**, 293.
90. Lai, V.W.M., Cullen, W.R., Ray, S., *Mar. Chem.*, 1999, **66**, 81.
91. Goessler, W., Maher, W., Irgolic, K.J., Kuehnelt, D., Schlagenhaufen, C., and Kaise, T., *Fresenius J. Anal. Chem.*, 1997, **359**, 434.
92. Hanaoka, K., Goessler, W., Kaise, T., Ohno, H., Nakatani, Y., Ueno, S., Kuehnelt, D., Schlagenhaufen, C., Irgolic, K.J., *Appl. Organomet. Chem.*, 1999, **13**, 95.
93. Goessler, W., Rudorfer, A., Mackey, E.A., Becker, P.R., and Irgolic, K.J., *Appl. Organomet. Chem.*, 1998, **12**, 491.
94. Dhandhukia, M.M., and Seshadri, K., *Phykos*, 1969, **8**, 108.
95. Lunde, G., *J. Sci. Food Agric.*, 1970, **21**, 416.
96. Tagawa, S., and Kojima, Y., *J. Shimonoseki Univ. Fish.*, 1976, **25**, 67.
97. Sanders, J.G., *Estuarine Coast Mar. Sci.*, 1979, **9**, 95.
98. Whyte, J.N.C., and Englar, J.R., *Botanica Marina*, 1983, **26**, 159.
99. Maher, W.A., and Clarke, S.M., *Mar. Pollut. Bull.*, 1984, **15**, 111.
100. Rao, C.K., Chinnaraj, S., Inamdar, S.N., and Untawale, A.G., *Ind. J. Mar. Sci.*, 1991, **20**, 283.
101. Raber, G., Francesconi, K.A., Irgolic, K.J., and Goessler, W., *Fresenius J. Anal. Chem.*, 2000, **367**, 181.
102. Lai, V.W.M., Cullen, W.R., Harrington, C.F., and Reimer, K.J., *Appl. Organomet. Chem.*, 1997, **11**, 797.
103. McSheehy, S., Pohl, P., Lobinski, R., and Szpunar, J., *Anal. Chim. Acta*, 2001, **440**, 3.
104. McSheehy, S., Pohl, P., Vélez, D., and Szpunar, J. *Anal. Bioanal. Chem.*, 2002, **372**, 457.
105. McSheehy, S., Pohl, P., Lobinski, R., and Szpunar, J. *Analyst*, **126**, 1055.
106. Francesconi, K.A., and Edmonds, J.S., *Rapid Commun. Mass Spectrom.*, 2001, **15**, 1641.

6 Organoarsenic Compounds in the Terrestrial Environment

DORIS KUEHNELT and WALTER GOESSLER

Institute of Chemistry, Karl-Franzens-University, Graz, Austria
The authors dedicate this chapter to Professor K.J. Irgolic, who died in a tragic accident in the Austrian Alps in July 1999.

6.1 INTRODUCTION

Since arsenobetaine was isolated for the first time from the tail muscle of the western rock lobster, *Panulirus longipes cygnus* George, in 1977 [1], arsenic compounds in the marine environment have been investigated intensively. The following organic arsenic compounds have been identified in marine ecosystems during the past three decades: methylarsonic acid (MA) (sea water, sediments, sediment pore water, algae), dimethylarsinic acid (DMA) (sea water, sediments, sediment pore water, algae, animals), trimethylarsine oxide (TMAO) (sediments, sediment pore water, animals), the tetramethylarsonium cation (TETRA) (animals), arsenobetaine (AB) (animals), arsenocholine (AC) (animals), arsenoriboses (algae, animals) (Figure 6.1) [2]. In addition to these 'common' organoarsenicals, a few other organoarsenic compounds have been detected in the marine environment [2] (see also Chapter 5). Although the total arsenic concentration in sea water is almost uniformly 0.5 to $2 \, \mu g \, dm^{-3}$ [2], marine algae and animals normally contain arsenic at the $mg \, kg^{-1}$ level, and some even show arsenic concentrations higher than $100 \, mg \, As \, kg^{-1}$ (dry mass). These arsenic concentrations are sufficiently high to allow the determination of arsenic compounds in marine organisms.

Arsenic concentrations in uncontaminated terrestrial organisms are generally lower than in marine organisms. For example, uncontaminated terrestrial plants usually contain 0.2 to $0.4 \, mg \, As \, kg^{-1}$. However, in plants from contaminated sites arsenic concentrations of up to several thousand $mg \, As \, kg^{-1}$ (dry mass) were observed [3]. The median arsenic concentration in organs and body fluids of unexposed humans was reported to be between 0.02 and $0.06 \, mg \, As \, kg^{-1}$ [4]. Higher concentrations may be found in muscle, lung, femur, skin, teeth, nails,

Organometallic Compounds in the Environment
Edited by P.J. Craig © 2003 John Wiley & Sons Ltd

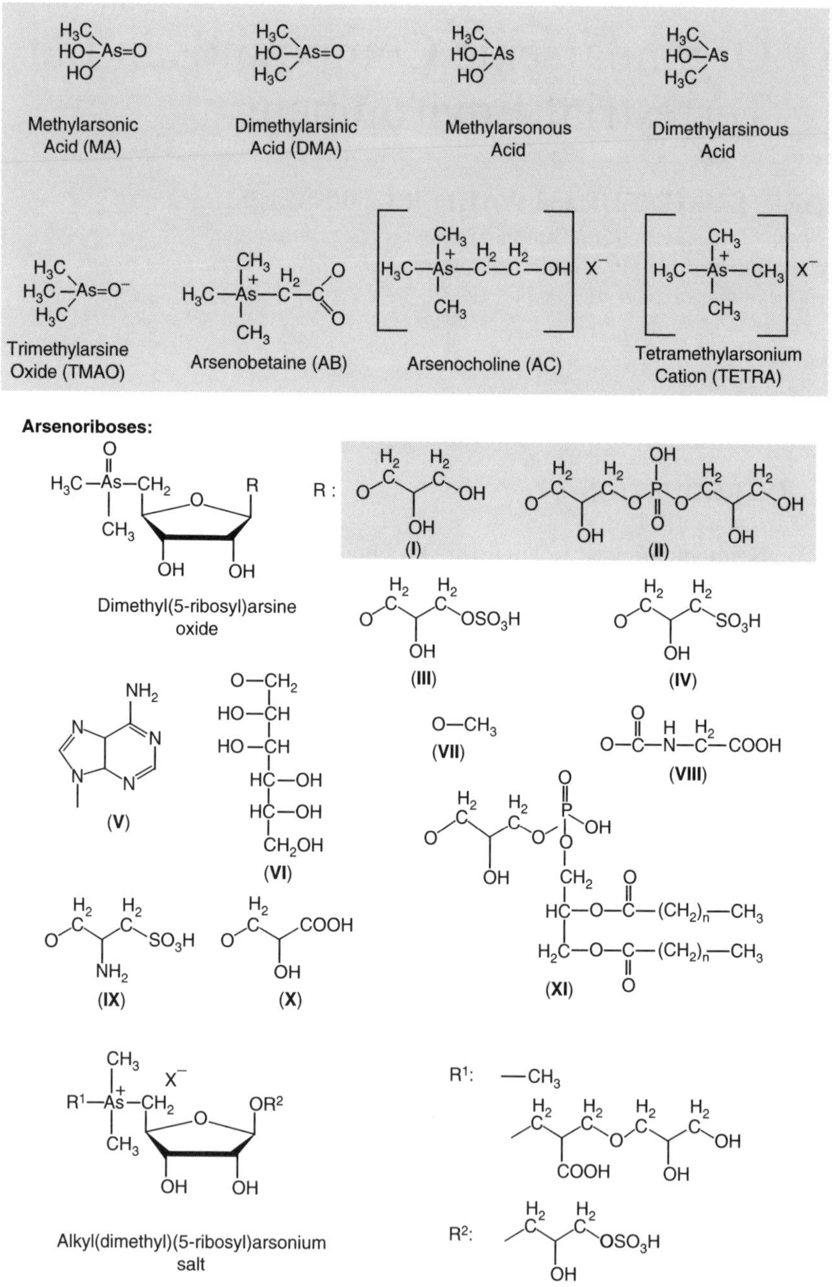

Figure 6.1 Organoarsenic compounds detected in the environment. Shaded compounds have been detected in the terrestrial environment (see also Chapter 5)

and especially hair [4], which can contain from a few to several hundred μg As kg^{-1} [5]. Blood and urine arsenic concentrations from unexposed humans are usually about 10 μg As dm^{-3} [6]. Meat of animals destined for human consumption was reported to contain arsenic in the sub-mg As kg^{-1} dry mass range. Exceptions were meat of swine (up to 6.3 mg As kg^{-1}) and poultry (up to 5.5 mg As kg^{-1}) [4]. These higher arsenic concentrations result from phenylarsonic acids added to swine and poultry feeds to enhance weight and feed efficiency [7].

Arsenic concentrations in fresh waters are strongly dependent on the geological composition of the drainage area as well as on the extent of anthropogenic input, which can cause arsenic concentrations of several hundred mg As dm^{-3}. The geometric mean of the arsenic concentration in European, and North and South American river water was reported to be 1.4 μg As dm^{-3} [3]. Arsenic concentrations in soils and sediments are also dependent on the geological conditions as well as on anthropogenic activities such as the use of pesticides, mining activities, or industrial processes. Average arsenic concentrations of uncontaminated soils were reported to be usually below 15 mg As kg^{-1}. In contaminated soils arsenic concentrations of more than a thousand mg As kg^{-1} can be reached [8]. Uncontaminated sediments usually contain less than 10 mg As kg^{-1} dry mass, whereas contaminated sediments can contain up to 10 000 mg As kg^{-1} dry mass [9].

The fact that many terrestrial samples from uncontaminated areas or from unexposed animals or humans contain arsenic at the sub-mg As kg^{-1} level has hampered arsenic speciation analysis in the terrestrial environment for a long time. The determination of arsenic compounds was restricted to material obtained from contaminated areas or samples of arsenic-exposed humans or animals. Non-quantitative extraction procedures and the distribution of the total arsenic concentration among several arsenic compounds require analytical methods with detection limits below 1 μg As dm^{-3} for the determination and quantification of arsenic compounds in uncontaminated terrestrial material. Nowadays, the availability of such analytical techniques, in combination with powerful separation systems, allows the investigation of arsenic compounds in samples with total arsenic concentrations at the sub-mg As kg^{-1} level. In recent years, knowledge about the occurrence of organoarsenic compounds in material of terrestrial origin has improved significantly, and the assumption that organic arsenic compounds in the terrestrial ecosystem are restricted to the 'simpler' methylated arsenic compounds (methylarsonic acid, dimethylarsinic acid and trimethylarsine oxide) has been revised.

6.2 ANALYTICAL METHODS FOR THE IDENTIFICATION AND QUANTIFICATION OF ORGANOARSENIC COMPOUNDS

For a long time arsenic analyses were restricted to the determination of total arsenic concentrations. Total arsenic concentrations do not provide information about the health risks posed by a material, since the toxicity of arsenic compounds varies significantly. Nowadays, chromatographic methods for the separation of arsenic compounds, in combination with the development of analytical methods with detection limits in the low or sub-μg As dm^{-3} range, allow the identification and quantification of arsenic compounds in environmental and biological samples, even if the total arsenic concentration is below 1 mg As kg^{-1}. For most of the analytical methods the arsenic compounds have to be present in solution. For gas-chromatographic (GC) analysis the arsenic compounds must be volatilized. The extraction procedures should ensure that the arsenic compounds are extracted quantitatively or at least close to quantitatively. The arsenic compounds must not be chemically converted during the extraction step. In some cases, one extraction procedure may not be sufficient to extract all arsenic compounds from a sample and a combination of different procedures has to be applied. Liquid samples like urine can be analyzed directly in most cases. Dilution or clean-up steps before analysis minimize matrix effects. Generally, the determination of arsenic compounds in environmental samples or their extracts consists of two sequential steps. First, the arsenic compounds have to be separated. Since it is not possible to separate all arsenic compounds identified until now within one chromatographic run, a combination of procedures has to be used, when all these compounds have to be determined. The second step is the detection of the separated arsenic compounds. During recent years a variety of hyphenated techniques for the determination of arsenic compounds has been developed (Figure 6.2). The techniques applied in the literature cited in Chapter 6.3–6.6 for the determination of arsenic compounds in terrestrial organisms will be discussed here briefly.

Several techniques for the separation of arsenic compounds are available (Figure 6.2). Hydride generation (HG) has frequently been applied. Arsenous acid, and arsenic acid, MA, DMA, and TMAO can be reduced to the corresponding volatile arsines AsH$_3$, CH$_3$AsH$_2$, (CH$_3$)$_2$AsH, and (CH$_3$)$_3$As by NaBH$_4$ in acidic solution. The arsines can be collected in a cold trap. Subsequent warming of the trap releases the arsines according to their different boiling points [10–38]. A better separation of the arsines can be achieved with GC [39–46]. AB, AC, TETRA, and arsenoriboses do not form volatile arsines under these conditions. In order for these arsenic compounds to be determined by HG, a digestion step must precede the treatment with NaBH$_4$. Digestion is performed with NaOH at 80 to 95 °C [18–26, 28, 30–35, 37, 38, 42, 47], microwave oxidation [48–51], UV-photooxidation [51–55], or pyrolysis and combustion in an oxygen atmosphere [56]. GC can also be applied directly to separate gaseous samples [34, 35, 45, 57]. Liquid chromatography (LC) is often used for the separation of arsenic compounds in liquids or liquid sample extracts [57–80]. HPLC offers powerful separation and only small amounts of sample are required. It can easily be connected to most detection systems commonly in use. Depending on the arsenic compounds to be determined, several separation procedures on anion-exchange [45, 46, 49, 54, 55, 62, 81–125], cation-exchange

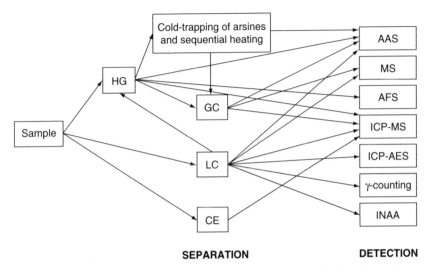

SEPARATION   **DETECTION**

Figure 6.2   Instrumental methods for the identification and quantification of arsenic compounds

Abbreviations:

	AAS	atomic absorption spectrometry
	AFS	atomic fluorescence spectrometry
	CE	capillary electrophoresis
	GC	gas chromatography
	HG	hydride generation
	HPLC	high performance liquid chromatography
	ICP-AES	inductively coupled plasma – atomic emission spectrometry;
	ICP-MS	inductively coupled plasma – mass spectrometry
	INAA	instrumental neutron activation analysis
	LC	liquid chromatography
	MS	mass spectrometry

[51–53, 86, 92, 95, 97–101, 106–108, 111, 119, 120, 122, 125–128] or reversed-phase [43, 48, 50, 53, 56, 61, 95–97, 102, 112, 125, 129–138] columns were developed. Capillary electrophoresis (CE) offers even better separation efficiency but this method is still suffering from high relative detection limits as a consequence of the small sample volumes applied into the capillary. Additionally, coupling the capillary to element specific detectors like inductively coupled plasma mass spectrometers (ICP-MS) is more difficult because of the low flow in CE.

The separation systems listed above can be coupled to various detectors. The cold-trapping device for arsines is commonly coupled to an atomic absorption spectrometer (AAS) [11–38]. The method can determine the different levels of arsenic methylation but it cannot unequivocally distinguish which compounds were originally present in the sample. Coupling of GC to AAS [39, 40, 43, 44, 46], mass spectrometry (MS) [34, 35, 41, 42], or ICP-MS [45] is used for the determination of volatile or volatilized arsenic compounds in environmental and biological samples. The detection of arsenic compounds after separation with LC can be performed with AAS [13, 49, 51–53, 58, 60, 63, 64, 66–68, 70,

71, 74, 79, 80, 82, 84, 90, 112, 129, 130, 139], MS [92], atomic fluorescence spectrometry (AFS) [48, 50, 54, 55, 102, 114, 121, 135–138], ICP-MS [14, 43, 45, 46, 54, 62, 81, 83, 85–89, 91–101, 103–105, 107–111, 113, 115–120, 122–126, 131–134], inductively coupled plasma-atomic emission spectrometry (ICP-AES) [61, 106, 127, 128], γ-counting [57, 59, 65, 69, 72, 73, 75–78, 140–142], if radioactively labeled arsenic compounds have to be determined, or instrumental neutron activation analysis (INAA) [61, 62, 74]. To achieve lower detection limits a HG step after the LC separation is sometimes employed [13, 48–54, 55, 58, 60, 63, 64, 66–68, 70, 71, 74, 79, 80, 82, 84, 90, 102, 112, 114, 121, 130, 131, 135–139]. Coupling of CE to ICP-MS is also used for the determination of arsenic compounds in environmental samples [143]. Typical detection limits for various techniques have been summarized elsewhere [144].

6.3 ORGANOARSENIC COMPOUNDS PRESENT IN THE TERRESTRIAL ENVIRONMENT

A variety of different terrestrial samples has been reported to contain organoarsenic compounds: freshwaters including underground, well, tap, river, pond and lake water as well as sediment pore water; soil and sediments; microorganisms; freshwater algae; fungi; lichens; green plants; urine, feces, blood, and tissues of animals; and human urine, blood, and tissues. Many of these publications report the presence of arsenic compounds in samples from arsenic-rich regions, or from exposed humans, or the analysis of arsenic compounds after a defined dosage of certain arsenic compounds to plants, animals or humans. This 'forced' uptake/incorporation and metabolism of arsenic compounds will also be included in the following sections. Since this chapter focuses on organoarsenic compounds, samples in which only inorganic arsenic was detected will not be considered (also in sections 6.4–6.6). The presence of inorganic arsenic in samples containing organoarsenic compounds will also not be explicitly noted. It should also be noted that many of the terrestrial arsenic compounds are the same as those found in the marine environment (Chapter 5).

6.3.1 WATER

In contrast to seawater, which almost uniformly shows arsenic concentrations of 0.5 to $2 \mu g \ dm^{-3}$ [2], the arsenic concentration in freshwater is strongly dependent on the geological characteristics of the drainage area as well as on anthropogenic input. Besides inorganic arsenic, organoarsenic compounds were also detected in water samples. In 1973 Braman and Foreback reported the presence of MA and DMA in one sample of river and well water, and in two samples of pond and lake water from the region of Tampa, Florida, USA [10]. The river water contained 48 % of the total arsenic as DMA and 14 % as MA. In the well water, 44 % of the total arsenic was present as DMA and 16 % as MA. The percentage of DMA in the pond water samples was 7.7 % and 58.5 %, and the percentage of MA 2.6 % and 11 % of the total arsenic. In lake water DMA accounted for 9 % in both samples, MA for 3.1 % and 12.6 %, respectively, of the total arsenic. Faust and coworkers analyzed water from a river in New Jersey, USA, contaminated by a pesticide plant upstream of the sampling site [145]. Besides inorganic arsenic, MA and trace amounts of DMA were found in the water. River water from an arsenic-rich hot spring environment in Japan was reported to contain 6.8 % of the total arsenic in the form of trimethylated arsenic, but no mono- or dimethylated arsenic [42]. DMA and the trivalent methylarsenic species monomethylarsonous acid and dimethylarsinous acid were detected in the water of Lake Biwa, Japan [16]. A detailed study of the seasonal variations of the arsenic compounds in this lake was carried out by Sohrin and coworkers [146]. They reported that DMA became the major arsenic compound within the euphotic zone of the water in summer. Mono-methylarsonous acid, dimethylarsinous acid, and MA were also detected in the water as minor arsenic compounds. The authors suggested that the seasonal variations of the arsenic species in the lake water largely depend on biological processes such as the metabolism of phytoplankton, the decomposition of organic matter by bacteria, or the microbial reduction of iron and manganese oxides in the sediments, and that eutrophication affects the arsenic species in the lake water. Pongratz reported the presence of trace amounts of DMA ($0.2 \, \mu g$ As dm^{-3}) in rain water samples collected in Wolfsberg, Austria [104]. Two unknown arsenic compounds, MA, and DMA at the sub $\mu g \, dm^{-3}$ level were detected in underground and tap water from a blackfoot disease area in Taiwan [131]. Some Indonesian river and tap water samples were reported to contain MA and DMA [114]. These two organoarsenic compounds were also present in freshwater from the Odiel river in Spain [54]. Traces of MA and DMA (< 1 % of total arsenic) were detected in a French thermal water sample [46]. Thus it seems that MA and DMA are the dominating organoarsenic compounds detected in natural freshwaters.

6.3.2 SOIL AND SEDIMENTS

Like in natural waters, arsenic concentrations in soils and sediments depend on the geological situation and anthropogenic input. Methylation and reduction of arsenic in soil is attributed to microbial activity. This assumption is supported by the fact that no evolution of arsines from sterilized soil treated with DMA occurred, whereas generation of arsines from non-sterilized soil treated with DMA was observed [15]. Organoarsenic compounds in soils and sediments can also be derived from organic matter. Little data on organoarsenicals in sediments is available in the literature. Organoarsenic compounds in sediments seem to be restricted to MA and DMA. Braman and Foreback reported the presence of these two organoarsenic compounds in a sedimentary rock sample [10]. MA, DMA, and sometimes TMAO were detected in the culture media of sediments incubated with arsenate or arsenite, and, additionally, the release of methylarsines was observed [3]. Faust and coworkers determined the arsenic compounds in river and lake sediments from a river drainage basin in New Jersey, USA and also detected MA and DMA in their samples [145]. More work has been carried out in determining arsenic compounds in soils because of the agricultural impact of arsenicals in fertilizers, herbicides, and pesticides. Depending on the microbial community, soil conditions and soil type, processes involving demethylation, methylation, and reduction can occur [3]. Gao and Burau investigated the transformation of arsenic compounds incorporated in a Sacramento silty clay soil after a 70 days' incubation period under controlled conditions [15]. The effects of different arsenic compounds (arsenite, arsenate, MA, sodium salt of DMA), arsenic concentration, soil moisture, soil temperature, and cellulose addition were studied. In the soil samples treated with inorganic arsenic no methylation was observed, whereas MA and DMA were partially demethylated to arsenate. Small amounts of MA were detected in the soil treated with DMA. The percentage of arsenate formed by demethylation decreased with increasing concentration of DMA added to the soil. Above a soil moisture value of $350 \, g \, H_2O \, kg^{-1}$ soil a significant increase of DMA demethylation was observed in DMA treated soil. DMA demethylation also increased with increasing temperature. The addition of cellulose to DMA treated soil decreased DMA demethylation. Generation of arsines was observed, when the soil was amended with arsenite, arsenate, MA, or DMA. The arsine evolution rate followed the order DMA > MA > arsenite = arsenate, but the arsines formed were not definitively identified. The formation of volatile organoarsenicals was also observed by several other working groups [3]. The influence of microbes on the biomethylation of arsenic in soil samples collected from a wood impregnating plant in Southern Finland was investigated by Turpeinen and coworkers [109]. Arsenate was the main arsenic compound in the four soil samples and the control soil, in which microbial activity had been inhibited by the addition of formaldehyde. MA was found in one sample. After 5 days of aerobic incubation, MA was detected in two, and DMA in three, of the four samples, whereas no methylarsenicals were found in the control soil. No organoarsenic compounds were formed when the soil samples were incubated anaerobically. In the water extract of one sample of industrially contaminated soil, the presence of MA and DMA was reported [46]. Pongratz analyzed the arsenic compounds present in soil that contained high arsenic concentrations as a result of natural geological processes [104]. Only arsenate could be detected before incubation, whereas arsenite, MA, DMA, and AB were found in addition to a high quantity of arsenate in water extracts of the soil after an aerobic incubation period of one week. The authors suggested that the AB derived from microorganisms, from where it was released during the shaking applied for the extraction of the arsenic compounds from the soil. This suggestion was supported by the fact that no AB could be detected in the soil pore water after incubation. Traces of AB were also detected in ant-hill material collected at a former arsenic roasting facility in Austria [98]. Additionally, DMA and traces of MA and TMAO were present. Since ant-hill material consists of soil particles and organic matter like the needles of trees the presence of the organoarsenic compounds probably derived from the organic part of the ant-hill material.

In general, the presence of MA and DMA in soils and sediments in connection with microbial activities seems to be well established, whereas the presence of AB is yet to be confirmed.

6.3.3 MICROORGANISMS

The ability of microorganisms to generate volatile arsines was established in 1933, when Challenger and coworkers reported their production by strains of *Penicillium brevicaule* growing on sterile bread crumbs containing various arsenic compounds [147]. The literature data on arsenic metabolism by microorganisms until 1989 was reviewed by Cullen and Reimer and will only be summarized briefly in this section [3].

Several organoarsines were volatilized when the mould *Scopulariopsis brevicaulis* was exposed to different organic arsenic compounds, e.g. arsine oxides R_3AsO were reduced to the corresponding arsines R_3As [3]. Methylation was also observed for a variety of compounds investigated. When *Scopulariopsis brevicaulis* was exposed to As_2O_3 or Na_2HAsO_4, methylation occurred and trimethylarsine was released [3]. Other fungi like some *Candida, Aspergillus, Penicillium,* or *Gliocladium* spp. were also found to release volatile organoarsines after exposure to various arsenic compounds [3]. Again the methylation of inorganic arsenic (As_2O_3) to trimethylarsine was observed in two species of *Aspergillus* and in *Candida humicola,* and in *C. humicola* arsenate methylation was dependent on the pH. In contrast, a number of *Penicillium* spp. were reported to form no volatile arsines from inorganic arsenic [3]. Different varieties of bacteria were also reported to produce methylarsines. When microorganisms were exposed to the herbicides NaH_2AsO_4 or $MeAsO(ONa)_2$, the volatile arsines $MeAsH_2$, Me_2AsH, and Me_3As were detected [3]. In addition to these methylation processes demethylation was also observed. Methylation and demethylation can occur simultaneously. For example the bacteria *Pseudomonas* sp., *Flavobacterium* sp., *Achromobacter* sp., and *Norcardia* converted $MeAsO(ONa)_2$, not only to methylarsines but also to arsenate [3].

Cullen and coworkers reported the metabolism of methylarsine oxide and sulfide, two possible intermediates in the methylation of inorganic arsenic, and of dimethylthioarsinite dervatives in microorganisms [57]. *Candida humicola, Scopulariopsis brevicaulis, Lactobacillus brevis, Streptococcus sanguis, Pseudomonas aeruginosa, Escherichia coli,* and *Bacillus subtilis,* all grown aerobically, anaerobically grown *Veillonella alcalescens* and *Fusobacterium nucleatum,* as well as rumen fluid, soil, compost, and ceca of mice were investigated. Volatile arsenic compounds in the headspace of the cultures as well as arsenic compounds in the culture medium and in extracts of the cells were determined. When cultures of *C. humicola, V. alcalescens, L. brevis,* or *S. brevicaulis* were exposed to methylarsine oxide, DMA was detected in the culture medium in addition to unchanged methylarsine oxide. Only small amounts of arsenic were volatilized. After incubation of *C. humicola* in the presence of methylarsine sulfide, methylarsine and trimethylarsine, but no dimethylarsine, were detected in the headspace of the culture. Dimethylarsine was identified in the headspace of cultures of *C. humicola* exposed to the glutathione derivative of dimethylthioarsinite. Trimethylarsine was released from cultures of *C. humicola* containing DMA, or the glutathione or cysteine derivatives of dimethylthioarsinite. No trimethylarsine was present, when the mercaptoethanol or thioacetic acid derivatives of dimethylthioarsinite were added to cultures of *C. humicola.* Methylarsine oxide was metabolized to DMA by cell-free extracts of *C. humicola* in the presence of glucose. Methylarsine oxide was converted to DMA and volatile arsines by rumen fluid, but no volatile arsines were generated from methylarsine sulfide. No metabolism of methylarsine oxide or sulfide was observed in soil or compost. In ceca of mice, methylarsine oxide was partially converted to arsenate and DMA.

In another study, arsenic compounds in the growth medium of *Apiotrichum humicola* (previously named *Candida humicola*) and *Scopulariopsis brevicaulis* after incubation with arsenite, arsenate, MA, or DMA were determined [40]. Arsenite, TMAO, and small amounts of DMA and MA were detected as arsenic metabolites in the growth medium, when *A. humicola* was exposed to arsenate. Within a 30 day incubation period the concentration of arsenite decreased and the concentration of TMAO concurrently increased. In the growth medium of *S. brevicaulis* arsenite, TMAO, and traces of DMA were present, but the concentration of arsenite did not decrease significantly and the concentration of TMAO increased slowly during 5 weeks of incubation. Arsenite was metabolized to TMAO, DMA, and traces of MA by *A. humicola* and to TMAO and traces of DMA by *S. brevicaulis*. No oxidation to arsenate was observed. MA was partially converted to DMA and TMAO by *A. humicola* and *S. brevicaulis*. No demethylation was observed. The only metabolite of DMA detected in both growth media was TMAO. No trimethylarsine was produced and arsenic was not accumulated in the cells during any of the experiments. TMAO seemed to be the main methylation product in both microorganisms. The authors discussed how these data support the pathway for the methylation of inorganic arsenic proposed by Challenger, which includes sequential reduction and oxidative methylation steps.

Apiotrichum humicola was also used for another study concerning arsenic biomethylation [41]. The incorporation of CD_3 from L-methionine with a deuterated methyl group into arsenic metabolites formed by this microorganism from arsenite, arsenate, MA, or DMA in the growth medium was investigated in order to verify that *S*-adenosylmethionine was a probable methyl donor in arsenic methylation. In the presence of arsenate, arsenite, DMA, and TMAO were formed. A considerable amount of DMA and TMAO was deuterated in the presence of L-methionine-*methyl-d₃*. Deuterated DMA and TMAO were also detected, when arsenite or MA were metabolized. Deuterated TMAO was formed from DMA as substrate. The addition of L-methionine-*methyl-d₃* did not enhance the methylation process in any of the experiments. The results strongly support the hypothesis that methionine or *S*-adenosylmethionine acts as the methyl donor in the biological methylation of arsenic in *A. humicola*.

Arsenic-tolerant bacteria identified as *Pseudomonas putida* were isolated from a contaminated culture of the freshwater alga *Chlorella* sp. [21]. When these bacteria were grown in a medium containing arsenate, trimethylated arsenic compounds were detected in the cells in addition to inorganic arsenic. When bacterial cells were transferred from the arsenic-containing growth medium to an arsenic-free growth medium, mono-, di-, and trimethylated arsenic compounds were excreted besides inorganic arsenic. Two other arsenic–tolerant bacteria, *Klebsiella oxytoca* and *Xanthomonas* sp., isolated in the same way as *P. putida*, were also grown in arsenic-spiked culture media [25]. In the cells of *K. oxytoca* inorganic arsenic as well as di-, tri- and traces of monomethylated arsenic compounds were detected after exposure to inorganic arsenic. MA was partially transformed to di-, and trimethylated arsenic compounds. When exposed to DMA, mono-, di- and trimethylated arsenic compounds were present in the cells and after exposure to AB, di- and trimethylated arsenic compounds were found. No inorganic arsenic was formed when the cells were exposed to organoarsenic compounds. Bioaccumulation experiments with the same compounds in *Xanthomonas* sp. resulted in the presence of mono-, di-, and trimethylated arsenic compounds in the cells with one exception: monomethylated arsenic after DMA exposure was detected only, when the cells had been pre-exposed to inorganic arsenic or DMA. Inorganic arsenic was only detected after exposure to inorganic arsenic.

Anaerobic microorganisms have also been investigated. Bright and co-workers studied arsenic methylation by anaerobic microbial consortia, namely anaerobic oligotrophs, nitrate dissimilators, iron, manganese, and sulfate re-ducers (with acetate and lactate as alternative carbon-sources), isolated from an arsenic-contaminated lake sediment in the Canadian sub-arctic [39]. After 35 days of incubation in an arsenate-containing medium, mono-, di- and trimethy-lated arsenic compounds were detected in the cultures of the anaerobic oligo-trophs and sulfate reducers, and mono- and trimethylated arsenic compounds in the cultures of the iron and manganese reducers. Arsine, methylarsine, and trimethylarsine were volatilized into the headspace of the culture vessels by the anaerobic oligotrophs, arsine and methylarsine by one species and arsine and trimethylarsine by the other species of sulfate reducer. Mono-, di-, and tri-methylated arsenic compounds were also detected in sediment pore water. Wickenheiser and coworkers investigated the arsenic metabolism in the anaer-obic methanogen *Methanobacterium formicicum* exposed to arsenate, because volatile arsines and methylated arsenic species are known to be formed in sewage treatment facilities and municipal landfills and this process is attributed to microbial activity [45]. In the culture medium, arsenite, MA, DMA and an unknown arsenic compound were detected as metabolites of arsenate. Arsine, methylarsine, di-, and trimethylarsine, and an unknown volatile arsenic compound were released into the headspace of the culture. The authors concluded that microbial activity is at least partially responsible for the release of arsenic compounds during waste treatment. In addition, the presence of arsenite as a metabolite in the cultures indicated that anaerobic bacteria also follow the methylation pathway proposed by Challenger. This suggestion is supported by the fact that MA in the culture medium appeared prior to DMA, and monomethylarsine was released into the gas phase prior to dimethylarsine, which itself was formed prior to trimethylarsine. Furthermore, the methylated arsenic compounds MA and DMA in the culture medium were formed before their corresponding arsines.

In a recent study, Koch and coworkers reported for the first time the presence of arsenoriboses in microbial mats [95]. In Canada four samples of microbial mats were collected at a hot spring environment with naturally elevated arsenic concentrations. The microorganisms forming the mats were not definitively identified. Besides inorganic arsenic, which was present in all samples, an orange microbial mat contained MA and DMA, another DMA, and a top layer microbial mat MA. A brown microbial mat contained arsenoribose **II** (Figure 6.1). In the top layer microbial mat, and in the orange microbial mats, arsenoriboses **I** and **II** (Figure 6.1) were present. The mats were thought to contain cyanobacteria. Since cyanobacteria are algae and marine algae are well known to contain arsenoriboses, they could—if present—be a possible source of the riboses in the mats. However, whether the riboses were formed in the mats or taken up from the surroundings cannot be deduced, but the presence of these compounds in the microbial mats gives new perspectives for future investigations into arsenic metabolism in microorganisms.

6.3.4   FRESHWATER ALGAE

Although algae are frequently classified as plants, they will be discussed separately because their structure is different from that of other plants. As mentioned above, marine algae were intensively investigated for arsenic compounds, and the organoarsenicals so far detected in these organisms are arsenoriboses and traces of MA and DMA. TETRA, AB, and AC have not been found in marine algae [2]. Although fewer publications on freshwater algae are available, there have been some studies on the arsenic metabolism in these organisms, most of them investigating algae growing in arsenic-spiked cultures. A variety of data on freshwater algae were reported by Maeda and coworkers [20, 24, 148]. Five arsenic-tolerant freshwater algae, *Chlorella vulgaris, Hydrocolium* sp., *Phormi-*

dium sp., *Nostoc* sp., and *Microchaete* sp., were isolated from microorganisms living in arsenic-polluted sites and cultured in media spiked with arsenate [20]. Cells were extracted and the arsenic compounds in the whole cells as well as in the protein and polysaccharide fraction and in the water- and lipid-soluble fractions were determined. In the whole cells of *C. vulgaris*, traces of mono-, di- and trimethylated arsenic compounds were detected. A trace of dimethylated arsenic compounds was found in *Nostoc* sp. In the protein and polysaccharide fraction of *C. vulgaris* a trace of monomethylated arsenic compounds was present. Dimethylated arsenic compounds were found in the lipid-soluble fractions of all five algae. When the lipid-soluble fraction of *Nostoc* sp. was investigated, the dimethylated arsenic compounds were found to be bound to lipids. No free dimethylated arsenic compounds were detected. The water-soluble fraction of *C. vulgaris* contained mono-, di-, and trimethylated arsenic compounds. Apart from dimethylated arsenic compounds in the water-soluble fraction of *Nostoc* sp., no methylarsenicals were identified in the water-soluble fractions of the other algae.

When *C. vulgaris* grown in arsenate-containing media was transferred into arsenic-free culture medium or pure water, excretion of dimethylated and trimethylated (traces only) arsenic compounds was observed [24]. No arsenic methylation occurred in cell homogenates of *C. vulgaris*, indicating that the enzymes in the homogenate alone are not able to methylate arsenate, and intact cells are required for arsenic methylation. In another experiment on the time-dependence of arsenic metabolism in *C. vulgaris* mono, di- and trimethylated arsenic compounds were detected in the cells and the culture medium. When *C. vulgaris* was grown in media spiked with MA, DMA, or AB, MA was partially metabolized to dimethylated and traces of trimethylated arsenic compounds; DMA and AB remained essentially unchanged, although traces of dimethylated arsenic compounds after the AB administration were detected. Recently, *C. vulgaris* dosed with arsenite ($50\,\text{mg dm}^{-3}$) was reported to contain mainly arsenate and low amounts of arsenite, mono-, and dimethylated arsenic [37].

An investigation of arsenic biomethylation and the excretion of arsenicals by *Phormidium* sp. in arsenate-spiked culture medium showed that mono-, di- and trimethylated arsenic compounds were formed by the cells and excreted into the culture medium [148]. The methylation of arsenic in the cells continued when they were transferred from arsenate-spiked to arsenic-free culture medium. Kaise and coworkers investigated a variety of organisms living in an arsenic-rich freshwater environment in Japan [42]. The green alga *Clodophora glomerata* and a diatom were found to contain more than 80% of the total arsenic in the dimethylated form, about 9% as inorganic arsenic, and about 3% as trimethylated arsenic compounds.

In the studies discussed above, arsenic compounds were determined by HG-AAS. In most cases the samples were digested by heating with sodium hydroxide to detect arsenic compounds like AB or arsenoriboses, that do not form volatile arsines with sodium borohydride. However, although this method provides information on the 'degree of methylation' of arsenic, the specific arsenic compounds present in the sample cannot be definitely identified.

Chlorella spp. grown in arsenate-spiked culture media were also investigated as a possible reference material for arsenic compounds by Goessler and co-workers [87]. The analysis of cell extracts of *Chlorella boehm* showed that a small amount of arsenate was metabolized to arsenite, MA and DMA. An unknown arsenic compound was also detected, when cadmium was present in the growth medium. Recent results indicate that this unknown arsenic compound is arsenoribose **II** (Figure 6.1). DMA was the only methylated arsenic compound excreted into the growth medium by *Chlorella boehm*, *Chlorella kessleri* and *Chlorella* 108.

In 1997 Lai and coworkers were the first to identify an arsenoribose in an algal sample of terrestrial origin. Ribose **I** (Figure 6.1) and DMA were detected as organoarsenicals in commercial products of *Nostoc* sp. purchased in food stores in China, Canada, and Hawaii [132]. Koch and coworkers investigated two samples of green algae collected at a hot spring environment in Canada [95]. Both algae contained arsenoriboses **I** and **II** (Figure 6.1). One of the samples additionally contained the organoarsenic compounds MA and DMA. The presence of arsenoriboses in algal samples of terrestrial origin indicates that freshwater algae probably behave similarly to marine algae, and arsenoriboses were simply not detected before because the analytical techniques formerly employed allowed no definite identification of the arsenic compounds. However, arsenoriboses **III** and **IV** (Figure 6.1), which are often found in marine algae, have not been identified in freshwater algae.

In a recent study, the freshwater alga *Closterium aciculare* isolated from Lake Biwa, Japan, was cultured in an arsenate-containing medium under axenic conditions and the medium was investigated for arsenic compounds [36]. During the exponential phase, arsenate was initially converted to arsenite, whereas methylarsenicals (mostly DMA, but also low concentrations of MA) accumulated in the medium in the stationary phase of the algal growth. In addition to MA and DMA methylarsonous acid and dimethylarsinous acid were also detected. This was the first report of methylarsenic(III) species in algae.

The freshwater algae *Chlorella vulgaris*, *Phormidium* sp., and *Nostoc* sp. were also investigated in experiments on the arsenic metabolism in freshwater food chains [18, 22, 23, 26, 37, 38, 42, 149]. These experiments will be summarized in Section 6.6 below.

6.3.5 HIGHER FUNGI

Total arsenic concentrations in higher fungi are dependent on the soil arsenic concentration as well as on their ability to accumulate arsenic. An example for an arsenic accumulator is *Laccaria amethystina*. Specimens of these fungi collected in central Slovenia were found to contain arsenic concentrations of between 109 and 200 mg As kg^{-1}, although no evidence of soil contamination with arsenic existed in this area [61]. Total arsenic concentrations and organoarsenic compounds detected in fungi (Ascomycetes, Basidiomycetes) are summarized in Table 6.1.

In the two kinds of ascomycetes, MA (*Sarcosphaera coronaria*) [62] or DMA (*Tarzetta cupularis*) [95] accounted for the major part of the extractable arsenic. In two samples of *Sarcosphaera coronaria* an additional chromatographic signal was present, but the authors did not further investigate if this signal was corresponding to DMA, TMAO, or AB. In *Tarzetta cupularis* an arsenoribose [arsenoribose I (Figure 6.1)] was detected for the first time in fungi.

Among the basidiomycetes Agaricales, Boletales, Cantharellales, Gomphales, Lycoperdales, Poriales, and Thelephorales were investigated. In the Agaricales AB was a major arsenic compound in the Agariceae (Table 6.1). Most of the Agariceae contained also MA and DMA. In *Leucocoprinus badhamii* AC and TETRA, which was also present in *Agaricus campester* [108], were detected. Unknown arsenic compounds were reported in *Agaricus elvensis, Agaricus lilaceps, Agaricus silvicola,* and *Leucocoprinus badhamii* [108]. In the Amanitaceae, which also belong to the order of Agaricales, a higher diversity in organoarsenic compounds was observed. All samples contained DMA, whereas MA was only detected in *Amanita magniverrucata* and *Amanita phalloides* [108]. In the latter a trace of AB was also found. This compound was also detected in *Amanita muscaria* [96, 97, 108]. Unknowns, AC, and TETRA were additionally reported in this fungus. The fact that the same organoarsenic compounds were detected in two samples of *Amanita muscaria* of different origin (one from Austria [96, 97], and one from a site not described in detail but not from Austria [108]) indicates that the same kind of fungus has the ability to form or accumulate the same arsenic species without any influence of the area where it grows. In *Amanita caesarea* TETRA was present in addition to DMA [108].

Both Coprinaceae (*Psathyrella candolleana, Coprinus comatus*) contained AB and DMA [125]. In *Psathyrella candolleana* MA and TETRA, and in *Coprinus comatus,* traces of AC and an unknown arsenic compound were detected. DMA was the only organoarsenic compound in *Entoloma rhodopolium* (order: Agaricales; family: Entolomataceae) [108]. This compound was also present in the Plutaceae investigated (Table 6.1) [95, 108], in *Volvariella volvacea* as a major arsenic compound [108]. In this fungus MA and AB were also detected.

The major arsenic compound in the Tricholomataceae was DMA (Table 6.1). Exceptions were the *Collybia* spp. [96], *Lyophyllum conglobatum,* and *Tricholoma pardinum,* in which AB was the dominant organoarsenic compound, and *Tricholoma sulphureum,* which contained mostly inorganic arsenic [108]. AB was also detected in *Laccaria* spp. (Table 6.1); in *Laccaria amethystina* only in one sample grown on highly arsenic-contaminated soil [101]. *Laccaria amethystina* was the only fungus in which TMAO was identified so far. An unknown arsenic compound was additionally reported in this fungus [101]. An unknown arsenic compound was also detected in *Collybia butyracea* [96]. In most

Table 6.1 Organoarsenic compounds in higher fungi (the classification of the fungi corresponds to the *NCU-3. Names in Current Use for Extant Plant Genera* [172])

Class	Order	Family	Fungus species	Total (mg As kg^{-1} dry mass)	MA	DMA	TMAO	AB	AC	TETRA	Riboses	Unknowns	Reference
Ascomycetes	Pezizales	Pezizaceae	*Sarcosphaera coronaria*	360, 332[a]	++			t(nr)					62
				2120	t(nr)		t(nr)	+(nr)					62
				15[b]	+(nr)	+	+(nr)	+(nr)					62
		Pyronemataceae	*Tarzetta cupularis*	t		+					t Ribose I		95
Basidiomycetes	Agaricales	Agaricaceae	*Agaricus abruptibulbus*	3.49	+	+		++					108
			Agaricus bisporus	1.00	+	+		++					108
			Agaricus campester	1.32	t	+		++					108
			Agaricus elvensis	2.43	+	t		++		+			108
			Agaricus fuscofibrillosus	2.68		+		++				+	108
			Agaricus haemorrhoidarius	8.2, 9.3[a]		++(nr)	++(nr)	++(nr)					62
			Agaricus haemorrhoidarius	8.2, 9.3[a]	t	+							62
			Agaricus lilaceps	1.78		+		++				+	108
			Agaricus macrosporus	3.32	++	+		++					108
			Agaricus placomyces	8.1, 9.2[a]	++			++					62
			Agaricus silvicola	6.2		+		++				+	108
			Agaricus subrutilescens	10.8	+	+		+					108
			Leucocoprinus badhamii	2.9	t	+		++	t	t		+	108
			Macrolepiota procera	0.42	t	t							108
		Amanitaceae	*Amanita caesarea*	0.5	t	t							108
			Amanita magniverrucata	0.50		t			+	+			108
			Amanita muscaria	3.1		t		++	+	+		++	108
			Amanita muscaria	21.9	t	+		+	+	+		++	96,97
			Amanita phalloides	0.55	t	t		t					108
			Amanita rubescens	0.1		+							108
		Coprinaceae	*Psathyrella candolleana*	13.6	+	+		+	t	+		t	125
			Coprinus conatus	410		+		++					125
		Entolomataceae	*Entoloma rhodopolium*	0.55		+							108
		Pluteaceae	*Pluteus cervinus*	t		t							95
			Volvariella volvacea	0.82	++	++		++					108
			Volvariella volvacea	1.06	t	++		t					108
		Tricholomataceae	*Collybia maculata*	• 30.0		t		++	t (not resolved)	t (not resolved)			96
			Collybia butyracea	10.9		t		++	t (not resolved)	t (not resolved)		+	96
			Laccaria laccata	0.66	t	++		t					108
			Laccaria laccata	4.26	t	++		+					108
			Laccaria amethystina	109 to 200	+	++							61
				3.4[b]		++	+(nr)	+(nr)					62
				40.5[b]	t	++	t(nr)	+(nr)					62
				23	t	t	t	t					101
				77	++	+	+	+					101
				1420	t	+		t				+	101
			Lyophyllum conglobatum	0.63		+		++					108

Order	Family	Species	mg As/kg							+ Ribose II	Ref
		Tricholoma inamoenum	0.38	t							108
		Tricholoma pardinum	0.63		++		++				108
		Tricholoma sulphureum	0.26		+						108
Boletales	Boletaceae	*Leccinum scabrum*	8.3	++	++			t		+	125
	Paxillaceae	*Paxillus involutus*	36	++	++			t		++	125
Cantharellales	Sparassidaceae	*Sparassis crispa*	1.03	t		t	++				108
			0.57	+		t	+	t		+	108
Gomphales	Gomphaceae	*Gomphus clavatus*	4.47	+	+	++	+	+			108
	Ramariaceae	*Ramaria pallida*	3.7	+	+	++	+ (not resolved)				108
Lycoperdales	Geastraceae	*Geastrum* sp	3.12	+	+	++					108
	Lycoperdaceae	*Calvatia excipuliformis*	0.72		+	++					108
		Calvatia utriformis	0.79		+	++					108
		Lycoperdon echinatum	1.23		+	++					108
		Lycoperdon perlatum	2.81		+	+					108
			0.23[b]	++(nr)	++(nr)	++(nr)	++(nr)				62
		Lycoperdon pyriforme	0.46	t	t	+	t				108
			1010	+	+	+	t	t		t	125
			nr	+	+	++	+	t			125
Poriales	Polyporaceae	*Albatrellus cristatus*	7.7	+	+	++	+	+		+	108
		Albatrellus ovinus	0.26	t	t	+	t	t		t	108
		Albatrellus pes-caprae	0.77			++					108
Thelephorales	Thelephoraceae	*Thelephora terrestris*	15.9	+	++(nr)	++(nr)	+	+			108
		Sarcodon imbricatum	23,24.6[a]	+	+(nr)	+(nr)	+	+			62
			0.9,0.86[a]								62

[a]: different analytical methods

[b]: mg As/kg wet mass

nr: not reported

t = trace (<1 % of extractable arsenic)

+ = minor compound (1–50 % of extractable arsenic)

++ = major compound (> 50 % of extractable arsenic)

nr = not resolved

Tricholomataceae MA was found (Table 6.1). In the *Collybia* spp. AC and/or TETRA were present, but the authors did not definitively distinguish between these two arsenic compounds [96].

In the Boletales DMA was the major arsenic compound (Table 6.1). In *Paxillus involutus* arsenoribose **II** (Fig. 6.1), two unknowns, and traces of TETRA were additionally reported [125]. This was the first report of this ribose in fungi. *Sparassis crispa* (order: Cantharellales, family: Sparassidaceae) was the only fungus in which AC was a major arsenic compound [108]. Additionally, AB and in one sample DMA, TETRA, and two unknowns were detected. AB was the major arsenic compound in the Gomphales, Lycoperdales, and Poriales (Table 6.1). In all fungi of these orders with the exception of *Albatrellus pescaprae* DMA was present [108]. MA was present with the exception of the *Calvatia* spp., *Lycoperdon echinatum*, *Albatrellus cristatus*, and one sample of *Lycoperdon pyriforme* [108, 125]. Both kinds of Gomphales additionally contained AC and TETRA (*Ramaria pallida* contained AC and/or TETRA), whereas unknown compounds were only found in *Gomphus clavatus* [108]. In *Lycoperdon pyriforme* collected at Yellowknife, NWT, Canada, AC, TETRA, and (in one sample) an unknown arsenic compound were additionally reported [125]. In the Poriales *Albatrellus cristatus* and *Albatrellus ovinus* AC, TETRA and unknowns were detected [108]. Among the Telephorales *Telephora terrestris* was reported to contain MA as the only organoarsenic compound [108]. In *Sarcodon imbricatum* AB, TETRA, DMA and MA (in one of two samples) were found [62].

A systematic study on the metabolism of different arsenic compounds in fungi was conducted by Slejkovec and coworkers [107]. The oyster mushroom *Pleurotus* sp. was grown on straw spiked with arsenate. At a concentration of $50\,mg\,As\,kg^{-1}$ straw, methylation occurred and MA was detected in the fruit bodies of the fungus after 55 days. At lower arsenate concentrations no methylation was observed. In a second experiment, mycelium of *Agaricus placomyces*, a fungus containing mainly AB under natural conditions, was cultured under sterile conditions on nutrient medium plates spiked with arsenite, arsenate, MA, DMA, AB, TMAO, or TETRA, and the mycelium and medium were analyzed for arsenic compounds after 39 days' incubation. AB and TETRA were accumulated by the mycelium. No methylation of inorganic arsenic was observed. MA was partially methylated to DMA. DMA, AB, and TETRA remained unchanged except for a small amount of arsenite (1% of total arsenic present) detected in the mycelium grown on the DMA-spiked medium. Traces of TMAO were metabolized to AB detected in the mycelium, and to DMA detected in both mycelium and growth medium, but more than 96% and almost 100% remained as TMAO in the growth medium and mycelium, respectively. The methylation of MA to DMA in a sterile culture of the mycelium, where no microorganisms are present, is a proof that mycelium is capable of methylating arsenic. The authors discussed that a growth period of 39 days could be too short to detect other methylation processes, or some methylation steps are taking place in the fruit bodies, which were not investigated in this study.

Generally, a large number of different organoarsenic compounds first detected in the marine environment have been identified in fungi. In 1995 AB was reported for the first time in fungi [62]. In the following years this organoarsenic compound was found to be contained in a variety of fungi, in many of them even as a major arsenic compound (Table 6.1). Arsenocholine, first detected in 1997 [96, 97], was identified in several fungi, but it does not seem to be as widespread as AB. TETRA was also reported less frequently than AB. In 1999 an arsenoribose was first detected in a fungus [95]. Whether arsenoriboses also are common arsenic compounds in fungi has to be confirmed in the future. The identification of organoarsenic compounds was favored by the fact that some fungi accumulate arsenic to the mg kg^{-1} level, even when they grow on non-contaminated soil, and therefore contain enough arsenic for analysis. However, if the organic arsenic compounds are formed by the fungi themselves or by microorganisms in the soil and then taken up by the fungi is not certain yet. The fact that the mycelium of *Agaricus placomyces* is able to methylate MA to DMA [107] shows that at least this step can occur in the fungus itself. The finding that AB and TETRA are accumulated from the culture medium into the mycelium of *Agaricus placomyces* indicates that organoarsenic compounds with low toxicity may be taken up and accumulated by the fungus, even if they are present only in low concentrations in the surrounding environment.

6.3.6   LICHENS

Only few data about arsenic compounds in lichens are available. Lichens are symbiotic organisms of algae and fungi. Therefore, they should combine the characteristics of these two forms of life and contain arsenic compounds known to be present in these organisms. The investigation of the arsenic compounds in organisms collected at the Canadian hot spring environment conducted by Koch and coworkers included samples of the lichens *Alectoria* sp. (three samples), *Bryoria* sp. (two samples), and *Cladonia* sp. [95]. Arsenoribose **II** (Figure 6.1) was the only organoarsenic compound detected in two samples of *Alectoria* sp. and one sample of *Bryoria* sp. *Cladonia* sp. contained traces of DMA and arsenoribose **I** as organoarsenic compounds. Five *Cladonia* sp. from Yellowknife, NWT, Canada, from an area with elevated arsenic levels were reported to contain the organoarsenic compounds AB (in all five samples), MA and DMA (in four of the samples each), TMAO (in three of the samples), TETRA and an unknown arsenic compound (in one of the samples each), and arsenoribose **I** (in two of the samples) [125]. In *Cladina* sp. from the same area DMA, AB, TMAO, and an unknown were found. An unidentified lichen contained MA, DMA, and traces of AB [125]. Arsenoribose **I** was identified in *Alectoria ochroleuca* and *Usnea articulata* from a former arsenic roasting facility in Austria [99]. In these lichens, AB, TMAO, and an unknown arsenic compound were also present. *Alectoria ochroleuca* additionally contained TETRA and traces of MA. TMAO was the major arsenic compound in extracts of *Usnea articulata*, AB in extracts of *Alectoria ochroleuca*. AB was thought to come from the fungus, while the algal component of the lichens was the probable source of arsenoribose **I**.

Therefore, all organoarsenic compounds identified in higher fungi so far were also identified in lichens. The only exception is AC which has never been detected in extracts of lichens until now.

6.3.7   GREEN PLANTS

For a long time organic arsenic compounds in terrestrial plants seemed to be restricted to the 'simpler' methylated compounds MA, DMA, and TMAO [3]. Since AB was reported for the first time in a fungus sample of terrestrial origin in 1995 [62], a variety of organoarsenic compounds formerly attributed only to marine organisms has been detected also in terrestrial green plants. As mentioned above, uncontaminated green plants usually contain 0.2 to 0.4 mg As kg^{-1} dry mass. With respect to non-quantitative extraction procedures and dilution due to the extraction, arsenic speciation analysis in these organisms requires analytical methods with detection limits at the low or sub μg dm^{-3} level. Green plants from contaminated areas, e.g. mine wastes, were found to contain arsenic up to the high mg kg^{-1} level [3]. Tomato leaves were reported to contain MA and DMA [3]. Tissue cultures of *Cantharantus roseus* are able to metabolize arsenate to a low extent to MA and DMA [3]. Aggett and Kadwani detected MA in the lakeweed *Largarosiphon major* collected at a region in New Zealand with elevated arsenic concentrations as a result of geothermal activity [58].

Recent work on arsenic compounds in green plants is rather limited. In 1998, Heitkemper and coworkers determined the arsenic compounds in baby-food carrots grown on contaminated soil in Texas [89]. MA and two unknown arsenic compounds were present in addition to inorganic arsenic, whereas only inorganic arsenic was detected in a control sample. Vela *et al.* reported MA as major arsenic species in four carrot samples [117]. In two of the samples an unknown arsenic compound was additionally detected. DMA was identified in rice [27, 116, 123, 150], beet and cane sugar, fruits, and fruit juices [27], as well as in apples [115]. The presence of MA was reported in rice [27, 150], apples [115], apple juice [27], and yams [150].

AB, TMAO, and TETRA at the low μg kg^{-1} dry mass level were reported in the green plants *Dactylis glomerata*, *Trifolium pratense*, and *Plantago lanceolata* collected from arsenic contaminated soil in Austria [151]. A trace of AC was detected in *Plantago lanceolata*. Mattusch *et al.* reported the presence of DMA in monocots (*Calamagrostis epigejos*, *Phleum pratense*, *Carex leporina*, *Agrostis capillaris*) and dicots (*Silene vulgaris*, *Polygonum persicaria*) grown on a reference soil mixed with mine tailing material [118]. MA was only detected in *Carex leporina*. AB was found in *Silene vulgaris, Polygonum persicaria, Calamagrostis epigejos*, and *Agrostis capillaris*, and AC and/or TMAO were found in *Silene vulgaris, Calamagrostis epigejos, Phleum pratense*, and *Carex leporina*. When *Holcus lanatus, Juncus effusus*, and an unidentified grass were grown on ferrous oxyhydrate precipitate formed in the seepage of the dam foot of the mine tailings, AB as well as AC and/or TMAO were detected in these plants. DMA was present in *Holcus lanatus* and the unidentified grass [118].

Koch and coworkers investigated the water soluble fraction of thirteen plants, most of them submergent, collected at a gold mining area in the North West Territories in Canada, where the mined ore is associated with arsenopyrite and arsenic is subsequently released into the environment during the smelting operation [152]. Methylated arsenic compounds were detected in *Hordeum jubatum* (MA, DMA), *Agrostis scabra* (MA, DMA), *Carex* sp. (MA, DMA, TETRA), *Bidens cernua* (MA, DMA, TETRA), *Typha latifolia* (MA, DMA, TETRA), *Lemna minor* (MA, DMA, TETRA), *Potomogetan richardsonii* (DMA, TETRA), *Myriophyllum* sp. (MA, DMA, TETRA), and *Drepanogladus* sp. (MA, DMA, TETRA). Arsenoriboses were present in some of the submergent plants, namely *Lemna minor* (arsenoriboses **I** and **II**, Figure 6.1), *Potomogetan richardsonii* (arsenoribose **I**), and *Myriophyllum* sp. (arsenoriboses **I** and **II**). Since these plants were growing in physical contact with algae, the arsenoriboses could have derived from algal sources. MA was the only organoarsenic compound in *Scirpus* sp. collected at a Canadian hot spring environment with naturally elevated arsenic concentrations in November 1996, DMA in *Scirpus* sp. collected in July 1997 [95]. These two organoarsenicals were also detected in *Thuja plicata*. In *Fumaria hygrometrica* collected in November, arsenoribose **I** was present. In one sample collected in July, DMA and arsenoribose **II** were additionally detected; in another sample collected in July, MA and DMA were identified in addition to arsenoribose **I**. DMA was the only organoarsenic compound present in *Erigeron* sp., TETRA in *Mimulus* sp. The rather surprising presence of arsenoriboses in green plants was confirmed by investigations into the arsenic compounds in twelve different green plant samples (leaves of the ferns *Asplenium viride* and *Dryopteris dilata*, the grass *Deschampsia cespitosa*, needles of the trees *Picea abies* and *Larix decidua*, leaves of the tree *Alnus incana*, as well as leaves of *Vaccinium myrtilis, Vaccinium vitis idaea, Rubus idaeus, Fragaria vesca, Achillea millefolium*, and *Equisetum pratense*) collected at a former arsenic roasting facility in Austria [99]. Apart from inorganic arsenic all samples contained TMAO. DMA was detected in all samples except *Larix decidua, Vaccinium vitis idaea*, and *Equisetum pratense*, MA in all samples except *Dryopteris dilata, Larix decidua*, and *Rubus idaeus*. TETRA was identified in the ferns and in *Fragaria vesca* and *Vaccinium vitis idaea*. Arsenoribose **I** (Figure 6.1) was detected in all green plants except *Larix decidua* in concentrations below 11 μg As kg^{-1} dry mass. Additionally, a trace of AB was present in *Deschampsia cespitosa*.

The identification of organoarsenicals like AB, AC, TETRA, or arsenoriboses in terrestrial green plants confirmed the theory, that 'complex' organoarsenic compounds are constituents of the terrestrial environment, although at much lower concentrations than in the marine environment. However, the major part of arsenic in terrestrial green plants is inorganic arsenic.

6.4 PRESENCE AND TRANSFORMATION OF ARSENIC SPECIES IN ANIMALS

6.4.1 NON-MAMMALS

Many investigations of arsenic compounds in non-mammals were conducted on freshwater animals. In most cases the uptake and transformation of arsenic spiked to the water was studied.

The uptake and biotransformation of arsenate from water by the planktonic grazer *Moina* sp. was investigated by Maeda and coworkers in two studies [22, 149]. In one experiment *Moina* sp. metabolized 50, 37, or 24%, respectively, of arsenate accumulated from water spiked with arsenate at concentrations of 0.1, 1, or 2 mg As dm^{-3}, respectively, to mono-, di- and traces of trimethylated arsenic compounds. Dimethylated arsenic compounds were the major form of organic arsenic [149]. In the other experiment *Moina* sp. metabolized 55% of the arsenate taken up from arsenate-spiked water containing 1 mg As dm^{-3} to dimethylated arsenic and traces to mono- and trimethylated arsenic compounds [22]. When the zooplanktonic grazer *Daphnia magna* was exposed to arsenite, it contained about 1% of the total arsenic as dimethylated arsenic and less than 1% as monomethylated arsenic [37].

Depending on the arsenate concentration in water (0.1 to 1.5 mg As dm^{-3}) the shrimp *Neocardina denticula* metabolized 9% (at a water arsenic concentration of 1 mg As dm^{-3}) to 32% (at a water arsenic concentration of 0.5 mg As dm^{-3}) of the arsenate accumulated from water to di- and trimethylated arsenic compounds [23]. Monomethylated arsenic compounds were only detected in traces. Kuroiwa and coworkers exposed *N. denticula* to arsenate, MA, DMA, or AB spiked to the water [18]. After exposure to arsenate, 10 to 30% of the arsenic in the shrimp were present as di- and trimethylated arsenic compounds. When MA was administered, 10 to 20% of the arsenic in the shrimp were as dimethylated arsenic compounds and 20 to 40% were inorganic arsenic. After DMA administration 10 to 60% of the arsenic in the shrimp was inorganic arsenic. When AB was added to the water to a final concentration of 1 mg As dm^{-3}, about 40% of the arsenic in *N. denticula* were dimethylated arsenic compounds and about 20% inorganic arsenic. At AB concentrations higher than 100 mg As dm^{-3} almost all arsenic was present as trimethylated arsenic. The authors discussed that *N. denticula* is capable of both methylation and demethylation of arsenic. Inorganic, mono-, di-, and trimethylated arsenic compounds were excreted into the water by *N. denticula* dosed with arsenate, MA, DMA, or AB. When *N. denticula* was exposed to arsenite (0.1, 0.5, or 1.5 mg dm^{-3}) 32%, 18%, or 7%, respectively, of the arsenic in the shrimp was dimethylated arsenic [37].

The freshwater prawn, *Macrobrachium rosenbergii*, was exposed to arsenate, MA, or DMA through the water by Kuroiwa and coworkers [19]. Arsenate was partially transformed to di-(about 1%) and trimethylarsenic (5 to 8%) compounds. Between 33 and 60% of MA accumulated from the water by the prawn was demethylated to inorganic arsenic, about 1.5% was methylated to dimethyl-arsenic compounds. When the prawn was kept in water containing DMA at concentrations of 100, 200, or 300 mg As dm^{-3} between 3% (300 mg As dm^{-3} water) and 23% (100 or 200 mg As dm^{-3} water) of the DMA taken up were demethylated to inorganic arsenic. Only at a DMA concentration of 300 mg As dm^{-3} monomethylated arsenic compounds (24% of total arsenic) were detected in the prawn. No further methylation of DMA was observed in the prawn. Kaise and coworkers investigated a variety of organisms including the freshwater prawn, *Macrobranchiura nipponense*, collected from an arsenic-rich freshwater environment in Japan [42]. Neither inorganic arsenic nor monomethylated arsenic compounds were found in the prawn. Dimethylated arsenic compounds accounted for 75%, and trimethylated arsenic compounds for 23% of the total arsenic in the prawn. Marsh snails (*Semisulcospira libertina*) collected at the same location also contained di- (27% of total arsenic) and trimethylated (62% of total arsenic) arsenic compounds. About 96%, 86%, and 58%, respectively, of the total arsenic in the larva of the dobsonfly (*Plotohermes grandis*), in the larva, and in the pupa of the caddis fly (*Stenopsyche marmorata*) were monomethylated and about 2%, 9%, and 41%, respectively, were trimethylated arsenic compounds.

Several species of freshwater fish were exposed to arsenate spiked to the water and the metabolites in the animals were studied. The goldfish, *Carassius carassius auratus*, accumulated arsenate from the water (0.5 or 1 mg As dm^{-3}) and methylated 3 or 28% of the arsenic [149]. The methylated arsenic was mainly present as trimethylated arsenic, but mono- and dimethylated arsenic compounds were also detected. When no arsenate was spiked to the water, 4% of the arsenic contained in the goldfish were in the methylated form (trimethylated arsenic compounds, traces of mono- and dimethylated arsenic compounds). About 9, 1, or 15% of arsenate taken up from the water (0.5, 1, or 10 mg As dm^{-3}) by the guppy *Poecilia reticulata* were metabolized to mono-, about 2% to di- and about 16, 12, or 7%, respectively, to trimethylated arsenic compounds [22]. When the guppy was kept in arsenic-free water 30% of the total arsenic in the fish was in the dimethylated form, 3% in the mono-, and 9% in the trimethylated form. About 13% (arsenic concentration in water 10 mg As dm^{-3}) to 33% (arsenic concentration in water 50 mg As dm^{-3}) of arsenate taken up from the water at concentrations of 10 to 70 mg As dm^{-3} by the carp *Cyprinus carpio* was present as mono-, di-, and trimethylated arsenic compounds in the remnants including skin, scale, bone, and fin [26]. In the gut, 24% (arsenic concentration in water 10 mg As dm^{-3}) to 73% (arsenic concentration in water 60 mg As dm^{-3}) of the total arsenic was in the mono-, di-, and trimethylated form. When the carp was kept in arsenic free water, less than 5% of the total arsenic in the remnants and gut were methylarsenic. No monomethylated arsenic was present in this case. In the meat 5% (arsenic concentration in water 10 mg dm^{-3}) to 41% (arsenic concentration in water 60 mg dm^{-3}) were present as mono-, di-, and trimethylated arsenic compounds, whereas the only methylated form of arsenic was trimethylated arsenic (10% of the total arsenic), when *C. carpio* was kept in arsenic free water. The killifish *Oryzias lapites* transformed about 20 to 40% of arsenate accumulated from arsenate-spiked water (1 to 20 mg As dm^{-3}) to mono-, di-, and trimethylated arsenic compounds, of which the monomethylated form was predominant [18]. A variety of different freshwater fish (*Plecoglossus altivelis, Oncorhynchus masou masou, Rhinogobius* sp., *Phoxinus steindachneri, Tribolodon hakonensis, Sicyopterus japonicus*) from the arsenic-rich Japanese freshwater environment mentioned above, were investigated by Kaise and coworkers [42]. Di- (0.8 to 76% of total arsenic in the fish) and trimethylated arsenic compounds (20 to 78% of total arsenic in the fish) but no inorganic or monomethylated arsenic compounds were detected in all fish species.

When the freshwater carnivorous fish *Tilapia mossambica* was exposed to arsenite in the water (0.1 or 1 mg dm^{-3}), a high percentage of the arsenic ($\sim$70%) in *T. mossambica* was trimethylated arsenic [37]. Traces of mono- and dimethylated arsenic were detected in this fish at an arsenite concentration of 0.1 mg dm^{-3} in the water. At an arsenite concentration of 5 mg dm^{-3}, the percentage of trimethylated arsenic decreased to 12% of the total arsenic. Mono-(5%) and dimethylated arsenic (8%) were also detected. When *T. mossambica* was exposed to arsenate (10 mg dm^{-3}) for 7 days, organic arsenic compounds in the mono-, di-, and trimethylated forms were detected in the liver, intestine, ovary, bone, muscle, gill, and eyes of the fish. In the liver all the arsenic was present as methylated arsenic [38]. In the brain, which showed the highest total arsenic concentration of all tissues investigated, no organoarsenic (with the exception of a trace of trimethylated arsenic) was found. When *T. mossambica* was fed on artificial diets containing arsenite ($\sim$ 1.4 mg kg^{-1}) or arsenate ($\sim$ 1.9 mg kg^{-1}) for 7 days, about 30% of the arsenic in the fish was trimethylated arsenic. Traces of dimethylated arsenic were reported in case of the artificial diet containing arsenite [38].

However, in none of the investigations summarized above, were the arsenic compounds present in the animals definitely identified. From the degree of methylation no conclusions can be made concerning parallels between marine and freshwater animals. In 1995 Shiomi and coworkers investigated the arsenic compounds in the freshwater fish species *Salmo gairdneri* (rainbow trout) and *Hypomesus nipponensis* (Japanese smelt), purchased from a market in Tokyo, by HPLC-ICP-AES, which allows definite identification of the arsenicals preent in the samples [128]. The major arsenic compound in the muscle of both ishes was identified as AB, accounting for more than 80 % of the total arsenic n the muscle of *S. gairdneri* and for about 60 % of the total arsenic in the muscle of *H. nipponensis*. Since *H. nipponensis* was a wild specimen, these results uggest the natural occurrence of AB in freshwater ecosystems. AB detected in *. gairdneri*, of which cultured specimens were analyzed, was mainly attributed o AB contained in the assorted feed. Biotransformation of arsenate, DMA, \B, TMAO, and AC after intraperitoneal injection in the carp *Cyprinus carpio* vas investigated [106]. Intraperitoneal injection was preferred to oral adminis-ration to avoid the transformation of arsenic compounds by bacteria in the gut)f the carp. Arsenic compounds were determined in the viscera and muscle of he fish. Arsenate was quantitatively reduced to arsenite within 72 hours. DMA ind AB remained unchanged. TMAO was partially metabolized to another irsenic compound, which was probably DMA. AC was partially transformed o AB. When catfish and rainbow trout were investigated by Schoof and co-vorkers in a market basket survey of arsenic in food, DMA was detected in the ·ainbow trout [27]. The arsenic compounds in muscle and kidney of freshwater ish (bream) stored in the German Environmental Specimen Bank were deter-nined by Slejkovec and coworkers [74]. Besides inorganic arsenic MA, DMA, ΓETRA, and TMAO and/or AB (the chromatographic separation applied was iot able to distinguish between AB and TMAO) were detected in these samples. Γhe same arsenic compounds were detected in the freshwater mussel, *Dreisena* ·olymorpha*, also investigated in this study.

The confirmation of AB in freshwater fish in recent years strongly indicates)arallels between marine and freshwater animals. Whether AB is taken up or ormed within the fish has not been ascertained yet. However, the results)btained for carp fed with different arsenicals did not indicate any methylation :apability by the fish. Since the arsenic compounds were administered intraper-toneally and not orally in this study in order to investigate their transformation)y the fish and not by bacteria in the gut of the fish, methylation of arsenic aken up orally with the food by bacteria living in the gut of this fish cannot be ·xcluded.

Only a few data are available on non-aquatic, non-mammalian animals. Ants :ollected from an anthill located at a former arsenic roasting facility in Austria vere reported to contain DMA and traces of MA and AB in addition to norganic arsenic [98]. In earthworms from six sites in Styria, Austria, with ioil arsenic concentrations above 5 mg kg^{-1}, DMA, AB, arsenoribose **I** and irsenoribose **II** (Figure 6.1) were identified in addition to inorganic arsenic and wo unknown arsenic compounds [86]. Zheng and coworkers reported the)resence of inorganic arsenic, MA, DMA, AB, and an unknown arsenic com->ound in earthworms collected in Styria, Austria [113]. The presence of AB and :specially of arsenoriboses in these two animals again indicates that 'complex')rganic arsenic compounds are more widespread in the terrestrial environment han was formerly believed.

6.4.2 MAMMALS

A variety of studies have been conducted on arsenic metabolism in mammal exposed to different arsenic compounds, sometimes in order to draw conclu sions concerning the behavior of arsenic compounds in humans. The mammal most often investigated were mice, rats, hamsters, and rabbits. Sometimes dogs pigs, and monkeys were used as test animals. The arsenic compounds present in urine after exposure of the animal to an arsenic compound were most fre quently determined. The advantage of urine is the fact that it is a liquid and can be analyzed directly in many cases. Feces, blood and several kinds of tissue were also investigated in some of the studies summarized below.

6.4.2.1 Mice

When inorganic arsenic (arsenite or arsenate) was administered orally or sub cutaneously to mice, DMA was the major arsenic compound (44 to 83% of the administered dose) detected in urine after 48 hours [76]. A low percentage (about 1%) was present as MA. About 8 to 41% of the administered dose was excreted unchanged as inorganic arsenic. Arsenite was methylated to a higher extent than arsenate. Arsenite was also metabolized to DMA after intravenous injection [141]. About 68% of the administered dose was excreted as DMA in the urine, 23% as inorganic arsenic. After intravenous injection of arsenate, more inorganic arsenic (52% of administered dose) than DMA (40% was detected in the urine. Besides inorganic arsenic, DMA was the main metabolite in the ultrafiltrated fractions of plasma, and liver, kidney and lung tissue. MA accounted for less than 1% of the total ultrafiltrable arsenic. The percentage of DMA calculated on the total arsenic concentration in tissue ultrafiltrates was highest in lung tissue (more than 90% DMA) and lowest in liver tissue (between 45 and 60% DMA). Arsenite exposure lead to highe DMA concentrations in the plasma ultrafiltrate than arsenate exposure. Inor ganic arsenic was better retained in the tissue than DMA. Therefore highe methylation rates resulted in faster excretion of the administered arsenic. In another study arsenite (0.3 to 3.4% of the administered dose), arsenate (1.3 to 27%), and DMA (2 to 9%) were detected in bladder urine of mice 1 hour after oral administration or intravenous injection of arsenite or arsenate [77]. MA accounted for less than 1%. Again arsenite was methylated more efficiently than arsenate. The reduction of administered arsenate to arsenite supports the consecutive reduction–oxidative methylation pathway also for arsenic bio methylation in mammals. The same arsenic compounds as in the urine were detected in the plasma, which contained about 1% of the administered dose.

Mice were also exposed to organic arsenic compounds and the metabolites o these compounds were investigated. DMA was excreted almost quantitatively and unchanged in the urine of mice after intravenous injection [141]. In anothe study, DMA orally administered to mice was excreted as DMA (56 and 24% of the administered dose), and a DMA complex (7.7 and 4.9%) in urine and feces during 48 hours after administration. TMAO (3.5% of the administered dose) was detected in the urine, indicating that DMA is not unequivocally the end point of methylation [44]. No demethylation was observed. Orally administered AB was not biotransformed in mice [78]. After oral or intravenous administration of AC about 40% of the administered dose was excreted as AB in the urine within 48 hours [153]. About 20% (intravenous administration) or less than 10% (oral administration) of the dose was excreted as AC via the urine. In the tissues, AB and arsenic in the phospholipid fraction were present. When arsenoriboses isolated from a marine red alga were dosed orally to mice, 86% of the dose was excreted as arsenoriboses in the feces, indicating no adsorption from the gastrointestinal tract [127]. MA, DMA, and AB, but no riboses, were detected in the urine after oral or intravenous administration, which led to the excretion of 13% and 92%, respectively, of the administered dose via the urine. Since the definite identity of the arsenoriboses in the alga was not determined, it can only be concluded that they are biotransformed, but the detailed pathway remains unclear.

Generally the main metabolite of inorganic arsenic in mice is DMA. MA is present in negligible amounts. DMA, which is less toxic than inorganic arsenic, is also excreted more rapidly than inorganic arsenic. The end point of arsenic methylation in mice is DMA in most cases. However, partial methylation of DMA to TMAO was reported. Demethylation of organic arsenic compounds in mice was not observed.

6.4.2.2 Rats

When arsenite or arsenate were orally administered to rats 13% or 2.3% of the administered doses, respectively, were excreted as inorganic arsenic, about 0.3% as MA, and about 4% as DMA via the urine after 48 hours [76]. Compared with mice, methylation was lower and, therefore, body retention was higher, since inorganic arsenic is excreted more slowly from the body than is methylated arsenic. When arsenite was dosed intraperitoneally, about 5% of the administered dose was detected in the urine within 48 hours, mainly as DMA [154]. Inorganic arsenic and small amounts of MA were also present. These compounds were also found in the ultrafiltrable fractions of the plasma, liver, and kidney. The urinary excretion of arsenic compounds before and after intraperitoneal administration of arsenite was studied by Buchet and Lauwerys [155]. Prior to administration of arsenite, mainly DMA and inorganic arsenic, but no MA, were excreted. After administration, DMA, MA, and inorganic arsenic were present. Additional administration of cofactors which are known to be involved in methylation processes resulted in no changes in the urinary excretion of arsenic compounds. A reduction of the hepatic GSH (reduced glutathione) level exceeding 90% caused decreased urinary excretion of MA and DMA and increased excretion of inorganic arsenic. Cornelis and De Kimpe also reported the presence of DMA besides lower amounts of MA, inorganic and protein-bound arsenic in rat urine after intraperitoneal injection of arsenate [65]. Gonzales and coworkers found that dichromate decreased total arsenic excreted via the urine but supported the methylation of orally administered arsenate in rats by favoring the reduction of arsenate to arsenite [68]. These results confirm that a reduction step is required when arsenate is biomethylated. When the effect of the arsenate dose was studied for high arsenate concentrations (160 μg kg^{-1} body weight) the feces became the major excretion route and enzymatic saturation resulted in a decrease of DMA in the urine.

Yoshida and coworkers investigated urinary excretion of arsenic compounds by rats before and after long-term oral administration of various arsenic compounds [111]. AB was the major arsenic compound in the urine of the unexposed rats after one week and seven months. Arsenate, MA, DMA and TMAO were also detected. In the urine of arsenite treated rats arsenite, TETRA, and two unidentified metabolites were also present. The concentrations of DMA and TMAO in the urine were markedly elevated after 1 week and after 7 months, and the concentration of arsenate after 7 months compared to the control rats. Several organic arsenic compounds were also administered to rats in this study. After exposure to MA, TETRA and the two unknown metabolites were detected in addition to the compounds found in the urine of the control rats. The concentrations of MA, DMA, and TMAO were elevated after 1 week and after 7 months. After DMA exposure the same compounds as after MA exposure were excreted via the urine. In this case DMA and TMAO were significantly elevated. In a recent study rats were exposed to DMA via drinking water [119]. A third unidentified metabolite was detected after 20 weeks of exposure. This metabolite was only found in the feces but not in the urine of the rats, whereas the other two unknowns were present in both urine and feces. Experiments with bacterial cultures isolated from the cecal content of rats indicated that intestinal bacteria are involved in the formation of these three unknowns [120]. When the rats were administered TMAO, TETRA and the two unknown metabolites were found compared with the control group [111]. The concentrations of MA, DMA and TMAO were higher than in the control group. In contrast to a previous work reporting no transformation of orally dosed AB in the rat [78], AB administration resulted in the detection of TETRA and the two unknown metabolites, but no MA was present. The concentration of TMAO was markedly elevated.

These results indicate that DMA is also not the end point of arsenic methylation in rats. Further methylation to TMAO and even TETRA can obviously occur. Since most of the older investigations into arsenic compounds in rats only included the 'common' arsenic compounds in urine, which are arsenite, arsenate, MA, and DMA, other organoarsenic compounds were probably not detected, although they could have been present. The elevated MA concenration in the urine after 7 months of DMA administration suggests that demethylation processes, possibly caused by intestinal microorganisms, can take place in addition to methylation processes.

Other studies conducted on the metabolism of DMA orally administered to rats confirmed the presence of arsenite, MA, and TMAO as metabolites in addition to DMA in the urine [91, 94]. In one work an unidentified metabolite (also after intraperitoneal administration) [110], in another work AB and two unidentified metabolites were also detected [126]. No MA but an unknown arsenic metabolite was detected by Chen and coworkers in feces and urine in addition to arsenite, DMA, and TMAO after oral long-term exposure of rats to DMA [83]. In a recent work Inoue and coworkers reported MA, DMA, TMAO, AB, TETRA, and two unknown arsenic metabolites to be present in rat urine after chronic exposure to DMA [92]. AB was also detected in control rats and, therefore, attributed to the food administered. TETRA was not present in the urine of the control rats. When AC was administered orally or intravenously to rats, AB was the major arsenic compound in addition to small amounts of AC detected in the rat urine [153]. Rats exposed to arsine gas excreted inorganic arsenic, MA, and DMA via the urine [14]. AB was not a major metabolite and was only detected at low concentrations when AB-free food was administered.

A variety of different arsenic compounds including unidentified species were detected, when rats were exposed to different arsenic compounds. Methylation even to TETRA was observed in some experiments. Additionally, demethylation of arsenic also occurred. Recently, the presence of methylarsonous acid in the bile of rats after injection of arsenite or arsenate was reported [121].

5.4.2.3 Hamsters

Most of the studies on arsenic metabolism in hamsters were conducted by Yamauchi and coworkers. When arsenic trioxide was administered orally to hamsters, inorganic arsenic, MA, and DMA were detected in the brain, kidneys, lungs, muscle, and skin [32]. In the liver trimethylated arsenic was additionally found. Inorganic arsenic and MA were present in the spleen, inorganic arsenic only in the hair. The concentration of MA was higher than of DMA. The concentration of MA increased prior to DMA. In a control group the only arsenic species in organs and tissues was inorganic arsenic. In the blood, inorganic arsenic, MA, DMA but no trimethylated arsenic compounds were present. About 50 % of the administered dose was excreted within 5 days via the urine mainly as inorganic arsenic and DMA, lower amounts also as MA and trimethylated arsenic compounds, which were also excreted by control animals and may have derived from the food. About 11 % of the administered dose was detected in the feces, but no trimethylated arsenic compounds were present. Arsenite, arsenate and DMA were excreted via the urine of hamsters administered arsenite, arsenate, As_2S_3, or $Pb_3(AsO_4)_2$ orall or intratracheally [73].

The metabolism of a variety of organoarsenic compounds in hamsters wa investigated. After oral administration of MA, DMA was detected in additio to MA in whole blood, plasma, brain, kidney, liver, lung, muscle, skin an spleen [33]. Inorganic arsenic was also present in whole blood, plasma, lung muscle, skin, and spleen and was the only form of arsenic in the hair of th hamsters but, just like trimethylated arsenic, it did not seem to derive from th MA administration when compared with control animals. Oral administratio of MA favored excretion via the feces, intraperitoneal administration favore excretion via the urine. MA (about 30 % of the administered dose), DM (between 0.4 and 8 % depending on dose), and trimethylated (2 % at a dose o 5 mg MA kg^{-1} body weight) were detected in the urine, MA (about 50 % of th administered dose) and DMA (1 % at a dose of 5 mg MA kg^{-1} body weight) i the feces after oral administration of MA. From the results it was conclude that MA remains almost unchanged when administered to hamsters.

When DMA was orally administered to hamsters, it was partially methylate to trimethylated arsenic compounds, detected in liver, lungs, kidneys, muscle brain, and spleen, whereas trimethylated arsenic compounds were only found i the liver of control animals [30]. MA was detected in the liver, brain, kidneys and lungs, but only in the liver and brain of the control group. Compared wit the control animals the concentration of inorganic arsenic in most organs an tissues did not vary significantly with the administration of DMA. However inorganic arsenic was slightly increased in the liver, lungs, kidneys, and muscle Inorganic arsenic, DMA and trimethylated arsenic compounds were detected i whole blood of control hamsters. In DMA-exposed hamsters, DMA, an trimethylated arsenic compounds were increased and MA was additionall present. The concentration of inorganic arsenic was slightly higher than in th controls. Within one day 45 % of the dose was excreted via the urine, whic contained inorganic arsenic, MA, DMA and trimethylated arsenic compounds DMA and trimethylated arsenic compounds were markedly increased com pared to controls. About 35 % of the dose was found in the feces, whic contained no trimethylated arsenic compounds; inorganic arsenic and M were slightly, DMA was markedly increased. The results indicate that DM is partially methylated in hamsters. It is not certain, whether demethylatio occurs, since MA and inorganic arsenic were not (or only slightly) increase The trimethylated arsenic compound in hamster urine after oral DMA adminis tration was identified as TMAO by Marafante and coworkers [44]. The majo arsenic compound in urine and feces was DMA. Also in this study no TMA was excreted via the feces. No inorganic arsenic was detected. AB orall administered to hamsters was rapidly absorbed through the digestive glan and excreted [156]. No MA or DMA was detected in organs, tissues, or whol blood and the concentration of inorganic arsenic was similar to controls. Onl the concentration of trimethylated arsenic compounds was increased indicatin to demethylation of AB in hamsters. After subtraction of the arsenic compounds excreted by control hamsters it was found that trimethylated arsenic compounds were the only form of arsenic elevated in urine and feces. Since the definite identity of trimethylated arsenic compounds was not determined, the conversion of AB to other trimethylated arsenic compounds cannot be excluded. Orally administered trimethylarsine was excreted via the expired air or via the urine, where it was present as TMAO [35]. No demethylation and no conversion to AB occurred. TMAO was rapidly excreted via the urine by hamsters without conversion [34]. Trimethylarsine was detected in the expired air. Orally administered gallium arsenide, a compound commonly used in the electronic industry, was mainly excreted via the feces. Intraperitoneally administered gallium arsenide was excreted poorly via both the urine and the feces [157]. Oral administration was followed by the appearance of inorganic arsenic, MA, and DMA in organs and tissues, but concentrations were low. DMA was the main metabolite in the urine.

Like in rats, DMA is not the end point of arsenic methylation in hamsters. DMA is further methylated to trimethylated arsenic, which was identified to be TMAO, but no *in vivo* formation of AB was observed. Methylation to TETRA reported to take place in rats was not found in hamsters. No demethylation of organic arsenic compounds in hamsters occurred.

5.4.2.4 Rabbits

In rabbits the transformation of inorganic arsenic was most intensively investigated. When arsenite was administered to these animals by intraperitoneal injection, about 80 % of the dose was excreted mainly as DMA via the urine during the first 4 days after administration [59, 154]. Inorganic arsenic, of which the highest amounts were detected in the first hours after administration, and low amounts of monomethylated arsenic compounds were also detected in the urine [59]. The same compounds were also present in the plasma and the ultrafiltrable fractions of lung [59], liver, kidney [59, 154], and plasma [154] of the rabbits. In the lung the highest percentage of DMA in combination with poor binding to proteins was found, whereas more inorganic arsenic was detected in liver and kidney. In these tissues significant binding of arsenic to proteins was reported, supporting the theory that inorganic arsenic interacts more efficiently with tissue than does methylated arsenic [59]. Arsenite intravenously injected in rabbits was excreted as DMA (which was the major arsenic compound) and inorganic arsenic via the urine [141]. The same compounds were also detected in ultrafiltrated plasma and tissue cytosols of liver, lung, and kidney. MA accounted for less than 1 % of the arsenic present. Again the percentage of DMA in the ultrafiltrate of lung tissue cytosol (74 % of total arsenic 1 hour after administration) was significantly higher than in the ultrafiltrates of plasma (9 %), liver tissue cytosol (12 %), and kidney tissue cytosol

(11%). Low dietary intake of methionine, choline, or proteins by rabbits exposed to arsenite by intravenous injection resulted in a significant decrease of DMA excretion via the urine and increased the tissue retention of arsenic, again indicating that inorganic arsenic is more efficiently retained in the body than methylated arsenic [142]. Methionine deficiency most strongly influenced the excretion of DMA, proving that methionine is a main source of methyl groups in mammals. Deficiency of methionine and choline decreases the concentration of S-adenosylmethionine and increases the concentration of S-adenosinehomocysteine, thereby inhibiting transmethylation reactions. The decrease of DMA excretion by a low protein diet showed that dietary proteins are the main source of methionine. The excretion of inorganic arsenic was only elevated by a low methionine diet. No significant difference in MA excretion (which was between 5 and 8% only) between the controls and the rabbits on the low choline, methionine, or protein diets was observed.

When arsenate was administered intravenously to rabbits, inorganic arsenic and DMA were excreted in almost equal amounts via the urine [141]. Excretion was faster than after arsenite administration. The arsenic compounds detected in the ultrafiltrates of plasma and of liver, lung and kidney cytosols were inorganic arsenic, DMA, and less than 1% MA. Like in the arsenite dosed rabbits the DMA concentration in the ultrafiltrate of lung cytosol was higher than in the ultrafiltrates of plasma, liver cytosol, and kidney cytosol. About 10% of an arsenate dose of 0.04 mg As kg^{-1} body weight was excreted as arsenite in the urine [77]. Arsenite appeared in the urine prior to DMA after arsenate administration, indicating a consecutive reduction–oxidative methylation pathway also for arsenic biomethylation in mammals. When arsenate was injected intravenously into rabbits arsenite was detected in the plasma 15 minutes after the injection, whereas DMA appeared after 1 hour [72]. When the methyltransferase activity was inhibited by periodate-oxidized adenosine (PAD), DMA first appeared in the plasma after 6 hours. PAD also delayed the appearance of DMA in the red blood cells. Arsenite, arsenate, and DMA were excreted via the urine, whereas after PAD treatment no DMA was present in the urine 6 hours after the arsenate administration. The urinary excretion of DMA during the 24 hours after administration was significantly lower in PAD treated rabbits. PAD also delayed the presence of DMA in several tissues.

Only a few investigations on the administration of organic arsenic compounds to rabbits are available. Intravenously injected DMA was excreted unchanged and almost quantitatively within the first day after administration, indicating the low body retention of this compound in contrast to inorganic arsenic [141]. Demethylation was not observed. AB was also not biotransformed by rabbits [78]. Like in mice and rats, AC was mainly transformed to AB in rabbits [153].

In contrast to observations made in the rat and in the hamster, DMA was the end point of arsenic methylation in rabbits. Trimethylated arsenic compounds were only detected when trimethylated compounds had been administered before. If further methylation cannot occur in rabbits or was simply not observed, because most of the studies only investigated the 'commonly present' arsenic compounds arsenite, arsenate, MA, and DMA, is not certain yet.

6.4.2.5 Other Mammals

Only a few data are available on mammals other than mice, rats, hamsters, and rabbits. Inorganic and dimethylated arsenic were detected in plasma and urine and monomethylated arsenic only in the urine of beagle dogs after intravenous injection of ^{74}As [75]. After intraperitoneal injection of arsenite, only inorganic arsenic was detected in urine and tissue of marmoset monkeys, indicating that these animals (in contrast to other mammals or humans) are not able to methylate arsenic [140]. No methylated arsenic compounds were found in the urine and plasma of chimpanzees intravenously administered arsenate [79]. Only low amounts of MA and DMA were detected in the urine of guinea pigs after intraperitoneal injection of arsenate [69].

Seaweed-eating sheep from Northern Scotland excreted DMA ($\sim$95% of total excreted arsenic), MA, arsenoribose I (Figure 6.1), the tetramethylarsonium cation, and an unknown arsenic compound via the urine, whereas DMA and inorganic arsenic were the only arsenic compounds excreted by sheep living entirely on grass [122]. The only arsenic compound in the blood serum of the seaweed-eating sheep was DMA. In a study investigating the arsenic compounds in various food products, AB was the major arsenic compound in fresh minced chicken [123]. DMA and MA were also detected. Spread chicken liver also contained low concentrations of AB and MA.

6.5 PRESENCE AND TRANSFORMATION OF ARSENIC SPECIES IN HUMANS

Human urine has been investigated most intensively, because arsenic compounds are present in solution already and no extraction step is necessary in most cases. Dilution or clean-up steps to avoid matrix effects can be required prior to analysis. Blood was also investigated in some studies. Most of the data reported refer to human subjects intoxicated with arsenic compounds through voluntary ingestion or through work-place or environmental exposure (such as arsenic contaminated drinking water). As in mammals, methylation of inorganic arsenic in humans is discussed as a form of detoxification. Methylated arsenic compounds are less reactive with tissue constituents and are, therefore, rapidly excreted. Typically, arsenic in human urine consists of 10–15% inorganic arsenic, 10–15% MA, and 60–80% DMA. Total arsenic concentrations in the urine of humans without arsenic exposure are below 20 µg As dm^{-3}. Consumption of seafood can increase the arsenic concentration in urine to more than 1 mg As dm^{-3} [158]. Since arsenic in seafood is mainly present as organic arsenic, like AB in marine animals or arsenoriboses in seaweed, the determination of total arsenic concentrations is not useful for the assessment of health risks. Not only total arsenic concentrations but also the distribution of arsenic compounds in urine are influenced by dietary habits.

During the last year, the role of arsenic biomethylation as a detoxification process has been questioned. This development was initiated by the identification of methylarsonous acid and dimethylarsinous acid (Figure 6.1) in human urine. Methylarsonous acid [135–137], and dimethylarsinous acid [135] were identified in the urine of Inner Mongolian persons consuming well water with total arsenic concentrations of 510 to 660 μg dm^{-3}. Urine samples of 41 persons were collected before and 0–2, 2–4, and 4–6 hours after the administration of sodium 2,3-dimercapto-1–propanesulfonate (DMPS), a chelating agent previously used to treat arsenic poisoning [135]. Methylarsonous acid was detected in 51 urine samples out of a total of 164 urine samples; dimethylarsinous acid was detected in two of the samples. The highest concentration of methylarsonous acid found in the urine samples was 240 μg dm^{-3}. Methylarsonous acid and dimethylarsinous acid occurred in the urine samples after DMPS treatment, but reduction of MA and DMA to these two compounds by DMPS was excluded, since incubation of MA and DMA with DMPS in urine did not result in reduction of MA or DMA. It was discussed that methylarsonous acid and dimethylarsinous acid are usually bound to proteins and glutathione in the body and that they are released when they become chelated by DMPS. Methylarsonous acid was also reported in the urine of persons consuming drinking water with elevated arsenic levels in Romania [138]. In this study methylarsonous acid was excreted without DMPS treatment. Analysis of the urine of 428 persons exposed to inorganic arsenic in the drinking water in West Bengal revealed that methylarsonous acid was present in the urine of 48 % of the subjects and dimethylarsinous acid in 72 % [124].

Methylarsonous acid and dimethylarsinous acid are intermediates in the pathway for the methylation of inorganic arsenic to MA and, further, to DMA proposed by Challenger [147]. The presence of methylarsonous acid and dimethylarsinous acid in human urine strongly support this proposed pathway. However, methylarsonous acid and dimethylarsinous acid were reported to be toxic, most probably even more toxic than inorganic arsenic [159–162]. Due to their presence in human urine the usual belief in arsenic methylation as a detoxification process has to be reconsidered.

The dietary intake of arsenic was studied in 35 healthy Japanese volunteers [163]. The mean dietary intake was 195 ± 235 μg As day^{-1} consisting of 76 % trimethylated, 17.3 % inorganic, 5.8 % dimethylated, and 0.8 % monomethylated arsenic compounds. In the urine of seven females out of the 35 volunteers, less inorganic arsenic, but more mono- and dimethylated arsenic compounds than ingested were excreted, supporting the assumption that inorganic arsenic is methylated in the body. Urinary and blood arsenic levels and compounds were determined in 56 volunteers without determination of the dietary arsenic intake. The mean percentage of total arsenic present as inorganic, mono-, di-, and trimethylated arsenic compounds was 6.7%, 2.2%, 26.7%, and 64.6%, respectively. The mean blood total arsenic concentration was composed of 9.6% inorganic, 14% di-, and 73% trimethylated arsenic compounds. No monomethylated arsenic was detected. Mean concentrations of inorganic arsenic, MA, and DMA in 148 healthy unexposed Italian men were reported to be $1.9\,\mu g$ As dm^{-3}, $1.9\,\mu g$ As dm^{-3}, and $2.1\,\mu g$ As dm^{-3}, respectively [60, 67]. About 60% of the total arsenic in urine was present in other organic forms not further determined [60, 67]. Inorganic arsenic, MA, and DMA were also determined in the urine of 137 Italian males divided into groups of persons never eating seafood, regular or intermittent fish consumers, and regular or intermittent consumers of fish and shellfish [13]. When total urinary arsenic concentrations for the persons living on seafood-free diet, for persons intermittently eating fish, regularly eating fish, intermittently eating fish and shellfish, and regularly eating fish and shellfish were compared, only the concentrations detected in the urine of the persons regularly eating fish, or fish and shellfish were significantly higher than those of non-seafood consumers. Only inorganic arsenic in the urine of persons regularly consuming fish and shellfish was significantly elevated in comparison with persons never eating seafood. When 15 Belgians abstinent from seafood for at least 5 days provided 24-hour urine samples the mean values amounted to 0.6, 0.5, 3.8, and $20\,\mu g$ As 24 hours^{-1} for inorganic arsenic, MA, DMA, and total arsenic, respectively [13]. Background levels of arsenic compounds in non-exposed Dutch inhabitants (61 subjects; 18 of them reported seafood consumption within 48 hours before sampling) ranged from below the detection limit to $17\,\mu g$ As dm^{-3} for MA and from below the detection limit to $212\,\mu g$ As dm^{-3} for AB. DMA concentrations were between 0.4 and $29.1\,\mu g$ As dm^{-3}. Inorganic arsenic was not detected because of higher detection limits for these species [105]. DMA (maximum concentration $39\,\mu g$ As g^{-1} creatinine) was the only arsenic compound detected in all urine samples of 40 unexposed adults from Glasgow, UK. Arsenite (maximum concentration $1.0\,\mu g$ As g^{-1} creatinine) and MA (maximum concentration $0.6\,\mu g$ As g^{-1} creatinine) were only detected in five and four of the samples, respectively [66].

Several studies have been conducted on urinary arsenic compounds in humans after ingestion of arsenic compounds. After oral administration of $NaAsO_2$ mainly inorganic arsenic was excreted via the urine during the first 4 hours, but this mode of excretion rapidly disappeared leading to the excretion of methylated derivatives, especially DMA, which became the major arsenic compound in urine about eight hours after ingestion [11, 12]. Daily oral ingestion of 125, 250, 500, or $1000\,\mu g$ $NaAsO_2$ for 5 days showed that inorganic arsenic and MA excretion in the urine during the first 24 hours after the last administration increased linearly with the administered dose, while the DMA concentration tended to level off at higher doses [164]. Acute intoxication with

As$_2$O$_3$ was studied in subjects who had made suicide attempts using this compound. Buchet and coworkers reported that inorganic arsenic was the major form of arsenic excreted in the urine of two persons during the first 3 days after the suicide attempt [11]. DMA was the major arsenical excreted from the fourth day on. In other reports the predominance of inorganic arsenic during the first days after intoxication and the subsequent predominance of DMA was also found [67, 71, 164]. The arsenic compounds excreted in the urine of a person after acute arsenic trioxide poisoning 37 to 109 hours after ingestion were arsenite (39.3% of the total excreted arsenic), arsenate (3.5%), MA (25.6%), and DMA (31.6%). Urinary arsenic levels in acute poisoning were found to be 30 to 2000 times higher than normal urinary arsenic levels [71]. In a recent study of arsenic compounds in blood and tissues (liver, kidneys, muscle, heart, spleen, pancreas, lung, cerebellum, brain, and skin) of a person who died of acute poisoning with arsenic trioxide, arsenite was reported to be dominant in blood and all tissues (47 to 85% of total arsenic) [129]. MA concentrations (10 to 32%) were higher than DMA concentrations (4 to 19%). MA and DMA were more prevalent in lipidic organs. Arsenate was only detected in liver, kidneys, and blood (2% of total arsenic). The highest total arsenic concentrations were found in the liver (147 mg kg^{-1} dry mass) and kidneys (26.6 mg kg^{-1} dry mass) emphasizing the role of these organs in arsenic detoxification. The total arsenic concentration was lowest in hemolyzed blood (0.422 mg kg^{-1} dry mass). In the urine of a person acutely poisoned with Na$_2$HAsO$_4$.7H$_2$O, arsenite was the dominant form of arsenic 45 to 70 hours after ingestion [71]. Arsenate was present at significantly lower concentrations than arsenite.

About 87% of the total arsenic excreted within 96 hours after oral administration of MA to volunteers was reported to be MA, 13% were methylated to DMA, when the urinary concentrations of the arsenic compounds were corrected for basal daily excretion of arsenic compounds [12]. DMA orally administered to volunteers was excreted unchanged. About 75% of the administered dose were excreted via the urine within 96 hours in both cases. Marafante and coworkers reported that, when DMA was orally administered to a volunteer, during the first 3 days after administration 4% of the dose was excreted as TMAO and 80% as DMA via the urine [44]. Results obtained for humans after ingestion of various synthetic arsenic compounds indicate that methylation of inorganic arsenic occurs in the body. Methylation can be understood as a detoxification mechanism, since methylated arsenic compounds are significantly less toxic than inorganic arsenic. Additionally, methylated arsenicals are excreted more rapidly from the body than inorganic arsenic, which interacts with tissue components. However, the role of arsenic methylation as a detoxification process has recently been questioned with the identification of methylarsonous acid and dimethylarsinous acid, toxic intermediates of the proposed methylation pathway, in human urine [124, 135–138]. Many authors reported DMA as the methylation end point in the human body. The results obtained by

Marafante and coworkers [44] indicate that methylation to trimethylated arsenic compounds can occur. The authors discussed that the DMA concentration in the liver, the main site of arsenic methylation [165], is the rate-limiting factor in the synthesis of TMAO. Since this concentration is not very high after ingestion of inorganic arsenic, no significant amounts of TMAO are formed from inorganic arsenic. Methylation of MA to DMA, but no demethylation of MA or DMA was observed in humans.

Environmental exposure to arsenic can derive from industrial impact, as well as from drinking water and from the diet. Several studies on humans exposed to arsenic released by industrial processes were conducted. Urinary arsenic excretion in some groups of arsenic-exposed industrial workers as well as chemists working with arsenic compounds were investigated. Workers in glass factories exposed to As_2O_3 (used as decolorizing agent) by skin contact or inhalation showed blood and urinary arsenic levels proportional to the intensity of exposure [67]. Inorganic arsenic, MA, and DMA levels in urine were elevated compared with a control group. The elevated urinary DMA concentration reflected arsenic accumulation in the body due to high exposure. At low exposure levels, no short-time accumulation of arsenic occurred and inorganic arsenic excretion is the best indicator of exposure. When the relative mean proportions of inorganic arsenic, MA, and DMA in urine of glass factory workers were compared with an unexposed control group, DMA decreased from 98% to about 70%, whereas MA increased from 0.7% to 15% and arsenite from 1.7% to about 14% [66]. These results, indicating elevated exposure to As_2O_3, also support the theory that the conversion of MA to DMA can be inhibited by high arsenite concentrations in the liver [166]. Arsenate was not detected in the urine of the control group and was below 3% in the urine of the glass factory workers. In the same study, total arsenic concentrations and species distribution in the urine of persons exposed to As_2O_3 by working on the production of semiconductors were similar to that of the control subjects [66]. Excretion patterns in workers employed in a timber treatment firm and consequently exposed to pentavalent inorganic arsenic as well as in workers of a chemical firm manufacturing arsenic compounds, were similar to those observed in glass workers, but urinary total arsenic concentrations were higher in the workers employed in the chemical firm. Workers in a sulfuric-acid-producing plant showed elevated urinary levels of inorganic arsenic after exposure to airborne As_2O_3 [167]. Workers exposed to coal-fly ash during boiler cleaning operations in a coal-fired power plant excreted slightly higher concentrations of inorganic arsenic and MA and markedly higher concentrations of DMA via the urine than before exposure [29]. In another study workers at an arsenic trioxide refinery excreted arsenite, arsenate, MA, and DMA via their urine, whereas in the urine of unexposed workers only DMA was detected at concentrations about 2.5 times lower than in exposed workers [130]. The sum of the concentrations of the four arsenic species in the urine of exposed workers was significantly higher than in the urine of those unexposed. Total urinary arsenic concentrations were also markedly elevated and the differences between the sum of the concentrations of inorganic arsenic, MA, and DMA, and the total urinary arsenic concentration after acid digestion, indicating the presence of trimethylated arsenic compounds, increased. Arsenic compounds in the urine of copper smelter workers showed a similar behavior, but the difference between total urinary arsenic and the sum of inorganic arsenic, MA, and DMA was similar to unexposed workers [130]. Residents living in the vicinity of arsenic emission sources also showed elevated urinary arsenic levels. For example, inhabitants of an area of past intense mining and smelting activity of arsenical ores showed elevated urinary levels of inorganic arsenic, MA, and DMA compared with a control population [93].

In 1997 the formation of AB in the human body was discussed for the first time. In a subject exposed to trimethylarsine via synthesizing arsenic compounds in a chemistry laboratory, a urinary increase of AB was reported [88]. In addition to elevated AB concentrations the concentrations of arsenate, MA, and DMA in the urine increased compared with urinary concentrations before exposure. The formation of AB after acute occupational intoxication of a galvanic factory worker by arsine was discussed by Apostoli and coworkers [81]. The excretion of arsenite, MA, and DMA increased markedly, and the excretion of arsenate and AB slightly, after intoxication. Further investigations showed that the AB in the urine could have not only derived from the diet. These results indicated that arsine and trimethylarsine can be metabolized to AB [81, 88] as well as demethylated to DMA, MA, and even inorganic arsenic in the human body [88].

Exposure to arsenic compounds by the diet, namely drinking water and food, was also investigated by several working groups. High arsenic concentrations in drinking water deriving from the geological situation cause serious health problems (skin cancer, blackfoot disease and others) in the population of several regions in the world, e.g. some parts of India, Bangladesh, or Taiwan, giving rise to investigations of the urinary arsenic compounds in persons consuming this water. Drinking water can also be contaminated with arsenic by industrial impact. Urinary arsenate and DMA concentrations were reported to be elevated, when arsenate-containing well-water was ingested [168]. When urinary arsenic metabolites of the inhabitants of two towns in Northern Chile, one of which had arsenate-contaminated drinking water ($600\,\mu g$ As dm^{-3}), were compared, the percentage of total urinary arsenic in the form of inorganic arsenic and MA was increased, whereas the percentage of DMA was decreased, and total urinary arsenic concentrations were about 10-times higher in persons living on contaminated tap water [17]. Differences were best described by an increased MA-to-DMA ratio for persons drinking the arsenic-contaminated water. When persons living on arsenic-contaminated drinking water were provided with water of lower arsenic content ($45\,\mu g$ As dm^{-3}) for 2 months, the percentage of inorganic arsenic and the MA-to-DMA ratio in the urine slightly decreased, but the magnitude of this decrease did not appear to be associated with the magnitude of the decrease in total arsenic [169]. Total urinary arsenic concentrations after the 2 months were still higher than in inhabitants of a town generally having low arsenic concentrations in the drinking water. No evidence of an exposure-based threshold for arsenic methylation in humans was found.

Inhabitants of a Mexican region with arsenic drinking water levels of about 400 μg As dm^{-3} had 20- to 95-times higher total urinary arsenic concentrations than unexposed controls [139]. The percentages of arsenic species calculated on total urinary arsenic concentrations were as follows: inorganic arsenic increased from 8.7 to 30.6% and MA from 6.8 to 11.3%, whereas DMA significantly decreased from 78.5 to 54.1% in the exposed group supporting the increase of the MA to DMA ratio reported by Hopenhayn-Rich and coworkers [17]. The increase of urinary inorganic arsenic and MA, and subsequent decrease of DMA, were positively correlated to the time of exposure. In the urine of persons, who already showed cutaneous signs of arsenicism, the proportion of MA increased and the proportion of DMA remained constant compared with exposed subjects without cutaneous signs. The authors concluded that high chronic arsenic exposure decreases the body's ability to methylate inorganic arsenic due to the depletion of cofactors and inhibition or saturation of enzymes and that, as in rat liver, high arsenite concentrations inhibit the conversion of MA to DMA. Therefore, when methylation of inorganic arsenic to MA becomes impaired, the higher arsenite concentrations inhibit the methylation of MA to DMA and the MA to DMA ratio in the urine increases.

Native women of four Andean villages in Northern Argentina with elevated levels of arsenic in drinking water were reported to excrete mainly inorganic arsenic and DMA via the urine [80]. Only a very low amount of MA (2.2% of total arsenic) compared with other studies (10 to 20% MA) was detected in the urine indicating the existence of genetic polymorphism in the control of the methyltransferase activity involved in the methylation of arsenic. Additionally, in contrast to other studies on arsenic exposed subjects, the percentage of DMA in the urine of women in the village with the highest arsenic concentration in drinking water (200 μg As dm^{-3}) was significantly elevated. In another study conducted on inhabitants of this region, the urine of children and women was also found to contain only low amounts of MA (average 2.1 to 3.6%) [63]. Compared with women (average of 25% of total urinary arsenic as inorganic arsenic, 74% as DMA) the percentage of inorganic arsenic was significantly higher (average 49%) and the percentage of DMA much lower (average 47%) in the urine of the children in one village, indicating a low general methylation efficiency and, therefore, higher sensitivity to arsenic-induced toxicity. Arsenic concentrations in cord blood of pregnant native Andean women living on drinking water with elevated arsenic levels (200 μg As dm^{-3}) was almost as high as in maternal blood (11 μg As dm^{-3}) [64]. The total arsenic concentration in the placenta was about five-times higher than in non-exposed women. Almost all arsenic in the blood of the mother and child, and 90% of the arsenic excreted via the urine by mothers in late gestation and newborns, was present as

DMA, indicating an increase of arsenic methylation associated with lower blood and higher urinary arsenic levels and a transfer of arsenic mainly in the form of DMA to the fetus during pregnancy. Lactating native Andean women exposed to arsenic via the drinking water ($200\,\mu g$ As dm^{-3}) showed elevated blood and urinary arsenic concentrations [170]. About 80% of the urinary arsenic was inorganic arsenic. The total arsenic concentration in breast milk was only slightly elevated and total urinary arsenic in nursing babies was low, suggesting protection of the child by no transfer of inorganic arsenic to breast milk under arsenic exposure.

In residents of the northeast coast of Taiwan no relation between the percentages of inorganic arsenic, MA, and DMA in the urine and arsenic concentrations in drinking water (below 1 to more than $300\,\mu g$ As dm^{-3}) was observed, whereas genetic polymorphisms of glutathione S-transferase (GST) M1 and T1, enzymes involved in arsenic methylation, were significantly associated with arsenic methylation [84]. Subjects with the null genotype of GST M1 showed a significant increase of the percentage of urinary inorganic arsenic compared with subjects having the non-null genotype. A positive correlation between the null genotype of GST T1 and the percentage of urinary DMA was reported. Elevated levels of urinary MA and DMA were reported in Taiwanese previously exposed to arsenic through the consumption of contaminated drinking water [90]. Data suggested that women had a higher methylation capability than men and aging diminishes this capability. Residents of the southwest coast of Taiwan suffering from severe health effects, namely blackfoot disease, due to arsenic exposure by drinking water prior to 1956, still showed significantly higher total urinary arsenic concentrations comprising elevated concentrations of inorganic arsenic, MA, and trimethylated arsenic compared with subjects from control regions [70]. MA and trimethylated arsenic were suggested to have probably derived from higher seafood consumption. This discussion may not be valid for MA, since MA was not found to be a common constituent of seafood in a recent study [27]. No increase in the DMA concentration was observed [70]. Among inorganic arsenic and DMA (the only arsenic species in hair and fingernails) inorganic arsenic was elevated in subjects from the blackfoot disease area. No significant difference between males and females was reported in this study.

Most studies on arsenic exposure by food focused on the consumption of seafood, which is known to contain arsenic at the mg kg^{-1} dry mass level. The major arsenic compound in marine animals is AB, whereas seaweed, an important part of the diet of the Japanese, mainly contains arsenoriboses [2]. Arsenic levels in food other than seafood is generally in the low $\mu g\ kg^{-1}$ dry mass range. Exceptions are rice, coffee extracts, spinach, and parsley [7]. Food products from animals raised on medicated feeds containing arsenicals, like poultry or pork, can also show elevated arsenic concentrations. The application of arsenic herbicides and pesticides in agriculture can lead to increased arsenic concentrations in food products.

Urinary arsenic metabolites after seafood consumption were investigated by several working groups. AB ($2500\,\mu g\;dm^{-3}$), TMAO ($21\,\mu g\;dm^{-3}$), TETRA ($30\,\mu g\;dm^{-3}$), DMA ($11\,\mu g\;dm^{-3}$), and arsenite ($20\,\mu g\;dm^{-3}$) were detected in the urine of a person, who ingested 250 mg of plaice meat [100]. No arsenic species were detected in the urine before seafood ingestion (detection limits for arsenite, arsenate, MA, DMA, and AB between 6 and $10\,\mu g$ As dm^{-3} undiluted urine). Urinary concentrations of DMA and AB were reported to increase two- to fivefold and 30- to 50-fold, respectively, after the consumption of codfish [88]. The urinary concentration of these two arsenic compounds also increased after the consumption of tuna fish [51, 105], shrimps [51, 103], mussels [103], and crabs [51]. Additionally, an unknown compound was detected in the urine of shrimp and mussel consumers [103]. AB (61 to $233\,\mu g\;dm^{-3}$) and an unknown compound were found in human urine after the consumption of a dinner containing prawns, clams, razor shell, and anchovy, or prawns and clams, or tuna, prawns and clams, whereas all arsenic compounds were below the detection limit (8 to $15\,\mu g$ As dm^{-3} for the various species) before seafood consumption [49]. In all these investigations the arsenic compounds originally present in the seafood ingested were not determined. Whether the arsenic compounds elevated in urine after seafood consumption are simply taken up, or are metabolic products of other compounds originally present in seafood cannot be ascertained. Since AB is well known to be the major arsenic compound in marine animals, it can be concluded that this arsenic compound is taken up from seafood and is mostly excreted unchanged via the urine. This behavior was also reported by Le and coworkers, who found that about 70 % of the AB ingested with a meal of shrimp were excreted unchanged via the urine within 37 hours [43, 133]. AB also was the major arsenic compound excreted after the consumption of crab meat [133]. No increase in urinary inorganic arsenic, MA, and DMA was reported for the consumption of shrimp mainly containing AB [43]. However, Yamauchi and coworkers found that 3 to 5 % of the trimethylated arsenic ingested by the consumption of prawn containing 98.8 % trimethylated arsenic, 0.96 % inorganic arsenic, and 0.14 % DMA was excreted as other species (inorganic arsenic, MA, DMA) via the urine, and discussed that small amounts of trimethylated arsenic were demethylated in the human body [31].

When arsenoriboses are ingested by humans through the consumption of seaweed or marine algae, excretion patterns different from those of AB were observed. After ingestion of powdered kelp, which contained arsenoriboses **I** and **II** (Figure 6.1) and an unknown arsenic compound as major arsenic compounds, DMA, arsenate, and several unknown metabolites were detected in urine after 23 hours; after 46 hours only DMA and arsenate were present [133]. Differences in the excretion patterns of the unknown arsenic compounds were reported when two individuals ingested the same kelp powder. Significant differences in total urinary arsenic excretion and in the excretion patterns of arsenic compounds was observed in nine volunteers after the consumption of the same amount of Nori, in which most of the arsenic is present as arsenoribose **I** [133]. Up to six unknown metabolites were reported in the urine in addition to DMA and arsenate, which had already been present prior to the ingestion of Nori. The concentration of DMA increased after the seaweed consumption. The authors discussed that arsenoriboses are less stable than AB and, therefore, are metabolized in the human body. The metabolism of arsenoriboses in different individuals, even when they are relatives or of the same gender or age, can differ significantly. In another experiment, five unknown arsenic compounds and an increase in the DMA concentration was observed, when a volunteer ingested Nori [50]. When commercially available, Yakinori containing two arsenoriboses was consumed, again no arsenoriboses but two unknown metabolites were detected in the urine and the urinary DMA concentration increased [48, 50]. When mussels containing AB and the arsenoriboses **I** and **II** as major arsenic compounds were consumed, AB was rapidly excreted into the urine [102]. As after seaweed consumption, no arsenoriboses but four unidentified metabolites were detected in the urine and the concentration of DMA increased. Since the mussels did not contain significant amounts of DMA or inorganic arsenic, DMA detected in the urine after mussel consumption was a metabolic product of the arsenoriboses.

Urinary DMA is a commonly used indicator for exposure to inorganic arsenic. Although it was reported that seafood consumption does not influence urinary inorganic arsenic, MA, and DMA concentrations, and the difference between total arsenic in urine and the sum of inorganic arsenic, MA, and DMA can be attributed to 'seafood-arsenic' [11, 60, 67], several cases of an increase of urinary DMA after seafood consumption have been described recently. Buchet and coworkers also reported an increase of urinary DMA after the consumption of mussels [13, 47]. A slight increase in urinary DMA was also observed after the ingestion of cod, whiting, sole [13], tuna and crab [82], and a significant increase after the consumption of mackerel and herring [82] thus indicating that seafood consumption can lead to falsified results in the monitoring of exposure to inorganic arsenic on the basis of the urinary DMA concentrations.

Data on arsenic compounds in human body liquids other than urine are limited. However, some investigations were conducted on arsenic compounds in blood. Two hours after the ingestion of prawns, trimethylated arsenic appeared in blood plasma and was excreted almost quantitatively after 24 hours [31]. The concentrations of inorganic arsenic and DMA, which were also detected before seafood ingestion, remained constant. Inorganic arsenic, DMA, and trimethylated arsenic were present in red blood cells prior to and after seafood ingestion and slightly increased after seafood ingestion. AB was detected in blood plasma (0.9 to $3.3\,\mu g$ As dm^{-3}), serum (1.6 to $4.6\,\mu g$ As dm^{-3}), and red blood cells lysate (5.7 to $10.1\,\mu g$ As dm^{-3}) of healthy male volunteers about 12 hours after fish consumption [134]. AB was detected in most predialysis and some postdialysis blood plasma samples of persons undergoing renal dialysis [85]. Arsenate, MA, and DMA were below the detection limit in all samples. Since no data on the diet were available, the source of AB present in the plasma was not clear. Zhang and coworkers also found AB and additionally DMA in the serum of uremic patients, whereas all arsenic compounds investigated were below the detection limit in the serum of a healthy control person [52], in serum samples of non-hemodialyses and hemodialysis patients with chronic renal disease [53], and in serum of patients on continuous ambulatory peritoneal dialysis [171]. In another study, DMA was the only arsenic compound present in the serum of uremic patients [112]. Inorganic arsenic was only detected in one single sample. Inorganic arsenic, MA, and DMA were below the detection limit (0.40 to $0.92\,\mu g$ As dm^{-3}) in serum samples from healthy controls. AB could not be detected, because HPLC-HG-AAS was employed without a further mineralization step. DMA was also reported in serum and erythrocytes of two leukemia patients after ingestion of a realgar-containing drug [55].

Summarizing investigations in humans, MA and DMA are well established organoarsenic compounds in the human body. The transformation of inorganic arsenic to MA and DMA is discussed as a detoxification mechanism, since MA and DMA are less toxic than inorganic arsenic and more rapidly excreted from the human body due to their lower affinity to tissue. However, the role of arsenic methylation as a detoxification process has recently been questioned. In one case the excretion of TMAO after incorporation of DMA was observed, indicating that DMA is probably not the end point in arsenic methylation in the human body. The excretion of elevated amounts of arsenate, MA, and DMA after exposure to trimethylarsine indicates that demethylation processes can also occur. Other organoarsenic compounds like AB or arsenoriboses are ingested, e.g. from seafood. AB is excreted unmetabolized from the body, whereas arsenoriboses, arsenic compounds incorporated mainly by the consumption of seaweed and mussels, cannot be detected in the urine for they are completely metabolized to unidentified arsenic compounds and DMA. Different metabolization patterns were observed in different individuals. In recent years the detection of AB in the urine of persons who did not ingest any seafood but were exposed to trimethylarsine or arsine, posed the question as to whether AB can be formed from other arsenic compounds in the human body. Future work is necessary to confirm these findings and to gain more information on processes involved in the metabolism of arsenic compounds in the human body.

6.6 THE METABOLISM OF ARSENIC COMPOUNDS IN FRESHWATER FOOD CHAINS

Several freshwater food chains were investigated for arsenic metabolism. They are discussed separately, since they include plants as well as animals and these organisms should be dealt with together to visualize the transformation of arsenic compounds in the food chain. The lowest link in the food chains investigated is always algae, which are the diet of several freshwater animals.

Maeda and coworkers conducted several studies on such food chains. The metabolism of arsenate in a food chain consisting of an autotroph (the alga *Chlorella vulgaris* Beijerinck var. *vulgaris*), a grazer (the zooplankton *Moina macrocopa*), and a carnivore (the goldfish *Carassius carassius auratus*) was investigated [149]. Arsenate was accumulated by *C. vulgaris*. but only about 1% of total arsenic in the cell was methylated, mainly to dimethylated arsenic compounds. Traces of mono- and trimethylated arsenic compounds were also detected. When *M. macrocopa* was fed with *C. vulgaris* grown in an arsenate-spiked medium, 17% of the total arsenic was transformed to mono-, di-, and traces of trimethylated arsenic compounds. When *C. carassius* was fed with *M. macrocopa*, 37% of the total arsenic was methylated, mainly to monomethylated arsenic compounds. The concentration of trimethylated was lower than the concentration of monomethylated arsenic compounds. The concentration of the dimethylated form was lowest. Generally, total arsenic accumulation decreased within the food chain, but the relative concentration of methylated arsenic increased successively.

The transformation of arsenate in the food chain *Chlorella vulgaris* Beijerinck var. *vulgaris* and *Phormidium* sp. (algae)–*Moina macrocopa*–*Poecilia reticulata* (carnivorous guppy) was also investigated by Maeda and coworkers [22]. About 5% of the arsenate accumulated from the culture medium into *C. vulgaris* was methylated to dimethylated and traces of mono- and trimethylated arsenic compounds. In the case of *Phormidium* sp. about 0.2% of the arsenate accumulated was transformed to di- and 0.1% to trimethylated arsenic compounds. Traces of monomethylated arsenic compounds were also detected in *Phormidium* sp. When *M. macrocopa* was fed on *C. vulgaris*, 12% of the arsenic in *M. macrocopa* was transformed to dimethylated arsenic, and traces only of mono- and trimethylated arsenic were formed. Monomethylated (8%), dimethylated (17%), and traces of trimethylated arsenic compounds were detected in *M. macrocopa*, when the diet consisted of *Phormidium* sp. *M. macrocopa* took up arsenic more efficiently from the diet than from water. *M. macrocopa* fed on *C. vulgaris* as diet for *P. reticulata* resulted in 82% tri-, 3% di-, and traces of monomethylated arsenic compounds in *P. reticulata*. After the intake of *M. macrocopa* grown in water spiked with arsenate (1 mg As dm^{-3}), 86% of tri- and traces of mono- and dimethylated arsenic compounds were detected in *P. reticulata*. Therefore, *P. reticulata* seems to be able to methylate mono- and dimethylated and perhaps inorganic arsenic to trimethylated arsenic compounds. Again the total arsenic concentration in the food chain decreased about one order of magnitude with each step, but the relative amount of methylated arsenic increased significantly.

The same results regarding the decrease of the total arsenic concentration as well as the relative increase of methylated arsenic compounds with an elevation in the trophic level were obtained for the food chain including the alga *Chlorella vulgaris* Beijerinck var. *vulgaris* (isolated from an arsenic-polluted environment), the planktonic grazer *Moina macrocopa* or the herbivorous shrimp

Neocardina denticula, and the carnivorous guppy *Poecilia reticulata* after the administration of arsenate [23]. Arsenate was accumulated by *C. vulgaris* and to a small extent biotransformed to mono-, di-, and trimethylated arsenic compounds, which were also excreted to the culture medium. *N. denticula* fed on *C. vulgaris* contained inorganic, di-, and trimethylated, but no monomethylated arsenic compounds. The relative concentrations of arsenic accumulated in the food chain *C. vulgaris–M. macrocopa–P. reticulata* were almost 100% inorganic arsenic in the alga, ~75% inorganic, ~25% dimethylated and less than 1% trimethylated arsenic compounds in the planktonic grazer, and 18% inorganic, 1% dimethylated, and 81% trimethylated arsenic compounds in the guppy. Monomethylated arsenic compounds were only present in trace amounts. A decrease in the total arsenic concentration and a relative increase of methylated arsenic compounds were also reported for the food chain *Nostoc* spp. (alga)–*N. denticula–Cyprinus carpio* (carp) with arsenate as substrate [26]. When *Nostoc* sp. was cultured in an arsenate-containing medium, less than 1% of the arsenate accumulated in the alga was transformed to di- and traces of monomethylated arsenic compounds. In the food chain arsenic compounds were distributed as follows: alga: 99% inorganic, traces of mono-, 1% dimethylated arsenic compounds; shrimp: 90% inorganic, 5% di-, and 5% trimethylated arsenic compounds; carp: 79% inorganic, traces of mono-, 8% di-, and 14% trimethylated arsenic compounds.

Kuroiwa and coworkers investigated the arsenate metabolism in the food chain *C. vulgaris–N. denticula–Oryzias lapites* (killifish) [18]. About 11% of the arsenic taken up from the culture medium was metabolized to mono- and dimethylated arsenic compounds by *C. vulgaris*. *N. denticula* fed on *C. vulgaris* contained 39% of the incorporated arsenic mainly as dimethylated, a small amount as trimethylated, and traces as monomethylated arsenic compounds. In *O. lapites* fed on *N. denticula* 80% of the total arsenic was present as trimethylated arsenic compounds. Only traces of mono-, or dimethylated arsenic compounds were detected in this organism. Kaise and coworkers investigated the transformation of arsenate in the food chain *C. vulgaris–Daphnia magna* (water flea) [42]. Arsenate was transformed to dimethylated arsenic compounds by *C. vulgaris*. The dimethylated arsenic compounds were partially metabolized to trimethylated arsenic compounds by *D. magna* fed on *C. vulgaris*.

Recently, Suhendrayatna and coworkers investigated food chains including *C. vulgaris, D. magna, N. denticula*, and the two carnivorous fish *Tilapia mosambica* and *Zacco platypus* [37]. When *C. vulgaris* cultured in a medium containing arsenite (50 mg dm^{-3}) for 7 days was fed to *D. magna* for 7 days, and *D. magna* then was fed to *N. denticula* for 7 days, total arsenic in the organisms decreased within the food chain and inorganic arsenic was predominant in all organisms. Mono-(2% of the total arsenic) and dimethylated arsenic (6%) were additionally detected in *C. vulgaris*. Traces of these species were found in *D. magna*. In *N. denticula* dimethylated arsenic (1.6%) was present. No trimethylated arsenic was found within this food chain. Similar results were obtained when *D. magna* fed on *C. vulgaris* was administered to *Z. platypus*. *Z. platypus* contained mostly inorganic arsenic and only traces of mono-, di-, and trimethylated arsenic. When *C. vulgaris* cultured in arsenite-containing medium was fed to *N. denticula* and *N. denticula* was fed to *T. mosambica*, inorganic arsenic was again predominant and total arsenic concentrations decreased within the food chain. *N. denticula* contained traces of mono- and dimethylated arsenic whereas *T. mosambica* contained trimethylated arsenic (3 % of the total arsenic). When *N. denticula* fed on *C. vulgaris* was administered to *Z. platypus*, traces of mono-, di-, and trimethylated arsenic were detected in this fish.

When the distribution of arsenic compounds in *T. mosambica* after administration of *N. denticula* fed on *C. vulgaris* was investigated, the total arsenic concentration was highest in the ovary of the fish and lowest in the intestine [38]. Inorganic arsenic was predominant in all tissues. Trimethylated arsenic was found in the liver, bone, brain, muscle, gill, and eyes. Traces of dimethylated arsenic were present in the liver. Monomethylated arsenic compounds were not detected in any of the tissues. Comparison with *T. mosambica* exposed to arsenate via the water revealed that arsenic was more evenly distributed in the organs than in the food chain experiment.

Unfortunately the methylated arsenic compounds were not definitely identified in any of the food chain experiments. Apart from the degree of methylation, no predictions on the identity of the arsenic compounds present in the organisms can be made. However, the decrease of the total arsenic concentration and the relative increase of methylated arsenic compounds with an increase of the trophic level was observed in most of the food chains investigated.

6.7   CONCLUSIONS AND FUTURE PROSPECTS

Due to the low arsenic concentrations in the terrestrial environment, the existence of 'complex' organoarsenicals like AB, AC, TETRA, or arsenoriboses was questioned for a long time, although these compounds are well known constituents of the marine environment. The development of new analytical methods, especially HPLC-ICP-MS with detection limits in the low μg As kg^{-1} (dry mass) range, created a tool that has allowed the detection of the arsenic compounds in the terrestrial environment. With the detection of AB in fungi in 1995 many of the compounds previously attributed to the marine environment have been discovered in the terrestrial environment, albeit at much lower concentrations.

The main advantages of HPLC-ICP-MS over HPLC-(HG)-AAS, a technique commonly used in the past for the determination of arsenic compounds, are the superior detection limits and that all arsenic compounds can be detected without an additional mineralization step. The mineralization of the arsenic compounds, necessary to obtain volatile hydrides for AB, AC, and TETRA, results in information loss unless the mineralization is performed after the chromatoraphic separation. Many publications, especially fundamental work on transformation of arsenic compounds in various organisms after exposure, suffer from the problem that the arsenic compounds are not unequivocally identified (just the degree of methylation is given) because HG-AAS was used for their determination.

Biomethylation to the 'simple' arsenic compounds MA, DMA, and TMAO is believed to occur via reduction of inorganic As(V) to As(III) and addition of methyl groups from S-adenosylmethionine. The biogenesis of arsenic-containing ribosides is believed to occur via reduction of DMA and addition of the adenosyl group to produce dimethylarsinyladenosine, a precursor of the ribosides. This key compound has been scarcely detected. Dimethylarsinylethanol and dimethylarsinylacetic acid, proposed as possible intermediates for the biogenesis of AB from dimethylarsinyl ribosides, have not been discovered in the terrestrial environment yet. Variability in toxic responses of organisms to various arsenic compounds is still not clear and might have its origin in specific dietary components. Important factors to consider are the influence of methionine, cysteine, and vitamin B_{12}, as well as the role of trace elements such as zinc and selenium. All these facts clearly indicate that our knowledge of the biochemical pathway of arsenic in the environment (terrestrial and marine) as well as the impact on human health is not fully understood thus far.

The improvement of our analytical techniques with respect to separation efficiency (involvement of CE), as well as the continuous improvement in the method detection limits will be a contribution from the analytical point of view. The examination of the arsenic metabolism from an interdisciplinary point of view (chemists, biologists, medical scientists, etc.) will help to solve this puzzle, of which many pieces have been already found, and finally show it as a whole picture.

5.8 REFERENCES

1. Edmonds, J.S., Francesconi, K.A., Cannon, J.R., Raston, C.L., Skelton, B.W., and White, A.H., *Tetrahedron Lett.*, 1977, **18**, 1543.
2. Francesconi, K.A., and Edmonds, J.S., *Adv. Inorg. Chem.*, 1997, **44**, 147, and references cited therein.
3. Cullen, W.R., and Reimer, K.J., *Chem. Rev.*, 1989, **89**, 713, and references cited therein.
4. Leonard, A. In *Metals and their Compounds in the Environment. Occurrence, Analysis and Biological Relevance*, E. Merian (Ed.), VCH, Weinheim, 1991, pp. 751–774, and references cited therein.
5. Wolfsperger, M., Hauser, G., Goessler, W., and Schlagenhaufen, C., *Sci. Total Environ.*, 1994, **156**, 235.
6. Iyengar, V., and Woittiez, J., *Clin. Chem.*, 1988, **34**, 474.
7. Nriagu, J.O., and Azcue, J.M. In *Food Contamination from Environmental Sources*, J.O. Nriagu and M.S. Simmons (Eds), John Wiley & Sons, New York, 1990, pp. 121–143 and references cited therein.

8. Smith, E., Naidu, R., and Alston, A.M., *Adv. Agron.*, 1998, **64**, 149, and references cited therein.
9. World Health Organization, *Environmental Health Criteria 18. Arsenic*, World Health Organization, Geneva, 1981, pp. 33–34 and references cited therein.
10. Braman, R.S., and Foreback, C.C., *Science*, 1973, **182**, 1247.
11. Buchet, J.P., Lauwerys, R., and Roels, H., *Int. Arch. Occup. Environ. Health*, 1980, **46**, 11.
12. Buchet, J.P., Lauwerys, R., and Roels, H., *Int. Arch. Occup. Environ. Health*, 1981, **48**, 71.
13. Buchet, J.P., Lison, D., Ruggeri, M., Foa, V., and Elia, G., *Arch. Toxicol.*, 1996, **70**, 773.
14. Buchet, J.P., Apostoli, P., and Lison, D., *Arch. Toxicol.*, 1998, **72**, 706.
15. Gao, S., and Burau, R.G., *J. Environ. Qual.*, 1997, **26**, 753.
16. Hasegawa, H., Sohrin, Y., Matsui, M., Hojo, M., and Kawashima, M., *Anal. Chem.*, 1994, **66**, 3247.
17. Hopenhayn-Rich, C., Biggs, M.L., Smith, A.H., Kalman, D.A., and Moore, L.E., *Environ. Health Perspect.*, 1996, **104**, 620.
18. Kuroiwa, T., Ohki, A., Naka, K., and Maeda, S., *Appl. Organomet. Chem.*, 1994, **8**, 325.
19. Kuroiwa, T., Yoshihiko, I., Ohki, A., Naka, K., and Maeda, S., *Appl. Organomet. Chem.*, 1995, **9**, 517.
20. Maeda, S., Wada, H., Kumeda, K., Onoue, M., Ohki, A., Higashi, S., and Takeshita, T., *Appl. Organomet. Chem.*, 1987, **1**, 465.
21. Maeda, S., Ohki, A., Miyahara, K., Takeshita, T., and Higashi, S., *Appl. Organomet. Chem.*, 1990, **4**, 245.
22. Maeda, S., Ohki, A., Tokuda, T., and Ohmine, M., *Appl. Organomet. Chem.*, 1990, **4**, 251.
23. Maeda, S., Ohki, A., Kusadome, K., Kuroiwa, T., Yoshifuku, I., and Naka, K., *Appl. Organomet. Chem.*, 1992, **6**, 213.
24. Maeda, S., Kusadome, K., Arima, H., Ohki, A., and Naka, K., *Appl. Organomet. Chem.*, 1992, **6**, 407.
25. Maeda, S., Ohki, A., Miyahara, K., Naka, K., and Higashi, S., *Appl. Organomet. Chem.*, 1992, **6**, 415.
26. Maeda, S., Mawatari, K., Ohki, A., and Naka, K., *Appl. Organomet. Chem.*, 1993, **7**, 467.
27. Schoof, R.A., Yost, L.J., Eickhoff, J., Crecelius, E.A., Cragin, D.W., Meacher, D.M., and Menzel, D.B., *Food Chem. Toxicol.*, 1999, **37**, 839.
28. Suhendrayatna, Ohki, A., Kuroiwa, T., and Maeda, S., *Appl. Organomet. Chem.*, 1999, **13**, 127.
29. Yager, J.W., Hicks, J.B., and Fabianova, E., *Environ. Health Perspect.*, 1997, **105**, 836.
30. Yamauchi, H., and Yamamura, Y., *Toxicol. Appl. Pharm.*, 1984, **74**, 134.
31. Yamauchi, H., and Yamamura, Y., *Bull. Environ. Contam. Toxicol.*, 1984, **32**, 682.
32. Yamauchi, H., and Yamamura, Y., *Toxicology*, 1985, **34**, 113.
33. Yamauchi, H., Yamamoto, N., and Yamamura, Y., *Bull. Environ. Contam. Toxicol.*, 1988, **40**, 280.
34. Yamauchi, H., Takahashi, K., Yamamura, Y., and Kaise, T., *Toxicol. Environ. Chem.*, 1989, **22**, 69.
35. Yamauchi, H., Kaise, T., Takahashi, K., and Yamamura, Y., *Fund. Appl. Toxicol.*, 1990, **14**, 399.
36. Hasegawa, H., Sohrin, Y., Seki, K., Sato, M., Norisuye, K., Naito, K., and Matsui, M., *Chemosphere*, 2001, **43**, 265.

37. Suhendrayatna, Ohki, A., and Maeda, S., *Appl. Organomet. Chem.*, 2001, **15**, 277.
38. Suhendrayatna, Ohki, A., Nakajima, T., and Maeda, S., *Appl. Organomet. Chem.*, 2001, **15**, 566.
39. Bright, D.A., Brock, S., Cullen, W.R., Hewitt, G.M., Jafaar, J., and Reimer, K.J., *Appl. Organomet. Chem.*, 1994, **8**, 415.
40. Cullen, W.R., Li, H., Hewitt, G., Reimer, K.J., and Zalunardo, N., *Appl. Organomet. Chem.*, 1994, **8**, 303.
41. Cullen, W.R., Li, H., Pergantis, S.A., Eigendorf, G.K., and Mosi, A.A., *Appl. Organomet. Chem.*, 1995, **9**, 507.
42. Kaise, T., Ogura, M., Nozaki, T., Saitoh, K., Sakurai, T., Matsubara, C., Watanabe, C., and Hanaoka, K., *Appl. Organomet. Chem.*, 1997, **11**, 297.
43. Le, X.C., and Ma, M., *J. Chromatogr. A.*, 1997, **764**, 55.
44. Marafante, E., Vahter, M., Norin, H., Envall, J., Sandstroem, M., Christakopoulos, A., and Ryhage, R., *J. Appl. Toxicol.*, 1987, **7**, 111.
45. Wickenheiser, E.B., Michalke, K., Drescher, C., Hirner, A.V., and Hensel, R., *Fresenius J. Anal. Chem.*, 1998, **362**, 498.
46. Guerin, T., Molenat, N., Astruc, A., and Pinel, R., *Appl. Organomet. Chem.*, 2000, **14**, 401.
47. Buchet, J.P., Pauwels, J., and Lauwerys, R., *Environ. Res.*, 1994, **66**, 44.
48. Le, X.C., Ma, M., and Wong, N.A., *Anal. Chem.*, 1996, **68**, 4501.
49. Lopez-Gonzalvez, M., Gomez, M.M., Palacios, M.A., and Camara, C., *Chromatographia*, 1996, **43**, 507.
50. Ma, M., and Le, X.C., *Clin. Chem.*, 1998, **44**, 539.
51. Sur, R., Begerow, J., and Dunemann, L., *Fresenius J. Anal. Chem.*, 1999, **363**, 526.
52. Zhang, X., Cornelis, R., De Kimpe, J., and Mees, L., *Anal. Chim. Acta*, 1996, **319**, 177.
53. Zhang, X., Cornelis, R., De Kimpe, J., Mees, L., Vanderbiesen, V., De Cubber, A., and Vanholder, R., *Clin. Chem.*, 1996, **42**, 1231.
54. Gomez-Ariza, J.L., Sanchez-Rodas, D., Giraldez, I., and Morales, E., *Talanta*, 2000, **51**, 257.
55. He, B., Jiang, G.B., and Xu, X.B., *Fresenius J. Anal. Chem.*, 2000, **368**, 803.
56. Momplaisir, G.M., Blais, J.S., Quinteiro, M., and Marshall, W.D., *J. Agric. Food Chem.*, 1991, **39**, 1448.
57. Cullen, W.R., McBride, B.C., Manji, H., Pickett, A.W., and Reglinski, J., *Appl. Organomet. Chem.*, 1989, **3**, 71.
58. Aggett, J., and Kadwani, R., *Analyst*, 1983, **108**, 1495.
59. Bertolero, F., Marafante, E., Edel Rade, J., Pietra, R., and Sabbioni, E., *Toxicology*, 1981, **20**, 35.
60. Buratti, M., Calzaferri, G., Caravelli, G., Colombi, A., Maroni, M., and Foa, V., *Intern. J. Environ. Anal. Chem.*, 1984, **17**, 25.
61. Byrne, A.R., Tusek-Znidaric, M., Puri, B.K., and Irgolic, K.J., *Appl. Organomet. Chem.*, 1991, **5**, 25.
62. Byrne, A.R., Slejkovec, Z., Stijve, T., Fay, L., Goessler, W., Gailer, J., and Irgolic, K.J., *Appl. Organomet. Chem.*, 1995, **9**, 305.
63. Concha, G., Nermell, B., and Vahter, M., *Environ. Health Perspect.*, 1998, **106**, 355.
64. Concha, G., Vogler, G., Lezcano, D., Nermell, B., and Vahter, M., *Toxicol. Sci.*, 1998, **44**, 185.
65. Cornelis, R., and DeKimpe, J., *J. Anal. At. Spectrom.*, 1994, **9**, 945.
66. Farmer, J.G., and Johnson, L.R., *Brit. J. Ind. Med.*, 1990, **47**, 342.
67. Foa, V., Colombi, A., Maroni, M., Buratti, M., and Calzaferri, G., *Sci. Total Environ.*, 1984, **34**, 241.

68. Gonzales, M.J., Aguilar, M.V., and Martinez, M.C., *Vet. Human Toxicol.*, 1995, **37**, 409.
69. Healy, S.M., Zakharyan, R.A., and Aposhian, H.V., *Mutat. Res.-Rev. Mutat.*, 1997, **386**, 229.
70. Lin, T.H., Huang, Y.L., and Wang, M.Y., *J. Toxicol. Environ. Health A*, 1998, **53**, 85.
71. Lovell, M.A., and Farmer, J.G., *Human Toxicol.*, 1985, **4**, 203.
72. Marafante, E., Vahter, M., and Envall, J., *Chem.-Biol. Interact.*, 1985, **56**, 225.
73. Marafante, E., and Vahter, M., *Environ. Res.*, 1987, **42**, 72.
74. Slejkovec, Z., Byrne, A.R., Smodis, B., and Rossbach, M., *Fresenius J. Anal. Chem.*, 1996, **354**, 592.
75. Tam, K.H., Charbonneau, S.M., Bryce, F., and Lacroix, G., *Anal. Biochem.*, 1978, **86**, 505.
76. Vahter, M., *Environ. Res.*, 1981, **25**, 286.
77. Vahter, M., and Envall, J., *Environ. Res.* 1983, **32**, 14.
78. Vahter, M., Marafante, E., and Dencker, L., *Sci. Total Environ.*, 1983, **30**, 197.
79. Vahter, M., Couch, R., Nermell, B., and Nilsson, R., *Toxicol. Appl. Pharmacol.* 1995, **133**, 262.
80. Vahter, M., Concha, G., Nermell, B., Nilsson, R., Dulout, F., and Natarajan, A.T., *Eur. J. Pharmacol. Environ. Toxicol. Pharmacol. Sect.*, 1995, **293**, 455.
81. Apostoli, P., Alessio, L., Romeo, L., Buchet, J.P., and Leone, R., *J. Toxicol. Environ. Health*, 1997, **52**, 331.
82. Arbouine, M.W., and Wilson, H.K., *J. Trace Elem. Electrolytes Health Dis.*, 1992, **6**, 153.
83. Chen, H., Yoshida, K., Wanibuchi, H., Fukushima, S., Inoue, Y., and Endo, G., *Appl. Organomet. Chem.*, 1996, **10**, 741.
84. Chiou, H.Y., Hsueh, Y.M., Hsieh, L.L., Hsu, L.I., Hsu, Y.H., Hsieh, F.I., Wei, M.L., Chen, H.C., Yang, H.T., Leu, L.C., Chu, T.H., Chen-Wu, C., Yang, M.H., and Chen, C.J., *Mutat. Res.-Rev. Mutat.*, 1997, **386**, 197.
85. Ebdon, L., Fisher, A., Roberts, N.B., and Yaqoob, M., *Appl. Organomet. Chem.*, 1999, **13**, 183.
86. Geiszinger, A., Goessler, W., Kuehnelt, D., Francesconi, K., and Kosmus, W., *Environ. Sci. Technol.*, 1998, **32**, 2238.
87. Goessler, W., Lintschinger, J., Szakova, J., Mader, P., Kopecky, J., Doucha, J., and Irgolic, K.J., *Appl. Organomet. Chem.*, 1997, **11**, 57.
88. Goessler, W., Schlagenhaufen, C., Kuehnelt, D., Greschonig, H., and Irgolic, K.J., *Appl. Organomet. Chem.*, 1997, **11**, 327.
89. Heitkemper, D.T., Flurer, R.A., and Barnes, B., Poster presented at the Winter Conference on Plasma-Spectrochemistry, January 5–10 1998, Scottsdale, USA, 1998.
90. Hsueh, Y.M., Huang, Y.L., Huang, C.C., Wu, W.L., Chen, H.M., Yang, M.H., Lue, L.C., and Chen, C.J., *J. Toxicol. Environ. Health A*, 1998, **54**, 431.
91. Inoue, Y., Kawabata, K., Takahashi, H., and Endo, G., *J. Chromatogr. A*, 1994, **675**, 149.
92. Inoue, Y., Date, Y., Sakai, T., Shimizu, N., Yoshida, K., Chen, H., Kuroda, K., and Endo, G., *Appl. Organomet. Chem.*, 1999, **13**, 81.
93. Kavanagh, P., Farago, M.E., Thornton, I., Goessler, W., Kuehnelt, D., Schlagenhaufen, C., and Irgolic, K.J., *Analyst*, 1998, **123**, 27.
94. Kawabata, K., Inoue, Y., Takahashi, H., and Endo, G., *Appl. Organomet. Chem.*, 1994, **8**, 245.
95. Koch, I., Feldmann, J., Wang, L., Andrewes, P., Reimer, K.J., and Cullen, W.R., *Sci. Total Environ.*, 1999, **236**, 101.

96. Kuehnelt, D., Goessler, W., and Irgolic, K.J., *Appl. Organomet. Chem.*, 1997, **11**, 289.

97. Kuehnelt, D., Goessler, W., and Irgolic, K.J., *Appl. Organomet. Chem.*, 1997, **11**, 459.

98. Kuehnelt, D., Goessler, W., Schlagenhaufen, C., and Irgolic, K.J., *Appl. Organomet. Chem.*, 1997, **11**, 859.

99. Kuehnelt, D., Lintschinger, J., and Goessler, W., *Appl. Organomet. Chem.*, 2000, **14**, 411.

00. Larsen, E.H., Pritzl, G., and Hansen, S.H., *J. Anal. At. Spectrom.* 1993 **8**, 557.

01. Larsen, E.H., Hansen, M., and Goessler, W., *Appl. Organomet. Chem.*, 1998, **12**, 285.

02. Le, X.C., and Ma, M., *Anal. Chem.*, 1998, **70**, 1926.

03. Lintschinger, J., Schramel, P., Hatalak-Rauscher, A., Wendler, I., and Michalke, B., *Fresenius J. Anal. Chem.*, 1998, **362**, 313.

04. Pongratz, R., *Sci. Total Environ.*, 1998, **224**, 133.

05. Ritsema, R., Dukan, L., Roig i Navarro, T., van Leeuwen, W., Oliveira, N., Wolfs, P., and Lebret, E., *Appl. Organomet. Chem.*, 1998, **12**, 591.

06. Shiomi, K., Sugiyama, Y., Shimakura, K., and Nagashima, Y., *Fisheries Sci.*, 1996, **62**, 261.

07. Slejkovec, Z., Byrne A.R., Goessler, W., Kuehnelt, D., Irgolic, K.J., and Pohleven, F., *Acta Chim. Slov.*, 1996, **43**, 269.

08. Slejkovec, Z., Byrne, A.R., Stijve, T., Goessler W., and Irgolic, K.J., *Appl. Organomet. Chem.*, 1997, **11**, 673.

09. Turpeinen, R., Pantsar-Kallio, M., Haeggblom, M., and Kairesalo, T., *Sci. Total Environ.*, 1999, **236**, 173.

10. Yoshida, K., Chen, H., Inoue, Y., Wanibuchi, H., Fukushima, S., Kuroda, K., and Endo, G., *Arch. Environ. Contam. Toxicol.*, 1997, **32**, 416.

11. Yoshida, K., Inoue, Y., Kuroda, K., Chen, H., Wanibuchi, H., Fukushima, S., and Endo, G., *J. Toxicol. Environ. Health A*, 1998, **54**, 179.

12. Zhang, X., Cornelis, R., De Kimpe, J., and Mees, L., *J. Anal. At. Spectrom.*, 1996, **11**, 1075.

13. Zheng, J., Goessler, W., and Kosmus, W., *Mikrochim. Acta*, 1998, **130**, 71.

14. Bohari, Y., Astruc, A., Astruc, M., and Cloud, J., *J. Anal. At. Spectrom.*, 2001, **16**, 774.

15. Caruso, J.A., Heitkemper, D.T., and B'Hymer, C., *Analyst*, 2001, **126**, 136.

16. Heitkemper, D.T., Vela, N.P., Stewart, K.R., and Westphal, C.S., *J. Anal. At. Spectrom.*, 2001, **16**, 299.

17. Vela, N.P., Heitkemper, D.T., and Stewart, K.R., *Analyst*, 2001, **126**, 1011.

18. Mattusch, J., Wennrich, R., Schmidt, A.-C., and Reisser, W., *Fresenius J. Anal. Chem.*, 2000, **366**, 200.

19. Yoshida, K., Kuroda, K., Inoue, Y., Chen, H., Date, Y., Wanibuchi, H., Fukushima, S., and Endo, G., *Appl. Organomet. Chem.*, 2001, **15**, 539.

120. Kuroda, K., Yoshida, K., Yasukawa, A., Wanibuchi, H., Fukushima, S., and Endo, G., *Appl. Organomet. Chem.*, 2001, **15**, 548.

121. Gregus, Z., Gyurasics, A., and Csanaky, I., *Toxicol. Sci.*, 2000, **56**, 18.

22. Feldmann, J., John, K., and Pengprecha, P., *Fresenius J. Anal. Chem.*, 2000, **368**, 116.

123. Zbinden, P., Andrey, D., and Blake, C., *Atom. Spectrosc.*, 2000, **21**, 205.

24. Mandal, B.K., Ogra, Y., and Suzuki, K.T., *Chem. Res. Toxicol.*, 2001, **14**, 371.

125. Koch, I., Wang, L., Reimer, K.J., and Cullen, W.R., *Appl. Organomet. Chem.*, 2000, **14**, 245.

126. Inoue, Y., Date, Y., Yoshida, K., Chen, H., and Endo, G., *Appl. Organomet. Chem.*, 1996, **10**, 707.

127. Shiomi, K., Chino, M., and Kikuchi, T., *Appl. Organomet. Chem.*, 1990, **4**, 281.
128. Shiomi, K., Sugiyama, Y., Shimakura, K., and Nagashima, Y., *Appl. Organomet Chem.*, 1995, **9**, 105.
129. Benramdane, L., Accominotti, M., Fanton, L., Malicier, D. and Valon, J.J., *Clin Chem.*, 1999, **45**, 301.
130. Hakala, E., and Pyy, L., *J. Anal. At. Spectrom.*, 1992, **7**, 191.
131. Hwang, C.J., and Jiang, S.J., *Anal. Chim. Acta*, 1994, **289**, 205.
132. Lai, V.W.M., Cullen, W.R., Harrington, C.F., and Reimer, K.J., *Appl. Organomet Chem.*, 1997, **11**, 797.
133. Le, X.C., Cullen, W.R., and Reimer, K.J., *Clin. Chem.*, 1994, **40**, 617.
134. Shibata, Y., Yoshinaga, J., and Morita, M., *Appl. Organomet. Chem.*, 1994, **8**, 249
135. Le, X.C., Lu, X., Ma, M., Cullen, W.R., Aposhian, H.V., and Zheng, B., *Anal Chem.*, 2000, **72**, 5172.
136. Le, X.C., Ma, M., Lu, X., Cullen, W.R., Aposhian, H.V., and Zheng, B., *Environ Health Perspect.*, 2000, **108**, 1015.
137. Aposhian, H.V., Zheng, B., Aposhian, M.M., Le, X.C., Cebrian, M.E., Cullen, W. Zakharyan, R.A., Ma, M., Dart, R.C., Cheng, Z., Andrewes, P., Yip, L., O'Malley G.F., Maiorino, R.M., Van Voorhies, W., Healy, S.M., and Titcomb, A., *Toxicol Appl. Pharmacol.*, 2000, **165**, 74.
138. Aposhian, H.V., Gurzau, E.S., Le, X.C., Gurzau, A., Healy, S.M., Lu, X., Ma, M. Yip, L., Zakharyan, R.A., Maiorino, R.M., Dart, R.C., Tircus, M.G., Gonzalez Ramirez, D., Morgan, D.L., Avram, D., and Aposhian, M.M., *Chem. Res. Tox icol.*, 2000, **13**, 693.
139. Del Razo, L.M., Garcia-Vargas, G.G., Vargas, H., Albores, A., Gonsebatt, M.E. Montero, R., Ostrosky-Wegman, P., Kelsh, M., and Cebrian, M.E., *Arch. Toxicol.* 1997, **71**, 211.
140. Vahter, M., Marafante, E., Lindgren, A., and Dencker, L., *Arch. Toxicol.*, 1982 **51**, 65.
141. Vahter, M., and Marafante, E., *Chem.-Biol. Interact.*, 1983, **47**, 29.
142. Vahter, M., and Marafante, E., *Toxicol. Lett.*, 1987, **37**, 41.
143. Van Holderbecke, M., Zhao, Y., Vanhaecke, F., Moens, L., Dams, R., and Sandra, P., *J. Anal. At. Spectrom.*, 1999, **14**, 229.
144. National Research Council, In *Arsenic in Drinking Water*, National Academy Press, Washington DC, 1999, pp. 35–41.
145. Faust, S.D., Winka, A.J., and Belton, D., *J. Environ. Sci. Health A*, 1987, **22** 209.
146. Sohrin, Y., Matsui, M., Kawashima, M., Hojo, M., and Hasegawa, H., *Environ. Sci. Technol.*, 1997, **31**, 2712.
147. Challenger, F., Higginbottom, C., and Ellis, L., *J. Chem. Soc.*, 1933, 95.
148. Maeda, S., Fujita, S., Ohki, A., Yoshifuku, I., Higashi, S., and Takeshita, T., *Appl. Organomet. Chem.*, 1988, **2**, 353.
149. Maeda, S., Inoue, R., Kozono, T., Tokuda, T., Ohki, A., and Takeshita, T. *Chemosphere*, 1990, **20**, 101.
150. Schoof, R.A., Yost, L.J., Crecelius, E., Irgolic, K., Goessler, W., Guo, H.R., and Greene, H., *Hum. Ecol. Risk Assess.*, 1998, **4**, 117.
151. Geiszinger, A., *PhD thesis*, Institute for Analytical Chemistry, Karl-Franzens-University, Graz, Austria, 1998.
152. Koch, I., Wang, L., Ollson, C.A., Cullen, W.R., and Reimer, K.J., *Environ. Sci. Technol.*, 2000, **34**, 22.
153. Marafante, E., Vahter, M., and Dencker, L., *Sci. Total Environ.*, 1984, **34**, 223.
154. Marafante, E., Bertolero, F., Edel, J., Pietra, R., and Sabbioni, E., *Sci. Total Environ.*, 1982, **24**, 27.

96. Kuehnelt, D., Goessler, W., and Irgolic, K.J., *Appl. Organomet. Chem.*, 1997, **11**, 289.
97. Kuehnelt, D., Goessler, W., and Irgolic, K.J., *Appl. Organomet. Chem.*, 1997, **11**, 459.
98. Kuehnelt, D., Goessler, W., Schlagenhaufen, C., and Irgolic, K.J., *Appl. Organomet. Chem.*, 1997, **11**, 859.
99. Kuehnelt, D., Lintschinger, J., and Goessler, W., *Appl. Organomet. Chem.*, 2000, **14**, 411.
00. Larsen, E.H., Pritzl, G., and Hansen, S.H., *J. Anal. At. Spectrom.* 1993 **8**, 557.
01. Larsen, E.H., Hansen, M., and Goessler, W., *Appl. Organomet. Chem.*, 1998, **12**, 285.
02. Le, X.C., and Ma, M., *Anal. Chem.*, 1998, **70**, 1926.
03. Lintschinger, J., Schramel, P., Hatalak-Rauscher, A., Wendler, I., and Michalke, B., *Fresenius J. Anal. Chem.*, 1998, **362**, 313.
04. Pongratz, R., *Sci. Total Environ.*, 1998, **224**, 133.
05. Ritsema, R., Dukan, L., Roig i Navarro, T., van Leeuwen, W., Oliveira, N., Wolfs, P., and Lebret, E., *Appl. Organomet. Chem.*, 1998, **12**, 591.
06. Shiomi, K., Sugiyama, Y., Shimakura, K., and Nagashima, Y., *Fisheries Sci.*, 1996, **62**, 261.
07. Slejkovec, Z., Byrne A.R., Goessler, W., Kuehnelt, D., Irgolic, K.J., and Pohleven, F., *Acta Chim. Slov.*, 1996, **43**, 269.
08. Slejkovec, Z., Byrne, A.R., Stijve, T., Goessler W., and Irgolic, K.J., *Appl. Organomet. Chem.*, 1997, **11**, 673.
09. Turpeinen, R., Pantsar-Kallio, M., Haeggblom, M., and Kairesalo, T., *Sci. Total Environ.*, 1999, **236**, 173.
10. Yoshida, K., Chen, H., Inoue, Y., Wanibuchi, H., Fukushima, S., Kuroda, K., and Endo, G., *Arch. Environ. Contam. Toxicol.*, 1997, **32**, 416.
11. Yoshida, K., Inoue, Y., Kuroda, K., Chen, H., Wanibuchi, H., Fukushima, S., and Endo, G., *J. Toxicol. Environ. Health A*, 1998, **54**, 179.
12. Zhang, X., Cornelis, R., De Kimpe, J., and Mees, L., *J. Anal. At. Spectrom.*, 1996, **11**, 1075.
13. Zheng, J., Goessler, W., and Kosmus, W., *Mikrochim. Acta*, 1998, **130**, 71.
14. Bohari, Y., Astruc, A., Astruc, M., and Cloud, J., *J. Anal. At. Spectrom.*, 2001, **16**, 774.
15. Caruso, J.A., Heitkemper, D.T., and B'Hymer, C., *Analyst*, 2001, **126**, 136.
16. Heitkemper, D.T., Vela, N.P., Stewart, K.R., and Westphal, C.S., *J. Anal. At. Spectrom.*, 2001, **16**, 299.
17. Vela, N.P., Heitkemper, D.T., and Stewart, K.R., *Analyst*, 2001, **126**, 1011.
118. Mattusch, J., Wennrich, R., Schmidt, A.-C., and Reisser, W., *Fresenius J. Anal. Chem.*, 2000, **366**, 200.
119. Yoshida, K., Kuroda, K., Inoue, Y., Chen, H., Date, Y., Wanibuchi, H., Fukushima, S., and Endo, G., *Appl. Organomet. Chem.*, 2001, **15**, 539.
120. Kuroda, K., Yoshida, K., Yasukawa, A., Wanibuchi, H., Fukushima, S., and Endo, G., *Appl. Organomet. Chem.*, 2001, **15**, 548.
121. Gregus, Z., Gyurasics, A., and Csanaky, I., *Toxicol. Sci.*, 2000, **56**, 18.
122. Feldmann, J., John, K., and Pengprecha, P., *Fresenius J. Anal. Chem.*, 2000, **368**, 116.
123. Zbinden, P., Andrey, D., and Blake, C., *Atom. Spectrosc.*, 2000, **21**, 205.
124. Mandal, B.K., Ogra, Y., and Suzuki, K.T., *Chem. Res. Toxicol.*, 2001, **14**, 371.
125. Koch, I., Wang, L., Reimer, K.J., and Cullen, W.R., *Appl. Organomet. Chem.*, 2000, **14**, 245.
126. Inoue, Y., Date, Y., Yoshida, K., Chen, H., and Endo, G., *Appl. Organomet. Chem.*, 1996, **10**, 707.

127. Shiomi, K., Chino, M., and Kikuchi, T., *Appl. Organomet. Chem.*, 1990, **4**, 281.
128. Shiomi, K., Sugiyama, Y., Shimakura, K., and Nagashima, Y., *Appl. Organomet. Chem.*, 1995, **9**, 105.
129. Benramdane, L., Accominotti, M., Fanton, L., Malicier, D. and Valon, J.J., *Clin. Chem.*, 1999, **45**, 301.
130. Hakala, E., and Pyy, L., *J. Anal. At. Spectrom.*, 1992, **7**, 191.
131. Hwang, C.J., and Jiang, S.J., *Anal. Chim. Acta*, 1994, **289**, 205.
132. Lai, V.W.M., Cullen, W.R., Harrington, C.F., and Reimer, K.J., *Appl. Organomet. Chem.*, 1997, **11**, 797.
133. Le, X.C., Cullen, W.R., and Reimer, K.J., *Clin. Chem.*, 1994, **40**, 617.
134. Shibata, Y., Yoshinaga, J., and Morita, M., *Appl. Organomet. Chem.*, 1994, **8**, 249.
135. Le, X.C., Lu, X., Ma, M., Cullen, W.R., Aposhian, H.V., and Zheng, B., *Anal. Chem.*, 2000, **72**, 5172.
136. Le, X.C., Ma, M., Lu, X., Cullen, W.R., Aposhian, H.V., and Zheng, B., *Environ. Health Perspect.*, 2000, **108**, 1015.
137. Aposhian, H.V., Zheng, B., Aposhian, M.M., Le, X.C., Cebrian, M.E., Cullen, W., Zakharyan, R.A., Ma, M., Dart, R.C., Cheng, Z., Andrewes, P., Yip, L., O'Malley, G.F., Maiorino, R.M., Van Voorhies, W., Healy, S.M., and Titcomb, A., *Toxicol. Appl. Pharmacol.*, 2000, **165**, 74.
138. Aposhian, H.V., Gurzau, E.S., Le, X.C., Gurzau, A., Healy, S.M., Lu, X., Ma, M., Yip, L., Zakharyan, R.A., Maiorino, R.M., Dart, R.C., Tircus, M.G., Gonzalez-Ramirez, D., Morgan, D.L., Avram, D., and Aposhian, M.M., *Chem. Res. Toxicol.*, 2000, **13**, 693.
139. Del Razo, L.M., Garcia-Vargas, G.G., Vargas, H., Albores, A., Gonsebatt, M.E., Montero, R., Ostrosky-Wegman, P., Kelsh, M., and Cebrian, M.E., *Arch. Toxicol.*, 1997, **71**, 211.
140. Vahter, M., Marafante, E., Lindgren, A., and Dencker, L., *Arch. Toxicol.*, 1982, **51**, 65.
141. Vahter, M., and Marafante, E., *Chem.-Biol. Interact.*, 1983, **47**, 29.
142. Vahter, M., and Marafante, E., *Toxicol. Lett.*, 1987, **37**, 41.
143. Van Holderbecke, M., Zhao, Y., Vanhaecke, F., Moens, L., Dams, R., and Sandra, P., *J. Anal. At. Spectrom.*, 1999, **14**, 229.
144. National Research Council, In *Arsenic in Drinking Water*, National Academy Press, Washington DC, 1999, pp. 35–41.
145. Faust, S.D., Winka, A.J., and Belton, D., *J. Environ. Sci. Health A*, 1987, **22**, 209.
146. Sohrin, Y., Matsui, M., Kawashima, M., Hojo, M., and Hasegawa, H., *Environ. Sci. Technol.*, 1997, **31**, 2712.
147. Challenger, F., Higginbottom, C., and Ellis, L., *J. Chem. Soc.*, 1933, 95.
148. Maeda, S., Fujita, S., Ohki, A., Yoshifuku, I., Higashi, S., and Takeshita, T., *Appl. Organomet. Chem.*, 1988, **2**, 353.
149. Maeda, S., Inoue, R., Kozono, T., Tokuda, T., Ohki, A., and Takeshita, T., *Chemosphere*, 1990, **20**, 101.
150. Schoof, R.A., Yost, L.J., Crecelius, E., Irgolic, K., Goessler, W., Guo, H.R., and Greene, H., *Hum. Ecol. Risk Assess.*, 1998, **4**, 117.
151. Geiszinger, A., *PhD thesis*, Institute for Analytical Chemistry, Karl-Franzens-University, Graz, Austria, 1998.
152. Koch, I., Wang, L., Ollson, C.A., Cullen, W.R., and Reimer, K.J., *Environ. Sci. Technol.*, 2000, **34**, 22.
153. Marafante, E., Vahter, M., and Dencker, L., *Sci. Total Environ.*, 1984, **34**, 223.
154. Marafante, E., Bertolero, F., Edel, J., Pietra, R., and Sabbioni, E., *Sci. Total Environ.*, 1982, **24**, 27.

155. Buchet, J.P., and Lauwerys, R., *Toxicol. Appl. Pharmacol.*, 1987, **91**, 65.
156. Yamauchi, H., Kaise, T., and Yamamura, Y., *Bull. Environ. Contam. Toxicol.*, 1986, **36**, 350.
157. Yamauchi, H., Takahashi, K., and Yamamura, Y., *Toxicology*, 1986, **40**, 237.
158. Vahter, M., *Clin. Chem.*, 1994, **40**, 679.
159. Styblo, M., Serves, S.V., Cullen, W.R., and Thomas, D.J., *Chem. Res. Toxicol.*, 1997, **10**, 27.
160. Lin, S., Cullen, W.R., and Thomas, D.J., *Chem. Res. Toxicol.*, 1999, **12**, 924.
161. Petrick, J.S., Ayala-Fierro, F., Cullen, W.R., Carter, D.E., and Aposhian, H.V., *Toxicol. Appl. Pharmacol.*, 2000, **163**, 203.
162. Mass, M.J., Tennant, A., Roop, B.C., Cullen, W.R., Styblo, M., Thomas, D.J., and Kligerman, A.D., *Chem. Res. Toxicol.*, 2001, **14**, 355.
163. Yamauchi, H., Takahashi, K., Mashiko, M., Saitoh, J., and Yamamura, Y., *Appl. Organomet. Chem.*, 1992, **6**, 383.
164. Buchet, J.P., and Lauwerys, R., *Appl. Organomet. Chem.*, 1994, **8**, 191.
165. Marafante, E., and Vahter, M., *Chem.-Biol. Interact.*, 1984, **50**, 49.
166. Buchet, J.P., and Lauwerys, R., *Arch. Toxicol.*, 1985, **57**, 125.
167. Offergelt, J.A., Roels, H., Buchet, J.P., Boeckx, M., and Lauwerys, R., *Brit. J. Ind. Med.*, 1992, **49**, 387.
168. Crecelius, E.A., *Environ. Health Perspect.*, 1977, **19**, 147.
169. Hopenhayn-Rich, C., Biggs, M.L., Kalman, D.A., Moore, L.E., and Smith, A.H., *Environ. Health Perspect.*, 1996, **104**, 1200.
170. Concha, G., Vogler, G., Nermell, B., and Vahter, M., *Int. Arch. Occup. Environ. Health*, 1998, **71**, 42.
171. Zhang, X., Cornelis, R., De Kimpe, J., Mees, L., and Lameire, N., *Clin. Chem.*, 1997, **43**, 406.
172. Greuther, W., Brummit, R.K., Farr, E., Kilian, N., Kirk, P.M., and Silva, P.C., (Eds.), *NCU-3 Names in Current Use for Extant Plant Genera*, Koeltz Scientific Books, Koenigstein, 1993.

7 Organoantimony Compounds in the Environment

PAUL ANDREWES

United States Environmental Protection Agency, Research Triangle Park, NC, USA

and

WILLIAM R. CULLEN

Department of Chemistry, University of British Columbia, Canada

7.1 INTRODUCTION

Antimony (atomic number 51, atomic mass 121.8) belongs to group 15 of the periodic table, immediately under arsenic. Antimony compounds are found in the environment with antimony possessing an oxidation state of +III or +V. The structures of some important common antimony compounds are shown in Figure 7.1. Although ubiquitous, antimony is one of the least abundant elements on Earth (ranked 62nd in crustal abundance [1]) and for this reason antimony in the environment has received comparatively little attention. Nevertheless, the United States Environmental Protection Agency (US-EPA) considers it a priority pollutant. The environmental chemistry of antimony is often discussed with respect to that of arsenic because both elements occur in the same group of the periodic table and are often found in association in the environment, however there are very significant differences in their chemistry [2].

Our knowledge of organoantimony compounds in the environment has greatly increased since the publication of the first edition of this work in 1986. This increase can be attributed to chemical curiosity and the application of recently developed analytical methodology, and also, in particular, to the examination of the 'toxic gas hypothesis' that has been advanced to account for sudden infant death syndrome (SIDS). The toxic gas hypothesis provided considerable impetus to investigate the production of organoantimony compounds by microorganisms. New analytical instrumentation, especially inductively coupled plasma mass spectrometry (ICP-MS), provided the sensitivity and element specificity to detect organoantimony compounds. In addition, the multi-element capability of ICP-MS allowed the detection of organoantimony compounds during investigations that may not have been directed specifically towards determining these species. Some of this material is covered in a recently published conference presentation [3].

Organometallic Compounds in the Environment
Edited by P.J. Craig © 2003 John Wiley & Sons Ltd

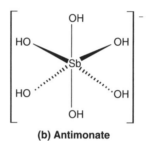

(a) Potassium antimony tartrate

(b) Antimonate

(c) Sodium stibogluconate

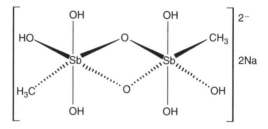

(d) Disodium-di-u-oxo-bis[trihydoxo methylantimonate(V)]
(sodium salt of monomethylstibonic acid)

Figure 7.1   Structures of some antimony compounds

Triethylstibine was the first organometallic compound of antimony to be synthesized (in 1850) [4]. Over 3000 organoantimony compounds are now reported in the literature. The methylantimony species are currently the most studied organoantimony compounds in the environment largely because, by analogy with arsenic, they are expected to be the major species present, and because analytical methods are available for their determination.

7.2   CHEMISTRY

7.2.1   TERMINOLOGY

The compound Me_3Sb is commonly named trimethylstibine or trimethylanti-mony. The terms methylantimony species, dimethylantimony species and tri-methylantimony species are often used to refer to any antimony compounds containing one, two and three methyl groups, respectively. Therefore, to avoid confusion, in this chapter, Me_3Sb is referred to as trimethylstibine, $MeSbH_2$ as methylstibine and Me_2SbH as dimethylstibine.

7.2.2   SYNTHESIS OF METHYLANTIMONY SPECIES

The preparation of dimethylstibinic acid was first reported as follows (Equations 7.1–7.3):

$$Me_2SbBr + Br_2 \xrightarrow{-20 \text{ to } -40\,°C} Me_2SbBr_3 \tag{7.1}$$

$$Me_2SbBr_3 + 3NaOR \xrightarrow{-20 \text{ to } -40\,°C} Me_2Sb(OR)_3 + 3NaBr \tag{7.2}$$

$$Me_2Sb(OR)_3 + 2H_2O \longrightarrow Me_2Sb(O)OH + 3ROH \tag{7.3}$$

The product, claimed to be pure, did not melt and is insoluble in common organic solvents, and, presumably, in water [5]. It now seems that the charac-terization of the material was suspect (H. A. Meinema, personal communication to W. R. Cullen) and other workers have been unable to reproduce this work [6, 7].

Methylstibonic acid can be prepared via two routes [8] (Equations 7.4 and 7.5):

$$n[MeSb(OR)_4]_2 + 2mH_2O \longrightarrow 2[MeSbO_2]_n.mH_2O + 4nROH \tag{7.4}$$

$$nMeSb(OMe)_2 + mH_2O_2 \longrightarrow [MeSbO_2]_n.mH_2O + 2nMeOH \tag{7.5}$$

The 'acid' is soluble in base to give the water soluble salt $Na_2[CH_3Sb(OH)_3O]_2$, the X-ray crystal structure of which has revealed that the anion is oxygen bridged [9] (Figure 7.1d).

Parris and Brinckman examined rates of quaternization of trimethylstibine with alkyl halides [10]. They proposed that this might be an important process because significant concentrations of alkyl halides are found in the ocean and the atmosphere. However, the rates of reaction were found to be very slow. It is significant that most antimony(III) species are not nucleophilic enough to react with methyl iodide to form methylantimony(V) derivatives. This reaction is at the heart of alkylarsenic chemistry and is the foundation of the Challenger pathway for the biological methylation of arsenic, Figure 7.2 (see Section 7.8.6) [11].

Trimethylstibine and the trimethylantimony(V) species $Me_3SbX_2(X = Cl, Br, I, OH)$ are readily accessed via Grignard reactions [12].

7.2.3   STABILITY OF TRIMETHYLSTIBINE

The environmental mobility of antimony is partially determined by the stability of trimethylstibine. Bamford and Newitt studied the gas phase interaction of trimethylstibine with oxygen [13]: they monitored the pressure change as oxygen was admitted into a bulb containing trimethylstibine. Some of the data are shown in Figure 7.3. In curve (1) the initial departure from linearity signifies slow reaction of the gases. Higher pressure of trimethylstibine results in faster reaction, curve (2). Parris and Brinckman used these data to estimate a rate constant for the oxidation [14]. To do this they had to make a number of unjustified assumptions including 'the reaction is homogenous and second order.' Like many other oxidations, this reaction is probably a chain process, and even though at high pressures trimethylstibine and oxygen can ignite, their interaction at low pressure cannot be predicted because it depends on chain breaking and making reactions, either of which can occur on surfaces. Parris and Brinckman report that in unagitated methanol solution, at high concentration, trimethylstibine reacts with air to afford mainly the oxide 'Me$_3$SbO' accompanied by ill-defined products, the result of limited antimony–carbon bond cleavage. In spite of the absence of characterized products their results suggested that the major product is a trimethylantimony(V) species. Craig et al. investigated the solid formed by the reaction of trimethylstibine with oxygen by using mass spectrometry. They found that extracts of the product with water, acetonitrile, and methanol contain trimethylantimony oxide and a range of cyclic and linear oligomers. The presence of tetramethylantimony species indicates that group migration can occur in the analytical system; however no dimethyl species were seen [15].

Extrapolating these results to environmental conditions is difficult but they seem to indicate that at low concentration the oxidation of trimethylstibine is slow and does not result in significant antimony–carbon bond cleavage. The fact that landfill gases can be collected in Tedlar bags one day and analyzed

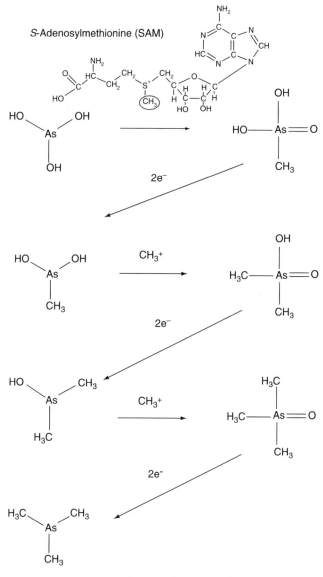

Figure 7.2 The mechanism of arsenic biomethylation initially proposed by Challenger [11] Reproduced with permission

successfully for trimethylstibine some days later speaks for the stability of the species [16].

Unfortunately, a number of authors have invoked the conclusions of Parris and Brinckman to explain why in some experiments trimethylstibine was not detected in the headspace of cultures of *S. brevicaulis*, and similar arguments

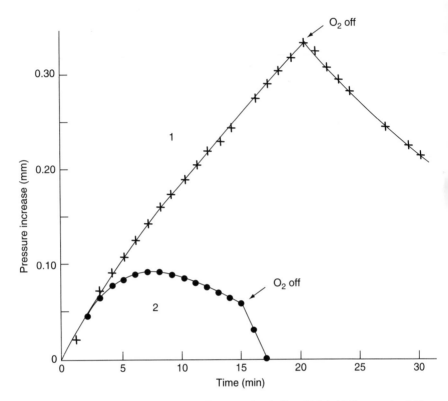

Figure 7.3   Pressure increase versus time for a reaction bulb, which initially contained 49 mm (curve 1) or 0.89 mm (curve 2) Me_3Sb and had oxygen bled in at the rate of 0.03 mm min^{-1} (pressure is mm H_2SO_4). From Reference [13] Reproduced with permission

have been used to dismiss trimethylstibine as an environmentally significant species. (see Section 7.8)

7.3   USES AND ENVIRONMENTAL DISTRIBUTION OF ANTIMONY COMPOUNDS

The concentration of antimony in the Earth's crust is 0.2 mg Sb kg^{-1}, which is about an order of magnitude less than that of arsenic (1.5 mg As kg^{-1}) [1]. However, significant deposits containing high concentrations of antimony as the ore stibnite (Sb_2S_3) are found in China, Bolivia, Russia and South Africa, and these deposits are mined at the rate of 140 000 metric tons antimony per year (1998) [17]. High antimony concentrations, up to 500 mg kg^{-1}, were found in soils in the Palatinate Forest, Germany, an area that had seen considerable mining activity in the past; however, the concentrations of antimony in the plants grown on the soil were not elevated. The authors attribute this to the low availability of the element, maximum 0.6%, as measured by extraction with ammonium nitrate. Plants grown in prepared soils that contained higher amounts of extractable antimony took up more of the element [18].

Historically antimony compounds have been used as cosmetics, medicines and as a poison (although not as popular as arsenic!). Potassium antimony tartrate (tartar emetic, Figure 7.1a) was once commonly used as an emetic and as an expectorant [19]. Trivalent antimony compounds were used in the treatment of leishmaniasis, but their use in modern medicine has been superseded by less-toxic pentavalent antimony compounds (such as sodium stibogluconate, Figure 7.1c, and meglumin antimonate) that are the preferred treatments for the tropical parasite leishmaniasis today [20]. Small quantities of trivalent antimony compounds are used in the treatment of schistosomiasis. Compounds derived from phenylstibonic acid have been used as trypanocidal drugs and the compounds Stibamine and Stibosan were once used as antisyphilitic drugs, but the discovery of penicillin superseded this application. The small quantity of antimony used in drugs is insignificant compared with the total world production of antimony, so this use is unlikely to have any environmental consequences.

Today the predominant use of antimony is as antimony trioxide, which is added to plastics, rubber, adhesives and textiles as a flame retardant. This use accounted for 56% of total antimony consumption in the United States in 1997 [21]. When antimony trioxide is incorporated into polymers it is tightly bound (in one study $< 0.5\%$ of the antimony in a PVC sample was released after 8 days extraction with domestic cleaning fluid [22]) and poses little environmental threat (see Section 7.8.2). However, the disposal of such products by incineration mobilizes significant quantities of antimony compounds. The average antimony concentration of a feed stream into municipal waste incinerators is approximately $42\,mg\,Sb\,kg^{-1}$ and over half of this is derived from fire retardants [23]. Fortunately, during incineration, scrubbers can effectively remove antimony from the flue gas.

Antimony compounds are also used in ceramics, glasses, paints, pigments and ammunition primers. High purity stibine is used as a dopant in the manufacture of semiconductors and trialkylstibines are used as catalysts in the polymerization of vinyl compounds.

Metallic antimony is alloyed with other metals for use in solder, bearings, ammunition, castings, cable coverings, sheet-and-pipe, and batteries. Antimonial lead is used as the grid metal in lead-acid storage batteries. This use is particularly significant because stibine (SbH_3) can be released into the atmosphere during the production of these batteries [24]. Stibine can also be released if batteries are overcharged.

In the United States, the Environmental Protection Agency (EPA) maintains a Toxics Release Inventory (TRI). Over a period of 7 years the reported release of antimony into the US environment was over 5000 metric tons, mostly from nonferrous metal smelting, and most of this emission was into the atmosphere. Other major inputs of antimony into the atmosphere include industrial dust and gases from the burning of fossil fuels. The global anthropogenic emission of antimony into the atmosphere has been estimated at 6100 metric tons per year. In contrast, the total natural global emission of antimony into the atmosphere has been estimated at 2600 metric tons per year with major sources including volcanic activity and sea salt spray [25]. Atmospheric concentrations of antimony range from 0.6 ng m^{-3} (rural areas) to 8000 ng m^{-3} (urban areas of North America) [26]. Information about the antimony species in airborne particulate matter (APM) is desirable for risk assessment reasons. Zheng et al. found that aqueous extracts of APM collected in Tokyo (56 μg g^{-1} Sb) contain mainly Sb(V) [27].

An average antimony concentration of 0.2 μg Sb L^{-1} is found in natural waters (rivers, lakes and oceans) although polluted waters can, for example, contain 700 μg Sb^{-1} L^{-1} [28].

7.4 TOXICOLOGY OF ANTIMONY COMPOUNDS

Because of their current, and previous, use in medicine, the effects of inorganic antimony(III) and antimony(V) complexes, such as sodium stibogluconate (Figure 7.1c), on humans are well documented [29]. Antimony(III) species are more toxic than antimony(V) species, and the most serious effects usually involve the liver or heart. The toxicological properties of antimony species are thought to arise from the high affinity of antimony for sulfur but the evidence for this is minimal. The mechanistic toxicology of antimony has been reviewed in comparison to that of arsenic [30].

Leonard and Gerber reviewed the literature on the mutagenicity, carcinogenicity and teratogenicity of antimony compounds and concluded: 'claims that antimony compounds have mutagenic properties are based on insufficient and not particularly relevant evidence [31].' They consider antimony to be less of a mutagenic risk than other metals such as arsenic, nickel and chromium.

The threshold limit value (TLV) for antimony, and its compounds, in workplace air, has been set at 0.5 mg Sbm^{-3}. The United States Food and Drug Administration (US-FDA) has set the limit for antimony in food at 2 mg Sb kg^{-1}. The US-EPA has set the maximum contaminant level (MCL) for antimony in drinking water at 6 μg SbL^{-1} (cf. the MCL for arsenic was, until very recently, 50 μg As L^{-1}, the new MCL is 10 μg As L^{-1} [32]). The European Union standard for antimony in drinking water is 10 μg Sb L^{-1}.

Little attempt has been made to evaluate the toxicity of organoantimony compounds. In 1939, Seifter performed experiments to determine the acute toxicity of trimethylstibine to animals and he concluded that: 'Trimethylstibine possesses no great or pronounced acute toxicity to animals [33]. Fungal toxicities of some diphenylantimony, triphenylantimony and trimethylantimony compounds were determined and only diphenylantimony oxinate, diphenylantimony acetonate and diphenylantimony chloride had EC$_{50}$ values (effective concentration to kill 50 % of the colonies) less than 30 mg Sb L^{-1} [34].

7.5 ANALYTICAL METHODS FOR ENVIRONMENTAL ANTIMONY SPECIATION

7.5.1 INTRODUCTION

A variety of techniques for performing inorganic antimony speciation, i.e. the determination of antimony(III) and antimony(V), have been described. In a recent review Smichowski et al. outlined methods for antimony speciation in waters at trace and ultratrace levels [35]. Selective hydride generation is currently the most popular technique for the speciation of antimony(III) and antimony(V) and has been successfully applied to a number of environmental samples.

In contrast, only a few techniques for organoantimony compound speciation have been described, the most popular being hydride generation gas chromatography (HG-GC) for solutions, and GC for gases. The main reasons for the paucity of analytical methods for organoantimony speciation are:

(i) The total antimony content in environmental matrices is relatively low.
(ii) There is a lack of accessible well-characterized compounds that could be used as standards for the development and validation of new (and in some cases, old) analytical methods.

7.5.2 SAMPLE PREPARATION AND STORAGE

In order to determine antimony species in biological materials, it is desirable to extract the compounds without modifying them. The most commonly used procedures initially involve rupturing the cells (freeze-dry and grind, or blend) and extracting the material with mild solvents. To accelerate the process heat, pressure, agitation and sonication are often used.

Dodd and coworkers sonicated homogenized plant samples in acetic acid for 1 hour and left the mixture to stand overnight [36]. Koch et al. sonicated homogenized plant material in methanol/water (50:50) for 20 minutes; extraction efficiencies were typically less than 25 %. Craig et al. extracted plant samples using three different procedures: suspension of dried material in water for 2 days; suspension of dried material in sodium hydroxide (10^{-5} M) for 3 days in sunlight; and overnight sonication in acetic acid (0.2 M) [37]. Extracts were then analyzed by using HG-GC–AAS: organoantimony species were detected only in the acetic acid extracts.

Slurries of homogenized samples can be directly analyzed by using hydride generation methodology. For instance, sediment can be freeze-dried, homogenized and then analyzed by using HG-GC–ICP-MS [38]. Koch and coworkers analyzed lysed fungal cells directly by using HG-GC–AAS [39].

The determination of organoantimony species in matrices that contain relatively large amounts of inorganic antimony species is facilitated if the inorganic antimony can be removed by using solid phase extraction (SPE) [40].

Little is known about the stability of organoantimony compounds in solution. Andreae and coworkers [41] reported that aqueous solutions of methylstibonic acid and dimethylstibinic acid are stable for several weeks, although details on how this conclusion was reached are not provided.

7.5.3   GAS CHROMATOGRAPHY

The volatile antimony compounds: stibine (BP $-17\,^{\circ}$C), methylstibine (BP $41\,^{\circ}$C), dimethylstibine (BP $61\,^{\circ}$C) and trimethylstibine (BP $81\,^{\circ}$C) are easily separated, although pure samples of methylstibine and dimethylstibine are difficult to acquire. Capillary columns or packed columns may be used; the separation is best at low temperatures. Packing materials that have been successfully used include Tenax-TA, Porapak-PS and SP-2100 (10 % on Chromosorb).

Volatile stibines in the environment can be collected by trapping *in situ* [42] or by remote trapping, after collecting the gases in Tedlar bags [16]. The trap, which functions as a crude GC column, is connected to a detector, and slowly heated. The stibines slowly desorb from the trap in order of increasing boiling point, so that separation is achieved. If this separation is not sufficient (usually because rapid sampling results in a broad distribution of stibines on the trap) the stibines can be cryofocused onto a second GC column [43].

Quantification of stibines is a problem because of the difficulty of working with gaseous standards at the appropriate concentrations, and because of the unavailability of reference samples of methylstibine and dimethylstibine. A method for semiquantification (reproducibility $\pm\,30\,\%$) has been described for GC–ICP-MS where an aqueous sample is used as the calibrant [44]. An internal standard (usually rhodium) is aspirated during the analysis. This procedure has been applied to determining the concentration of trimethylstibine in gases from landfills, sewage and hot-springs [43].

AAS can be used as a detector for analyzing samples that contain large quantities of organoantimony compounds. However, the use of an ICP-MS as a detector gives significantly lower detection limits and allows monitoring of antimony isotopes. Ion trap and quadrupole mass spectrometers have also been used as detectors, these allow compound verification and can also be used to monitor incorporation of isotope labels when performing mechanistic studies. The incorporation of a time of flight mass spectrometer as detector (GC-ICP-time of flight-MS) has allowed the investigation of the possibility of isotope fractionation of antimony through bivolatilization processes [45].

7.5.3.1   Hydride Generation

Braman *et al.* were the first to report a technique where inorganic antimony species were reduced to stibine for subsequent determination by spectral emission [46]. Nakashima then demonstrated that careful control of the pH at which derivatization is performed allows the determination of total antimony and antimony(III). Antimony(V) may then be determined by difference [47]. Andreae and coworkers inserted a GC column between the reaction vessel and the AAS detector (i.e. HG-GC–AAS) and showed that it is possible, in principle, to determine methylstibines that are formed from unknown precursors as a result of hydride-forming reactions [41]. Samples of methylstibonic acid and dimethylstibinic acid were used as precursors in this work but their purity is suspect. The synthesis of these compounds is difficult (see Section 7.2.2). Furthermore, results from hydride generation need to be interpreted with care because artifact peaks can be formed during the reaction. For example, when solutions of pure trimethylantimony dichloride, trimethylantimony dihydroxide, or dimethylantimony dihydroperoxychloride were reacted with $NaBH_4$ all compounds formed four stibines (SbH_3, $MeSbH_2$, Me_2SbH and Me_3Sb) [7]. Dodd et al. claimed that preconditioning the analytical system with the reagents $NaBH_4$ and 4-M acetic acid eliminated demethylation [36], but other workers could not reproduce this work [39]. The production of artifact peaks is enhanced in both batch and semi-continuous hydride generation systems as the pH is decreased, and is enhanced in some analytical matrices [39]. However, Craig et al. report that rigorous exclusion of oxygen minimizes artifact peak production when performing HG-GC–AAS even at pH 1 [37].

One general disadvantage of hydride generation methodology is that information on the original species, from which the hydrides are formed, is lost [48]. Despite these problems hydride generation has proven useful for revealing that methylantimony species exist in the environment.

7.5.4 LIQUID CHROMATOGRAPHY

Hydride generation is inadequate for determining anything other than the presence of inorganic or methylated antimony species in solution. Other organoantimony compounds, which might be expected to be found in the environment, either would not yield volatile hydrides or would not form hydrides at all. It is clear that if the environmental chemistry of antimony is to be advanced, liquid chromatography methods need to be developed. The use of liquid chromatography and an appropriate detector has enabled the determination of a diverse range of organoarsenic compounds in the environment [49].

HPLC has been used to separate antimony(III) and antimony(V) species with various degrees of success [28, 50–58]. However, application of these techniques to organoantimony species has proven difficult because of a lack of synthetic standards. The only readily available methylantimony standards are Me_3SbX_2(X = Cl, Br, OH), and Me_3SbO, and it is likely that all these compounds hydrolyze to the same species in aqueous solution.

Antimony(V) and trimethylantimony species can be separated by using anion exchange HPLC under alkaline conditions (2 mM KOH was the mobile phase that provided the best separation) [50, 51]. However, the trimethylantimony species eluted in the solvent front, so the use of this procedure is limited, and even with ICP-MS as the detector, detection limits ($\sim 0.8\,\mu g\,L^{-1}$) were still too high for the method to be of practical use. Detection limits were significantly reduced ($\sim 5\,ng\,L^{-1}$) when a 'metal-free' chromatographic system coupled to an ultrasonic nebulizer was employed [51]. Cation exchange did not prove successful for determining Me_3SbCl_2 [50], even though there is evidence that Me_3SbCl_2 hydrolyses to $[Me_3SbOH]^+$ in aqueous solution [51]. The species does not elute and may be lost because of precipitation.

Ulrich used anion exchange HPLC with a mobile phase of phthalic acid (pH 5) to separate antimony(V), antimony(III) and Me_3SbO [54, 55]. Similar methodology is described by Krachler and Emons who employed HG-AAS for detection [59]. Citric acid complexes with both antimony(III) and antimony(V) and resulting species separate on a PRP-X100 anion exchange column (EDTA/phthalic acid as mobile phase) [60].

Antimony speciation has been performed by using capillary electrophoresis (CE) coupled to ICP-MS [61, 62] The best method used Na_2HPO_4/NaH_2PO_4, 20 mM, pH 5.6 as the background electrolyte, with sodium hydroxide (100 mM) and acetic acid (1 %, pH 2) as the stacking electrolytes. The concentration detection limit for all antimony species was approximately $0.1\,\mu g\,L^{-1}$. Trimethylantimony dichloride gave two peaks, and potassium antimony tartrate gave three peaks in the electropherogram under alkaline conditions.

7.5.5   MASS SPECTROMETRY

Antimony has two naturally occurring isotopes, ^{121}Sb and ^{123}Sb, with abundances of 57.3 % and 42.7 % respectively, and these characterize the fragmentation pattern of antimony compounds.

A number of researchers have reported the mass spectra of methylstibines for both ion trap [16, 63] and quadrupole mass spectrometers [36, 64, 65]. The fragmentation patterns obtained from quadrupole mass spectrometers are much simpler than from ion traps. An example of the later is shown in Figure 7.4a.

Lintschinger et al. reported the electrospray ionization mass spectrum of aqueous trimethylantimony dichloride, which is a commonly used standard for the analysis of organoantimony compounds in the environment [51]. They concluded that Me_3SbCl_2 hydrolyzes in water (pH 7), most probably forming the species $[Me_3SbOH]^+$. A similar conclusion was reached by Zheng et al. who claim that whereas negative ion electrospray mass spectra are more suitable for the identification of inorganic antimony species, positive ion spectra afford better results from organic species. The ion Me_3SbOH^+ is characteristic of trimethylantimony(V) species [66].

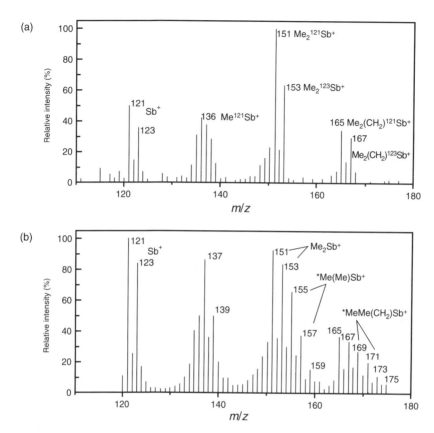

Figure 7.4 Mass spectra of trimethystibine obtained by using CGC-MS (ion trap): (a) trimethylstibine containing no labeled methyl groups, (b) trimethylstibine that was obtained by hydride derivatization of medium from a culture of *S. brevicaulis* that had been incubated with $^{13}CD_3$-L-methionine and potassium antimony tartrate ($^*Me = {}^{13}CD_3$). From Reference [93] Chemosphere, Vol 41, Cullen et al. p 1717 with permission from Elsevier Science © 2000

7.6 INORGANIC ANTIMONY COMPOUNDS IN THE ENVIRONMENT

A review of antimony in the environment, with emphasis on the inorganic species in natural waters has been recently published [67]. The oxidation state of antimony is important with respect to the possible formation of organoantimony compounds, because the formation of antimony–carbon bonds probably involves oxidative addition reactions to antimony(III) species. This is the situation for arsenic (Figure 7.2 and Chapter 1, Section 13). Thus, biomethylation of antimony(V) would occur only after reduction of antimony(V) to antimony(III). In the case of arsenic, microbial reduction of arsenic(V) to arsenic(III), with or without further methylation is commonly found [49]. The biological reduction of inorganic antimony(V) species has not been demonstrated.

As expected from the thermodynamic data, antimony(V) predominates in oxygenated marine and fresh waters, and the concentration ranges from 0.025 to 0.29 μg Sb L^{-1} (average 0.2 μg Sb L^{-1} [41, 68–71].

There are few reports of antimony speciation in other matrices. One study examined antimony speciation in soils [56]. Only small amounts of the total antimony in the soil were readily extracted and antimony(III) was not detected in any of the extracts even though the contamination was from an antimony(III) producing industry.

Aqueous extracts of airborne particulate matter APM collected in Tokyo (56 μg g^{-1}Sb) contain mainly Sb(V) (89.5 % of the extracted antimony) but some trimethyl antimony species are present (0.5 %) in addition to some other unidentified antimony species (6.6 %) (anion exchange and size exclusion HPLC –ICP-MS) [27].

7.7   ORGANOANTIMONY COMPOUNDS IN THE ENVIRONMENT

7.7.1   VOLATILE ORGANOANTIMONY COMPOUNDS

Volatile organoantimony compounds have been detected in a variety of environments, often in association with a miscellany of other volatile organometallic compounds.

The antimony oxide deposits in biogas burners reveal the presence of volatile antimony species in sewage and landfill fermentation gas [72]. Direct evidence for volatile antimony species in such sites in central Germany was first provided by injection of gas samples into an ICP-MS [73]. Trimethylstibine was detected in landfill and sewage gas by using GC–ICP-MS [42, 74]. The initial identification of trimethylstibine was based on retention time comparisons. Subsequently, confirmation was made by using GC–MS (ion trap) to analyze landfill and sewage gas from Canadian sites [16]. The concentration ranges of trimethylstibine in German fermentation gases were estimated as 0.62–15 μg Sb m^{-3} and 23.9–71.6 μg Sb m^{-3} for sewage and landfill gas, respectively. Similar concentrations were found in Canadian fermentation gases.

Condensed water samples, obtained from the outlet of the landfill gas collection pipeline, were found to contain methylantimony, dimethylantimony and possibly trimethylantimony and triethylantimony species by using HG-GC–ICP-MS [74]. Similarly, standing water on a landfill site was shown to contain methylantimony, dimethylantimony and trimethylantimony species by using HG-GC–ICP-MS [39] (on a wet day, evolution of gas can often be seen as a constant stream of bubbles passing through puddles on top of a landfill).

The environmental impact of trimethylstibine, or any other volatile metal-(loid) species that is released from landfills, and other methanogenic environments, has not been investigated, but it has been suggested that metal(loid) gas evolution may be significant with respect to chronic poisoning and global cycling of the elements [73]. However, the apparently low toxicity of trimethyl-stibine [33] makes it unlikely that this species is a hazard to landfill workers, although other volatile metal(loid) species may cause problems.

Trimethylstibine and dimethylstibine were detected, by using GC–ICP-MS analysis, above (and within) algal mats growing in the Meager Creek hot springs(BC, Canada) [43]. Methylantimony species have been detected in geothermal waters (see Section 7.7.2).

Feldmann et al. described a cryotrapping GC-ICP-MS method for the determination of volatile metal species in anaerobic environments. Unfortunately the antimony species react with the NaOH used to clean up the gas stream [75].

7.7.2 ORGANOANTIMONY COMPOUNDS IN WATERS AND SEDIMENT

The first organoantimony compounds to be detected in the environment were found in natural waters. Andreae et al. derivatized antimony compounds in natural waters by using sodium borohydride [41]. The derivative compounds: stibine, methylstibine and dimethylstibine were detected by using GC–AAS and it was unjustifiably concluded that the waters contain *methylstibonic* and *dimethylstibinic* acid. Although their work indicated that methylantimony species are present in natural waters, the inorganic functional groups in the species remain unknown, as do the oxidation states.

It is now known that the experimental conditions (30 mM HCl) used by Andreae and coworkers are likely to produce artifacts (methyl groups redistribute during the hydride generation process so that SbH_3, $MeSbH_2$, Me_2SbH and Me_3Sb can form from a pure methylantimony compound) [7, 39]. Furthermore the reference compounds that were used were impure (see Section 7.2.2), and there is no guarantee that they were not also subject to rearrangement upon hydride generation. Nevertheless, the generation of methylstibines from pure inorganic antimony compounds, during hydride generation is very unlikely, so the work of Andreae and coworkers indicates that these natural waters must contain some methylantimony species of indeterminate structure and oxidation state.

The following 'quantitative' results were reported by Andreae et al. [41] for river and sea waters and similar results were obtained by Bertine and Lee (using the methods of Andreae et al. on sea water from Saanich Inlet, BC, Canada) [69]. However, it should be realized that these data must be treated with caution in view of the aforementioned problems. Methylantimony species were the only organoantimony species detected in rivers (0.5–1.8 ng Sb L^{-1}). Methylantimony (0.8–12.6 ng Sb L^{-1}) and dimethylantimony (0.6–3.2 ng Sb L^{-1}) species were detected in seawater. No trimethylantimony species were detected during these studies. The methylantimony compounds typically comprised 10 % of the total antimony present. Because there is no known anthropogenic discharge of these types of compounds into natural waters they most probably arise from biological methylation. It was also reported, by Andreae *et al.* [41], that methylantimony species were present in pure cultures of marine planktonic algae (details on the methodology used were not given) but, in a subsequent paper, Andreae and Froelich stated that methylantimony compounds were not detected in marine phytoplankton (extracts were prepared by grinding the algae with 2-M HCl and analyzed by using HG-GC–AAS) [68].

In 1984 Andreae and Froelich [68] presented an impressive survey of arsenic, antimony and germanium speciation in the Baltic Sea, which is one of the most polluted marine areas on the planet. Deep marine basins that are often anoxic are characteristic of the Baltic. Samples were collected on a cruise in 1981 from five points along the axis of the Baltic Sea and the Gulf of Finland. Samples were taken from 10 to 235 m, depth mostly in 10-m increments. The samples were analyzed by using HG-GC–AAS. Methylantimony species were detected throughout the water column and were more abundant than dimethylantimony species that were only detected occasionally [68]. Slightly higher levels of methylantimony species were found in the surface layers for all five sample-points. Also, for two sample-points slightly higher levels of methylantimony species were observed in the anoxic zone. Andreae and Froelich suggested that the methylantimony species were produced by bacteria rather than algae, because no methylantimony species were detected in marine phytoplankton [68] or macro-algae [76], and production of methylantimony species is not closely tied to the eutrophic zone (as it is for arsenic).

Methylantimony species were identified in sediment from a West German river, by using HG-GC–ICP-MS [38], although the methodology used for hydride derivatization could have resulted in artifact formation. Rough quantification was performed on the observed species by using interelement calibration and the concentration ranges reported were 0.2–9.8, 0.1–1.2, and 0.1–0.9 μg Sb kg^{-1} for MeSbH$_2$, Me$_2$SbH, and Me$_3$Sb respectively. In some samples, an additional peak was observed in the gas chromatogram. This peak was attributed (based on boiling point calculations) to triethylstibine.

Hirner *et al.* measured methylantimony species in geothermal waters obtained from various New Zealand locations by using HG-GC–ICP-MS [43]. Methylantimony, dimethylantimony and trimethylantimony species were detected as the corresponding stibines. The highest concentration of 31 ng Sb kg^{-1} was for a dimethylantimony species present in Frying Pan Lake, Waimangu, New Zealand. However, the reliability of these numbers is questionable because (a) the possibility of demethylation in the analytical system was not examined; (b) methylantimony species in the blanks had to be corrected for; and (c) rough quantification was made by using interelement calibration.

Koch *et al.* studied (HG-GC-AA) two antimony rich environments in Canada: Yellowknife, NWT, a gold mining region, and the Meager Creek hot springs in British Columbia. Methylated species and Sb(III) were found in water samples from both environments although inorganic Sb(V) was the major species, up to 300 ppm in mine effluent. One hot spring sample contained Sb(III) and trimethylantimony species [77].

7.7.3 ORGANOANTIMONY COMPOUNDS IN SOIL

Methanol–water (1:1) extracts of highly polluted soil samples from Bitterfeld, Germany, were found to contain trace amounts of an antimony species having the same retention time as trimethylantimony oxide by using HPLC–ICP-MS [55].

Michalke and Schramel used CE–ICP-MS to obtain evidence for a trimethyl-antimony species in the liquid phase (obtained by centrifugation at 25 800 g) of a sewage sludge and a fouling sludge [61]. A small peak in the electropherogram had the same retention time as a standard of trimethylantimony dichloride.

7.7.4 ORGANOANTIMONY COMPOUNDS IN PLANTS AND ANIMALS

Samples of pondweed (*Potamogetan pectinatus*) from antimony-impacted lakes near Yellowknife, Canada, were extracted with acetic acid (0.2 M) and the extracts were analyzed by using HG-GC–MS (quadrupole) [36]. The extracts contained methylantimony, dimethylantimony, and trimethylantimony species. When a pure standard of $Me_3Sb(OH)_2$ was analyzed by using this analytical system, only a single peak was observed (Me_3Sb) indicating that the analytical conditions used did not cause disproportionation. The pondweed samples contained between 48 and 68 mg total Sb kg^{-1} dry weight; however, quantification of the organoantimony species was not reported.

Plant samples (liverwort and moss) from the former Louisa antimony mine, Eskdale, Scotland were extracted with acetic acid (0.2 M) and the extracts were analyzed by using HG-GC–AAS [37]. The extracts contained methylantimony and dimethylantimony species, but trimethylantimony species were not detected. As a rough estimate, the plant samples contained between 101 and 186 ng extractable dimethylantimony g^{-1} dry weight. The total antimony content of the plant material was between 8.9 and 57.3 µg Sb g^{-1} dry weight. In a similar study [77], methanol/water was used for the extraction of biota samples and the extraction efficiencies were typically low but ranged from 0.7% for *Funaria hygrometricia* (190 ppm Sb dry weight) to 95% for *Typha latifolia* (0.2 ppm Sb) both from Yellowknife. Inorganic Sb(V) was the main species in the extracts. The authors suggested that one abundant moss, *Drepanocladus* sp., could be used as a laboratory standard for dimethylantimony species.

Methylantimony species were found for the first time in an animal, the snail *Stagnicola* sp. from Yellowknife. A headspace HG-GC–MS method was used to confirm some of the speciation results [77].

7.8   BIOLOGICAL PRODUCTION OF ORGANOANTIMONY COMPOUNDS IN THE LABORATORY

7.8.1   INTRODUCTION

Ever since the remarkable discovery that cultures of microorganisms can bio-methylate arsenic, selenium and tellurium compounds [11], it has been pre-sumed that the same would be true for antimony. Indeed, Barnard found evidence that *Penicillium notatum* produces a volatile antimony compound (he suggested that it was trimethylstibine) when cultured with potassium antimo-nate or the monosodium salt of phenylstibonic acid [78]. The tests used (Marsh and Gutzeit tests) on the gas-trapping solution, through which volatile products were bubbled, had low sensitivity and were not specific.

As mentioned in the introduction, interest in the ability of microorganisms to transform antimony has recently received considerable attention because of a possible association of this process with sudden infant death syndrome.

7.8.2   ANTIMONY, MICROORGANISMS AND SUDDEN INFANT DEATH SYNDROME

Sudden infant death syndrome (SIDS), also known as cot death, is defined as 'unexpected deaths, in sleeping quarters, of apparently healthy infants'. Richardson has proposed a 'toxic gas hypothesis' as a possible explanation of SIDS. The hypothesis is that microorganisms growing on infants' cot bedding material, that could contain phosphorus, arsenic, or particularly antimony (used as a fire retardant) are able to transform these compounds to toxic volatile species. Inhalation of the resulting phosphines, arsines and stibines, by infants can result in acute or chronic poisoning [79]. An extensive summary of the literature on the toxic gas hypothesis is provided in Reference 80. The hypothesis placed emphasis on the filamentous fungus *Scopulariopsis brevicau-lis*, which is well known for its ability to produce the supposedly toxic trimethy-larsine from a variety of arsenic substrates [11, 49].

Richardson used colorimetric methods to identify gases produced by microorganisms growing on samples taken from cot mattresses. He concluded that phosphines, arsines or stibines, or mixtures of gases, were produced [79]. Most other investigators suggest that the observed color changes could have been the result of the generation of gases [81] such as dimethyl sulfide [82].

One early study with cultures of *S. brevicaulis* or *Phaeolus schweinitzii* (a wood rotting fungus that is known for its ability to biomethylate arsenic) failed to show that these fungi could biovolatilize potassium antimony tartrate or antimony trioxide, even though arsenic trioxide was readily biovolatilized

[83]. In this study silver nitrate impregnated paper strips were suspended in the culture headspace in the expectation that they would trap any evolved arsines and stibines. None were found by using HG-AAS.

Gates and coworkers incubated various aerobic and anaerobic microorganisms, including *S. brevicaulis* and six *Bacillus* sp., with antimony(III) compounds. Mercuric chloride impregnated papers were set up to trap any volatile species and the papers were later analyzed using mass spectrometry [82]. No trimethylstibine was produced under aerobic conditions and it was not produced by some facultative anaerobes or anaerobes cultured from cot mattresses. However, some anaerobic cultures from a pond sediment did produce trimethylstibine.

7.8.3 PRODUCTION OF TRIMETHYLSTIBINE BY MIXED ANAEROBIC CULTURES

The microbiological production of trimethylstibine was first unambiguously demonstrated using anaerobic cultures obtained by inoculating antimony-rich medium with soils [64]. Anaerobic cultures from a sewage plant soil and an automobile body shop soil produced trimethylstibine, which was detected by using GC–chemiluminescence and GC–MS. Trimethylstibine was produced from both potassium antimony tartrate and potassium hexahydroxyantimonate substrates. However, the production of trimethylstibine by these cultures was highly variable (the concentration of trimethylstibine in culture headspace ranged from 566 to 4897 parts per billion by volume after 1 month of incubation; quantification was made by using a standard solution of trimethylstibine). There was no evidence for any other volatile antimony compounds in the headspace gases of these cultures.

As part of a larger study described in the previous section, Gates and co-workers prepared 56 anaerobic cultures, containing potassium antimony tartrate, using mud from a pond as the inoculum [82]. Three of these cultures produced trimethylstibine, which was identified by using mass spectrometry (quantification was not performed).

Jenkins *et al.* performed a similar study of anaerobic antimony-enriched cultures, but they used a range of soils obtained in the United Kingdom [84]. Volatile antimony species were purged from the cultures and trapped on a GC column (OV-101, 3%) and analyzed by using GC–AAS and GC–MS. The antimony substrate used was potassium antimony tartrate. The generation of trimethylstibine, the only volatile antimony compound, was highly variable. Rough estimates of the quantities of trimethylstibine in the culture vials ranged from 40–3450 ng, i.e. 0.03–2.30% of the elemental antimony substrate after 5–8 weeks of incubation.

7.8.4   MICROBIOLOGICAL PRODUCTION OF ORGANOANTIMONY COMPOUNDS BY AEROBIC CULTURES

7.8.4.1   Scopulariopsis brevicaulis

Because of its long association with the biomethylation of arsenic, the potential for *S. brevicaulis* to biovolatilize antimony to stibines has received much attention [11, 49]. There is little evidence that this fungus is a major part of the microbial population on cot bedding material [81] yet this is the organism that has been singled out by proponents of the toxic gas hypothesis for SIDS as the major culprit. As discussed above, many attempts to produce volatile antimony compounds from *S. brevicaulis* were unsuccessful [82, 83]. It was suggested that trimethylstibine may have been lost by rapid oxidation [65, 85, 86]; however, the analytical methodology used was not very sensitive.

Jenkins *et al.* showed that *S. brevicaulis* (IMI 17297) could biomethylate potassium antimony tartrate, antimony trioxide and antimony(V) compounds by using a combination of aerobic and anaerobic incubations [65, 85]. Trimethylstibine in the headspace was trapped on Tenax-TA during an anaerobic incubation (that followed an aerobic phase), and identified following desorption into a MS (GC–MS). However, trimethylstibine was not detected by GC–MS during aerobic incubation, although mass transport of antimony was observed. They concluded that the trimethylstibine trapped during the anaerobic incubation was a product arising from aerobic growth because *S. brevicaulis* does not grow in the absence of oxygen. Subsequently Andrewes *et al.* unequivocally established that trimethylstibine is produced by *S. brevicaulis* (ATCC 7903), cultured under entirely aerobic conditions in medium containing potassium antimony tartrate, by using GC–ICP-MS analysis of headspace gases [87]. The headspace gases were continuously purged from the culture and trapped at $-78\,°C$ on SP-2100 (10 % on Chromosorb). This methodology avoids any problems that might arise as a result of instability of trimethylstibine at room temperature. Craig *et al.* reported similar results but reproducibility was poor [86].

Trimethylstibine is the predominant volatile antimony compound detected in headspace gases of *S. brevicaulis* cultures but other stibines have been detected in some cultures [40]. *S. brevicaulis* was grown in media containing both arsenic and antimony in order to probe the possible interference of one element on the biomethylation of the other. The trimethyl derivatives of both elements were quantified (HG-GC-AAS) to reveal that the biomethylation of antimony can be enhanced by the presence of arsenite whereas antimony(III) can inhibit the biomethylation of arsenic(III). The same study showed that arsenic is much more readily biomethylated than antimony by factors of at least 500 [88].

Phaeolus schweinitzii, a wood rotting fungus, is relatively efficient in methylating antimony(III) species. Both di- and trimethylated products are found in the culture medium with the later predominating with a maximum yield of 0.4 % from potassium antimony tartrate ($1000\,mg\,Sb\,L^{-1}$). Inefficient biomethylation of antimony(V) was observed. In contrast, the yield of trimethylarsenic species was 20% following incubation of *P schweinitzii* with sodium arsenite (1 mg As L^{-1}) [89].

7.8.4.2 Bacteria

Dimethylantimony species were detected in the cell pellet of an antimony-resistant soil isolate, *Aeromonas* sp., and in the culture supernatent of *Flavobacterium* sp. (HG-GC-AA). Provision of inorganic arsenic(III) alongside antimony(III) at an equivalent concentration enhanced the formation of involatile methylantimony species, but high relative concentrations of arsenic(III) retarded the formation of the antimony metabolites. These authors suggest that biomethylation of antimony is a fortuitous process and is not a primary resistance mechanism for the bacteria [90]. Antimony biomethylation by bacteria under aerobic conditions could account for the presence of methylantimony species in cot mattress inner foam [91].

7.8.5 MICROBIOLOGICAL PRODUCTION OF ORGANOANTIMONY COMPOUNDS BY ANEROBIC BACTERIA

Michalke *et al.* found that *Methanobacterium formicicum* catalyses the anaerobic formation of stibine, methylstibine, dimethylstibine, and trimethylstibine from cultures containing 0.1 mM or less antimony(III) (GC-ICP-MS analysis). At 0.5 mM antimony(III) trimethylstibine is the only product from *M. formicicum* as it is from other microflora that are involved in anaerobic digestion of sewage sludge: *M. barkeri, M. thermoautotrophicum, Clostridium collagenovorans, Desulfovibrio vulgaris,* and *D. gigas.* As part of the same study the head space gas above incubated sewage sludge samples contained trimethylstibine and trimethylbismuthine as the dominant species with concentrations up to 80 ng per liter in spite of the data showing that antimony and bismuth (see also Chapter 9) were at relatively low concentration (Sb1.7 mg kg^{-1} (dw), As 15.2 mg kg^{-1} (dw)) in the sludge [62].

7.8.6 THE MECHANISM OF ANTIMONY BIOMETHYLATION AND THE PRODUCTION OF SOLUBLE ORGANOANTIMONY SPECIES

The redox potentials for arsenic and antimony are similar so it is feasible that the mechanism of arsenic biomethylation first proposed by Challenger [11] also applies to antimony (Figure 7.2). According to this mechanism biomethylation is thought to proceed by the oxidative addition of a methyl carbocation to an arsenic(III) species. The methylarsenic(V) product is then reduced in a two electron reduction and the process is repeated. Subsequent work revealed the source of the methyl carbocation in this mechanism to be S-adenosylmethionine (SAM, Figure 7.2) [49]. Evidence for this mechanism also includes the identification of the expected arsenic(V) intermediates in cultures [92].

Similar evidence has now been obtained from studies involving antimony as follows:

(i) Antimony(III) compounds are much more readily biomethylated than antimony(V) compounds. Antimony(V) is not reduced in cultures of *S. brevicaulis*) [40, 65].

(ii) Both dimethylantimony and trimethylantimony species are found in the medium of cultures of *S. brevicaulis*. The identification of dimethylstibine and trimethylstibine produced by hydride derivatization of the medium was made by using GC–AAS [40], GC–ICP-MS [40] and GC–MS (ion trap) [63]. Dimethylantimony species and trimethylantimony species were present at approximately equal concentration and the concentrations of these antimony species were between 0.8 and 7.1 μg Sb L^{-1}. The presence of dimethylantimony species at such high concentrations (relative to trimethylantimony species) cannot be attributed to trimethylstibine oxidation (see Section 7.2.3).

(iii) When $^{13}CD_3$-L-methionine was added to cultures of *S. brevicaulis* and potassium antimony tartrate, the methylantimony species produced contained approximately 50 % labeled methyl group [63]. L-Methionine is the precursor for S-adenosylmethionine, hence it is likely that the source of the methyl carbocation for antimony biomethylation is SAM. Label incorporation was measured by using GC–MS (ion trap) (Figure 7.4). The dimethyl and trimethyl antimony species in the culture medium can be separated by using anion exchange chromatography, and isotope incorporation studies reveal that CD_3-labels from methionine are incorporated to approximately the same extent into the methylarsenic and methylantimony species [93].

7.9 ANTIMONY AND ALGAE

Kantin examined antimony species in green and brown macro-algae (seaweed) by using HG-GC–AAS and did not detect methylantimony species [76]. Andreae and Froelich performed similar measurements on marine phytoplankton with similar results (extracts were prepared by grinding the algae with 2-M HCl and analyzed by using HG-GC–AAS, but extraction efficiencies were not determined) [68].

Benson *et al.* found evidence for antimony metabolites in the marine diatom *Thalassiosira nana* [94]. The diatom was cultured in $^{125}SbCl_3$, the cells harvested and the suspension extracted with methanol. The R_F values for 2D paper ^{125}Sb radioautograms were similar to those they had previously assigned to compounds that were assumed to be the arsenosugars isolated by Edmonds and Francesconi [95]. However, these antimony species have not been further characterized.

Chlorella vulgaris, a fresh water alga, was found rapidly to bioaccumulate antimony(III). When the alga was transferred to an antimony-free medium it excreted antimony as 40% antimony(V) and 60% antimony(III). The authors suggested that this oxidation is a means of detoxification. The antimony in the cells was shown to be associated with proteins of molecular weight of approximately 40 000 [96].

Antimony speciation in an enclosed spring phytoplankton bloom was monitored by using HG-AAS. No change in antimony speciation was observed, i.e. the only significant species detected throughout the course of the experiment was antimony(V) [97].

7.10 ANTIMONY METABOLISM IN MAMMALS

Arsenic ingested by humans and other mammals is rapidly excreted in the urine, and a large percentage of the arsenic excreted in urine is in the form of methylarsenic species [32]. In contrast, very few studies of the human metabolism of antimony have been performed; this is probably because cases of antimony poisonings are rare. The usual four antimony species were detected in human urine by using hydride generation methodology, but the methylated precursors were in very low abundance relative to inorganic antimony [98].

Antimony compounds are poorly absorbed from the gastrointestinal tract [29]. The concentration of antimony in body fluids of an adult woman who had ingested an unknown amount of antimony trisulfide was monitored over 160 hours [99]. The concentration of antimony recovered from the urine was the same whether the sample was analyzed directly or pretreated (wet digestion or dry ash) and it was suggested that this is because there was no *in vivo* biomethylation.

A population of Sprague-Dawley rats was administered a single intraperitoneal or intravenous dose of antimony trichloride, and the concentration of antimony in feces and urine was monitored over 96 hours. There was no evidence for antimony biomethylation *in vivo*. The methodology used, in attempting to identify methylantimony compounds, included hydride generation, ion exchange chromatography and various mineralization procedures. However, no detection limits were reported for these procedures. An interesting observation was made that antimony (as antimony trichloride) completely inhibits the biomethylation of arsenic by rat liver cytosol [100].

7.11 CONCLUSIONS/FUTURE DIRECTIONS

The only organoantimony species that have been positively identified in the environment are methylantimony species. This is probably because the analytical methodologies developed to date can only be applied to these species. (The reported detections of ethyl antimony compounds that may be the result of trans-ethylation from ethyllead compounds, should be confirmable by using HG/GC methodology). The exact nature of these methylantimony species in biota and in extracts is yet to be established.

It is likely that antimony biomethylation proceeds via a mechanism similar to that of arsenic. If this is the case then, like arsenic, other more complex organoantimony compounds such as stibiosugars and stibiobetaine may be formed in analogous ways.

The first isolation of arsenobetaine and the arsenosugars was made by using a natural-products approach; milligram quantities of arsenic species were isolated from kilograms of substrate, by using column chromatography, and characterization was made by using NMR and X-ray structure analysis [44]. Once the arsenic species were isolated they could be used as standards for determination by using HPLC–MS (ICP-MS, ESIMS, etc.). Unfortunately, this approach is less viable for antimony because the concentrations of antimony species in biological compartments are much lower. In the absence of antimony standards, which are difficult to prepare, identification techniques must be developed that do not require prior knowledge of structure. HPLC (or CE) –MS^n with picogram or lower detection limits is the most promising. Fortunately, the characteristic isotope distribution pattern of ^{121}Sb and ^{123}Sb will make an MS approach slightly easier.

7.12 REFERENCES

1. Cox, P.A., in *The Elements: Their Origin, Abundance and Distribution*, Oxford University Press, Oxford, 1989.
2. *Chemistry of Arsenic, Antimony and Bismuth*, N. C. Norman (Ed.), Blackie Academic and Professional, London, 1998.
3. Craig, P.J., Forster, S.N., Jenkins, R.O., Miller, D. P., Ostah, N., Smith, L.M., and Morris, T.-A. *Chem. Processes Mar. Environ. Meeting Date 1998*, 2000, 265.
4. Lowig, C., and Schweizer, E., *Justus Liebigs Ann. Chem.*, 1850, **75**, 315.
5. Meinema, H.A., and Noltes, J.G., *Journal of Organometallic Chemistry*, 1972, **36**, 313.
6. Rheingold, A., *Synth. React. Inorg.*, 1978, **8**, 453.
7. Dodd, M., Grundy, S.L., Reimer, K.J., and Cullen, W.R., *Appl. Organomet. Chem.* 1992 **6**, 207.
8. Wieber, M., and Walz, J., *Z. Naturforsch.*, 1990, **45b**, 1615.
9. Wieber, M., Simonis, U., and Kraft, D., *Z. Naturforsch.*, 1991, **46b**, 139.
10. Parris, G.E., and Brinckman, F.E., *J. Org. Chem.*, 1975, **40**, 3801.
11. Challenger, F., *Chem. Rev.*, 1945, **36**, 315.
12. Morgan, G.T., and Davies, G.R., *Proc. R. Soc., Ser. A*, 1926, **110**, 523.
13. Bamford, C.H., and Newitt, D.M., *J. Chem. Soc.*, 1946, 695.
14. Parris, G.E., and Brinckman, F.E., *Environ. Sci. Technol.*, 1976, **10**, 1128.
15. Craig, P.J., Forster, S.A., Jenkins, R.O., Lawson, G., Miller, D., and Ostah, N., *Appl. Organomet. Chem.* 2001, **15**, 527.

16. Feldmann, J., Koch, I., and Cullen, W.R., *Analyst*, 1998, **123**, 815.
17. Carlin, J.F., *USGS Mineral Commodity Summary: Antimony*, US Geological Survey, Reston, VA, USA, 1999.
18. Hammel, W., Debus, R., and Steubing, L., *Chemosphere*, 2000, **41**, 1791.
19. McCallum, R.I., *Proc. R. Soc. Med.*, 1997, **70**, 756.
20. Berman, J.D., *Clin. Infect. Dis.*, 1997, **24**, 684.
21. *USGS Minerals Yearbook*; Volume I. *Metals and Minerals*, US Geological Survey, Reston, VA, USA, 1997.
22. Jenkins, R.O., Craig, P.J., Goessler, W., and Irgolic, K.J., *Hum. Exp. Toxicol.*, 1998, **17**, 138.
23. Van Velzen, D., Langenkamp, H., and Herb, G., *Waste Manage. Res.*, 1998, **16**, 32.
24. Kentner, M., Leinemann, M., Schaller, K.H., Welte, D., and Lehnert, G., *Internat. Arch. Occu. Environ. Health*, 1995, **67**, 119.
25. Clarke, L.B., and Sloss, L.L., in *Trace Elements*, 1992, Vol. 49.
26. Schroeder, W.H., Dobson, M., Kane, D.M., and Johnson, N.D., *J. Air Pollut. Control Assoc.*, 1987, **37**, 1267.
27. Zheng, J., Ohata, M., and Furuta, N., *Analyst*, 2000, **125**, 1025.
28. Smichowski, P., Madrid, Y., Guntinas, M.B.D., and Camara, C., *J. Anal. Atom. Spectrom.*, 1995, **10**, 815.
29. *Martindale: The Complete Drug Reference*, K. Parfitt (Ed.), Pharmaceutical Press, 1999.
30. Gebel, T., *Chem. Biol. Interact.*, 1997, **107**, 131.
31. Leonard, A., and Gerber, G.B., *Mutat. Res.: Rev. Genetic Toxicol.*, 1996, **366**, 1.
32. *Arsenic in Drinking Water*, National Academy Press, Washington, DC, USA, 1999, (and update 2001).
33. Seifter, J., *J. Pharmacol. Exp. Ther.*, 1939, **66**, 366.
34. Burrell, R.E., Corke, C.T., and Goel, R.G., *J. Agric. Food Chem.*, 1983, **31**, 85.
35. Smichowski, P., Madrid, Y., and Camara, C., *Fresenius J. Anal. Chem.*, 1998, **360**, 623.
36. Dodd, M., Pergantis, S.A., Cullen, W.R., Li, H., Eigendorf, G.K., and Reimer, K.J., *Analyst*, 1996, **121**, 223.
37. Craig, P.J., Forster, S.N., Jenkins, R.O., and Miller, D., *Analyst*, 1999, **124**, 1243.
38. Krupp, E.M., Grumping, R., Furchtbar, U.R.R., and Hirner, A.V., *Fresenius J. Anal. Chem.*, 1996, **354**, 546.
39. Koch, I., Feldmann, J., Lintschinger, J., Serves, S.V., Cullen, W.R., and Reimer, K.J., *Appl. Organomet. Chem.*, 1998, **12**, 129.
40. Andrewes, P., Cullen, W.R., Feldmann, J., Koch, I., Polishchuk, E., and Reimer, K.J., *Appl. Organomet. Chem.*, 1998, **12**, 827.
41. Andreae, M.O., Asmode, J., Foster, P., and Van't dack, L., *Anal. Chem.*, 1981, **53**, 1766.
42. Feldmann, J., and Hirner, A.V. *Internat. J. Environ. Anal. Chem.*, 1995, **60**, 339.
43. Hirner, A.V., Feldmann, J., Krupp, E., Grumping, R., Goguel, R., and Cullen, W.R., *Org. Geochem.*, 1998, **29**, 1765.
44. Feldmann, J., *J. Anal. Atom. Spectrom.*, 1997, **12**, 1069.
45. Haas, K., Feldmann, J., and Wennrich, R.H.J. *Fresenius' J. Anal. Chem.*, 2001, **370**, 587.
46. Braman, R.S., Justen, L.L., and Foreback, C.C., *Anal. Chem.*, 1972, **44**, 2195.
47. Nakashima, S., *Analyst*, 1980, **105**, 732.
48. Howard, A.G., *Appl. Organomet. Chem.*, 1997, **11**, 703.
49. Cullen, W.R., and Reimer, K.J., *Chem. Rev.*, 1989, **89**, 713.
50. Lintschinger, J., Koch, I., Serves, S., Feldmann, J., and Cullen, W.R., *Fresenius J. Anal. Chem.*, 1997, **359**, 484.

51. Lintschinger, J., Schramel, O., and Kettrup, A., *Fresenius J. Anal. Chem.*, 1998, **361**, 96.
52. Sayama, Y., Fukaya, T., and Kuno, Y., *Bunseki Kagaku*, 1995, **44**, 569 *Chem. Abs.* 1995: 690098.
53. Zhang, X.R., Cornelis, R., and Mees, L., *J. Anal. Atom. Spectrom.*, 1998, **13**, 205.
54. Ulrich, N., *Fresenius J. Anal. Chem.*, 1998, **360**, 797.
55. Ulrich, N., *Analytica Chimica Acta*, 1998, **359**, 245.
56. Lintschinger, J., Michalke, B., Schulte-Hostede, S., and Schramel, P., *Internat. J. Environ. Anal. Chem.*, 1998, **72**, 11.
57. Guy, A., Jones, P., and Hill, S.J., *Analyst*, 1998, **123**, 1513.
58. Lindemann, T., Prange, A., Dannecker, W., and Neidhart, B., *Fresenius J. Anal. Chem.*, 1999, **364**, 462.
59. Krachler, M., and Emons, H., *J. Anal. At. Spectrom.*, 2000, **15**, 281–285.
60. Zheng, J., Iijima, A., and Furuta, N., *J. Anal. At. Spectrom.*, 2001, **16**, 812.
61. Michalke, B., and Schramel, P., *J. Chromat. A*, 1999, **834**, 341.
62. Michalke, B., *J. Anal. At. Spectrom.*, 1999, **14**, 1297.
63. Andrewes, P., Cullen, W.R., Feldmann, J., Koch, I., and Polishchuk, E., *Appl. Organomet. Chem.*, 1999, **13**, 681.
64. Gurleyuk, H., Van Fleet-Stalder, V., and Chasteen, T.G., *Appl. Organomet. Chem.*, 1997, **11**, 471.
65. Jenkins, R.O., Craig, P.J., Goessler, W., Miller, D., Ostah, N., and Irgolic, K.J., *Environ. Sci. Technol.*, 1998, **32**, 882.
66. Zheng, J., Takeda, A., and Furuta, N., *J. Anal. At. Spectrom.*, 2001, **16**, 62.
67. Filella, M., Belzile, N., and Chen, Y.-W., *Earth-Science Reviews*, 2002, **57**, 125.
68. Andreae, M.O., and Froelich, P.N., *Tellus*, 1984, **36B**, 101.
69. Bertine, K.K., and Lee, D.S., in *Trace Metals in Seawater, NATO Conference Series, Series IV: Marine Sciences*, C.S. Wong, E. Boyle, K.W. Bruland, J.D. Berton, and E.D. Goldberg (Eds), Plenum Press, New York, 1983.
70. Cutter, L.S., Cutter, G.A., and San Diego-McGlone, M.L.C., *Anal. Chem.*, 1991, **63**, 1138.
71. Takayanagi, K., and Cossa, D., *Water Research*, 1997, **31**, 671.
72. Glindemann, D., Morgenstern, P., Wennrich, R., Stottmeister, U., and Bergmann, A., *Environ. Sci. Pollut. Res.*, 1996, **3**, 75.
73. Hirner, A.V., Feldmann, J., Goguel, R., Rapsomanikis, S., Fischer, R., and Andreae, M.O., *Appl. Organomet. Chem.*, 1994, **8**, 65.
74. Feldmann, J., Grumping, R., and Hirner, A.V., *Fresenius J. Anal. Chem.* 1994, **350**, 228.
75. Feldmann, J., Naels, L., and Haas, K., *J. Anal. At. Spectrom.*, 2001, **16**, 1040.
76. Kantin, R., *Limnol. Oceanogr.*, 1983, **28**, 165.
77. Koch, I., Wang, L., Feldman, J., Andrewes, P., Reimer, K.J., and Cullen, W.R., *Int. J. Environ. Anal. Chem.*, 2000, **77**, 111.
78. Barnard, K., Ph D Thesis, University of Leeds, Leeds, UK, 1947.
79. Richardson, B.A., *J. Foren. Sci. Soc.*, 1994, **34**, 199.
80. Limerick, S., *Expert Group to Investigate Cot Death Theories: Toxic Gas Hypothesis*, Department of Health, UK, 1998.
81. Warnock, D.W., Delves, H.T., Campell, C.K., Croudace, I.W., Davey, K.G., Johnson, E.M., and Sieniawska, C., *Lancet*, 1995, **346**, 1516.
82. Gates, P.N., Harrop, H.A., Pridham, J.B., and Smethurst, B., *Sci. Total Environ.*, 1997, **205**, 215.
83. Pearce, R.B., Callow, M.E., and Macaskie, L.E., *FEMS Microbiol. Lett.*, 1998, **158**, 261.
84. Jenkins, R.O., Craig, P.J., Miller, D.P., Stoop, L.C., Ostah, N., and Morris, T.A., *Appl. Organomet. Chem.*, 1998, **12**, 449.

85. Jenkins, R.O., Craig, P.J., Goessler, W., and Irgolic, K.J., *Hum. Exp. Toxicol.*, 1998, **17**, 231.
86. Craig, P.J., Jenkins, R.O., Dewick, R., and Miller, D.P., *Sci. Tot. Environ.*, 1999, **229**, 83.
87. Andrewes, P., Cullen, W.R., and Polishchuk, E., *Appl. Organomet. Chem.*, 1999, **13**, 659.
88. Andrewes, P., Cullen, W.R., and Polishchuk, E., *Environ. Sci. Technol.*, 2000, **34**, 2249.
89. Andrewes, P., Cullen, W.R., Polishchuk, E., and Reimer, K.J., *Appl. Organomet. Chem.*, 2001, **15**, 473.
90. Jenkins, R.O., Foster, S.N., Craig, P.J., and Goessler, W., *ICEBAMO, Schielleiten, Austria, Abstract PII-19*, 2001.
91. Jenkins, R.O., Morris, T.-A., Craig, P.J., Goessler, W., Ostah, N., and Wills, K.M., *Human and Experimental Toxicol.*, 2000, **19**, 1
92. Cullen, W.R., Li, H., Hewitt, G., Reimer, K.J., Zalunardo, N., *Appl. Organomet. Chem.*, 1994, **8**, 303.
93. Andrewes, P., Cullen, W.R., and Polishchuk, E., *Chemosphere*, 2000, **41**, 1717.
94. Benson, A.A., *The Biological Alkylation of Heavy Elements*, Royal Society of Chemistry, London, 1987, p. 135.
95. Edmonds, J.S., and Francesconi, K.A., *Nature*, 1981, **289**, 602.
96. Maeda, S., Fukuyama, H., Yokoyama, E., Kuroiwa, T., Ohki, A., and Naka, K., *Appl. Organomet. Chem.*, 1997, **11**, 393.
97. Apte, S.C., and Howard, A.G., *Marine Chem.*, 1986, **20**, 119.
98. Kresimon, J., Grueter, U.M., and Hirner, A.V., *Fresenius' J. Anal. Chem.*, 2001, **371**, 586.
99. Bailly, R., Lauwerys, R., Buchet, J.P., Mahieu, P., Konings, J., *Br. J. Ind. Med.*, 1991, **48**, 93.
100. Buchet, J.P., Lauwerys, R., *Arch. Toxicol.*, 1985, **57**, 125.

8 Organosilicon Compounds in the Environment

A.V. HIRNER, D. FLASSBECK

Institut für Umweltanalytik, University of Essen, Essen, Germany

and

R. GRUEMPING

GFA, Otto Hahn str. 22, Muenster, Germany

8.1 INTRODUCTION

Silicon is a main component of the solid earth [1]. More than 99 % of the silicon in the earth resides in the mantle, silicates range from 35 to 85 % in igneous rocks, and about 10^{10} tonnes of silicates are involved in the global geochemical cycle. In biological systems silicon occurs mainly as silica, silicic acid and silicates. Because it may also bind to oxygen and nitrogen in naturally occuring organic compounds, many researchers since the 1950s have looked for associations of silicon with organic matter [1, 2]. Until 1996 naturally occuring organosilicon species have not been reported [3]. Thus, silicones in the environment are usually assumed to be of anthropogenic origin. However, in the course of recent culture experiments using the freshwater diatom *Navicula pelliculosa* and substrates enriched in ^{15}N and ^{29}Si, evidence for the formation of a nitrogen-containing organic hexavalent silicon complex was found using ^{29}Si NMR [4].

Anthropogenically produced silicones typically contain Si–O–Si bonds (siloxanes). They have had a significant impact as speciality materials for consumer and industrial products since their commercial introduction in 1943. More than 80 % of the commercial products are based on the polymer system of the polydimethyl-siloxanes (PDMS) described by the formula $Me_3SiO(SiMe_2O)_nSiMe_3$. Because of the organic substituents in silicones—usually methyl, but among others also H, alkyl, aryl, trifluoropropyl, amino-, epoxy-, polyether-, and mercaptoalkyl groups—PDMS contain Si–C bonds and thus resemble organometallic compounds. Analytical and environmental aspects of organosilicon compounds have been extensively reviewed, especially during the last 15 years [2, 3, 5–7]. However, in the most recent review [7] just two of 25 authors reviewing the environmental chemistry of organosilicon materials are not employed by the silicone industry, notably a $6 billion industry worldwide. It is therefore

Organometallic Compounds in the Environment
Edited by P.J. Craig © 2003 John Wiley & Sons Ltd

possible to debate the balanced or objective views and interpretations of the subject/material presented.[†]

There are three organosilicon classes with potential significant environmental fate and transport characteristics leading to noteworthy environmental loading: volatile methylsiloxane (VMS), polydimethylsiloxane (PDMS), and polyether-methylsiloxane (PEMS). As environmental aspects of the latter have not often been discussed, PEMS will not be included in this article and this topic is reviewed elsewhere [8]. VMS with molecular weights < 600 amu (atomic mass units) have very low water solubility, relatively high vapor pressures and Henry constants (> 3) causing these materials to partition to the atmosphere with atmospheric lifetimes of 10–30 days. PDMS, however, have molecular weights > 600 amu and are non-volatile liquids insoluble in water, and thus partition onto solids (e.g. sediments, sewage sludge). 'Down-the-drain' use of PDMS (recycling is not yet realized) is estimated to be 14 million kg year^{-1} (data from US silicone industry) with a yearly increase of approximately 7% from 1991 to 1995. The annual US production of cyclic VMS (cVMS) in 1993 was 153 000 tonnes, of which approximately 20 000 tonnes had been used in personal care products as carriers or emollients.

Organosiloxane (silicone) resins are highly crosslinked siloxane systems and are more silica-like in properties than are linear polydimethylsiloxane fluids [9]. The resins are frequently reacted with organic systems (e.g. polyester) to form copolymers (Figure 8.1). Polysilane and related copolymer synthesis reactions have been reviewed [10]. Because silicone resins and elastomers used in sealants, caulks, and some surface coatings are solids in the PDMS class, they are treated as immobile and persistent in the environment, following several studies [11, 12], in which no evidence of significant biotic or abiotic degradation in landfills has been observed. About 89 thousand metric tons of silicone elastomers were produced or imported in the US in 1993 representing about 50% of all organosilicon products and 69% of all those becoming either landfilled or incinerated [9].

Silicone structures are commonly represented by the symbols D, M, T, and Q (Figure 8.1). The backbone units of PDMS with silicon atoms having two shared bonds to oxygen and two methyl substituents are symbolized by D. Thus, the basic PDMS structure is MD$_n$M (n from 10 to > 10000). PDMS fluids show viscosities ranging from 0.65 centistokes (cSt) for hexamethyldisi-loxane (L2) to over 60 000 cSt for some silicone greases, and they are classified in groups of 0.65–20 cSt, 50–100 cSt and 5000–250 000 cSt having low, medium and high viscosity, respectively; those with viscosities greater than 500 000 cSt are classified as gums.

Linear (abbreviation L) and cyclic VMS (abbreviation D) are characterized by the number of Si atoms in the molecule, as is indicated for L2 to L4 and D3 to D6 in Figure 8.1. Silanols as the environmentally most important silicone degradation products are also included in the Figure.

[†]The authors of the present review are independent university researchers and have not been involved in litigations and do not receive support from the silicone industry.

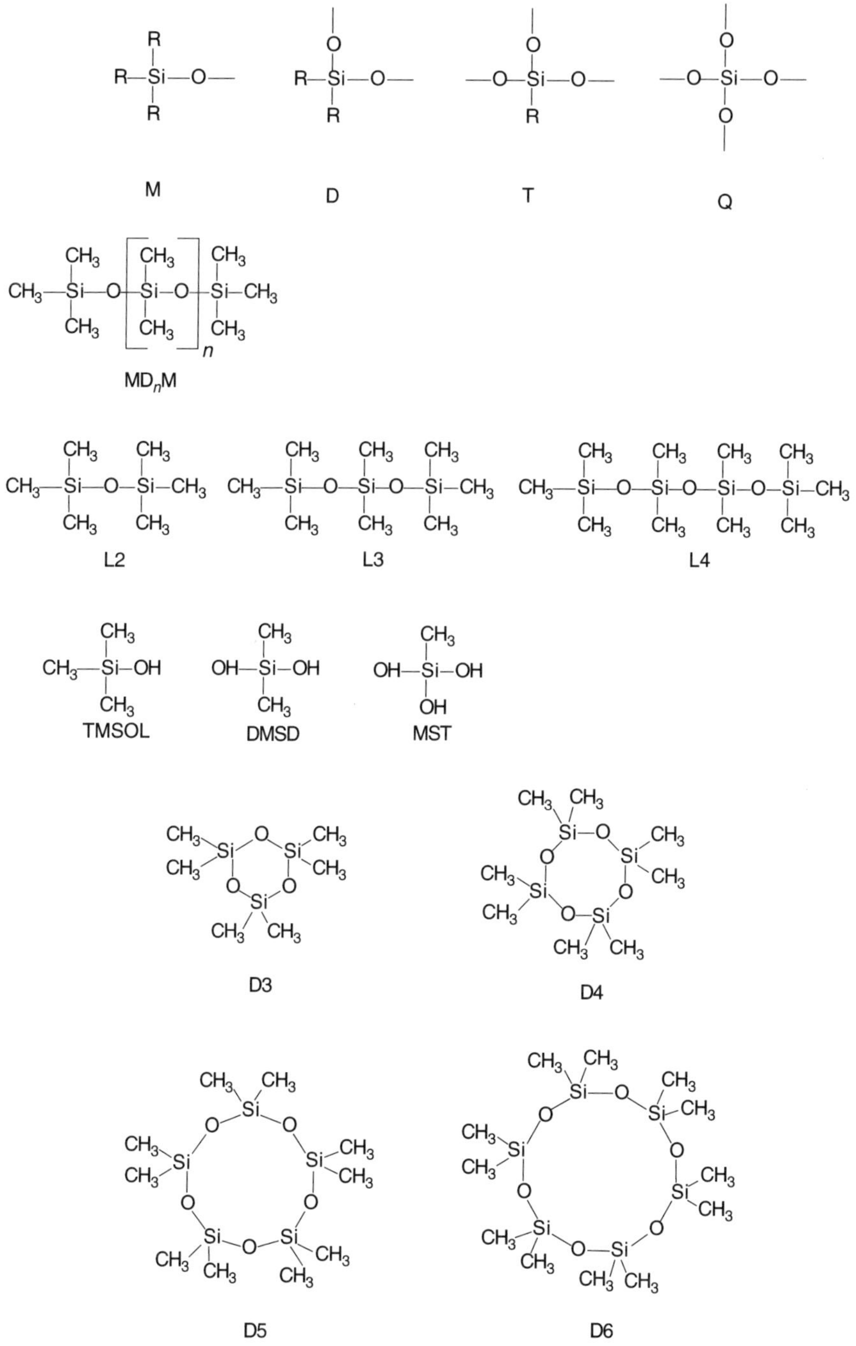

Figure 8.1 Environmental important silicones, their structural units and degradation products

The higher electronegativity of oxygen causes a charge movement of the Si–O bond resulting in an ionic character of this bond. Interaction of the occupied p orbitals of oxygen with the empty d orbitals of silicon occurs. This results in a shorter bonding length of approximately 0.163 nm compared with theoretical calculations, as well as a deviation of the bonding angle from tetrahedral. With a bonding energy of 470 kJ mol^{-1} the Si–O bond is very stable, leading to resistance of the siloxane structure against homolytic cleavage (see Chapter 1); however, ionic attack is easily possible. With a bonding energy of 336 kJ mol^{-1} the Si–C bond is slightly lower than a C–C bond [13, 14]. Data of dissociation energies of silicon compounds are often controversial [15]: for example, for the dissociation of the Si–C bond in tetramethylsilane, values of 271 to 374 kJ mol^{-1} are given, which may be compared with 356 kJ mol^{-1} for a paraffinic C–C bond. However, because of a significant ionic character, heterolytic cleavage (by a nucleophilic attack on Si or an electrophilic attack on C) of a Si–C bond is faster than for a C–C bond. Cation binding to C (especially H) and anion binding to Si (like O) decrease the polarization and stabilize the molecule. Consequently, PDMS are very stable compounds.

8.1.1 POLYDIMETHYLSILOXANES (PDMS)

PDMS tend to be stable and inert in the presence of heat, chemicals, and UV radiation [9]. Their low surface tension and physical properties are relatively insensitive to temperature change over a range of −50 to +150 °C; near 200 °C formaldehyde will evolve. PDMS can be modified to formulate a wide range of products with tailored hydrophobicity and durability [16]. Therefore, silicones find application in many diverse markets, such as aerospace, automotive, construction, electrical, electronics, medical materials, performance chemicals and coatings, personal care, household care, healthcare, food, paper and textiles. Silicones have broad utility in textile processing and finishing, antifoams for fabric and carpet dyeing, and coatings. They are used as softeners, wetting agents and water repellents, and in sewing operations. Mainly based on the assumption that they are biologically inert and toxicologically not relevant, highly polymeric silicones are used in medical equipment, and for implantation [13]. Of special importance are silicone breast implants (SBIs) consisting of an elastomer shell containing dimethyl and vinylmethyl siloxane gel copolymer networks [17].

PDMS fluids comprise the vast majority of organosilicon releases to wastewater [9]. Because of their physical properties, PDMS are effectively removed from the aqueous phase by adsorption onto sludge during sewage treatment. The ultimate fate of PDMS will therefore depend on the sludge disposal route. Incineration results in complete degradation of the siloxane polymer to carbon dioxide, water, and silica (SiO_2). Sludge-associated PDMS incorporated into soil during soil amendment will be expected to undergo soil-catalyzed degradation and ultimately become converted to silica, water, and carbon dioxide [18–24].

Similar soil-catalyzed degradation will also be expected to occur if sludge-borne PDMS is landfilled. Traces of sludge-borne PDMS may also enter the sediments of receiving waters from wastewater treatment plant (WWTP) effluents. If a specific country or region does not have wastewater treatment facilities, there is the possibility that organosilicon materials could enter the aquatic compartment. This could occur either through liquid discharge or from dumping sludge-borne silicones. Eventually slow hydrolysis of PDMS in sediments will occur.

Low PDMS levels detected in fish tissue confirm the conclusion that PDMS is unlikely to bioaccumulate [25]. While PDMS fluids have a high n-octanol/water partition coefficient, suggesting the ability to partition preferentially into cells and organisms, the high molecular weight of PDMS ($>10\,$cSt viscosity) prevents the fluid from bioaccumulating [14, 26–28]. The dominant pathway for PDMS dispersal in the environment is through sludge amendment of soil. A smaller fraction of PDMS that is not removed during wastewater treatment will be introduced into the aquatic environment as a component of suspended solids, which will settle onto bottom sediments. Summarized, 25–30 % of the down-the-drain PDMS directly flows to land application, 70–75 % enters wastewater treatment [11]. Of the latter 97 % are sorbed to sludge, 3 % are discharged to surface waters and will eventually end up in sediments.

8.1.2 VOLATILE METHYLSILOXANES (VMS)

The most important oligomeric alkylsiloxanes in either cyclic or linear configurations are also called 'volatile methylsiloxanes' or VMS [9]. These find application in PDMS manufacturing, antiperspirant/deodorants, hair and skin care products (cosmetic and pharmaceutical industry). 13 %, or about 20 000 tonnes, of VMS are used in personal care applications, of which 92 % (mainly as $(Me_2SiO)_{4-5}$), partitions to the air [9]. An emerging application for oligomeric methylsiloxanes is their use as solvents to replace ozone forming VOCs and ozone depleting CFCs.

Volatile methylsiloxane materials (VMS) are siloxanes with relatively low molecular weight ($<600\,$amu) and high vapor pressure [29]. A representative VMS is octamethylcyclotetrasiloxane (OMCTS) or D4. VMS fluids are used in the personal care industry, and are designated as 'cyclomethicones' by the Cosmetic, Toiletry, and Fragrance Association (CTFA). D4 has a relatively high molecular weight (296 amu) for a volatile compound, but low for a siloxane. D4 has relatively high vapor pressure (1.33 hPa) and low water solubility (74 and 33 μg L^{-1} in fresh and saltwater, respectively [30], which results in a high Henry constant (H); estimates of H for the D4 range from 3 to more than 17. The presence of humic acid or seawater increased H by a factor of 10 over fresh water under the same test conditions, indicating that environmental cosolutes will increase H even further. VMS molecules could be small enough to pass through biological membranes [16]. D4 shows no apparent biodegradation

[31], and a bioconcentration factor of 12 400 [32]. D4 will readily volatilize from water and would equilibrate to a half-life ranging from 3 hours to 6 days in rivers and streams. The fate of D4 was evaluated in a water/sediment system with active microflora, using radiolabeled ^{14}C-OMCTS [31]; there was no observed degradation of OMCTS in the test system based on measured ^{14}CO$_2$ evolution. Gas-phase VMS compounds can be removed from the troposphere by wet and dry deposition. Ignoring the latter, gas-phase reaction with the OH radical is calculated to be the only important loss process for methylsiloxanes in the troposphere, leading to lifetimes sufficiently long for transport over regional and longer distances (> 1000 km).

8.1.3   SILANOLS

Organosilanols are monomeric organosilicon compounds with at least one hydroxyl group. Silanols have high aqueous solubilities and low values of the Henry constant, indicating lower volatilization rates compared to siloxanes [33]. The octanol–water partition coefficients for linear and cyclic siloxanes are higher than those of silanols, which also correlates with higher bioconcentration factors.

The monomeric methylsilanols $Me_xSi(OH)_{4-x}(x = 1$–$3)$ are the fundamental building blocks of the silicone industry [34]. They are also the predominant transient intermediates in the environmental degradation of silicones. Silanols are generally unstable and tend to condense in acidic or alkaline environments to produce organosiloxanes. $Me_2Si(OH)_2$ can only be kept in a stable state over a long period under special, acid- and base-free conditions. The tendency of silanols to self-condense is reversed in dilute aqueous solutions. The condensation is a reversible reaction, which produce water and is driven in reverse by excess water if an appropriate catalyst is present. PDMS hydrolysis is a source of silanols in the environment [34]: PDMS will hydrolyze in dry soil to $Me_2Si(OH)_2$ and its low oligomer diols, and to a much lesser extent, Me_3SiOH (from end groups) and $MeSi(OH)_3$ (from any branching). Thus, most silanols generated in the environment will consist of $Me_2Si(OH)_2$ and its low oligomer diols. Dimethylsilanediol and methylsilanetriol are very water soluble organosilicon compounds, which are not easily extracted from water partitioning between water and ethyl acetate. The water solubility of the methylsilanols, combined with their low to moderate vapor pressures, indicates that they will be found in both aqueous and atmospheric compartments of the environment. To the extent that silanols form chemical bonds to mineral and humic materials, they will also be bound up in soils and sediments. Because their bonds are subject to hydrolysis, by interaction with water originally bound silanols may be released from their substrates [18, 19, 21, 23]. Methylsilanols are expected to degrade photolytically in the atmosphere and in water to silica and carbon dioxide. It was demonstrated that DMSD is oxidatively demethylated to methylsilanetriol and then to silicate by exposure to light in aqueous solutions containing nitrate or nitrite [35]. Reactivity of silanols to hydroxyl radicals is not limited to aqueous media. Trimethylsilanol is more reactive towards atmospheric hydroxyl radicals than are methylsiloxanes or methylsilanes [36]. The tropospheric half-life for trimethylsilanol was estimated to be about 2.5 days. Application of ^{14}C-PDMS to soils resulted in the production of $^{14}CO_2$, but only under conditions that led to the hydrolysis of PDMS to water soluble silanols [19]. Soils that had never been allowed to dry out demonstrated greatly reduced PDMS hydrolysis, producing 3 % silanol after 6 months as compared with over 50 % in the samples allowed to dry out [19]. Cultures able to degrade ^{14}C-DMSD to $^{14}CO_2$ were *Arthrobacter* and *Fusarium oxysporum* Schlechtendahl. Biodegradation and photolytic pathways are both oxidative demethylations, which led to sequential formation of silanol functions to replace the methyl groups, and ultimately to mineralization to SiO_2, CO_2 and H_2O [9]. For the low molecular weight diols $HO(SiMe_2O)_nH(n < 10)$, the aqueous solubility increases as n decreases, with $Me_2Si(OH)_2$ being a very water soluble material.

8.2 ANALYTICAL PROCEDURES AND TECHNIQUES

Total silicon content of organic extracts of solid/liquid samples can be determined by many analytical techniques, e.g. atomic emission spectrometry (AES) [37–39]; a similar example is the determination of PDMS in aqueous samples by inductively coupled plasma atomic emission spectrometry (ICP/AES) of organic sample extracts [40]. Although measurements of organic extracts using AAS and ICP techniques may be potentially influenced by coextraction and detection of inorganic silicates, the potential for overestimating PDMS concentrations using non-specific methodologies was found to be small [41]. Analytical methods have still not been validated by an international standard scheme; standard reference materials (SRMs) for quality control in silicon analyses are urgently required, especially in sensitive areas like blood analysis [42]. Probably the best option for certification of silicon contents in reference materials is the application of isotope-dilution high resolution mass spectrometry (ICP-HRIDMS) [43].

To obtain species-specific informations, however, analyses for silicones are usually performed by gas chromatography mass spectrometry (GC/MS) of the organic extract. Thus it is possible to analyse PDMS and its degradation products in aqueous environmental matrices [44–48]. Figure 8.2 gives an overview.

The most important instrumental techniques for the analysis of PDMS, VMS and its degradation products are indicated in Figure 8.2, and are discussed in the following sections. With exception of the total silicon determinations described above, all methods are species specific.

Proton nuclear magnetic resonance spectrometry (^1H NMR) enables quantification of organosilicon species (detection limit 0.05 ppm) with no interference from inorganic silicon (silicates). The signal for the methyl protons on silicon appears in a region free from proton interference from other non-silicon species.

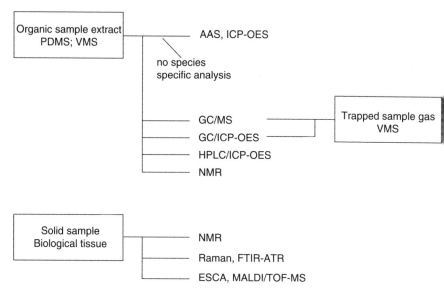

Figure 8.2 Instrumental techniques for the analysis of silicones in environmental and biological samples

Both FIA-ICP (FIA = flow injection analysis) and GPC-ICP (GPC = gel permeation chromatography) make the assumption that all detected silicon is in the form of dimethylsiloxy units, therefore the detected silicon is reported as PDMS using the appropriate conversion factor. Proton NMR detects dimethylsiloxy units only and is not subject to interference from silicate. Thus ICP and NMR report on the same PDMS concentrations when analyzing samples that contain PDMS only. When analyzing environmental samples, however, the nature of the contaminant is unknown. The ideal approach would be to analyze the environmental samples using both techniques.

8.2.1 ANALYSIS OF PDMS IN ENVIRONMENTAL SAMPLES BY GC TECHNIQUES

The analytical chemistry of silicones (mostly on PDMS) is reviewed by Smith [5] and Carpenter and Gerhards [6]. Examining PDMS levels in the environment requires validated extraction and detection methods. The Silicones Environmental, Health, and Safety Council (SEHSC), The Centre Europeen des Silicones (CES), and The Silicone Industry Association of Japan (SIAJ) have developed methods for the sampling, extraction, and detection of PDMS in environmental samples, including soil, sediment, sludge, and water. Wastewater treatment influent, effluent, and sludge are collected as 24-hour, 3-day composite samples.

PDMS extraction from soil, sediment, sludge, and water is performed by Soxhlet and ultrasonic extraction with diethyl ether, chloroform, petroleum ether or hexane [5]. The extraction of PDMS from soil, including sludge-amended soil, sediment, and sewage sludge is carried out using four sequential extractions with tetrahydrofuran (THF); there is no significant difference between the SEHSC and CES extraction methods. In a two-stage soil extraction, methyl isobutyl ketone (MIBK) is used to remove non-polar siloxanes, then acetonitrile or water is used to remove polar degradation products, such as low molecular weight silanols. In respect to airborne particulate matter three sequential Soxhlet extractions were applied (cyclohexane, followed by dichloromethane, followed by acetone). Inorganic silicon (silicates) is not co-extracted by organic solvents used for PDMS extraction, and does not interfere with the determination of organosilicon levels in environmental samples.

The pyrolysis of PDMS yields a series of cyclic siloxanes, which can be separated and detected by GC/MS [49]. The majority of the PDMS is converted to the cyclic trimer D3 upon pyrolysis; selective monitoring of the D3 parent ions in the mass spectrum allows detection of trace quantities of PDMS [49]. Pyrolysis vapors are collected on activated charcoal and extracted with tetrachloromethane prior to analysis.

Pellenbarg *et al.* [50] report both quantitative and qualitative analysis of silicones in sediment-like materials using ^{29}Si NMR, however a detection limit of only 45 ppm was realised for a 9-hour experiment.

Size exclusion chromatography (SEC, identical with GPC) with refractive index detection is used to characterize high molecular weight silicones [51]; however, refractive index detection lacks the sensitivity needed for environmental samples. Low detection limits for organosilicon compounds can be received by SEC coupled with ICP/AES [40, 52, 53]. Capillary supercritical fluid chromatography (SFC) [54] for separation of a series of organosiloxanes with ICP detection, and desorption chemical ionization mass spectrometry (DCI-MS) [55] for determining the molecular weight of PDMS oligomers, were used. Protonated molecular ions of the oligomeric species are produced without fragmentation, thus enabling the determination of molecular weights up to several thousand amu. Time-of-flight secondary ion mass spectrometry (TOF-SIMS) was applicable up to 10 000 amu. SFC with ion mobility detection to separate and detect a series of 70 PDMS oligomers has been described [56].

Gradient polymer elution chromatography (GPEC) coupled to an evaporative light scattering detector was developed for the characterization of copolymers based on their chemical composition distribution [57]. GPC/ICP methodology has been utilized for measuring PDMS levels in the environment [41].

Mahone *et al.* [48] describe a GPC/MS method for measuring and characterizing silicones in industrial and municipal wastewaters, surface and ground waters, foods and biological tissues. The determination of water-soluble organosilanol species is accomplished by the acid-catalyzed trimethylsilylation of the silanol functionality with L2 followed by gas–liquid chromatographic assay of the resulting L2 extract. Water-insoluble components are determined by subjecting the sample to acidic predigestion before the addition of the L2-derivatizing and extracting reagent. The possibility of dimethyl- or mono-methylsilicon artifacts arising from inadvertent trace cleavage of methyl groups from the L2 can be determined and/or precluded by the use of hexaethyldisilox-ane or perdeutero-L2 when appropriate.

8.2.2 ANALYSIS OF VMS AND SILANOLS IN ENVIRONMENTAL SAMPLES BY GC TECHNIQUES

Methods have been developed for the sampling, extraction, and detection of VMS in biogas and for the analysis of silanols in environmental samples [6]. VMS in biogas from WWTP and landfill is typically sampled by passage through a tube packed with sorbent materials (activated charcoal, XAD resin, or solvents); gas flow rates range from 0.5 to 1 L minute^{-1} for 10 to 60 minutes. To reduce the amount of water in the biogas stream, the gas is typically passed through a water trap prior to its concentration to the sorbent. For extracting VMS from the sorbent (by shaking or ultrasonification), a number of solvents have been used, among them hexane, cyclohexane, MIBK, and chloroform.

For silanol extraction solid samples are spiked with an aqueous solution of ^{14}C-DMSD, and water samples are spiked by the addition of pure ^{14}C-DMSD. The diol may actually be covalently bound to the soil [19, 20], so conventional extraction techniques are ineffective for removing it. Three sequential extrac-tions with water were somewhat effective for hydrolyzing these proposed diol–soil bonds to free up the diol. Calcium chloride can also be used to desorb silanols from soil [19, 20]. Aqueous extracts were frozen until analyzed, to prevent hydrolysis of oligomeric diols to DMSD [45].

By using headspace techniques and subsequent GC/FID analysis VMS could be determined in aqueous matrices [58]. The detection of both linear diols and cyclic siloxanes using GC/MS with electron impact ionization is reported [59]. Cyclics showed high detection sensitivity (1 pg), diols detection limits from 100 pg to 1 ng. Preconditioning of GC columns was done by bis(trimethylsilyl) acetamide. The electron impact (EI) spectra of the cyclics bear close resem-blance to their corresponding open chain diols. To distinguish cyclics from diols in EI mode, one could rely on retention time differences by GC. For positive identification of silanols and cyclics, GC/MS under positive chemical ionization (PCI) and 90:10 CH_4/NH_3 were used. The volatile species methoxy methylsi-lane, TMSOL, L2, D3 and D4 were detected in atmospheric samples by GC/MS [60].

In a method similar to techniques for voltile metal(loid) species, VMS gases are trapped at $-80\,^{\circ}C$ (acetone/liquid nitrogen mixture) and analysed by LTGC/ICP-AES (LT = low temperature) [61]. Another group [62] determined siloxanes in landfill and sewage gases by GC separation and simultaneous detection by MS and AES. By using gas canisters (stainless steel containers, internal surfaces electropolished), multiple analysis was possible. To clean, the used canisters were evacuated and flushed with dry nitrogen. After six cycles of evacuating and pressurizing very low blank values were obtained; heating of the canisters during the cleaning process was not necessary. Real samples were analyzed generally within 36 h. Even for samples stored for 10 days recovery was >85%. Relative standard deviations were <5%.

8.2.3 ANALYSIS OF ENVIRONMENTAL SAMPLES BY HPLC TECHNIQUES

DMSD in aqueous soil extracts was fractionated by HPLC on a polymeric C_{18} column as stationary and water/methanol as mobile phase [63]; non-polar siloxanes and polar silanols can be detected down to the low μg-range [64]. On-line coupling of HPLC separation systems with element selective detectors enable sensitive and selective determination of organosilicon compounds in complex matrices. When coupling with ICP the tolerance of the plasma for different solvents is very important [65]. A short flexible tubing (made of polytetrafluoroethylene PTFE or PEEK) connects the output of the HPLC column with the pneumatic (Meinhardt or cross-flow-type), thermospray or high pressure nebuliser [52, 66–71]. Micro-HPLC techniques using micro-bore columns together with special nebulizers are characterized by low flow rates (10 to 100 μL min^{-1}) and thus drastically reduce the solvent load for the ICP [72–74].

Dorn and Skelly-Frame [75] report a combination of high performance liquid chromatography (HPLC) for separation and inductively coupled plasma-atomic emission spectroscopy (ICP-AES) for detection. HPLC-ICP was operated in two ways: size exclusion or gel permeation chromatography (SEC or GPC) was utilized for the analysis of non-polar, high molecular weight PDMS polymers. By GPC-ICP the detection limit for PDMS was 0.1 ppm, while by column bypass direct injection (flow injection analysis) into ICP (FIA-ICP) 5 ppb was reached. Reversed-phase HPLC (RP-HPLC) was applied for the analysis of polar, low molecular weight silanols. Detection limit was 0.1 ppm, inorganic silicate was resolved from low molecular weight silanediols and hence provided no interference. Because silanols were prone to codensation in the presence of glass surfaces, polypropylene vials were used. This RP-HPLC/ICP-AES method is based on a C_{18} microbore column, a water/acetonitrile gradient and flow rates of 0.2 mL min^{-1} together with auxiliary gas flow of 0.2 to 1 L Ar min^{-1}; e.g. for DMSD the absolute detection limit is 3.75 ng Si.

As the silanol TMSOL has hydrophilic as well as volatile properties, this compound can be analysed by HPLC/ICP-AES [76] as well as by LTGC/ICP-AES [61]; the results of both methods were comparable, but the latter showed significantly lower detection limits.

ICP-MS as detector is usually characterized by extraordinary element specificity and the lowest detection limits. However, for masses below 80 amu, isobaric mass interferences play a significant role leading to mass coincidence of silicon isotopes with molecular ions (e.g. N_2^+, CO^+, CN^+) at mass 28, 29, and 30. It is only possible to detect at mass 44, 46, and 47 the cluster ions SiO^+ and $SiOH^+$ formed in cooler parts of the plasma. Because of this effect the detection limit for silicon is only in the lower µg-range compared to the higher fg-range for metal(loid)s [77].

8.2.4 ANALYSIS OF BIOLOGICAL SAMPLES

During the last few years several element-specific methods for the investigation of silicon in biological substances have been developed [78–80]. Silicon species have been measured in biological tissues by GC with flame ionization detection (FID), infrared spectroscopy (IR), and NMR. Electrothermal atomization or graphite furnace atomic absorption spectrometry (ETAAS and GFAAS, respectively) and inductively coupled plasma atomic emission spectrometry (ICP-AES) are the most frequently used techniques in laboratories for the determination of traces of the element silicon in biological samples [3]; flame atomic absorption spectrometry (FAAS) obviously lacks sensitivity [81]. For microscopic silicon analysis, usually the X-rays produced in SEM (scanning electron microscopy) are quantified by EDXA (energy-dispersive X-ray analysis) or electron-probe X-ray spectrometry (EPXRS) [82–87]. However, on the base of these determinations in most cases it is unclear as to whether the detection of silicones in the body of implant recipients is significantly related to any levels of silicone originating from the implants. Thus, although the implantation of polysiloxanes has been used in cosmetic and prosthetic medicine for several decades, there has been a lack of specific methodologies for monitoring any leaching of siloxanes from the implants or their possible migration to distant sites.

Gas chromatography and GC/MS are appropriate for the molecular identification and structure determination of silicones and for the sensitive determination of trace amounts of these species in biological materials [59]. Separation (using high-resolution capillary columns) together with identification on the basis of retention times and quantification of silicones by GC, usually with flame ionization detection (FID), is a common procedure. GC/MS is the most sensitive and selective technique for the quantification of specific silicone species [3, 59]. The characterization of silicones is possible through a combination of the structural information given by EI (electron impact) fragmentation patterns and the molecular mass information provided by the chemical ionisation method (CI) [59]. Silanols must be derivatized before introduction into the GC column [88]. One of the important problems in GC/MS systems are major sources of background silicone through PDMS stationary phases which are generally applied for GC separation, and also through components such as septa and seals made of silicone rubber or silicone diffusion pump fluids [89]. Usage of anhydrous magnesium sulfate for drying of the THF extract can avoid D4 formation by reduction of water traces within the stationary phase of the column [90]. A sensitivity of $0.2\,\text{ng}\,\text{mL}^{-1}$ was achieved by the purge-and-trap GC/MS method described for the determination of D4 in water [91], a detection limit of $0.1\,\text{ng}$ was reached when analysing PDMS associated with airborne particles by pyrolysis GC/MS [92].

Several studies have focused on the extraction of low molecular weight silicones from biological matrices by using different chromatographic equipment such as HPLC, GC/AES, and GC/MS [93–96]. The siloxane recoveries from mouse liver homogenates spiked with varying amounts of PDMS, containing linear and cyclic polydimethylsiloxanes, extracted with ethyl acetate and analyzed by GC/MS and GC/AES were found to be greater than 90% [93]. A study for detecting radioactively labeled D4 and its metabolites from plasma and blood revealed a concentration range between $21–2100\,\mu\text{g}\,\text{mL}^{-1}$; a lower range between $50–500\,\text{ng}\,\text{mL}^{-1}$ was found by extracting blood and plasma in the presence of glass beads [95]. After single extraction of these materials with THF the efficiency was 79–92% for the higher concentrations, depending on the material, and 82–102% for the lower concentration range. However, the monitoring of siloxanes by this analytical technique in real biological samples works only well if silicone is radioactively labeled.

Both D4 and its water-soluble polar metabolites from blood plasma, lung, liver, fat, urine, and feces were extracted with THF; for detection HPLC with a C_{18} column and a water/acetonitrile mobile phase was applied [90, 95]. In another study D4 from blood plasma was also extracted with THF and the extract analyzed by GC/MS; recovery was >90% and detection/quantification limits $3/11\,\text{ng}\,\text{g}^{-1}$ [96]. Cyclic siloxane species were analysed in blood serum by GC/MS at concentrations down to $0.1\,\mu\text{g}\,\text{mL}^{-1}$ using heptane/methanol extraction followed by water elution [97]. By mass fragmentography the biotransformation products (DMSD, other diols and phenol) could be identified as their trimethylsilyl derivatives; to confirm the source for DMSD, radioactive labelled substrates were used.

Low molecular weight silicones were analysed using GC/AES and GC/MS with detection limits of 80 and $10\,\text{pg}\,\mu\text{L}^{-1}$, respectively [93]. While GC/AES yields information related to organosilicon levels, GC/MS provides specific information related to the molecular structure of these compounds. The analytes were separated on a 30-m long capillary column. The AES is based on microwave-induced plasma (MIP) and diode array spectrophotometric detection of element-specific atomic emissions monitoring silicon atomic emission at 251.6 nm. For quantifying the dimethylpolysiloxane level of the samples, the MS was operated in the SIM (selected ion monitoring) mode on fragments of 73, 147, 221, 281, 295, and 369 amu. The samples were extracted with ethyl acetate for 1 h with continuous shaking. Subsequently the samples were centrifuged at

13 000 rpm for 5 min, and the ethyl acetate layer was used for the analysis. Test spike experiments with mouse liver homogenates yielded recovery rates higher than 90%. While no significant low molecular weight peaks corresponding to linear PDMS could be found in extracts from breast implants, by the described method it was possible to detect D3 to D7. Flassbeck et al. [98] developed a sensitive method for detection and quantification of D4 to D6 in plasma and blood using GC/MS with detection limits of one picogram. There were no blank values found when running hexane extracts; extraction efficiency was 80–90%.

Kennan et al. [80] describe a silicone-specific technique (modified aqueous silanol functionality test) [48, 99] involving microwave digestion of samples in acid solution rapidly to break down the biological matrix while hydrolyzing silicones to monomeric species. The resulting monomeric silanol species were then capped with trimethylsilyl groups, extracted into L2, and analyzed by GC. In serum detection limits for silicone species were below $0.5 \, \mu g$ of Si mL^{-1}. Compared with the latter, ^{29}Si NMR with a detection limit of only $64 \, \mu g \, mL^{-1}$ is far less sensitive.

Because of its ability to distinguish between silicone and silicones and the specificity of its response, ^{29}Si NMR may be regarded as the technique of choice for structural differentiation [100]. Although NMR is not usually used in trace analysis because of its lower sensitivity as compared with other techniques, a combination of in vivo ^{1}H NMR localized spectroscopy and in vitro ^{29}Si NMR spectroscopy was developed to investigate the migration and chemical modification of silicones in animal models and humans [101]. The results have shown that silicone migrates from breast implants to local and distant sites, and degrades in vivo (probably by the action of macrophages and monocytes). Earlier investigations by light microscopy had demonstrated microscopic droplets beyond the implant capsule and energy-dispersive X-ray analysis of the tissue samples showed a silicone peak, but by these techniques as well as by transmission electron microscopy, it was not possible to distinguish between silica, silicone, and silicone's various degrees of chemical functionality.

Silicon levels reported by Garrido et al. [102] were 4–5 orders of magnitude higher than levels reported by atomic spectroscopy techniques [103]. As a result, the reported high levels in blood have been challenged [103, 104], and the original authors have since published an erratum stating that the signal-to-noise ratios in their spectra could not be accurately determined [105]. Furthermore, Taylor and Kennan [103] also questioned other results from this group [101, 102, 106, 107] because S/N ratios were not reported.

Silicones can be very easily identified by their IR spectra because of the high specifity of the spectral patterns of organic groups attached to silicon. UV spectrometry is not a very good tool for identifying silicones because of their lack of absorption of UV radiation, except for aromatic substituents. Raman spectrometry and attenuated total reflectance (ATR) IR spectrometry are very attractive techniques for the characterization of biological materials; for example, a laser Raman probe was used to detect silicones in the lymph node [109], and by FTIR-ATR detection protein adsorption on silanized surfaces [110], especially *in vivo* adsorption of blood components on silicone rubber [111] and silicone membranes [112] was investigated.

Raman and FT-IR microscopy have been used to identify silicone in capsular tissue and in regional lymph nodes of women with breast implants [113–116]. A limitation of IR and Raman spectroscopy is that these methods do not give quantitative information. Extraction in conjunction with GC techniques was used to follow migration of D3 to D7 following injection in mice [94].

Electron spectroscopy for chemical analysis (ESCA) can be used to detect silicone in skin in a specific, highly sensitive, and quantitative manner [117]; this technique is more sensitive than the light microscopic evaluation.

Relatively new [3] techniques like MALDI/TOF-MS (matrix-assisted laser desorption/ionization (MALDI) time-of-flight mass spectrometry) and electrospray MS may permit the measurement of intact molecular ions of high mass biopolymers.

Generally, accidental contamination by environmentally omnipresent silicone compounds can occur during all steps of the analytical procedure [3]: silicon-containing dust in the laboratory atmosphere, stopcock grease, lubricants in syringes, pump fluids, fluids for glassware treatment, silicone rubber for different tubing, sealing, septa or O-ring elements and self-adhering labels are only some of the potential sources of silicones in laboratories. In gas chromatographic analysis, for example, leakage of chromatographic columns at temperatures above 200 °C takes place. Further, it is essential to avoid contamination during sampling procedures. When collecting biological samples, some commonly used items such as talc-powdered gloves or silicone-lubricated syringes must be absolutely avoided. It is important to control the purity of available reagents, solvents, and standards, and to use 'silicone-decontaminated' plasticware; sample preparation should be done under Class 10 or 100 (clean room) laboratory conditions or in cabinets under laminar air flow [42].

Molecules such as silanols are especially sensitive and unstable, undergoing condensation even under mild conditions, so that their determination becomes a particularly difficult analytical task. The low heat of vaporization and subsequent evaporation of low molecular mass siloxanes contribute to their losses. Silicones have a strong affinity for adsorption on glass and all siliceous materials. They also absorb atmospheric moisture [89].

If a sample contains trace amounts of components having $(CH_3)_3Si$ moieties, a hydrolytic cleavage of the $Si–CH_3$ bond can occur, especially in the presence of certain metals (e.g. Fe, Al, and Ca) [48]. Organic solvents can extract liposoluble silicone compounds but also possibly some inorganic silicon species in colloidal form.

8.3 ENVIRONMENTAL CONCENTRATIONS AND PROCESSES

8.3.1 PDMS IN SOIL, SEDIMENT AND WASTE

The general high stability of PDMS in the environment enables its long-range transport over continents and oceans. Thus PDMS has become a ubiquitous substance and is reported as an important component of Arctic aerosols [118].

The most important emission path for PDMS into environmental systems are 'down-the-drain' applications via wastewaters. Because of their hydrophobic nature about 97 % of the PDMS in wastewater become sorbed to sewage sludge particles [11].

While concentrations of silicones in surface waters do usually not exceed $54 \,\mu g \, L^{-1}$, at WWTPs in North America PDMS levels were between 87 and $374 \,\mu g \, L^{-1}$ at plants influenced primarily by discharge from households [41]. A PDMS concentration of $714 \,\mu g \, L^{-1}$ was measured at a plant receiving effluent from an industrial PDMS source, up to $20 \, mg \, kg^{-1}$ in filter cake and up to $6290 \, mg \, kg^{-1}$ in sewage sludge. The very low water solubility of PDMS, coupled with a very high octanol/water partition coefficient leads to very tight absorption of PDMS onto organic matter during the sewage treatment process, eventually leading to PDMS concentrations up to the lower $g \, kg^{-1}$ range in sewage sludge; no biodegradation occurs during sewage treatment [11, 22]. Concentration ranges for PDMS in sewage sludge were given from 284 to $1593 \, mg \, kg^{-1}$ for the US [41], and from 65 to $701 \, mg \, kg^{-1}$ for Germany [37]. Waters leaving the WWTPs contain residual PDMS often below the detection limit, the highest measured concentration being $13 \,\mu g \, L^{-1}$ [38, 40, 41]. Fifty-seven water samples from bays in Japan contained $< 2.5 \,\mu g \, L^{-1}$ PDMS [40]; slightly higher concentrations were found in surface waters of Chesapeake and Delaware Bays [38]. Significant levels of PDMS in surface waters were only detected in the proximity of heavily industrialized areas. However, concentrations of trace organics in the surface microlayer can be greater by a factor of ten or more than concentrations in the bulk water, leading to measurable concentrations of PDMS in surface microlayer samples.

Deposition of sewage sludge, together with dumping of industrial PDMS products, leads to PDMS transfer into soil. PDMS concentrations in agricultural sludge-amended soil range up to $10 \, mg \, kg^{-1}$ [41], and from Spain 3 to $340 \, mg \, kg^{-1}$ silicones in special waste deposits are reported [119]. In Germany a maximum of $0.63 \, mg \, kg^{-1}$ organosilicon material was found in soil, and $1.38 \, mg \, kg^{-1}$ in forest soil [37]; the latter was interpreted to be derived from atmospheric cVMS.

Although usually within the lower $mg \, kg^{-1}$ range [37, 38, 40, 41], PDMS concentrations in marine sediments range up to $126 \, mg \, kg^{-1}$ [41]. Highest levels of PDMS were associated with high organic content sediments in depositional areas receiving significant anthropogenic inputs. Organosilicon contents of sediments of the Potomac river, Chesapeake Bay and New York Bight ranged up to 96, 36 and 50 mg kg^{-1} [2, 38, 120]. Sediments from New York Bight represent PDMS loadings that would occur from off-shore sludge disposal, which was practiced up to the late nineteen-seventies [38]. At more than 90 % of the locations, PDMS concentrations in sediment are less than 26 mg kg^{-1}, and soil concentrations are less than 17 mg kg^{-1}. Thus, in the opinion of many scientists [11] PDMS does not appear to pose a risk to the terrestrial or benthic environments because measured environmental concentrations for both soil and sediment are much lower than measured effect concentrations for representative terrestrial or benthic organisms.

PDMS was measured in the surface sediments of marine and freshwater areas heavily impacted by municipal wastewater discharge to illustrate worst-case situations in the US [121]. Although the concentrations found ranged up to 309 mg kg^{-1}, the authors concluded that concentrations in worst-case sediments (including those from areas having poor wastewater treatment) were less than NOEC (no-observed-effect-concentration) values established in laboratory studies using sediment-dwelling organisms.

Observed levels of silicone in sediment cores range from 0.01 to over 1 mg kg^{-1} [122], and vertical distribution in the sediment column generally increases upwards, with the highest silicone content in the most contemporary sediments. Concentration spikes are associated with major, documented storm events. The utility of silicones as tracers for the movement of surficial sediments in a variety of aquatic systems has been demonstrated [38, 40, 120]. Pellenbarg and Tevault [123] described the existence of a sedimentary siloxane horizon in a selected sediment column; sediments deposited prior to approximately 1945 showed no siloxane; siloxanes have come into widespread use only since the early 1950s.

8.3.2 VMS IN AIR, SOIL, SEDIMENT AND WASTE

Because of their hydrophobic and sorption properties as well as their stability, high molecular silicone products like PDMS show low mobility and are preferentially sorbed onto solid surfaces, where they become enriched with time. However, this is not true for linear and cyclic VMS. These materials are fluids of low viscosity, and high vapor pressure. Thus these volatile substances are transported from condensed phases into the gaseous phase, i.e. the atmosphere. Up to 10 μg m^{-3} D3 and 20 μg m^{-3} D4 were determined in the industrial area of Guangzhou (China) [124]; similar maximal concentrations were found for D4 and D5 in indoor air. The dominance of D4 in industrial areas and waste treatment facilities was caused by industrial activities and used personal care and consumer products.

While one part of VMS is volatilized directly during application (e.g. in deodorants or perfumes), the other one is emitted from sewage waters into the atmosphere. The concentration of D4 in surface waters should be low

($< 1 \mu g \ L^{-1}$) and transient [125]. With a half-life of 16 days [36, 125, 126], D4 and other VMS would not be expected to build up in the atmosphere, but instead undergo constant degradation by hydroxyl radicals (see Chapter 1). Environmental fate analyses predicted a worst-case D4 concentration in STP (sewage treatment plant) of $27 \mu g \ L^{-1}$, based on 100% loading from consumer products [29]. A more realistic estimate of $9.5 \ ng \ L^{-1}$ was produced by back calculation from measured sludge concentrations [29]. Mean measured D4 concentrations in STP effluents ranged from 0.06 to $0.41 \mu g \ L^{-1}$ [29]. Assuming 50% bioavailability of total aquatic D4 and a conservative dilution factor of 3 as a realistic scenario [29], calculated concentrations of free or dissolved D4 in surface waters range from 0.01 to $0.068 \mu g \ L^{-1}$; worst-case sediment concentrations would be expected to range from 50 to $343 \mu g \ kg^{-1}$. Altogether the described scenario would imply a chronic margin of safety of 65 to 440 [30]. This risk assessment, although generally applicable to any geographic area, is designed to represent the US, with documented use rates for D4 and a standard type of WWTP.

D4 in sewage water and its treatment was also targeted in a model study postulating a nearly complete volatilization of D4 out of the sludge matrix into the atmosphere [125]. Concentration ranges of D4 in treated waste water (0.39–$0.66 \mu g \ L^{-1}$), in surface water (10^{-5}–$10^{-8} \mu g \ L^{-1}$) and in the atmosphere ($9.7 \ ng \ m^{-3}$) were calculated. Laboratory experiments demonstrated that D4 introduced in a sediment/water test system was nearly completely volatilized in the absence of biological degradation processses [31]. At five US sewage plants, the input water contained between 0.64 and $7.09 \mu g$ D4 L^{-1}, the output between 0.06 and $0.41 \mu g$ D4 L^{-1} [125]; in Germany corresponding concentrations for D3 to D6 were $34.5 \mu g \ L^{-1}$ (input) and $0.9 \mu g \ L^{-1}$ (output), respectively. For waste deposit waters average concentrations of $0.3 \mu g \ L^{-1}$ were reported [44], beside many organic contaminants in the Elbe river and its tributaries, D3 to D6 could also be identified [127].

Just 1.5% of the VMS in waste water are sorbed by sewage sludge [125]. In the US, sewage sludge from < 0.21 to $0.48 \ mg \ kg^{-1}$ D4 (dry weight) was analysed [125]. In primary sludge (1.2% dry substance) of a German sewage plant cVMS were found in a total concentration of $290.5 \mu g \ L^{-1}$; in mature sludge (0.59% dry substance) this concentration was lower by a factor of 4.5 [44].

Temperatures in the mesophile range (30 to 40 °C) together with a slight overpressure are apparently sufficient to produce significant volatilization of VMS from sewage sludge as shown by the fermenter gas: in Germany $28 \ mg \ m^{-3}$ of Si were detected in sewage gas on the average [44], and especially D4 and D5 were identified [61]. In Singapore respective contents of $> 4 \ mg \ m^{-3}$ were measured, and the air in the sewage plant area was loaded with up to $0.2 \ mg \ m^{-3}$ siloxanes [128]. Because of the relatively high concentrations, it is suggested that part of the VMS is generated by anaerobic degradation of PDMS polymers [128].

VMS not only exist as the original material, but at least to some extent as degradation products of PDMS polymers within domestic waste deposits. For example, one cubic meter (m^3) waste gas contains up to 66 mg VMS (Table 8.1), half a meter above the surface the atmospheric VMS concentrations are lower by more than one order of magnitude [128].

TMSOL, L2, D4, and D5 were found in the concentration range of mg m^{-3} in waste gases [61, 62, 129]. In sewage gas, TMSOL, L2–L4, and D3–D6 were identified. In contrast to landfill gas, the most volatile compounds TMSOL and L2 were found only in low concentrations, indicating that the sources for silicon in landfill gas and sewage gas are different. The cyclic species D4 and D5, however, were determined in the mg m^{-3} range, leading to a total silicon content similar to landfill gas.

During the combustion of landfill gas siloxanes are converted into microcrystalline silicon dioxide, which contributes to the abrasion of the surfaces within the combustion chamber. Given a daily gas production rate of 30 000–40 000 m^3 at each of the sewage plants sampled, several hundred grams of siloxanes per day are combusted causing severe problems in regard to the lifetime of the gas engines and revealing the necessity for gas pretreatment with effective siloxane elimination [130]. Effective relevant processes for the removal of siloxanes are chemical degradation by acid-catalyzed rupture of silicon–oxygen bonds at elevated temperatures, and adsorption onto activated charcoal or silica beds [47, 131].

VMS can not only be found at selected point sources like sewage plants or waste deposits, but also in ambient air (diffuse sources). In living areas far away from sewage plants and waste deposits cVMS were found up at to 17 μg m^{-3} [60, 128]; for the continent of Northern America an atmospheric D4 concentration of about 10 ng m^{-3} was estimated [125]. Indoor air is influenced by silicone outgassing of building materials and furnishing, or air conditioning [132, 133]. Organosilicon compounds including D3 and a volatile silanol species could be identified in the breath of 62 non-smokers from Chicago [134].

Table 8.1 VMS contents of environmental gases

Study	Samples		
	Waste gases (mg m^{-3})	Sewage gases (mg m^{-3})	Diffuse sources (μg m^{-3})
Martin et al. [44]	25	40–73	
Arendt and Kohl [129]	< 0.1–55	< 0.1	
Grümping et al. [61]	0.7–66	8.5–12	
Schweigkofler and Niessner [62]	6.6		
Schröder [128]		> 4	17–4000

8.3.3 PERSISTENCE OF SILICONES IN THE ENVIRONMENT: PDMS

PDMS is not expected to biodegrade during wastewater treatment: ^{14}C-PDMS showed no [14]CO$_2$ generation or volatile organic product formation after 70 days of incubation [135]. Matsui *et al.* [136] also showed that PDMS was resistant to biodegradation, essentially all of the PDMS partitioned onto the sludge biomass. These results show that PDMS behaves as an inert material during wastewater treatment, which does not impact the process and will be primarily associated with sludge following treatment. While it was stated that PDMS will proceed inertly through the active composting process, Spivack and Dorn [45] reported the presence of DMSD, a hydrolysis product of PDMS degradation, in composted sludge. The presence of DMSD in composted sludge provides direct evidence of PDMS hydrolysis, because PDMS hydrolysis is the only known source of DMSD in the environment [18, 19]. In general, PDMS will degrade both during outside storage of compost, and after compost is mixed with soil [137].

There is strong evidence for PDMS degradation occuring under strictly anaerobic conditions [61, 76]: TMSOL was determined in waste deposit gases in the mg m^{-3} range, and in leachates in the μg L^{-1} range together with DMSD in the mg L^{-1}-range. This is in accordance with the PDMS structure (see Figure 8.1) consisting of n intermediate D blocks (which are degraded to DMSD) and two end capping M blocks (which are degraded to TMSOL) in the proportion of several thousands to one.

Within aerobic environments, it is well known that PDMS will degrade in some agricultural soils [18, 19, 138–140] predominantly by hydrolyzation through random scission of its Si–O–Si backbone [141]. PDMS degradation mechanisms involve also biological degradation by bacteria and/or fungi, which takes place after an initial abiotic reaction initiating depolymerization [139]. Simultaneous with the formation of water desorbable products a decrease in molecular weight of extracted material was observed. DMSD was identified as the major PDMS degradation product [20].

When PDMS loaded sludge is amended to soil, PDMS is transferred from an aqueous to an essentially non-aqueous environment, where PDMS undergoes hydrolysis and rearrangement in contact with soil (primary degradation products are D4 and D5). HPLC-GPC revealed a significant change in the molecular weight from around 7000 to about 160. The presence of TMSOL was confirmed by GC/MS [18].

PDMS is hydrolyzed and rearranged on low-moisture soils to low-molecular-weight oligodimethylsiloxane-α, ω-diols hydrolyzing to DMSD. Addition of ^{14}C-PDMS to soils resulted in the production of [14]CO$_2$ under conditions that first led to the hydrolysis of PDMS to water-soluble silanols [20]. The production of [14]CO$_2$ followed a period of soil drying and rewetting in which the drying period led to the hydrolysis of the PDMS to DMSD, and the rewetting led to biological activity and degradation of DMSD.

PDMS hydrolysis was faster in ultisols and oxisols than in alfisols at the same soil–water potential [23]. The PDMS hydrolysis rates are primarily controlled by soil moisture and to a lesser extent by soil type. PDMS hydrolysis for all soils is thus expected to be most rapid at the surface, where the soil is most likely to dry, and during periods of hot, dry weather. In the field PDMS degrades by 50% in 4 to 10 weeks (degradation is faster for lower concentrations) [18, 142]. DMSD as the main degradation product was detected in most samples at <5% of original PDMS, indicating biodegradation and volatilization of DMSD as shown in laboratory experiments demonstrating that PDMS can be hydrolyzed in 1 to 2 weeks in dry soil [19, 138].

Soil factors such as clay content and clay type are very important in determining the degradation rates of PDMS [24]; other materials were also found to have a catalytic effect on hydrolytic PDMS degradation [45]. Clay minerals varied substantially in their catalytic activity: kaolinite, beidellite, and nontronite were the most, and goethite and allophane the least, active [24]. Clays were sequentially extracted first with 0.01-M $CaCl_2$ solution, next THF, and finally 0.1-M HCl to separate the different silicone species using a protocol obtained by modifying the procedure of Lehmann et al. [21]. The $CaCl_2$ solution was used to obtain water-extractable, monomeric, and small oligomeric silanols. The THF extracted the remaining water-insoluble polymer residue, and the HCl solution extracted the bound silanols [24].

Although PDMS is usually regarded as being immobile in the sedimentary environment [11], actual sediments show PDMS concentrations decreasing with depth in sediment cores, with the onset of detection corresponding to the introduction of silicone technology in the middle of this century. Following a year of incubation under aerobic conditions, 5–10% of the PDMS had been hydrolyzed to DMSD [19]. Biologically mediated hydrolysis of PDMS may have been partially responsible for the formation of DMSD in the sediment.

8.3.4   PERSISTENCE OF SILICONES IN THE ENVIRONMENT: VMS

VMSs in the atmosphere are not persistent, but are removed by physical processes [29] and transformed by several chemical reactions. Among the chemical processes is the photolytic and oxidative degradation of VMS in the gaseous phase including reactions with hydroxy and nitrate radicals as well as ozone [36, 125, 126, 143]; e.g. cVMSs react with OH radicals in the air with atmospheric lifetimes of 10 to 15 days. Similar results apply for L2 to L4 leading, among others, to silanol and siloxanol species [145]. VMS degradation reactions by OH and O_2 can be summarized [29] as in Figure 8.3.

The atmospheric gas-particle partitioning of D5 and its hydroxylated compound, 1-hydroxynonamethylcyclopentasiloxane (D4OH) was investigated [144]. The effect of temperature is considered to be the key parameter that affects the partitioning coefficients; relative humidity is also an important

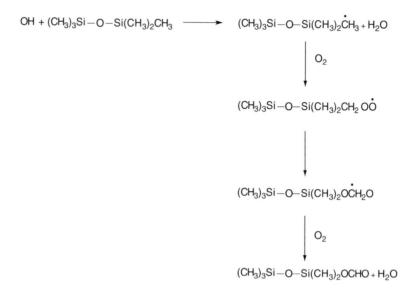

Figure 8.3   Degradation mechanism [29]

parameter affecting D4OH partitioning. In particular, partitioning to the particle phase is increased by two orders of magnitude in going from 25 to 0°C. D5 has recently been identified as a possible replacement candidate for chlorofluorocarbons and other solvents of environmental concern. From relevant rate constants and the atmospheric OH concentration a lifetime in rural atmospheres of approximately 7 to 9 days, and in polluted atmospheres of about one day could be estimated. If D5 resides on fine particles and does not react, its lifetime in the atmosphere will be 1–2 weeks [144].

Once cVMS fluids enter the aquatic system, they partition rapidly to the atmosphere due to their high volatilities. However, up to 46 % of the cVMSs entering a WWTP will be sorbed by sludge. Altogether about 95 % of the D4 in the wastewater influent was removed by municipal activated sludge treatment processes [125]. A model prediction indicated that about 49 % of the D4 was removed by sludge adsorption and 38 % was removed by volatilization in municipal activated sludge systems. Also in a pilot plant system, primary sludge removal and volatilization in the aeration basin were found to contribute approximately equally to the removal of D4 and D5 [146]. When experimentally investigating the microbial degradation of D4 into DMSD in sewage sludge, the degradation rate was highest after approximately 100 days, small > 150 days and neglegible < 20 days [147].

The degradation of cVMS is a multistep process [148], initiated by ring-opening hydrolysis of the cyclics to form linear oligomeric siloxane diols, followed by further hydrolysis of these oligomeric diols to the monomer DMSD. All oligomeric siloxane diols were unstable in air-dried soil and ultimately hydrolyzed to DMSD within a few hours to a week, depending on the humidity. When the soil was rewetted to water saturation, some of the intermediates continued to hydrolyze to DMSD, and some are converted back to cVMS, which then evaporated from wet soil. cVMS degradation was more significant than loss by volatilization in soils with low moisture levels. At high humidity the degradation slowed, while volatilization was accelerated and became the predominant process in regulating the cVMS removal from soils. These findings demonstrate that cVMS fluids are unlikely to persist in any soil within a wide range of moisture conditions.

8.3.5 PERSISTENCE OF SILICONES IN THE ENVIRONMENT: SILANOLS

Sequential oxidative demethylation is the mechanism responsible for the formation of siloxanols in the environment [34]. When exposed to atmospheric conditions volatile methylsilicon compounds undergo oxidative degradation involving hydroxyl radicals at a rate comparable with that of ordinary hydrocarbons [36]; for TMSOL the tropospheric lifetime is about 3 days. The mechanism of oxidative degradation can be summarized in Figure 8.4 [36].

Photolytic oxidation also degrades siloxanols in the aqueous phase [35, 149]. DMSD which would formally be expected to be the ultimate hydrolysis product from PDMS derivatives, can be demethylated on irradiation with UV light in aqueous solution. Altogether time and exposure to soils and sunlight will convert oganosilicon compounds finally to carbon dioxide, water and silicic acid [150].

DMSD is very water soluble (solubility $142\,mg\,L^{-1}$ at 25°C) [33], and is therefore available to microorganisms. As a result of biological activity the methyl groups of DMSD will be oxidized, eventually leading to silicic acid $(Si(OH)_4)$ as ultimate degradation product. DMSD is relatively volatile and, depending on the soil type, appreciable quantities are volatilized from the soil into the atmosphere [63]; the atmospheric lifetime of DMSD should be similar to that of TMSOL (2 to 9 days). There are sufficient levels of nitrates and nitrites in natural waters to catalyze the photolytic oxidative demethylation of DMSD. However, average concentrations in soil solutions are low (i.e. at or below the level of detection of $1–2\,\mu g\,L^{-1}$); even in soils receiving repeated applications of sewage sludge [41], DMSD can be regarded as a transient intermediate, before being biologically degraded to silicic acid and carbon dioxide.

DMSD biodegradation was confirmed and a pathway determined by Sabourin et al. [151], using ^{14}C-labelled DMSD. Biodegradation of DMSD is initiated by the oxidation of a methyl group, which converts $HO—Si(CH_3)_2$ $—OH$ into $CH_3—Si(OH)_3$. Thus biodegradation of DMSD led to the intermediate product methylsilanetriol (MST) and finally to silica [152]. Microorganisms

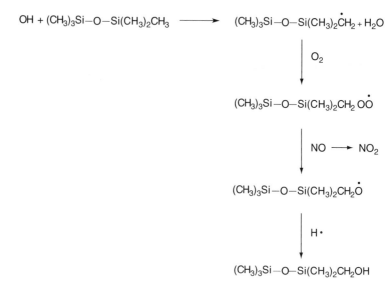

Figure 8.4   Oxidative degradation of siloxane involving hydroxyl radicals
Reproduced with permission from Ref. 36

isolated from soil are apparently able to mineralize DMSD completely in liquid culture when a primary carbon source is provided. MST is more strongly bound to soil than is DMSD, indicating that MST is more likely to sorb to soil, thus decreasing its bioavailability.

The activity of microbial degradation is determined by many parameters like temperature, water content, availability of nutrients and oxygen or pH [34, 63]. The fungus *Fusarium oxysporum* Schlechtendahl and bacteria of the type *Arthobacter* cometabolized DMSD to CO_2 probably in an aerobic pathway not further described by the authors [151].

Because in the environment an equilibrium between siloxanols, DMSD and water will quickly be established, in aqueous solutions DMSD will be the preferred species [45]. Silanols show low stability, and rapidly condense within acid and alkaline milieus [5, 34], therefore isolation and handling of these substances in environmental samples is very difficult. Numbers of hydroxyl groups and size of organic ligands influence stability; while large organic ligands exhibit a stabilizing effect, an increasing number of hydroxyl groups leads to decreasing stability [153]. Glass ware should be avoided, because condensation on glass surfaces is catalyzed by alkaline species [34]; sample storage at low temperatures is recommended. Frozen pure crystalline DMSD in polypropylene vials could be stored up to 8 weeks without self condensation effects [59, 75]. The stabilty of DMSD in aqueous solutions is also influenced by dilution; while in a 2 % DMSD standard solution dimeriation started immediately, in a $100\,mg\,L^{-1}$ DMSD solution condensation products could not be detected after 1 year standing time [59].

Besides solubility in water, silanols like TMSOL, and to a lesser degree DMSD, show significant volatilisation potential [29, 63], which inside soil is dependent upon several parameters such as soil structure and humidity [34, 63]. Especially at low soil humidity volatilization processes lead to significant decrease of DMSD in soil [63]; 8 weeks after incubation about 30% of DMSD was recovered in aqueous extracts and a comparable amount bound onto soil particles.

8.3.6   TRANSFER OF ALKYL GROUPS TO OTHER SPECIES

An environmentally important question—especially in the context of this book—is concerned with possible indirect contributions of silicones to the generation of element and metal(loid) organic compounds when acting as a donor of alkyl groups. While *in vivo* methyl transfer from silicones to metal (loid)s was described as an enzymatic catalysed process [154], abiogenic transfer of a methyl group from short and long chain silicones to inorganic mercury to yield monomethylmercury was also demonstrated experimentally [155]. Laboratory experiments showed that D4 and D5 stimulate the microbial production of trimethylbismuth (TMBi) [156]; TMBi is a volatile Bi species that has been identified in environmental gases [157], (Chapter 9) and *Methanosarcina barkeri* is a cobalamin-producing archaea commonly found in sewage [158]; Bi and PDMS are present in domestic waste as result of their use in household products. Experiments with pure cultures of *M. barkeri* spiked with Bi produced no detectable quantities of TMBi. When D4 or D5 was added, however, TMBi was produced at levels comparable to those observed in sewage sludge. Experiments with deuterated D4 and analysis of mass fragmentograms received by electron ionization GC/MS indicated that the methyl groups of the TMBi do not originate from the D4 or its degradation products. This points to an indirect influence of D4 on the process of methyltransfer, possibly by modification of the bacterial cell membranes. Thus, generally, organosilicon compounds may transfer alkyl and aryl groups to a variety of metal derivatives (e.g. mercury, lead and thallium) not only in spill situations, but also in the presence of trace amounts of metal(loid)s by direct or indirect routes [156].

   Influence of silicones onto biomethylation processes implies degradation of PDMS to smaller and hence more mobile and reactive substances as well as possible invasion of VMS into cells. That these scenarios occur in nature has already been shown in the preceding text, and is summarized as follows: in soil PDMSs are aerobically hydrolyzed to DMSD, partly transformed back to cVMS, and eventually biodegraded to carbon dioxide (CO_2) and silica (SiO_2) [148, 151]. Furthermore, within anaerobic environments (e.g. waste deposits) PDMS will be degraded [61, 76]. Microbial transformation of D4 into DMSD within sewage sludge could be demonstrated experimentally [147].

8.4   INTERACTIONS WITH ORGANISMS

Because silicones are widely used as inert and persistent materials in medicine, it is very important to know about possible interactions between these compounds and biological systems, i.e. to investigate the bioactivity of PDMS and VMS. Especially in connection with the question on safety of silicone breast implants, this topic is still the subject of tremendous current controversies among scientists, and will be summarized in this section.

8.4.1   COLONIZATION, DEGRADATION AND TRANSPORT OF SILICONES IN ORGANISMS

Pseudomonas putida and *Pseudomonas fluorescens* grow on various silicones (solutions containing L3, L7 and silicone oils) [159]. Organosilicon compounds including PDMS are biodegraded by enzymatically catalyzed hydrolytic bond scission, a mechanism also responsible for the biological degradation of other organic species. Silicone based catheters were often colonized by microorganisms leading to urethra infections [160]. Silicone breast implants can also be attacked by microbes, e.g. by *Staphylococcus epidermis* [161]. Voice prostheses are made of medical grade silicone rubber and can be subjected to rapid surface colonization and degradation by mixed biofilms of bacteria and yeast [162, 163]. Bands corresponding to C—O—Si, N—Si and Si—H were present in the infrared spectra of both the cell wall and and fluid material when the bacterium *Proteus mirabilis* was cultured in a silicon-containing medium [164].

As described for environmental systems, the siloxane bond can be hydrolyzed by exposure to water, creating two silanol groups. This reaction is highly reversible, and accelerated greatly in both directions by the presence of acid or base catalysts [17], and also occurs within normal physiological ranges ($37°C$ and mixed aqueous environment). Thus, methylsilanes undergo hydroxylation *in vivo* [2]; ingested siloxanes and silanes are broken down to silanols in the highly acidic environment of the stomach. Therefore, originally liposoluble silicone compounds could be excreted as water soluble silanols via the kidneys [3]. The hydroxylation reactions could be simple chemical reactions rather than enzyme-promoted processes. The only *in vivo* reaction at silicon so far reported that may be enzyme catalyzed is the direct transfer of methyl groups from silicon without prior oxidation [95]. Consequently various diols ($MeSi(OH)_3$, $Me_2Si(OH)_2$, $MeSi(OH)_2$—O—$SiMe_2OH$, $HOMe_2Si$—O—Si Me_2OH, H—$(OSiMe_2)_3$—OH) were found as urinary metabolites in rats by HPLC radiochromatogram after THF extraction [95].

Sixty-five genera of microorganisms were found not to be able to degrade PDMS in cosmetic ingredients [165]. However, there are several observations arguing against inertness of PDMS [14, 164, 166]. Microbial metabolization of low molecular weight siloxanes is postulated [14], and described [164]; silicon–

oxygen (Si—O) as well as silicon–carbon (Si—C) bonds can be enzymatically broken. Among other effects long-term environmental aging of a silicone rubber (monitored by MALDI/TOF-MS) may lead to degradation of the base polymer to a lower average molecular weight material, and to degradation leading to the production of cyclic compounds like D4 [166].

It was suggested that silicone oils (PDMS) migrate to distant sites along tissue planes [167] or are carried by macrophages to regional lymph nodes [168]. Polysiloxanes were injected in rats and were found together with partially hydrolyzed polysiloxanes and silica in lymph nodes [169]. Following subcutaneous administration of 1 mL of dimethylpolysiloxane into mice, accumulations of silicone-containing cells in the regional lymph nodes and formation of generalized foci of fat atrophy appeared as early as 2 weeks post injection [170]. Mice were injected with either breast implantate distillate composed primarily of D3 to D7 or a PDMS oil [94]. Subsequently low molecular weight silicones were found in brain, heart, kidney, liver, lymph nodes and other organs. While highest levels of cyclosiloxanes were found in the mesenteric lymph nodes, ovaries, and uterus, linear siloxanes were highest in the brain, lungs, and mesenteric lymph nodes. The observation that linear siloxanes accumulate preferentially in the brain can only be explained on the base that these species are passing the blood/brain barrier.

The most abundant siloxane oligomers extracted from implant gels are cyclic components (D4 to D6). The lower molecular weight soluble fraction (sol) can also largely pass through the elastomer shell of the implant; this sol can be picked up by the synovium-like cells (macrophages), and transported to tissues, lymph nodes and other locations, including the liver [17]. A number of studies suggest that the capsule as well as the blood is a suitable transport system for the distribution of this finely divided silicone. This is inferred from demonstration of increased levels of silicone in blood and capsules of women with implants, and its detection in distant organs. A case of extensive granuloma formation secondary to implant rupture, not only in the immediate vicinity of the breast, but also distant from the breast was reported [171]; thin gel migrated through subcutaneous planes as far as the groin.

Skin biopsy specimens were obtained from a patient who had received injections of silicone gel in his nose and chin [117]. Surface silicone concentrations were determined by ESCA in chin (21%), nose (19%), malar region (3%), and inner arm (below detection limit). Light microscopy revealed homogeneous 'globules' consistent with silicone in the chin and nose sections only; the malar region and inner arm sections showed no evidence of silicone.

Eller et al. [172] report a case in which intravitreal silicone oil migrated along the intracranial portion of the optic nerve and into the lateral ventricles of the brain after the repair of a retinal detachment secondary to cytomegalovirus retinitis. Common late complications of intravitreous silicone injection are cataract, glaucoma, and retinal changes [173].

8.4.2 TOXICOLOGY AND IMMUNOLOGY OF SILICONES

Some of the physico-chemical properties which make organosilicon materials so useful commercially also make their toxicity and ecotoxicity testing difficult using conventional methods [174]. Thus, although the toxicology and immunology of silicones has been extensively reviewed [2, 3, 175–177], many aspects of this field are still controversial and divergent interpretations exist. At this point it must also be mentioned that many silicone products contain additives (e.g. UV stabilizers, softeners, flame retardants, biostabilizers) potentially leading to health effects [178].

In the course of pertinent research during the last decades, increasing evidence was found that PDMS generally show very low toxicity. Earlier experiments with monkeys and dietary studies with humans gave no indications of any acute toxic effect of silicones [2]; while low molecular weight silicones were excreted unaltered via expiration and urination, high molecular compounds were not absorbed and were excreted in the faeces. During testing, daphnia apparently died of suffocation rather than poisoning. Reproduction of exposed birds was normal [2].

Low molecular weight linear and cyclic silicones are commonly used as inert bases in medical and cosmetic preparations. Since personal care products may come into contact with eyes and skin or be inhaled, both manufacturing and regulatory bodies require that the ingredients in such products be absolutely safe and have no troublesome effects even when used over a prolonged period of time. Therefore, this class of silicones has undergone even more elaborate testing than other silicones in both animal and human subjects. As with other silicones, it was stated that they generally show very low toxicity, and no indications of carcinogenicity or mutagenicity [2]. There were only a few exceptions like silatranes (rodenticide) and mival (herbicide).

PDMS fluids and formulations (Dow Corning 200 fluids 300/350 cSt containing up to 12% low molecular weight cyclic PDMS) were studied in certain biological systems to evaluate the possible environmental impact of these materials [135]. These siloxanes were not detectably degraded by sewage microorganisms, as shown in a study with ^{14}C labeled PDMS. The very low toxicity of these materials to daphnia, fresh water fish, marine species, mallard ducks, bobwhite quail, and domestic chickens, and their non-accumulation in the flesh and eggs of chickens and in the flesh of fish, minimizes concern regarding their potential to cause environmental damage. There have been no reproductive effects observed for VMS [29]. Marcus does not believe there is any valid experimental evidence that silicone elicits specific immunity in humans or in experimental animals [176].

Extensive work by the Dow Corning group on the health and environmental effects of silicone materials, including ecotoxicity testing of PDMS fluids and silicone oils, was carried out from 1970 to 1990. While there were just a few indirect troublesome effects, and generally silicones were rated as essentially inert, a cyclosiloxane copolymer producing androgen depressant activity in rats, rabbits and monkeys, and reducing fertility in female rats and rabbits, was reported [179].

D4 showed a 'functional' solubility in freshwater and sea water of 14–30 μg L^{-1} and 6–9 μg L^{-1}, respectively [180]; its bioconcentration factor ranged between 7000 and 13 400 [32]. For rainbow trout, the 14-day LC$_{50}$ for D4 is about 10 μg L^{-1}, and the NOEC (no observed effect concentration) < 4μg L^{-1} [29, 30]. From the mortalities observed in a chronic *Daphnia magna* study, a 21-day LOEC of 15 μg L^{-1} was determined. These values are higher than the highest maintained exposure levels, taken as the aqueous solubility limit of D4 in natural diluted waters. It is argued, however, that conditions maintained during laboratory studies were not representative of ambient environmental conditions found in natural waters, and exposures achieved during the laboratory studies would be expected significantly to exceed estimated environmental exposures both in concentration and in duration [31]; thus, D4 in sediments should not be a risk to organisms associated with them [181]. For PDMS polymers in terrestrial as well as aquatic ecosystems significantly higher NOEC are reported (250 to 2300 mg kg^{-1}) [11]; for the PDMS degradation product DMSD the NOEC for aquatic organisms is > 10 mg kg^{-1}.

The question is if the generalizations described above can really be made on the basis of the experimental and analytical data presented. A detailed toxicology/immunology implies differentiation between low (VMS) and high molecular weight silicones (PDMS), additionally considering the biological effects already discussed (transport and degradation within biological systems).

PDMS oligomers containing more than 14 siloxane units could not be detected in fish; probably because of their molecular size, these compounds could not pass biological membranes [26, 27]. A correlation between molecular size and biological resorption has been known for several decades [14], linear and cyclic siloxanes with up to six siloxane units show significant biological activity. It has been demonstrated that intake of D4 leads to resorption in the digestion tract [28].

While a 20 cSt PDMS silicone oil was inactive, D4 was found to have an adjuvant effect ascribed to its inflammatory properties [175]. There is some early evidence that certain low molecular weight siloxanes are toxic to cells in tissue cultures; low molecular weight siloxanes produce lethal effects on B-lymphocyte-derived target cells *in vitro* and permeabilize the plasma membranes at lower sub-lethal concentrations [182].

Silicone gel distillate containing D3 to D6 was found to be very toxic and all the mice injected with 35 g kg^{-1} died within 5–8 days; they developed inflammatory lesions of the lung and liver as well as liver cell necrosis [183]. D4 alone showed a LD$_{50}$ of 6–7 g/kg, being about as toxic as carbon tetrachloride or trichloroethylene.

For a long time it has been speculated that certain low molecular weight silicones like cyclosiloxanes may have potent biological activities that mimic estrogens [14, 184]. D4 was shown to have adjuvant activity when mixed with Dow Corning 360 medical grade silicone fluid [185]. These experiments high-lighted the unexpected activities of cyclosiloxanes. Brawer [186] concludes that it is now clear that high molecular weight silicones are neither chemically nor biologically inert. In this context it should be noted that pharmacologically active silicone, considered at one time for use as an oral estrogen, consists of a 2, 6-*cis*-biphenyl-substituted D4. Although L3 and D4 were nominated as candidates for carcinogenity testing [187], no significant chromosomal aberrations have yet been observed [188].

Proteins have hydrophobic regions that would bind strongly to siloxanes and silanols [17]. At the surface of PDMS droplets, hydrolysis and oxidative changes take place, leading to a surface that is a crosslinked silanol-containing composition. This composition is similar to that of silica that had been hydro-lyzed to a form which would have a significant biological effect.

Among the most surface-active of the blood proteins are fibrinogen and fibronectin. They play a key role in inflammation. Lymphocytes and macro-phages are much more effective against chronic persistent agents and therefore are especially relevant in the biological reaction to silicones. Fibrinogen, an adhesive plasma protein operating at interfaces, bound to droplets of PDMS renders an adhesive potential to the surface of the droplets, a potential that may have relevance to the biological processing of the polymer [189]. Some reports have implicated contaminated and aged silicones as the cause of inflammatory reactions [190].

Dimethylsilicones are not themselves immunogenic but, when emulsified with immunogenic proteins, lead to higher immune responses; i.e. silicones behave as adjuvants similar to mineral oils. Thus, silicone–protein complexes are poten-tially immunogenic [191]. D4 can induce denaturation and conformational changes in fibrinogen and fibronectin and it can be expected that protein molecules that have undergone denaturation or conformational change induced by D4 may act as antigens and stimulate the immune system to generate antibodies, ultimately resulting in autoimmune disease [192].

Silicone compounds possess the potential for creating antigen–antibody binding sites. Silicone surfaces provide a major electronegative static charge, and electrostatic forces are known to be important for immunochemical binding [3]. Moreover, hydrogen bonding (through side groups of silicones) and hydro-phobic interactions will also be important in antigen–antibody binding. How-ever, interactions of silicones with proteins is a issue with more theoretical assumptions than experimental confirmation; alternatively it was suggested that silicones are only unspecifically adsorbed to serum proteins [193].

Because silicones in the biological environment reveal numerous degradative reactions and surface interactions that produce bioreactive substances, Kossovsky [177] argues that silicones are toxic organometals. According to this author, implanted silicone prostheses and medical devices are associated with various local and systemic host inflammatory reactions, and associated with a form of autoimmune disease [191, 194]. These studies are criticized by Marcus [176] who claims a number of methodologic and conceptual defects in the work of Kossovsky *et al.* [177, 191, 194].

Several authors have expressed the opinion that silicones can be broken down *in vivo* to inorganic silica [101, 195]. Crystalline silica has been associated with severe inflammatory reactions, and is known to cause lung and kidney problems (e.g. lung fibrosis and Balkan nephropathy) [196, 197]; hence there is concern that this degradation might occur. Last but not least, because of the presence of fumed silica in the implant elastomer, the immunology of silica may also be considered when discussing the safety of silicone breast implants (SBIs).

Historically, the origin of safety and hygiene issues regarding organosilicon materials was the listing of organosilicon compounds in the 'Grey List' of marine wastes by the London Treaty of 1972 [198]. It was recognized that these substances are persistent, and thus may accumulate in the environment. In Japan, monitoring of organosilicon compounds occured in 1989 and 1993, when PDMS and D4 were tested for their ecological impact [199]. This work included growth inhibition tests on algae and terrestrial plants, and no data indicating any problem was obtained. However, it was stated that discarded oil and waste plastics made of organosilicon compounds should not be discharged into the ocean [198].

In Germany materials are classified into water pollution classes (Wasserge-fährdungsklassen WGK) [200]. PDMS and VMS are classified as WGK 1 (slightly hazardous in water). In the Netherlands, organic silicon compounds are listed in class C9 as hazardous waste in 'The Hazardous Waste Designation Decree' [200].

8.4.3   SILICONE BREAST IMPLANTS

The biocompatibility of silicones has come into question, as evidenced by the extensive silicone breast implant litigation occuring primarily in the US [201]. Relatively little is known about the physical and chemical interactions between silicone polymers and biomolecules.

Silicones were not introduced into the human body until the 1940s. For the past 50 years, the medically relevant silicones have been synthesized mostly from starting blocks of D4. The D4 cyclic is cracked open and polymerized by a catalyst to approximately 90% equilibrium, favoring the polymer (PDMS oil). The silicone-gel-filled breast implant was first developed in 1963 as a substitute for the former standard direct intramammary PDMS injections. From then, silicone-gel-filled breast implants have been used extensively for reconstruction and augmentation mammoplasty during the last three decades, and have become one of the most controversial applications of polysiloxanes in medicine.

The great range of viscosity of liquid and cross-linked silicone polymers and the elasticity of silicone rubbers made it possible to fabricate materials that mimic the consistency of tissues ranging from bone to mammary tissue. This has brought silicones into wide usage, and probably the largest group of persons involved are the recipients of silicone-gel-filled breast implants (SBIs). It has been estimated that more than a million US and Canadian women have been implanted with SBIs over the last two decades [3, 202]. Implants are not stable indefinitely, and after a period of years silicone materials may ultimately leak into tissues ('silicone bleeding') [203]. Issues of considerable controversies are whether implants evoke disease, and how the immune system responds to silicones.

Silicone gel breast implants consist of a rubber-like silicone elastomer envelope (shell), which varies in thickness from 0.075 to 0.75 mm, enclosing a known volume of silicone gel ranging from 80 to 800 cm^3. The elastomer casing of SBIs is made of highly crosslinked silicones, and liquid silicones within can slowly escape [204–206]. Tissue reactions and interfacial fibrous capsule formation may occur, and the elastomer casing may erode [203, 207]. Besides PDMS (viscosity usually 1000 ct) gels also contain lower molecular weight unpolymerized silicones that are derived from the polymerization reactions. The latter includes a series of linear and cyclic compounds ranging from 3 to >300 siloxy units; D4 is a typical example for these compounds. Generally, gel consists of a mixture of low (6000–19 000 amu) and high molecular weight (300 000–400 000 amu) components.

Of the first-generation implants (1963 to 1972), 96% remained intact after 14 to 28 years *in situ*, of the second-generation implants (1973 to mid 80s) that were explanted from 1992 to 1998, 73% had disrupted [203]. Diagnosis of disruption is difficult. Mammography is helpful only if there has been movement of silicone gel into breast tissue. Silicone migration was observed in only 4% of second-generation implants removed from 1992 to 1998. After disruption, none of the following were elevated above the levels seen in control women without implant exposure: serum autoantibodies, blood and serum silicon, and the incidence of breast cancer, autoimmune disease or any other medical disease. It was thus stated that there is no evidence to support the existence of any 'novel' or 'atypical' syndrome associated with gel implants [203].

Second-generation implants had a thin gel and a thin wall (0.13 mm). Since 1978, it has been recognized that low molecular weight silicones can be expected to diffuse out of all clinically intact gel implants [171, 208]. Although the implant remained clinically intact, globules of silicone have accumulated within the surrounding capsular tissue, after diffusing ('bleeding') through the intact elastomeric shell. Following most implant disruptions, the free silicone is contained within the capsule or can migrate into adjacent breast tissue. Up to 10 mg day^{-1} low molecular weight silicones were leaking through the intact implant outer shell into a lipid-rich medium per 250 g of implant at 37 °C [204]; this could lead to significant accumulation within lipid-rich tissues. It should also be noted that substances like blood or bacterial toxins were observed to diffuse through the silicone shell into the implant [209–211].

While mean silicon levels in breast tissue are $9980\,mg\,kg^{-1}$ in implant patients, controls show only $90\,\mu g\,kg^{-1}$ [212]. Capsule silicon levels were about 10^6-times higher than those in blood (see following section); capsular silicon levels are not significantly influenced by implant status [213–215]. Solvent extraction of the envelope and the gel has revealed 30 different linear and cyclic components, some of which have adjuvant activity in animals [212].

It is believed that macrophages take up polysiloxanes at the implantation site and transport them to local lymph nodes and distant sites. Silicon of undetermined chemical form has been found in macrophages [85, 216]. Macrophages can degrade polysiloxanes by generating reactive oxygen metabolites that attack the methyl protons, leading to the substitution of methyl groups by hydroxyl groups [106]. The degradation of PDMS in the lymph nodes parallels that of D4, the main difference being that linear high molecular weight polysiloxanes are less reactive and less cytotoxic than oligomers. It is likely that after phagocytosis of the polymers by macrophages, the silicon-containing metabolites are released into the lymphatic system and transported to organs (i.e. liver, kidneys) and then excreted.

These results demonstrate that all polysiloxanes and extracts from silicone-gel-filled implants are biotransformed in the lymph nodes, but high molecular weight polymer degrades at a slower rate than oligomers [169].

Investigations indicate that accumulation of silicones and silicone elastomer degradation products in human organs and other tissue can occur [217]. For example, chronic liver abnormalities of haemodialyzed patients were a result of accumulation of silicone particles in liver, the particles originating from siliconized blood pumps [217]. Several cases of pneumonitis or acute lung injury following the accidental direct intravascular injection of silicone fluid have been reported [210]. By element fingerprinting it could be demonstrated that silicone in the sputum was identical with that in ruptured implants [218].

In respect to silicone injections there appear to be more patients with unsatisfying final results and suffering complications than there are content patients [219]. Thus fluid PDMS has not proved to be a proper material for plastic reconstructive and aesthetic treatment in the head and neck region [219].

The symptoms experienced by women with breast implants ranged in severity from minor aching joints to debilitating pain and chronic fatigue [220]. Other symptoms include hair loss, stomach irritability, rashes, fevers, and allergic reactions to certain foods and chemicals found in, for example, perfumes and household cleansers. The existence of a relationship between implants and disease has remained the subject of both public and scientific debate (e.g. literature cited by Zimmermann [220]). By January 1992, so many implant recipients had reported experiencing implant-related complications and health effects to the Food and Drug Administration (FDA) that the FDA restricted the use of the devices [221]. Silicone elastomers *in vivo* may provoke an inflammatory response, fibrosis and (in the long term) calcification, and generate immune reactions, specifically various types of connective tissue disease;

considerable controversy exists as to the nature and magnitude of these effects [3]. The question of the safety of breast implants and their possible adverse effects on health is the subject of numerous discussions: on the one side it is stated that breast implants are safe [222], on the other side support groups and information networks for breast implant recipients have created home pages in the internet including bibliographies listing the latest scientific and epidemiological studies on the risks associated with these devices [223]. While there is still unresolved dispute concerning increased or reduced breast cancer risk among women with implants, excesses of cervical and vulvar cancers were observed compared with the general population [224].

Summarizing, there is still concern about the safety of silicones, triggered by case reports of autoimmune disease in women with breast implants, although controlled epidemiological studies have so far failed to establish an association between silicone breast implants and autoimmune disease [212, 225–227].

In April of 1996, Dow Corning Corporation entered into a Memorandum of Understanding [228] with the EPA, not only agreeing to conduct a comprehensive, multimillion dollar program for health and safety testing of six siloxanes, but in addition, committing to substantial product stewardship measures.

8.4.4   SILICON AND SILICONES IN BODY FLUIDS AND TISSUES

There is an estimation that a human assimilates 9–14 mg of the element silicon daily [3]. Important amounts of silicon are incorporated into foods such as vegetables, grain, rice and dairy products, and in beverages such as beer. A variety of sources of VMS and PDMS exist in the everyday diet within personal surroundings (e.g. body care products), and from environmental sources [229]. At least part of this silicon and silicone will be transported in the blood and excreted via urination. Urine silicon levels exceed blood serum levels 20 to 100-fold and have ranged from 2 to 52 mg L^{-1} [196, 230]. While these authors interpreted these differences to reflect different dietary (particularly vegetable) intakes of silicon, they also argued that silicone released from gel implants may also be rapidly excreted in the urine.

A controversy exists about possible contributions of leaking SBIs to VMS to body fluids; while Garrido et al. [101] discussed their evidence (obtained by NMR) that the silicone gel itself can be degraded in vivo into products that appear in blood plasma, Peters et al. [231] did not regard NMR measurements of silicone to be a useful method because of low sensitivity of detection. According to the latter authors, because of the large variability among patients, the use of silicon measurements in blood, serum, breast milk, or implant capsule tissue has no clinical role in the effective monitoring of implant leakage in women with silicone breast implants.

Because of the various sources of Si in blood, including inorganic silicon species, silicon levels cannot be assumed to be a proxy measurement for silicone.

Comparing implanted women to controls, mean silicon levels were not significantly different in breast milk (55 ± 35 and $51 \pm 31 \mu g$ L^{-1}, respectively) or in blood (79 ± 87 and $104 \pm 112 \mu g$ L^{-1}, respectively) [232] (note the higher values in blood controls compared with implant blood). Silicon levels are 10-times higher in cow's milk ($709 \mu g$ L^{-1}) and even higher in infant formulas ($4.4 mg$ L^{-1}). Lugowski et al. [233] measured silicon in whole blood, plasma, breast milk and infant formula and their methyl isobutyl ketone (MIBK) extracts (assumed to represent the silicone fraction). About 35 % of blood Si was found in plasma. The extractable species of silicon from blood, plasma, breast milk and formula are 0.3 to 16 % of total silicon. Mean concentrations of silicon (total/extractable in μg L^{-1}) in whole blood (79/6), plasma (57/8), breast milk (25/4) and formula (2930/9) were determined. In another study [229], breast milk levels of implant patients were $59 \pm 34 \mu g$ kg^{-1}, and of controls $51 \pm 31 \mu g$ kg^{-1}; infant formula mean was $4.4 mg$ kg^{-1}. No statistical significant difference between silicon in breast milk of nursing mothers with silicone breast implants and control nursing mothers, and no statistically significant difference in whole blood silicon levels between control and implant patients, was observed.

The results of measurements of the silicon content of human blood are summarized in Table 8.2. Concentrations from $2 \mu g$ L^{-1} up to $5 mg$ L^{-1} are reported, indicating variations due to different dietary and environmental influences; there does not appear to exist a 'normal' level. Additionally, differences in measurement techniques including sampling errors may contribute to the spread of data [229]; improving measurement techniques seem to result in lower concentrations.

In several studies no differences between the blood silicon concentrations in women with silicone breast implants and controls were found [42, 79]. Flassbeck et al. [98] measured low molecular weight silicone in plasma and blood of women who are or were exposed to silicone-gel-filled implants, and controls. However, D3–D6 were not detectable in plasma and blood of controls and even in some patients' plasma and blood. In contrast, D3 and D4 could be found unambiguously in most of the plasma and blood samples of women with silicone implants and even in blood of women who had undergone reimplantation up to 5 years ago. D3 varied between $6–12 \mu g$ L^{-1} (plasma) and $20–28 \mu g$ L^{-1} (blood), whereas the concentration range of D4 was $14–50 \mu g$ L^{-1} (plasma) and $79–92 \mu g$ L^{-1} (blood) (Figure 8.5).

Another potential for human exposure to cyclic siloxanes is by the respiratory route (antiperspirants, hair spray). Low levels (about 1 ppm) of D4, a material commonly found in personal care products, were detected in indoor air samples [238]. As hundreds of individuals are exposed to D4 at the workplace, an Industrial Hygiene Guideline of 5 to 10 ppm time-weighted average over an 8-hour shift has been established. The respiratory intake und uptake of D4 has been measured [96]. After 1 h exposure to 10 ppm D4 ($122 \mu g$ L^{-1}) the mean D4 intake of the volunteers was $137 \pm 25 mg$ (12 % of total exposure) exhibiting D4 plasma concentrations of $79 \pm 5 \mu g$ L^{-1}. The major routes

Table 8.2 Silicon content of human blood

Reference	Sample	Silicone implant patients (μg Si L^{-1})	Control group (μg Si L^{-1})
Marco-Franco et al. [234]	Serum		10–250
Roberts et al. [196]	Plasma		140
D'Haese et al. [235]	Serum		161
Bercowy et al. [236]	Blood		300–5000
Bercowy et al. [236]	Plasma		400–4000
Teuber et al. [237]	Serum	60–870	60–350
Jackson et al. [79]	Serum	60–340	60–720
Jackson et al. [79]	Plasma	70–340	210
Peters et al. [78]		39	25
		9–81[a]	
		10–121[b]	
Lugowski et al. [229]	Blood		39[a]
			38[b]
Flassbeck et al. [98]	Plasma	2–26[c]	n.d.
Flassbeck et al. [98]	Blood	n.d.–42[c]	n.d.

[a] Implantation of silicon breast implants before 1986 (second implant generation)
[b] Implantation of silicon breast implants after 1986 (third implant generation)
[c] Calculated silicon values based on analysis of cyclic siloxanes
n.d. Not detectable

for excretion were urine and expired air, the latter process being characterized by a half-life of $5\frac{1}{2}$ hours. No immunotoxic or proinflammatory effects of respiratory exposure of the volunteers to D4 were found on the basis of a panel of immune tests to screen for immunotoxicity and for an adjuvant effect [239].

Based on the described data, a comparison of different exposure paths can be established taking D4 concentrations in blood as a reference (Figure 8.5).

Experimental results plotted are typical static concentrations for implant patients (between graphs A_{min} and A_{max}) [98] and the concentration decline following the inhalation experiment (graph B) [96]. For comparison, workplace reference data are also included in the figure (graphs C and D): because of a lack of kinetic data, as a first approximation extrapolations were made linear, and no differentiation between blood and plasma were accounted for.

It can easily be seen that the D4 load caused by leaking implants may significantly exceed usual inhalation standards. Compared with these scenarios, inhalation of ambient air inside of domestic waste deposits seem to be of subordinate importance.

From Table 8.3 it seems to be clear that silicon content in body tissues in the vicinity of silicone implants are usually two to three orders of magnitude higher than comparable tissue samples from persons without implants. In a similar recent study in Essen, fat, muscular and capsular tissues contained 9 to 73 μg g^{-1} Si; extractable Si calculated from D4 to D6 determinations was substantially lower (20 to 1048 ng g^{-1} Si) [98].

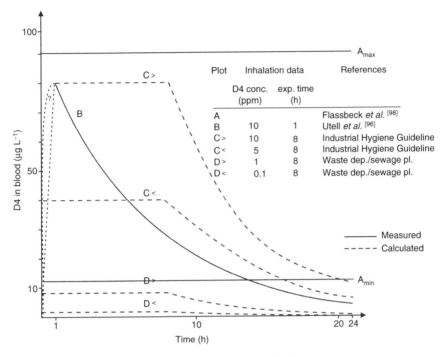

Figure 8.5   D4 in human blood (selected exposure scenarios)

8.5   SUMMARY AND OUTLOOK

Silicones are persistent materials at first instance for situations they were made for (e.g. PDMS elastomer and gum sealings in buildings). Driven mainly by analytical progress in elemental speciation, during the last two decades it became evident that, in general, PDMS exhibits limited stability and will be degraded within months to years in the environment and in contact with biological media. Particularly in air, soil and waste deposits, PDMS and VMS will be transformed to silanols and eventually be mineralized. Taking these observations into account, the listing of organosilicon compounds in the 'Grey List' of marine wastes by the London Treaty of 1972 [198] should be re-evaluated.

VMSs are the substances essentially responsible for the mobility of silicones in the environment and in organisms. Because of important properties like molecular size and structure, volatility or diffusional and lipophilic behaviour, these organosilicon compounds show similarities to metal(loid) organic species. Thus they may penetrate biological membranes (cell membranes, blood/brain barrier), and influence processes like biomethylation.

Table 8.3 Silicon content in breast tissue and capsular tissue of control group and patients with silicone gel breast implants

	Tissue	Si concentration (μg g^{-1})	
		Control group	Patients with implants
Evans et al. [213]	Capsular		370–14000
	Breast	0.5–6.8	
Evans et al. [213]	Breast		290–1600^d
			200–760^e
Lugowski et al. [229]	Capsular^a	0.235	25047^b
			13388^c
Schnur et al. [214]	Capsular		33–58800^d
			410–45532^e
	Breast	4.0–446.0	7–2995^d
			5–30333^e
Peters et al. [78]	Capsular		37–88703^d
			3235–65396^f
			1762–152387^e
	Breast	0.025–0.742	
Thomsen et al. [240]	Breast		6800^e
			5600^g
			1400^d

^{a} Extractable silicon content in tissue
^{b} Implantation of silicone-gel-filled implants before 1986
^{c} Implantation of silicone-gel-filled implants after1986
^{d} Intact silicone-gel-filled implants
^{e} Ruptured silicone-gel-filled implants
^{f} Silicone bleed
^{g} Possibly broken silicone-gel-filled implants

In respect to human health, the main problem is not that adverse health effects of silicones are proven but that, on the contrary, there does not exist a scientifically valid evaluation for the safety of these compounds. Concerning the latter, responsible care would imply that injection and implantation of these materials in human bodies should be avoided whenever possible.

Among the open questions to investigate further the environmental and health safety of silicones are topics of analytical, fundamental, and applied concern as listed.

Analytical research demand: What levels do low molecular weight silicones reach in various tissues with time, and can they exhibit toxicity or biological activity at these concentrations?

- Representative surveys of silicone levels in blood of women with implants and their children should be made, compared to controls. Studies on the effects of silicones on reproduction have not been done, the exposure level to babies is not known, and the effect of silicones on the developing fetus is not understood.

- Existence, concentrations as well as association and transformation of silanols *in vivo*—further data is required.

Fundamental research demand:
- Influence of silicones on environmentally important biological processes—further information is needed.
- Knowledge of the potential of D4 and other VMSs to enter intracellular compartments, and interact with enzymes and DNA.

Applied research demand:
- Proposal of a VMS toxicology based on threshold VMS levels in blood needs to be applied to various exposure scenarios in personal and workplace environments.

All these important questions need to be addressed before we can assess, for example, whether or not silicone breast implants truly pose a public health hazard. Also the European Parliament came to the conclusion that further research will be needed before the safety of implants can finally be evaluated (Report A5-0186/2001).

8.6   REFERENCES

1. Wedepohl, K.H. (Ed.) *Handbook of Geochemistry*, Vol. II-2, Springer-Verlag, Berlin, 1973.
2. Jarvie, A.W.P., In *Organometallic Compounds in the Environment—Principles and Reactions*, P.J. Craig (Ed.), Longman Group Ltd, Harlow, 1986, pp. 229–53.
3. Cavic-Vlasak, B.A., Thompson, M., and Smith, D.C., *Analyst*, 1996, **121**, 53R.
4. Kinrade, S.D., Gillson, A.M.E., Knight, C.T.G., *J. Chem. Soc. Dalton Trans.*, 2002, 307.
5. Smith, A.L., *The Analytical Chemistry of Silicones*, John Wiley & Sons, New York, 1991.
6. Carpenter, J.C., and Gerhards, R. In *Organosilicon Materials*, O. Hutzinger (Ed.), *Handbook of Environmental Chemistry* 3H, Springer-Verlag, Berlin, 1997, pp. 27–51.
7. Chandra, G., *Organosilicon Materials*, O. Hutzinger (Ed.), *Handbook of Environmental Chemistry*, Springer-Verlag, Berlin, 1997.
8. Powell, D.E., and Carpenter, J.C. In *Organosilicon Materials*, O. Hutzinger (Ed.), *Handbook of Environmental Chemistry*, Springer-Verlag, Berlin, 1997 pp. 225–39.
9. Allen, R.B., Kochs, P., and Chandra, G., In *Organosilicon Materials*, O. Hutzinger (Ed.), *Handbook of Environmental Chemistry*, Springer-Verlag, Berlin, 1997 pp. 1–25.
10. Richter, R., Roewer, G., Böhme, U., Busch, K., Babonneau, F., Martin, H.P., Müller, E., *Appl. Organomet. Chem.*, 1997, **11**, 71.
11. Fendinger, N.J., Lehmann, R.G., Mihaich, E.M., In *Organosilicon Materials*, O. Hutzinger (Ed.), *Handbook of Environmental Chemistry*, Springer-Verlag, Berlin, 1997, pp. 181–223.
12. Oldfield, D., and Symes, T. *Polymer Testing*, 1996, **15**, 115.

13. Moretto, H.H., Schulze, M., and Wagner, G. In *Ullmann's Encyclopedia of Industrial Chemistry*, B. Elvers, S. Hawkins, W. Russey, and G. Schultz (Eds) Verlag Chemie, Weinheim, 1993, pp. 57–93.

14. LeVier, R.R., Chandler, M.L., and Wendel, S.R. In *Biochemistry of Silicones and Related Problems*, G. Bendz and I. Lindquist (Eds), Plenum Press, New York, 1978, pp. 473–513.

15. Pawlenko, S., *Organosilicon Chemistry*, Walter de Gruyter, New York, 1986.

16. Chandra, G., *Text Chem. Color*, 1995, **27**, 21.

17. Lane, T.H., and Burns, S.A. In *Immunology of Silicones*, M. Potter, and N.R. Rose (Eds) Springer-Verlag, Berlin, 1996, pp. 4–23.

18. Carpenter, J.C., Cella, J.A., and Dorn, S.B., *Environ. Sci. Technol.*, 1995, **29**, 864.

19. Lehmann, R.G., Varaprath, S., and Frye, C.L., *Environ. Toxicol. Chem.*, 1994, **13**, 1061.

20. Lehmann, R.G., Varaprath, S., and Frye, C.L., *Environ. Toxicol. Chem.*, 1994, **13**, 1753.

21. Lehmann, R.G., Varaprath, S., Annelin, R.B., and Arndt, J.L., *Environ. Toxicol. Chem.*, 1995, **14**, 1299.

22. Lehmann, R.G., Frye, C.L., Tolle, D.A., and Zwick, T.C., *Wat. Air Soil Pollut.*, 1996, **87**, 231.

23. Lehmann, R.G., Miller, J.R., Xu, S., Singh, U.B., and Reece, C.F., *Environ. Sci. Technol.*, 1998, **32**, 1260.

24. Xu, S., Lehmann, R.G., Miller, J.R., and Chandra, G., *Environ. Sci. Technol.*, 1998, **32**, 1199–206.

25. Annelin, R.B., and Frye, C.L., *Sci. Tot. Environ.*, 1989, **83**, 1.

26. Bruggemann, W.A., Weber-Fung, D., Opperhuizen, A., Van der Steen, J., Wijbenga, A., and Hutzinger, O., *Toxicol. Environ. Chem.*, 1984, **7**, 287.

27. Opperhuizen, A., Damen, H.W.J., Asyee, G.M., Van der Stehen, J.M.D., *Toxicol. Environ. Chem.*, 1987, **13**, 265.

28. Firmin, R., and Raum, A.L.J., *Rev. Int. Océanogr. Méd.*, 1987, **85–86**, 46.

29. Hobson, J.F., Atkinson, R., Carter, W.P.L. In *Organosilicon Materials*, O. Hutzinger (Ed.) *Handbook of Environmental Chemistry*, Springer-Verlag, Berlin, 1997, pp. 137–79.

30. Hobson, J., and Silberhorn, E., *Environ. Toxicol. Chem.*, 1995, **14**, 1667.

31. Kent, D., Fackler, P., Hartley, D., and Hobson, J., *Environ. Tox. Water Qual.*, 1996, **11**, 145.

32. Fackler, P., Dionne, E., Hartley, D., and Hamelink, J., *Environ. Toxicol. Chem.*, 1995, **14**, 1649.

33. Mazzoni, S.M., Roy, S., and Grigoras, S. In *Organosilicon Materials*, O. Hutzinger (Ed.) *Handbook of Environmental Chemistry*, Springer-Verlag, Berlin, 1997, pp. 53–81.

34. Spivack, J.L., Pohl, E.R., and Kochs, P. In *Organosilicon Materials*, O. Hutzinger (Ed.) *Handbook of Environmental Chemistry*, Springer-Verlag, Berlin, 1997, pp. 105–35.

35. Buch, R.R., Lane, T.H., Annelin, R.B., and Frye, C.L., *Environ. Toxicol. Chem.*, 1984, **3**, 215.

36. Sommerlade, R., Parlar, H., Wrobel, D., and Kochs, P., *Environ. Sci. Technol.*, 1993, **27**, 2435.

37. Siebert, F., *Verteilung von Siliconen in der Umwelt*, Dissertation, University of Heidelberg, 1988.

38. Pellenbarg, R.E., *Environ. Sci. Technol.*, 1979, **13**, 565.

39. Parker, R.D., *Fresenius J. Anal. Chem.*, 1978, **292**, 362.

40. Watanabe, N., Nagase, H., and Ose, Y., *Sci. Tot. Environ.*, 1988, **73**, 1.

41. Fendinger, N.J., McAvoy, D.C., Eckhoff, W.S., and Price, B.B., *Environ. Sci. Technol.*, 1997, **31**, 1555.
42. Lugowski, S.L., Smith, D.C., Lugowski, J.Z., Peters, W., and Semple, J. *Fresenius J. Anal. Chem.*, 1998, **360**, 486.
43. Klemens, P., and Heumann, K.G., *Fresenius J. Anal. Chem.*, 2001, **371**, 758.
44. Martin, P., Ellersdorfer, E., and Zeman, A., *Korrespondenz Abwasser*, 1996, **43**, 1574.
45. Spivack, J.L., and Dorn, S.B., *Environ. Sci. Technol.*, 1994, **28**, 2345.
46. Wörmann H, Stuth J, Schulz J 1993 *GIT Fachz Lab* **37**: 400–3
47. Huppmann R, Lohoff H.W., Schröder, H.F., 1996 *Fresenius J. Anal Chem* **354**: 66–71
48. Mahone, L.G., Garner, P.J., Buch, R.R., Lane, T.H., Tatera, J.F., Smith, R.C., and Frye, C.L., *Environ. Toxicol. Chem.*, 1983, **2**, 307.
49. Kleinert, J.C., and Weschler, C.J., *Anal. Chem.*, 1980, **52**, 1245.
50. Pellenbarg, R.E., Bellama, J.M., and Meyer, S.R., *Appl. Organomet. Chem.*, 1991, **5**, 79.
51. Becker, M.A., and Steinmeyer, R.D. In *The Analytical Chemistry of Silicones*, A.L. Smith (Ed.) John Wiley & Sons, New York, 1991, pp. 255–303.
52. Hausler, D.W., and Taylor, L.T., *Anal. Chem.*, 1981, **53**, 1223.
53. Hausler, D.W., and Taylor, L.T., *Anal. Chem.*, 1981, **53**, 1227.
54. Forbes, K.A., Vecchiarelli, J.F., Uden, P.C., and Barnes, R.M., *Anal. Chem.*, 1990, **62**, 2033.
55. Vincenti, M., Pelizzetti, E., Guarini, A., and Costanzi, S. *Anal. Chem.*, 1992, **64**, 1879.
56. Morrissey, M.A., Siems, W.F., and Hill, H.H. Jr, *J. Chromatogr.*, 1990, **505**, 215.
57. Verhelst, V., and Vandereecken, P., *J. Chromatogr.*, 2000, **A 871**, 269.
58. Merten, H., and Richter, M., *Chem. Lab. Biotech.*, 1996, **47**, 170.
59. Varaprath, S., and Lehmann, R.G. *Environ. Poly. Degrad.*, 1997, **5**, 17.
60. Pelizzari, E.D., Bunch, J.E., Berkley, R.E., and McRae, J., *Anal. Chem.*, 1976, **48**, 803.
61. Grümping, R., Mikolajczak, D., and Hirner, A.V., *Fresenius J. Anal. Chem.*, 1998, **361**, 133.
62. Schweigkofler, M., and Niessner, R. *Environ. Sci. Technol.*, 1999, **33**, 3680.
63. Lehmann, R.G., and Miller, J.R., *Environ. Toxicol. Chem.*, 1996, **15**, 1455.
64. Cassidy, R.M., Hurteau, M.T., Mislan, J.P., and Ashley, R.W., *J. Chromatogr. Sci.*, 1976, **14**, 444.
65. Boorn, A.W., and Browner, R.F., *Anal. Chem.*, 1982, **54**, 1402.
66. Al-Rashdan, A., Heitkemper, and D., Caruso, J.A., *J. Chromatogr. Sci.*, 1991, **29**, 98.
67. Gast, C.H., Kraak, J.C., Poppe, H., and Haessen, F.J.M.J., *J. Chromatogr.*, 1979, **185**, 549.
68. Heine, D.R., Denton, M.B., and Schlabach, T.D., *J. Chromatogr. Sci.*, 1985, **23**, 454.
69. Irgolic, K.J., Stockton, R.A., and Chakraborti, D., *Spectrochim. Acta.*, 1983, **38B**, 437.
70. Roychowdhury, S.B., and Koropchak, J.A., *Anal. Chem.*, 1990, **62**, 484.
71. Weber, G., and Berndt, H., *Chromatographia.*, 1990, **29**, 594.
72. Dunemann, L., and Begerow, J., *Kopplungstechniken zur Elementspeziesanalytik*, Verlag Chemie, Weinheim, 1995.
73. Tarr, M.A., Zhu, G., and Browner, R.F. *Anal. Chem.*, 1993, **65**, 1689.
74. Gjerde, D.T., Wiederin, D.R., Smith, F.G., and Mattson, B.M., *J. Chromatogr.*, 1993, **640**, 73.
75. Dorn, S.B., Skelly-Frame, E.M., *Analyst.*, 1994, **119**, 1687.

76. Grümping, R., and Hirner, A.V., *Fresenius J. Anal. Chem.*, 1999, **363**, 347.

77. Feldmann, J., *Erfassung flüchtiger Metall-und Metalloidverbindungen in der Umwelt mittels GC/ICP-MS*, Dissertation, University of Essen, 1995.

78. Peters, W., Smith, D., Lugowski, S., McHugh, A., and Baines, C., *Ann. Plast. Surg.*, 1995, **34**, 343.

79. Jackson, L.W., Dennis, G.J., and Centeno, J.A. In *Metal Ions in Biology and Medicine*, Vol. 5, Ph. Collery, V.N. deBraetter, L. Khassanova, and J.C. Etienne (Eds), John Libbey Eurotext, Paris, 1998, pp. 33–38.

80. Kennan, J.J., McCann Breen, L.L., Lane, T.H., and Taylor, R.B., *Anal. Chem.*, 1999, **71**, 3054.

81. Lytle, N.W. In *The Analytical Chemistry of Silicones* A.L. Smith (Ed.), Wiley, New York, 1991, pp. 471–83.

82. Austin, J.H. In *Biochemistry of Silicon and Related Problems*, G. Bendz, and I. Lindqvist (Eds), Plenum Press, New York, 1978, pp. 255–68.

83. Wickham, M.G., Rudolph, R., and Abraham, J.L. *Science*, 1978, **199**, 437.

84. Silver, R.M., Sahn, E.E., Allen, J.A., Sahn, S., Greene, W., Maize, J.C., and Garen, P.D., *Arch. Dermatol.*, 1993, **129**, 63.

85. Greene, W.B., Raso, D.S., Walsh, L.G., Harley, R.A., and Silver, R.M. *Plast. Reconstr. Surg.*, 1995, **95**, 513.

86. delRosario, A.D., Bui, H.X., Petrocine, S., Sheehan, C., Pastore, J., and Ross, J.S. *Ultrastruct. Pathol.*, 1995, **19**, 83.

87. Raimondi, M.L., Sassara, C., Bellobono, I.R., and Matturi, L., *J. Biomed. Mater. Res.*, 1995, **29**, 59.

88. Steinmeyer, R.D. In *Analytical Chemistry of Silicones*, A.L. Smith (Ed.), Wiley, New York, 1991, pp. 255–81.

89. Smith, A.L., and Parker, R.D. In *Analytical Chemistry of Silicones*, A.L. Smith (Ed.) Wiley, New York, 1991, pp. 71–95.

90. Varaprath, S., Seaton, M., McNett, D., Cao, L., and Plotzke, K.P., *Intern. J. Environ. Anal. Chem.*, 2000, **77**, 203.

91. Moore, J.A., and Bujanowski, V.J. In *Analytical Chemistry of Silicones*, A.L. Smith (Ed.), Wiley, New York, 1991, pp. 82–4.

92. Wechsler, C.J., *Sci. Tot. Environ.*, 1988, **73**, 53.

93. Kala, S.V., Lykissa, E.D., and Lebovitz, R.M., *Anal. Chem.*, 1997, **69**, 1267.

94. Kala, S.V., Lykissa, E.D., Neely, M.W., and Lieberman, M.W., *Am. J. Pathol.*, 1998, **152**, 645.

95. Varaprath, S., Salyers, K.L., Plotzke, K.P., and Nanavati, S., *Anal. Biochem.*, 1998, **256**, 14.

96. Utell, M.J., Gelein, R., Yu, C.P., Kenaga, C., Geigel, E., Torres, A., Chalupa, D., and Gibb, F.R., *Toxicol. Sci..*, 1998, **44**, 206.

97. Vessman, J., Hammar, K.G., Lindeke, B., Strömberg, S., LeVier, R., Robinson, R., Spielvogel, D., and Hanneman, L. In *Biochemistry of Silicones and Related Problems*, G. Bendz and I. Lindquist (Eds), Plenum Press, New York, 1978, pp. 535–60.

98. Flassbeck, D., Pfleiderer, B., Grümping, R., and Hirner, A.V., *Anal. Chem.*, 2001, **73**, 606.

99. Baker, J.L., LeVier, R.R., and Spielvogel, D.E., *Plast. Reconstr. Surg.*, 1982, **69**, 56.

100. Taylor, R.B., and Parbhoo, B., and Fillmore, D.M. In *Analytical Chemistry of Silicones*, A.L. Smith (Ed.), Wiley, New York, 1991, pp. 347–419.

101. Garrido, L., Bogdanova, A., Cheng, L.L., Pfleiderer, B., Tokareva, E., Ackerman, J.L., and Brady, T.J. In *Immunology of Silicones* M. Potter and N.R. Rose (Eds), Springer-Verlag, Berlin, 1996, pp. 49–58.

102. Garrido, L., Pfleiderer, B., Jenkins, B.G., Hulka, C.A., and Kopans, D.B., *Magn. Reson. Med.*, 1994, **31**, 328.

103. Taylor, R.B., and Kennan, J.J., *Magn. Reson. Med.*, 1996, **36**, 498.
104. Macdonald, P., Plavac, N., Peters, W., Lugowski, S., and Smith, D., *Anal. Chem.*, 1995, **67**, 3799.
105. Garrido, L., Pfleiderer, B., Jenkins, B.G., Hulka, C.A., and Kopans, D.B., *Magn. Reson. Med.*, 1998, **39**, 689.
106. Garrido, L., Pfleiderer, B., Papisov, M., and Ackerman, J.L., *Magn. Reson. Med.*, 1993, **29**, 839.
107. Pfleiderer, B., Xu, P., Ackerman, J.L., and Garrido, L. *J. Biomed. Mater. Res.*, 1995, **29**, 1129.
108. Pfleiderer, B., Ackerman, J.L., and Garrido, L., *Magn. Reson. Med.*, 1993, **30**, 534.
109. Abraham, J.L., and Etz, E.S., *Science.*, 1979, **206**, 716.
110. Deng, X.M., Castillo, E.J., and Anderson, J.M., *Biomat.*, 1986, **7**, 247.
111. Seifert, L.M., and Greer, R.T., *J. Biomed. Mater. Res.*, 1985, **19**, 1043.
112. Kennedy, J.H., Ishida, H., Staikoff, L.S., and Lewis, C.W., *Biomater. Med. Dev. Art. Org.*, 1978, **6**, 215.
113. Frank, C.J., McCreery, R.L., Redd, D.C.B., and Gansler, T.S., *Appl. Spectrosc.*, 1993, **47**, 387.
114. Hardt, N.S., Yu, L.T., LaTorre, G., and Steinbach, B., *Mod. Pathology.*, 1994, **7**, 669.
115. Kidder, L.H., Kalasinsky, V.F., Luke, J.L., Levin, I.W., and Lewis, E.N., *Nat. Med. (NY).*, 1997, **3**, 235.
116. Ali, S.R., Johnson, F.B., Luke, J.L., and Kalasinsky, V.F., *Cell. Mol. Biol.*, 1998, **44**, 75.
117. Haycox, C.L., Leach-Scampavia, D., Olerud, J.E., and Ratner, B.D., *J. Am. Acad. Dermatol.*, 1999, **40**, 719.
118. Wechsler, C.J., *Atmos. Environ.*, 1981, **15**, 1365.
119. Navarro, A., Rosell, A., Villanueva, J., and Grimalt, J.O., *Chemosphere.*, 1991, **22**, 913.
120. Pellenbarg, R.E., *Mar. Pollut. Bull.*, 1982, **13**, 427.
121. Powell, D.E., Annelin, R.B., and Gallavan, R.H., *Environ. Sci. Technol.*, 1999, **33**, 3706.
122. Pellenbarg, R.E., DeCarlo, E.C., Boyle, M.E., and Lamontagne, R.A. *Appl. Organomet. Chem.*, 1997, **11**, 345.
123. Pellenbarg, R.E., and Tevault, D.E., *Environ. Sci. Technol.*, 1986, **20**, 743.
124. Wang, X.M., Lee, S.C., Sheng, G.Y., Chan, L.Y., Fu, J.M., Li, X.D., Min, Y.S., and Chan, C.Y., *Appl. Geochem.*, 2000, **16**, 1447.
125. Mueller, J.A., DiToro, D.M., and Maiello, J.A. *Environ. Toxicol. Chem.*, 1995, **14**, 1657.
126. Atkinson, R., *Environ. Sci. Technol.*, 1991, **25**, 863.
127. Franke, S., Hildebrandt, S., Schwarzbauer, J., Link, M., and Francke, W., *Fresenius J. Anal. Chem.*, 1995, **353**, 39.
128. Schröder, H.F., *Biochemisch schwer abbaubare organische Stoffe in Abwässern und Oberflächenwässern*, Habilitation Thesis, Technische Hochschule, Aachen, 1997.
129. Ahrendt, G., and Kohl, E.G., *Trierer Berichte zur Abfallwirtschaft*, Vol. 9, G. Rettenberger (Ed.), Economica Verlag, Bonn, 1996, pp. 9–20.
130. McCarthy, T.M., *Wasser-Abwasser*, 1998, **139**, 204.
131. Schweigkofler, M., and Niessner, R., *J. Hazard. Mat.*, 2001, **B83**, 183.
132. Engström, K., Nyman, L., Haikola, M., and Saarni, H. In *Healthy Buildings '88, International Conference 3*, Swedish Council for Building Research (Ed.), Stockholm, 1988, pp. 333–37.
133. Ezeonu, I.M., Price, D.L., Simmons, R.B., Crow, S.A., and Ahearn, D.G., *Appl. Environ. Microbiol.*, 1994, **60**, 4172.

134. Krotoszynski, B.K., and O'Neill, H.J., *J. Environ. Sci. Health: Environ. Sci. Engin.*, 1982, **A17**, 855.
135. Hobbs, E.J., Keplinger, M.L., and Calandra, J.C., *Environ. Res.*, 1975, **10**, 397.
136. Matsui, S., Murakami, T., Sasaki, T., Hirose, Y., and Iguana, Y., *Prog. Wat. Tech.*, 1975, **7**, 645.
137. Lehmann, R.G., Smith, D.M., Narayan, R., Kozerski, G.E., and Miller, J.R., *Compost Sci. Utilization*, 1999, **7**, 72.
138. Buch, R.R., and Ingebrigtson, D.N., *Environ. Sci. Technol.* 1979, **13**, 676.
139. Stevens, C., *J. Inorg. Biochem.*, 1998, **69**, 203.
140. Griessbach, E.F.C., and Lehmann, R.G., *Chemosphere*, 1999, **38**, 1461.
141. Xu, S., *Environ. Sci. Technol.*, 1998, **32**, 3162.
142. Lehmann, R.G., Miller, J.R., and Kozerski, G.E., *Chemosphere*, 2000, **41**, 743.
143. Abe, Y., Butler, G.B., and Hogen-Esch, T.E. *J. Macromol. Sci. Chem.*, 1981, **A16**, 461.
144. Latimer, H.K., Kamens, R.M., and Chandra, G., *Chemosphere* 1998, **36**, 2401.
145. Markgraf, S.J., and Wells, J.R., *Int. J. Chem. Kinet.*, 1997, **29**, 445.
146. Parker, W.J., Shi, J., Fendinger, N.J., Monteith, H.D., and Chandra, G., *Environ. Toxicol. Chem.*, 1999, **18**, 172.
147. Grümping, R., Michalke, K., Hirner, A.V., and Hensel, R., *Appl. Environ. Microbiol.*, 1999, **65**, 2276.
148. Xu, S., *Environ. Sci. Technol.*, 1999, **33**, 603.
149. Anderson, C., Hochgeschwender, K., Weidemann, H., and Wilmes, R., *Chemosphere*, 1987, **16**, 2567.
150. Chandra, G., Maxim, L.D., and Sawano, T. In *Organosilicon Materials*, O. Hutzinger (Ed.), *Handbook of Environmental Chemistry*, Springer-Verlag, Berlin, 1997, pp. 295–319.
151. Sabourin, C.L., Carpenter, J.C., Leib, K., and Spivack, J.L., *Appl. Environ. Microbiol.*, 1996, **62**, 4352.
152. Sabourin, C.L., Carpenter, J.C., Leib, K., and Spivack, J.L., *Environ. Toxicol. Chem.*, 1999, **18**, 1913.
153. Cella, J.A., and Carpenter, J.C. *J. Organomet. Chem.*, 1994, **480**, 23.
154. Craig, P.J. (Ed.), *Organometallic Compounds in the Environment—Principles and Reactions*, Longman Group Ltd, Harlow, 1986.
155. Nagase, H., Ose, Y., and Sato, T., *Sci. Tot. Environ.*, 1988, **73**, 29.
156. Wickenheiser, E.B., Michalke, K., Hirner, A.V., Hensel, R., and Flassbeck, D. In *Metal Ions in Biology and Medicine*, Vol. 6, J.A. Centeno, Ph. Collery, G. Vernet, R.B. Finkelman, H. Gibb, and J.C. Etienne (Eds), John Libbey Eurotext, Paris, 2000, pp. 120–22.
157. Feldmann, J., Krupp, E.M., Glindemann, D., Hirner, A.V., and Cullen, W.R., *Appl. Organomet. Chem.*, 1999, **13**, 1.
158. Gadd, G., *FEMS Microbiol. Rev.*, 1993, **11**, 297.
159. Wasserbauer, R., and Zadák, Z., *Folia Microbiol.*, 1990, **35**, 384.
160. Baldassarri, L., Gelosia, A., Fiscarelli, E., Donelli, G., Mignozzi, M., and Rizzoni, G., *J. Mat. Sci. Mat. Med.*, 1994, **5**, 601.
161. Doebke, M.K., Svahn, J.K., Vastine, V.L., Landon, B.N., Stein, P.C., and Parsons, C.L., *Ann. Plast. Surg.*, 1995, **34**, 563.
162. Eerenstein, S.E.J., Grolman, W., and Schouwenburg, P.F., *Clin. Otolaryngol.*, 1999, **24**, 398.
163. Neu, T.R., Van der Mei, H.C., Busscher, H.J., Dijk, F., and Verkerke, G.J., *Biomater.*, 1993, **14**, 459.
164. Heinen, W., *Biochemistry of Silicon and Related Problems—Proceedings of the Noble Foundation Symposium*, Vol. 40, Plenum Press, New York, 1978, pp. 129–46.

165. Yanagi, M., and Onishi, G., *J. Soc. Cosmet. Chem.*, 1971, **22**, 851.

166. Wolf, C.J., Brandon, H.J., Young, V.L., Jerina, K.L., and Srivastava, A.P. In *Immunology of Silicones*, M. Potter and N.R. Rose (Eds), Springer-Verlag, Berlin, 1996, pp. 25–37.

167. Raso, D.S., and Greene, W.B., *Ultrastruct. Pathol.*, 1997, **21**, 263.

168. Hardt, N.S., Emery, J.A., Steinbach, B.G., Latorre, G., and Caffee, H., *Int. J. Occup. Med. Toxicol.*, 1995, **4**, 127.

169. Pfleiderer, B., Moore, A., Tokareva, E., Ackerman, J.L., and Garrido, L., *Biomat.*, 1999, **20**, 561.

170. Ben-Hur, N., Ballantyne, D.L., Rees, T.D., and Seidman, I., *Plast. Reconstr. Surg.*, 1967, **39**, 423.

171. Capozzi, A., DuBou, R., and Pennisi, V.R., *Plast. Reconstr. Surg.*, 1978, **62**, 302.

172. Eller, A.W., Friberg, T.R., and Mah, F. *Am. J. Ophthalmol.*, 2000, **129**, 685.

173. Ni, C., Wang, W.J., Albert, D.M., and Schepens, C.L., *Arch. Ophthalmol.*, 1983, **101**, 1399.

174. Stevens, C., and Annelin, R.B. In *Organosilicon Materials*, O. Hutzinger (Ed.), *Handbook of Environmental Chemistry*, Springer-Verlag, Berlin, 1997, pp. 83–103.

175. Potter, M., and Rose, N.R. (Eds) *Immunology of Silicones*, Springer-Verlag, Berlin, 1996.

176. Marcus, D.M., *Arthritis Rheumat.*, 1996, **39**, 1619.

177. Kossovsky, N., *Appl. Organomet. Chem.*, 1997, **11**, 353.

178. Reller, A., Braungart, M., Soth, J., and von Uexküll, O., *Gaia*, 2000, **9**, 13.

179. Palazzolo, R.J., McHard, J.A., Hobbs, E.J., Fancher, O.E., and Calandra, J.C., *Toxicol. Appl. Pharmacol.*, 1972, **21**, 15.

180. Sousa, J.V., McNamara, P.C., Putt, A.E., Machado, M.W., Surprenant, D.C., Hamelink, J.L., Kent, D.J., Silberhorn, E.M., and Hobson, J.F., *Environ. Toxicol. Chem.*, 1995, **14**, 1639.

181. Kent, D.J., McNamara, P.C., Putt, A.E., Hobson, J.F., and Silberhorn, E.M., *Ecotoxicol. Environ. Safety*, 1994, **29**, 372.

182. Felix, K., Janz, S., Pitha, J., Williams, J.A., Mushinski, E.B., Bornkamm, G.W., Potter, M. In *Immunology of Silicones* M. Potter and N.R. Rose (Eds), Springer-Verlag, Berlin, 1996, pp. 93–9.

183. Lieberman, M.W., Lykissa, E.D., Barrios, R., Ou, C.N., Kala, G., and Kala, S.V., *Environ. Health Perspect.*, 1999, **107**, 161.

184. Hayden, J.F., and Barlow, S.A., *Toxicol. Appl. Pharmacol.*, 1972, **21**, 68.

185. Teuber, S.S., Yoshida, S.H., and Gershwin, E., *West. J. Med.*, 1995, **162**, 418.

186. Brawer, A.E., *Medical Hypotheses*, 1998, **51**, 27.

187. Doeltz, M.K., Mackie, M., Rich, P.A., Lent, D., Sigman, C.C., and Helmes, C.T. *J. Environ. Sci. Health*, 1984, **A19**, 27.

188. Vergnes, J.S., Jung, R., Thakur, A.K., Barfknecht, T.R., and Reynolds, V.L.L., *Environ. Mol. Mutagen.*, 2000, **36**, 13.

189. Whitlock, P.W., Clarson, S.C., and Retzinger, G.S., *J. Biomed. Mater. Res.*, 1999, **45**, 55.

190. Semanowitz, V.J., and Orentreich, N., *J. Dermatol. Surg. Oncol.*, 1977, **3**, 597.

191. Kossovsky, N., Heggers, J.P., and Robson, M.C. *J. Biomed. Mat. Res.*, 1987, **21**, 1125.

192. Sun, L., Alexander, H., Lattarulo, N., Blumenthal, N.C., Ricci, J.L., and Chen, G., *Biomat.*, 1997, **18**, 1593.

193. Wrobel, K., Gonzales, E.B., and Sanz-Medel, A., *Analyst*, 1995, **120**, 809.

194. Kossovsky, N., Zeidler, M., and Chun, G., *J. Appl. Biomater.*, 1993, **4**, 281.

195. Batich, C., DePalma, D., Marotta, J., Latoree, G., and Hardt, N. In *Immunology of Silicones*, M. Potter and N.R. Rose (Eds), Springer-Verlag, Berlin, 1996, pp. 13–23.

196. Roberts, N.B., and Williams, P., *Clin. Chem.*, 1990, **36**, 1460.
197. Hokosawa, S., and Yoshida, O., *Int. Urol. Nephrol.*, 1990, **22**, 373.
198. *www.londonconvention.org/London_Convention.htm*
199. Miyakawa, Y. In *Organosilicon Materials*, O. Hutzinger, (Ed.), *Handbook of Environmental Chemistry*, Springer-Verlag, Berlin, 1997, pp. 283–93.
200. Wischer, D., and Stevens, C. In *Organosilicon Materials*, O. Hutzinger (Ed.), *Handbook of Environmental Chemistry*, Springer-Verlag, Berlin, 1997, pp. 267–81.
201. Bartzoka, V., McDermott, M.R., and Brook, M.A., *Adv. Mater.*, 1999, **11**, 257.
202. Cook, R.R., and Perkins, L.L. In *Immunology of Silicones*, M. Potter and N.R. Rose (Eds), Springer-Verlag, Berlin, 1996, pp. 419–25.
203. Peters, W., *Can. J. Plast. Surg.*, 2000, **8**, 54.
204. Lykissa, E.D., Kala, S.V., Hurley, J.B., and Lebovitz, R.M., *Anal. Chem.*, 1997, **69**, 4912.
205. Bergman, R.B., and van der Ende, A.E., *Brit. J. Plast. Surg.*, 1979, **32**, 31.
206. Yu, L.T., LaTorre, G., Marotta, J., Batich, C., and Hardt, N.S., *Plast. Reconstr. Surg.*, 1996, **97**, 756.
207. Kumangai, V., Shiokawa, Y., Medsger, T.A., and Rodan, G.P., *Arthritis Rheumat.*, 1984, **27**, 1.
208. Barker, D.E., Retsky, M.I., and Schultz, S., *Plast. Reconstr. Surg.*, 1978, **61**, 836.
209. Dalal, M., Cooper, M., and Munnoch, D.A., *Plast. Reconstr. Surg.*, 2000, **105**, 2270.
210. Chen, N.T., Butler, P.F., Hooper, D.C.C., and May, J.W., Jr, *Ann. Plast. Surg.*, 1996, **36**, 337.
211. Niazi, Z.B., Salzberg, C.A., and Petro, J.A., *Plast. Reconstr. Surg.*, 1996, **98**, 1323.
212. Peters, W., Smith, D., Lugowski, S., McHugh, A., MacDonald, P., and Baines, C. In *Immunology of Silicones*, M. Potter and N.R. Rose (Eds), Springer-Verlag, Berlin, 1996, pp. 39–48.
213. Evans, G.R.D., Netscher, D.T., Schustermann, M.A., Kroll, S.S., Robb, G.L., Reece, G.P., and Miller, M.J., *Plast. Reconstr. Surg.*, 1996, **97**, 1207.
214. Schnur, P.L., Weinzweig, J., Harris, J.B., Moyer, T.P., Pelty, P.M., Nixon, D., and McConell, J.P., *Plast. Reconstr. Surg.*, 1996, **98**, 798.
215. Beekman, W.H., Feitz, R., van Diest, P.J., and Hage, J.J., *Ann. Plast. Surg.*, 1997, **38**, 441.
216. Winding, O., Christensen, L., Thomsen, J.L. *et al.*, *Scand. J. Plast. Reconstr. Surg.*, 1988 **22**, 127.
217. Hunt, J., Farthing, M.J., Baker, L.R., and Crocker, P.R., *Gut*, 1989, **30**, 239.
218. James, S.E., Tarr, G., Butterworth, M.S., McCarthy, J., and Butler, P.E., *J. R. Soc. Med.*, 2001, **94**, 133.
219. Pitanguy, I., Mayer, B., Mariz, S., and Salgado, F., *Laryng. Rhinol. Otol.*, 1988, **67**, 72.
220. Zimmermann, S.M., *Silicone Survivors*, Temple University Press, Philadelphia, 1998.
221. Kessler, D.A., *New Engl. J. Med.*, 1992, **326**, 1713.
222. Angell, M., *Science on Trial: The Clash of Medical Evidence and the Law in the Breast Implant Case*, W.W. Norton, New York, 1996.
223. *http://ts-tech.com/sbi-info*
     www.implantinfo.com
     http://199.45.69.176/tony/keeling/index.html
224. Brinton, L.A., Lubin, J.H., Burich, M.C., Colton, T., Brown, S.L., and Hoover, R.N., *Ann. Epidemiol.*, 2001, **11**, 248.
225. Hochber, M.C., and Perlmutter, D., *Curr. Top. Microbiol. Immunol.*, 1995, **210**, 411.

226. Perkins, L.L., Clark, B.D., Klein, P.J., and Cook, R.R., *Ann. Plast. Surg.*, 1995, **35**, 561.
227. Wang, O., *Regul. Toxicol. Pharmacol.*, 1996, **23**, 74.
228. Hatcher, J.A., and Slater, G.S. In *Organosilicon Materials*, O. Hutzinger (Ed.), *Handbook of Environmental Chemistry*, Springer-Verlag, Berlin, 1997, pp. 241–66.
229. Lugowski, S.L., Smith, D.C., Bonek, H., Lugowski, J.Z., Peters, W., and Semple, J., *J. Trace Elem. Med. Biol.*, 2000, **14**, 31.
230. Belliveau, J.F., Griffiths, W.C., Wright, G.C., and Tucci, J.R., *Ann. Clin. Lab. Sci.*, 1991, **21**, 328.
231. Peters, W., Smith, D., and Lugowski, S., *Ann. Plast. Surg.*, 1999, **43**, 324.
232. Semple, J.L., Lugowski, S.J., Baines, C.J., Smith, D.C., and McHugh, A., *Plast. Reconstr. Surg.*, 1998, **102**, 528.
233. Lugowski, S.L., Smith, D.C., Bonek, H., Semple, J., and Peters, W. In *Metal Ions in Biology and Medicine*, Vol. 6, J.A. Centeno, Ph. Collery G. Vernet, R.B. Finkelman, H. Gibb and J.C. Etienne (Eds), John Libby Eurotext, Paris, 2000, pp. 422–24.
234. Marco-Franco, J.E., Torres, V.E., Nixon, D.E., Wilson, D.M., James, E.M., Bergstralh, E.J., and McCarthy, J.T., *Clin. Nephrol.*, 1991, **35**, 52.
235. D'Haese, P.C., Shaheen, F.A., Huraib, S.O., Djukanovic, L., Polenkovic, M.H., Spasovski, G., Shikole, A., Schurgers, M.L., Daneels, R.F., Lamberts, L.V., Van Landeghem, G.F., and de Broe, M.E., *Nephrol.*, 1995, **10**, 1838.
236. Bercowy, G.M., Vo, H., and Rieders, F., *J. Anal. Toxicol.*, 1994, **16**, 46.
237. Teuber, S.S., Saunders, R.L., Halpern, G.M., Brucker, R.F., Conte, V., Goldman, B.D., Winger, E.E., Wood, W.G., and Gershwin, M.E., In *Immunology of Silicones*, M. Potter and N.R. Rose (Eds), Springer-Verlag, Berlin, 1996, pp. 59–65.
238. Shields, H.C., Fleischer, D.M., and Weschler, C.J., *Indoor Air*, 1996, **6**, 2.
239. Looney, R.J., Frampton, M.W., Byam, J., Kenaga, C., Speers, D.M., Cox, C., Mast, R.W., Klykken, P.C., Morrow, P.E., and Utell, M.J., *Toxicol. Sci.*, 1998, **44**, 214.
240. Thomsen, J.L., Christensen, L., Nielsen, M., Brandt, B., Breiting, V.B., Felby, S., and Nielsen, E., *Plast. Reconstr. Surg.*, 1990, **85**, 38.

9 Other Organometallic Compounds in the Environment

JÖRG FELDMANN

Department of Chemistry, University of Aberdeen, Scotland

9.1 ORGANOBISMUTH COMPOUNDS IN THE ENVIRONMENT

9.1.1 INTRODUCTION AND TERMINOLOGY

Bismuth is the heaviest element in group 15 of the periodic table and has chemical similarities to antimony and arsenic. It is a metal whereas the other elements are metalloids. Bismuth occurs in oxidation states of +III and +V of which +III is the more stable form (lone electron pair effect). This is in contrast to arsenic and also antimony, which occur in the environment mostly in the pentavalent state. Bi(V) is a strong oxidizing agent ($E^0 + 2.07V$). Bismuth is rather rare (0.1–0.2 ppm abundance in the earth's crust) [1] and occurs as bismuth ochre (α-Bi_2S_3), bismuth oxide (Bi_2O_3) and also as native bismuth (Bi^0) and forms a rather insoluble oxychloride (BiOCl) in seawater, which leads to its low concentration of about 0.02 $\mu g\ L^{-1}$ in seawater. Because of its insolubility and its limited abundance, bismuth has not been a concern of environmental legislation, despite considerable use.

The compound Me_3Bi is correctly named as trimethylbismuthine or trimethylbismuth. The abbreviation Me_3Bi will be used throughout this chapter. The term methylated bismuth compounds refers to MeBiX and Me_2BiX-compounds, in which X is not determined.

9.1.2 USES AND ENVIRONMENTAL DISTRIBUTION

The very low toxicity of ordinary inorganic bismuth compounds permits their use in very sensitive areas such as the cosmetic and pharmaceutical industries. The chemical industry also uses bismuth, as bismuth molybdophosphate, as a catalyst for the production of acrylonitrile and other intermediate products of acrylic fibres and various plastic products. The major applications are in cosmetic products as pigments in eye shadow and lipsticks (BiOCl) and as

Organometallic Compounds in the Environment
Edited by P.J. Craig © 2003 John Wiley & Sons Ltd

bismuth citrate in hair dye [2]. The pharmaceutical industry produces bismuth subgallate for infectious diseases, bismuth germanium oxide as contrast substances and bismuth subcitrate for ulcer therapy [3]. Anti-ulcer drugs contain bismuth(III) subcitrate, which occurs as a polymer [4]. In addition to the ulcer healing process of bismuth, compounds such as 2-chloro-1,3-dithia-2-bismolane and 1,2-[bis(1,3-dithia-2-bismolane)thio]ethane also show growth inhibition of the bacteria *Helicobacter pylori*, which plays a major role in the pathogenesis of peptic ulcer disease [5, 6]. When Bi(III) is absorbed it is probably transported as a glutathione complex to form a strong complex with metallothionein and transferrin [7, 8]. Pharmaceutical products such as 'Pepto-bismol' (Proctor & Gamble) contain bismuth subsalicylate in a rather high concentration of 1–2%. This product tends to prevent heartburn and enjoys worldwide distribution. All these substances are finally disposed to wastewater and increase the amount of bismuth in sewage sludge. For example, some data demonstrate a fairly high amount of bismuth, in the range of $5 \, mg \, kg^{-1}$ dry weight in sewage sludge, which is higher than the amounts of antimony and arsenic [9].

Recently an extensive review of the synthesis of organobismuth compounds in pharmacology has been published by Sadler and coworkers [10]. Bismuth is also used in fusible alloys and as a dopant in semiconductors, and Me_3Bi is used for the growth of bismuth telluride (Bi_2Te_3) thin films by metal organic compound vapor deposition (MOCVD) [11].

These products are discarded in municipal and industrial waste and end up on waste deposits. To calculate how much bismuth is actually discarded into waste, the world production of bismuth of about 5000 tons per year (1990) has to be considered. Of this, only 50% goes to the pharmaceutical industry [12]. However, the amount of bismuth used is expected to increase in the future because of progress in the substitution of lead in gun bullets, of organotin compounds as antifouling paints and also by the use of bismuth vanadate for high-performance lead-free pigments [13].

9.1.3   TOXICOLOGY OF ORGANOBISMUTH COMPOUNDS

The main hazard of bismuth is chronic exposure during routine medical therapy, when bismuth is orally administrated. The lowest dose of inorganic bismuth published, as lethal to humans, is $221 \, mg \, kg^{-1}$ [14]. LD_{50} values are known for the following bismuth species [15]: (i) BiOCl, $22\,000 \, mg \, kg^{-1}$ (rats, oral administration); (ii) TMB, $484 \, mg \, kg^{-1}$ (rabbits, oral administration); (iii) TMB, 182 $mg \, kg^{-1}$ (rabbits, subcutaneous); (iv) TMB, $11 \, mg \, kg^{-1}$ (rabbits, intravenous).

Acute toxicity of Me_3Bi is orders of magnitude higher than the inorganic bismuth form. Furthermore it can be absorbed through skin contact or by inhalation. However, little is known about its chronic toxicity at low concentrations.

Some toxicological data for Me_3Bi were obtained when it was investigated for its antiknock properties in petrol [16]. Sollmann and Seifter [15] examined the toxicological effects on various animals. In addition, toxicological data from accidental exposure of two subjects were reported [15]. A peculiar garlic or turnip taste was noticed which was more pronounced than the odour; itching and burning of the eyes were also mentioned. If the exposure was prolonged, a feeling of warmth developed in the exposed area accompanied by sweating and followed by headache. Dizziness and faintness developed as a first stage of anesthesia. The concentration of Me_3Bi in the air was unknown, however the 24-hour urinary output was measured to be as high as 820 μg bismuth. The urine from a second person, who felt symptom free, contained 380 μg bismuth. Urinary output from non-contaminated individuals is about 12.5 ± 29 μg per day [17].

Fatal dosages of Me_3Bi to rabbits and rats were examined and it appeared that it is not dependent on the kind of mammal used. The average fatal dose of an intravenous application of TMB was found to be $10\,mg\,kg^{-1}$ body weight. The animals (rabbits, rats) died after 10 min exposure to atmospheric Bi (310 mg Bi as Me_3Bi per m^3) or 45 min to $186\,mg\,m^{-3}$, respectively, of which an unknown proportion was oxidized before inhalation. However, it should be noted that no recent research has been done on the toxicological effects of Me_3Bi.

An elegant way to estimate toxicity is to use microorganisms rather than higher animals. Bioassays using bioluminescent bacteria offer a cheap, easy-to-measure and rapid way of determining toxicity. Currently, the Microtox® assay, which uses the naturally luminescent *Photobacterium phosphoreum*, is widely accepted as an international benchmark for ecotoxicity testing. Toxicity is assessed by quantifying the difference in light production between an uncontaminated blank and an actual sample. The main difficulty of this assay is that Me_3Bi is usually administered in its gaseous form. Therefore, partitioning through the phase boundary of liquid–gaseous systems has to occur. To immobilize the bacteria, 10 ml of resuscitated cells in 0.1-M potassium nitrate (KNO_3) are mixed with the same amount of a 0.1-M sodium algenate solution. When this mixture is added in small drops to a 0.1-M calcium chloride ($CaCl_2$) solution, jelly-like algenate beads form small drops. It has been shown that a Me_3Bi concentration in air of about 100 ppm (v/v) minimizes the luminescence of *Photobacterium phosphoreum* to 85% within 20 min, whereas 1000 ppm (v/v) reduces the luminescence to 39% [18]. Although, it is not clear if Me_3Bi or a degradation product is the active substance, it shows a concentration dependent toxic effect to the bacterium.

9.1.4   SYNTHESIS OF ORGANOBISMUTH STANDARDS

The easiest way to prepare Me_3Bi is by a Grignard reaction with bismuth triiodide (BiI_3) in diethyl ether (Et_2O) (Equation 9.1).

$$3 \text{ Me}_3\text{MgI} + \text{BiI}_3 \longrightarrow (\text{Me}_3)_3\text{Bi} + 3 \text{ MgI}_2 \qquad (9.1)$$

The bulk product has to be stored in a glass ampoule to avoid oxidation, although it is not pyrophoric. However, if the standard is to be used frequently, it can be stored in a glass flask under nitrogen in a freezer ($-20°C$, in the dark). Me_3Bi is volatile and can penetrate silicon septa, which can easily be identified by the formation of orange–yellow bismuth oxide (Bi_2O_3, formed by the immediate oxidation of Me_3Bi in air) [18].

The mono- and dimethylbismuth dihalides can be prepared by the mixture of 1:2 or 2:1 of Me_3Bi with BiX_3, respectively [19]. The compounds undergo redistribution reactions to form MeBiX_2 or Me_2BiX.

The instability of Me_3Bi standards is much more pronounced than trimethylarsine and even trimethylstibine. Their thermodynamics indicate that for the formation of TMB, the enthalpy of formation is largely endothermic due to the very weak carbon–bismuth bond (see Chapter 1). Stability of Me_3Bi in air has been measured at concentrations of 100 ng Bi as Me_3Bi in 1 L air in Tedlar bags [20]. The Tedlar bags were kept in the dark and continuously sampled with subsequent measurements of the Me_3Bi concentration in air over a period of 8 hours. After 7 hours a remarkable recovery rate of 65% was determined. Hence it can be concluded that low concentrations of Me_3Bi are fairly stable in air. The difference between bulk and low concentration stabilities is discussed in Chapter 1.

9.1.5 ANALYTICAL METHODS FOR ENVIRONMENTAL BISMUTH SPECIATION

Me_3Bi has a boiling point of 109 °C, and shows characteristic mass fragments of m/z 254 (35%, $(\text{CH}_3)_3\text{Bi}^+$), m/z 239 (87%, $(\text{CH}_3)_2\text{Bi}^+$), m/z 224, (77%, CH_3Bi^+), m/z 209 (100% Bi^+) using EI in a quadrupole MS. The chemical shift in ^{1}H-NMR using a 400 MHz-NMR Inova-400 'californium' shows a singlet at +1.08 ppm, whereas the ^{13}C-NMR spectrum shows a peak at -5.8 ppm [18].

It is well known that bismuth forms a rather unstable hydride, bismuth trihydride (BiH_3). The hydride however can still be used to enhance the transport of bismuth to the flame or plasma for AAS or ICP [21]. Recently it was suggested that sodium tetraethylborate (NaBEt_4) be used for ethylation derivatization in order to obtain better detection limits for bismuth in urine samples [22]. However, very few speciation approaches have been reported as yet [23, 24]. Volatile bismuth species can be separated by cryotrapping GC and bismuth as a heavy mono-isotopic metal can be detected very specifically and sensitively with ICP-MS. Packed column (10% SP2100 on Supelcoport) [23, 24] as well as capillary GC (Ultimetall CP-Sil 5CB 25 m × 0.53 mm, 1 μm Chrompack) [18, 20] were used for separation of volatile bismuth compounds. It was shown that

BiH_3, $MeBiH_2$, Me_2BiH, and Me_3Bi could be separated by a short packed column system [24, 25]. The species were identified by their element-specific detection and their retention times in comparison with standards. Me_3Bi has also been identified in sewage gas by GC-fractionation using a short packed column (10 % SP2100 on Chromosorb) followed by GC-MS (ion-trap) [23]. The packed column was used as a clean-up method before the fractions were sequentially and manually injected onto the capillary column (two-dimensional separation) followed by the detection of the molecular fragments in an electron impact MS. Using the ion trap, a secondary fragmentation had been performed to confirm the identity of individual primary fragments of Me_3Bi in sewage gas.

The identification of partially methylated bismuth compounds in water, sediment or soil was investigated using hydride generation of water at pH 2 or of slurries of soil and sediment adjusted to pH 2 [24, 25]. The identification of the ICP-MS signals (m/z 209) were compared with those achieved when hydride generation had been performed using a Me_3Bi/BiX_3 standard. According to Akiba and Yamamoto [19], inorganic bismuth and Me_3Bi are in equilibrium with partially methylated bismuth compounds. However, the method has not been validated and it is not known how quantitative the hydride generation method is. Rearrangements like those observed for antimony are possible [26]. The results of this method have to be interpreted as operationally defined fractions of methylated bismuth compounds. Up until now reliable, validated methods for non-volatile bismuth species, for which the integrity of the species is guaranteed, have not been available.

9.1.6 OCCURRENCE IN THE ENVIRONMENT

Volatile compounds of arsenic, and to a lesser extent of antimony, have been reported to be produced by certain microorganisms since the late nineteenth century for As and since the 1990s for Sb (see Chapter 7). Knowledge about organobismuth compounds in the environment however is very limited. A review of methylated bismuth compounds in the environment has recently summarized the findings by Feldmann and coworkers [24]. Despite the general similarity, bismuth shows major differences from arsenic and antimony. Bismuth occurs as a trivalent cation or as a bismuthoxide cation, whereas arsenic and antimony are metalloids and occur mostly as their pentavalent anions arsenate and antimonate, respectively. A look at the thermodynamic data of the organometallic compounds shows that the standard formation enthalpy for Me_3Bi ($+192\,kJ\,mol^{-1}$) is very high compared to $+13\,kJ\,mol^{-1}$ for trimethylarsine (Me_3As), and $+33\,kJ\,mol^{-1}$ for trimethylstibine (Me_3Sb) [27]. Me_3Bi is less stable than Me_3As or Me_3Sb but also less stable than Et_3Bi (triethylbismuth) [27]. Bismuth is more electropositive and is a metal, therefore the average dissociation energy of the metal–carbon bond is low for Me_3Bi ($143\,kJ\,mol^{-1}$) compared with $229\,kJ\,mol^{-1}$ for Me_3As and $216\,kJ\,mol^{-1}$ for

Me₃Sb [28]. This means that a lot of energy is involved in the formation of Me₃Bi and that the bismuth–carbon bond is very weak and can easily be cleaved (e.g. by oxygen). Although lower concentrations of an unstable compound in the gas phase are more stable than expected from the thermodynamical data, a determined half-life of more than 7 hours of Me₃Bi in air (0.1 μg Bi L^{-1}) is surprising (see Figure 9.1) [18, 20] (see Chapter 1 for a general discussion of these aspects.)

Me₃Bi is found to be abundant in gases released from municipal waste deposits and sewage-sludge fermentation tanks. The concentration of Me₃Bi in sewage gas is in general higher than the concentration measured for landfill gas. Feldmann *et al.* [24] reported data from eight different fermentation tanks (0.17–24.2 μg m^{-3}) and six different landfill sites (0.0002–0.892 μg m^{-3}). They estimated that 16–281 kg Me₃Bi per year is released from landfill sites worldwide, whereas 30–4980 kg Me₃Bi per year is generated in fermentation tanks from sewage treatment plants worldwide.

The migration of landfill gas from an old landfill site (> 20 years), which contained significant amounts of municipal waste, was studied by taking soil air sampled from bore holes [29]. The gas was cryotrapped onsite and was the subject of GC-ICP-MS analysis. The gas in the core area contained a Me₃Bi concentration between 0.01–0.4 μg m^{-3}, whereas the soil air at the edge of the site contained 0.015–0.034 μg m^{-3}. These concentrations are of the same order of magnitude as the concentration in the landfill gas. This is in contrast with the measured bore hole concentrations of other volatile metal(loid) compounds (e.g. tetramethyl tin, trimethyl stibine), which have been found to be orders of magnitude below the levels detected in landfill gas. Condensed water in gas pipelines on a landfill site did not contain any kind of methylated bismuth, either before or after hydride generation [30]. However, it has been claimed that landfill leachate contained significant amounts of methylated bismuth (0.2 ng L^{-1}) after hydride generation [29].

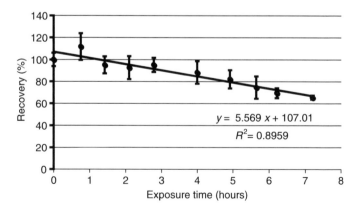

Figure 9.1 Stability of Me₃Bi(100 μg as Bi m^{-3}) in moisturized air at room temperature

Hirner and coworkers [31] investigated the occurrence of methylated bismuth compounds in soils from contaminated sites using a hydride generation method directly on the suspension at pH 2. They claimed to find 110 ng Bi as Me_3Bi per kg soil from a former gas station site and only 0.1 and 0.3 ng kg^{-1} from a former industrial and coal mining/processing site, respectively. A sample from a mature waste dump (>25 years) was shown to contain 40 ng Bi as Me_3Bi per kg sample and in addition they found partially methylated bismuth compounds as their hydrides ($MeBiH_2$; 4 µg kg^{-1}, Me_2BiH; 4 ng kg^{-1}). In another soil sample an unidentified volatile bismuth species has been detected after hydride generation [32]. This species has a higher boiling point than Me_3Bi.

9.1.7 BIOLOGICAL PRODUCTION OF ORGANOBISMUTH COMPOUNDS IN THE LABORATORY

Fermentation experiments using four sediments from contaminated sites in Germany were mixed 1:1 with an anaerobic pond sludge. This sludge was then mixed 1:1 with a nutrient solution for methanogenic bacteria [24]. This medium was stored at 30 °C in the dark for 2 weeks. Production of Me_3Bi was determined using cryotrapping GC-ICP-MS. Every culture produced considerable amounts of Me_3Bi, between 0.076–48 ng Bi as Me_3Bi per kg sludge. A correlation to the total bismuth in the sludge or to the available fraction of total bismuth (acidified sludge suspension was derivatized with sodium borohydride ($NaBH_4$) and the subsequent BiH_3 was quantified) failed. Furthermore, anaerobic cultures of sewage sludge ($E_h < -200$ mV, pH 6–8) containing 2.9 mg Bi kg^{-1}, 4.6 mg As kg^{-1}, 0.7 mg Sb kg^{-1}, produced, in addition to methane and carbon dioxide, volatile metal(loid) compounds. More Me_3Bi was produced than either trimethylated arsine or stibine [33].

Recently, it was shown that besides Me_3Bi, BiH_3 and the partially methylated bismuth compounds (Me_2BiH and $MeBiH_2$) are also generated from a culture of *Methanobacterium formicicum* [34]. The conversion rate of the spiked 1 µM $Bi(NO_3)_3$ to Me_3Bi was, at 2.6 ± 1.8%, relatively high. The other bismuthines only appeared at the end of the growth phase in a small proportion. The researchers found that methylcobalamin ($MeCoB_{12}$) can act as a methyl donor (with and without biota), whereas S-adenosylmethionine (SAM) does not transfer a methyl group to bismuth. It is not surprising that no oxidative methylation occurs with bismuth by the addition of CH_3^+ from SAM since bismuth is very difficult to oxidize to $Bi(V)$, in contrast to arsenic or antimony. One could suggest that the methyl group is probably transferred as a carbanion (CH_3^-), i.e. from CH_3CoB_{12}. This strong nucleophile can substitute one complexing ligand. A proposed mechanism for the successive substitution of the hydroxide ligands by carbanions is quite likely (Figure 9.2).

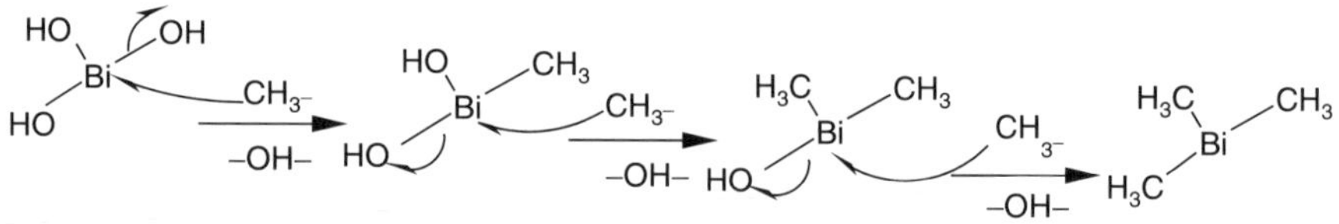

Figure 9.2 Proposed mechanism for bismuth methylation; consecutive SN_2 reactions

9.2 ORGANOTHALLIUM COMPOUNDS IN THE ENVIRONMENT

9.2.1 INTRODUCTION AND TERMINOLOGY

Thallium is the heaviest element of group 13 of the periodic table and has a crustal abundance of 3×10^{-5} mass-%, a factor of almost 8-times less abundant than arsenic. It exists in the environment in two stable oxidation states $+I$ and $+III$. In the presence of chloride $Tl(I)$ forms insoluble thallium(I) chloride (TlCl).

The compound Me_3Tl is trimethylthallium (TMTl). There are two methylated thallium species ($MeTl^{2+}$ and Me_2Tl^+) of interest.

9.2.2 USES AND ENVIRONMENTAL DISTRIBUTION

Thallium and its compounds are used in industry in acid resistance alloys, deep temperature measurement thermometers, in crystals highly permeable for IR light, in low-melting glass and in the semiconductor industry [35]. However, the main anthropogenic emission of thallium takes place when thallium-containing pyrites are roasted to produce iron.

9.2.3 SYNTHESIS OF ORGANOTHALLIUM STANDARDS

The synthesis of methylated thallium species was described by Gilman and Jones more than 50 years ago [36]. Thallium iodide (TlI) (0.073 mol) is mixed with 25 mL Et_2O and 0.08 mol methyl iodide (MeI) and stirred at room temperature, while 0.15 mol of methyl lithium (MeLi) in 110 mL ether was added dropwise into the solution over 6–7 h [37]. The brownish solution contains trimethyl thallium, Me_3Tl (Equation 9.2).

$$3MeLi + 3TlI \longrightarrow Me_3Tl + 2Tl + 3LiI \qquad (9.2)$$

Me_3Tl is stable in the dark under nitrogen or argon, but in diffuse light it decomposes. The extrapolated boiling point is 147 °C, and in bulk it will burn spontaneously in the open air to give thallium(III) oxide (Tl_2O_3). The solution can be distilled under vacuum to obtain Me_3Tl.

For the synthesis of DMTl, an ether solution of Me_3Tl was added to HI in order to obtain DMTl-I in a high yield (Equation 9.3).

Table 9.1 Other organometallic compounds in the environment

Elem	Medium	Species	Concentration	Analytical Method	Reference
Bi	Landfill gas	Volatile Bi	0.2–6.5 ng m^{-3}	CT-ICP-MS	40
	Landfill gas	Me$_3$Bi	312–892 ng m^{-3}	CT-GC-ICP-MS	30
	Sewage gas	Me$_3$Bi	2.7–1060 ng m^{-3}	CT-GC-ICP-MS	148
	Landfill gas	Me$_3$Bi		CT-GC-ICP-MS	144
	Sewage gas	Me$_3$Bi		CT-GC-MS and CT-GC-ICP-MS	23
	Sewage gas	Me$_3$Bi	1665–$24\,237$ ng m^{-3}	CT-GC-ICP-MS	29
	Sewage gas	Me$_3$Bi	3–$24\,237$ ng m^{-3}	CT-GC-ICP-MS	24
	Landfill gas	Me$_3$Bi	0.2–927 ng m^{-3}	CT-GC-ICP-MS	24
	Soil	Me$_2$Bi—X MeBi—X$_2$ Me$_3$Bi	4 4000 0.1–100 ng kg^{-1}	Slurry-HG-GC-ICP-MS	32
	Media of anaerobic cultures isolated from sediments	MeBi—X$_2$	8 ng kg^{-1}	Slurry-HG-GC-ICP-MS	24
	Headspace of anaerobic sediment cultures	Me$_3$Bi	0.084–48 ng kg^{-1} sediment 10–2031 ng m^{-3}	CT-GC-ICP-MS	29
	Landfill leachate	MeBi—X$_2$, BiH$_3$, MeBiH$_2$, Me$_2$BiH, Me$_3$Bi	0.2 ng L^{-1}	HG-GC-ICP-MS	29
	Methanobac. formicium	Bi-metallothionein		CT-GC-ICP-MS	34
	Cytosols of cells				7
Cd	Sewage gas	Volatile Cd	4–7 ng m^{-3}	CT-GC-ICP-MS	47
	Seawater/South Atlantic	MeCd$^+$	0.5–0.8 ng L^{-1}	Voltammetry	50
	Seawater/Spitzbergen	MeCd$^+$	0.47–0.67 ng L^{-1}	Voltammetry	51
	Media of Psychrotrophic bacteria	MeCd$^+$	0.56–1.3 ng L^{-1}	Voltammetry	50
	Pseudomonas sp.	Volatile Cd		Headspace trapping followed by AAS	54

(continues)

Table 9.1 Other organometallic compounds in the environment

Elem	Medium	Species	Concentration	Analytical Method	Reference
Ge	Ocean (Pacific)	MeGe—X$_3$	16 ng L^{-1}	HG-GC-AAS	119
	Pristine freshwater	MeGe—X$_3$	< 0.2 ng L^{-1}		
		Me$_2$Ge—X$_2$	< 0.5 ng L^{-1}		
	Contaminated river	MeGe—X$_3$	1.1 ng L^{-1}	HG-GC-AAS	123
		Me$_2$Ge—X$_2$	1.1 ng L^{-1}		
	Ocean	MeGe—X$_3$	9.7 ng L^{-1}		
		Me$_2$Ge—X$_2$	8.7 ng L^{-1}		
	Sediments	MeGe—X$_3$	0.3 µg kg^{-1}	Slurry-HG-GC-ICP-MS	75
	Hot spring water	MeGe—X$_3$	0.02–0.06	HG-GC-ICP-MS	124
		Me$_2$Ge—X$_2$	0.02–0.03		
		Me$_3$Ge—X	0.04–0.17 µg kg^{-1}		
Ni	Sewage gas	Ni(CO)$_4$	500–1000 ng m^{-3}	CT-GC-ICP-MS	145
Mn	Air (car park)	CMT	5.5–7.4 ng m^{-3}	SPME-GC-MIP	100
		MMT	8.8–22.9 ng m^{-3}		
	Air (mean atmospheric level in Canada)	MMT	5 ng m^{-3}		99
Mo	Landfill gas	Mo(CO)$_6$	200–300 ng m^{-3}	CT-GC-ICP-MS	144
	Sewage gas	Mo(CO)$_6$	3000–3600 ng m^{-3}	CT-GC-ICP-MS	145
	Landfill gas	Mo(CO)$_6$		CT-GC-ICP-MS	29
Po	Sea sediment	Volatile Po			143
Te	Landfill gas	Volatile Te	1.2–5.5 ng m^{-3}	CT-ICP-MS	40
	Landfill gas	Me$_2$Te	48–75 ng m^{-3}	CT-GC-ICP-MS	30
	Landfill gas	Me$_2$Te	12–1430 ng m^{-3}	CT-GC-ICP-MS	73
	Sewage gas	Me$_2$Te	44–47 ng m^{-3}	CT-GC-ICP-MS	29
	Sewage gas	Me$_2$Te	259–887 ng m^{-3}	CT-GC-ICP-MS	48

	Source	Species	Concentration	Method	Ref.
	Soil gas	Me$_2$Te	10–403 ng m^{-3}	CT-GC-ICP-MS	7
	Soil	Me$_2$Te		Slurry-HG-GC-ICP-MS	29
	Headspace of anaerobic sediment culture	Me$_2$Te	100 µg m^{-3}	CT-GC-ICP-MS	32 31
	Media of *Acremonium falciforme* and *Penicillium citrinum*	Me$_2$Te Me$_2$Te$_2$		GC-MS and GC-fluorine-induced chemiluminescence	68
	Media of *Pseudomonas fluorescens* and *Pencillium chrysogenum*	Me$_2$Te		GC-MS and GC-fluorine-induced chemiluminescence	69
Tl	Anaerobic culture of freshwater sediment	Me$_2$Tl$^+$	2600 ± 300 ng L^{-1}	Complexation/ oxidation/ extraction PTI-IDMS	39
	Seawater	Me$_2$Tl$^+$	0.4–3.2 ng L^{-1}	Complexation/ oxidation/ extraction PTI-IDMS	38
	Anaerobic river sediment	Me$_2$Tl$^+$			41
W	Landfill gas	W(CO)$_6$	5–10 ng m^{-3}	CT-GC-ICP-MS	144
	Sewage gas	W(CO)$_6$	10–15 ng m^{-3}	CT-GC-ICP-MS	145

$$Me_3Tl + HI \longrightarrow Me_2TlI + CH_4 \qquad (9.3)$$

The white precipitate is stable and can easily be separated from the solution. Me_2Tl^+ is isoelectronic to dimethylmercury and is the most stable methylated thallium compound.

9.2.4 ANALYTICAL METHODS FOR ENVIRONMENTAL THALLIUM SPECIATION

Not much is known about the occurrence of methylated thallium species in the environment because of the lack of a suitable analytical technique and the instability of most of the species. Methylated thallium compounds do not form stable hydrides. Therefore hydride generation methodology using $NaBH_4$ is not applicable. Recently, Schedlbauer and Heumann [38] described a method that can distinguish between Me_2Tl^+ and inorganic thallium. This method is based on the selective oxidation of total thallium to Tl(III) with an aqueous solution of bromine. Uncharged inorganic thallium can easily be extracted using methyl isobutyl ketone (MIBK) while Me_2Tl^+ is ionic and stays in the aqueous layer. For quantification, species-specific addition of isotopically labelled $Me_2^{203}Tl^+$ was used in order to follow the concept of isotope dilution methodology. Further clean up steps were introduced before isotope specific detection was performed using PTI-MS (positive thermal-ionization mass spectrometry). Although it is unknown how other species would be fractionated with this method, it is impressive that Me_2Tl^+ can be quantified in a more than 500-fold excess of inorganic thallium. Detection limits of 0.4 ng Tl L^{-1} were achieved with a sample volume of 500 mL using a preconcentration step. This step includes the complexation of inorganic thallium and Me_2Tl^+ with sodium diethyldithiocarbamate [NaDDTC]. The anionic complexes formed will be adsorbed onto a strong anion exchanger Dowex AG1-X8 and can be extracted with HNO_3 (2 M) [39].

9.2.5 OCCURRENCE IN THE ENVIRONMENT

Although there have been some attempts to detect volatile thallium compounds in landfill and sewage gases, no volatile compounds have been detected [40]. The instability of Me_3Tl may be the reason of the lack of occurrence. The bond enthalpy of Tl-C (92 kJ mol^{-1}) shows by far the lowest energy in the M—C bond compared with other permethylated species such as trimethylarsine, trimethylstibine or even trimethyl bismuthine [27]. Very limited information is available on occurrence in water, soil and sediment. The first modern detection of methylated thallium has just been reported [38]. Surface seawater was collected on two cruises through the Atlantic ocean. From 20 samples taken in the South

Atlantic only four samples contained measurable Me_2Tl^+, whereas three samples from ten show detectable Me_2Tl^+ concentrations above the detection limits of $0.5\,ng\,L^{-1}$ $(0.5-3.2\,ng\,L^{-1})$. It seems that the values appear where higher bioactivity has been detected either by chlorophyll c content or by dimethyl sulfide (Me_2S) production. One depth profile has been measured in the South Atlantic so far. The small concentration in the surface water of $0.59\,ng\,L^{-1}$ increases up to above $2.5\,ng\,L^{-1}$ Me_2Tl^+ at a depth between 40 and 200 m which is known as the biologically highly active zone. It is surprising that even at depths of 400 and 1000 m, detectable levels of Me_2Tl^+ have been measured. Total thallium shows a similar but not so pronounced profile with a maximum concentration of about $4.0\,ng\,L^{-1}$. In general however, the Me_2Tl^+ level does not correlate with the total amount of dissolved thallium in seawater, varying between 3–38 %. No organothallium compounds have yet been detected in biological tissue.

9.2.6 BIOLOGICAL PRODUCTION OF ORGANOTHALLIUM COMPOUNDS IN THE LABORATORY

It was suggested by Huber and coworkers that inorganic thallium may be oxidized and stabilized in the trivalent form (III) as Me_2Tl^+ [41, 42]. Huber and Kirchmann [42] and recently Schedlbauer and Heumann [39] were able to detect the production of Me_2Tl^+ in anaerobic cultures of incubated freshwater sediments. However, an anaerobic culture amended with $0.47\,ng\,mL^{-1}$ Tl(I) did not produce any Me_2Tl^+, whereas the duplicate culture amended with $0.024\,\mu g\,Tl\,L^{-1}$ transferred more than 10 % Tl(I) into Me_2Tl^+ within 30 days in the dark at $30\,°C$ [39]. Attempts to produce Me_2Tl^+ by aerobic cultures of sewage sludge amended with thallium nitrate $(TLNO_3)$ from concentrations 0.024 and $0.54\,\mu g\,mL^{-1}$ failed to produce any Me_2Tl^+.

9.3 ORGANOCADMIUM COMPOUNDS IN THE ENVIRONMENT

9.3.1 INTRODUCTION AND TERMINOLOGY

Cadmium is rare in the earth's crust $(3 \times 10^{-5}$ mass- %) and occurs naturally as greenockite (CdS) or otavite $(CdCO_3)$. The only stable oxidation state is Cd(II). The compound Me_2Cd is dimethylcadmium whereas $MeCd^+$ is generally described as methylcadmium.

9.3.2 USES AND ENVIRONMENTAL DISTRIBUTION

Cadmium is always associated with zinc and therefore it is found at all zinc production or processing sites. It is highly mobile in soil under slightly acidic conditions. It is a precursor for galvanic and electrochemical operations. The occurrence of inorganic cadmium in the environment is covered extensively in numerous other works and is not discussed here. Despite the knowledge of the distribution of cadmium in water, soil, sediment and air, it is surprising that only a few studies have been published about the occurrence of organometallic cadmium compounds in the environment. On the other hand, although used in the semiconductor industry, Me_2Cd is very unstable in the presence of water and it is very unlikely that it is detectable in the environment.

9.3.3 TOXICOLOGY OF ORGANOCADMIUM COMPOUNDS

Nothing is known about the toxicity of $MeCd^+$ or Me_2Cd. However, it should be assumed that parallels can be drawn with organomercury compounds, since $MeCd^+$ also shows a certain stability in the environment as does $MeHg^+$ (see Section 9.3.6).

9.3.4 SYNTHESIS OF ORGANOCADMIUM STANDARDS

Me_2Cd is a volatile colorless liquid with a boiling point of 106 °C. In bulk, it is inflammable in air and reacts vigorously with water. Me_2Cd can be synthesized from Cd^{2+} by a Grignard reaction in absolute diethyl ether [43]. A yield of more than 90 % of Me_2Cd can be achieved by the reaction of MeLi with cadmium chloride ($CdCl_2$) in tetrahydrofurane/ether at room temperature [44]. Methyl-cadmium can be synthesized from Me_2Cd using cadmium iodide (CdI_2). These two species are in the equilibrium with the monomethyl species (Equation 9.4).

$$Me_2Cd + CdI_2 \rightleftharpoons 2\ MeCdI \qquad (9.4)$$

The equilibrium constant was determined to be 110 [45]. Substoichiometric amounts of $CdCl_2$ have been added to double-distilled Me_2Cd in ether under nitrogen. The precipitated methylcadmium chloride can be isolated and dried under nitrogen [46].

9.3.5 ANALYTICAL METHODS FOR ENVIRONMENTAL CADMIUM SPECIATION

In general, the differences in volatility or polarity of the different metal species is used when metal species are separated by chromatographic methods, but these have not been used for cadmium speciation. The first published approach used the different oxidation behaviour of methylcadmium and free cadmium ion [46]. Using differential pulse anodic stripping voltammetry

(DPASV) the voltammogram showed a difference of 0.112 V for both species ($MeCd^+$: -0.759 V, Cd^{2+}: -0.871 V), which is enough to be able to determine both species simultaneously in seawater if the media had been adjusted to pH 8. Large amounts of lead do not affect the speciation method. Complexes of Cd(II) with humic acid do not interfere with the potential at which $MeCd^+$ is oxidized. Another method for distinguishing both species uses the ability of $MeCd^+$ to form a 1:1 non-charged complex with sodium NaDDTC, which can be extracted by n-hexane leaving only Cd(II) in the aqueous solution.

9.3.6 OCCURRENCE IN THE ENVIRONMENT

Using cryotrapping gas chromatography coupled to ICP-MS, gas samples from a sewage-sludge fermentation tank have been analyzed for volatile metal(loid) compounds. At 113 m/z one peak (^{113}Cd) was detected. The calculated boiling point was 73 °C, which is significantly different from 106 °C for Me_2Cd [47]. No positive identification of a volatile cadmium has therefore yet been reported.

The latest findings that cadmium distribution in the ocean shows similarities to those of nutrients have raised awareness to the possibility of organocadmium compounds occurring in the environment. Land and Morel [48] claim that cadmium has a biochemical role in the biosynthesis of an enzyme in a seawater alga. Heumann and coworkers [49] demonstrated elegantly the use of DPASV for the monitoring of $MeCd^+$ in seawater, in particular in the polar regions. The first evidence of $MeCd^+$ in polar and non-polar ocean water as well as in Arctic freshwater samples has been reported [49]. A depth profile at 47 °S, 15 °W in the South Atlantic showed levels of $MeCd^+$ between 0.5 and 0.8 ng L^{-1} down to a depth of 200 m [50]. A good correlation to Me_3Pb^+ and to chlorophyll a concentration was measured, peaking at a depth around 30 m. In the Arctic summer, concentrations of $MeCd^+$ of between 0.47–0.67 ng L^{-1} have been measured in surface seawater of the Kongsfjord of Spitzbergen in less than a half of the sample taken [51]. Overall it seems that $MeCd^+$ shows certain stability in seawater, which might be expected, due to the similarity to the mercury counterpart ($MeHg^+$).

9.3.7 BIOLOGICAL PRODUCTION OF ORGANOCADMIUM COMPOUNDS IN THE LABORATORY

Very little is known about the biotransformation of cadmium into organocadmium compounds, although it was demonstrated 20 years ago that methylcadmium can be produced if $MeCoB_{12}$ reacts in an aqueous solution with inorganic cadmium [52]. Indirect evidence of a volatile methylated cadmium species was described by Huey [53], who detected methylmercury when inorganic mercury was exposed to a volatile cadmium species. Cultures of heterotrophic saprophyllic *Pseudosomas* sp. bacteria isolated from a polluted water reservoir amended with $0.7–14\,\mu g$ Cd L^{-1} release volatile cadmium species [54]. The head-space was sampled by electrostatic accumulation. However, it has to be said that it could not be determined if the accumulated cadmium species was a truly volatile cadmium species or a cadmium containing aerosol, since only total cadmium was analyzed in the gas samples by electrostatic decomposition using direct GFAAS. A similar experiment was described by Robinson and Kiesel [52], who trapped a volatile cadmium species from the headspace of a reactor in which a saturated $CdCl_2$ solution reacted with $MeCoB_{12}$ at $37\,°C$ and pH 9.6.

Pongratz and Heumann [50] demonstrated that bacteria are able to generate $MeCd^+$ (analysed by DPASV). Psychrotrophic bacteria collected in East-Antarctica were inoculated in 20 g glucose, 20 g peptone and 250 mL Antarctic seawater. Seven cultures were inoculated and stored at 4 and 25 °C. Production of $MeCd^+$ showed a positive correlation to Me_3Pb^+ production. Significant production of $MeCd^+$ was identified after 116 h at 4 °C in five cultures, but all seven cultures showed production of $MeCd^+$ after 165 h incubation time in concentrations between 0.56 and 1.3 ng L^{-1}. However, only four cultures from the initial seven showed significant production of $MeCd^+$ ($0.49–0.96$ ng L^{-1}) after 68 h. These results were interpreted as the biological production of $MeCd^+$ by bacteria from the polar region [50]. Furthermore, a good correlation be-tween the concentration of organic carbon as an indicator for biomass and the $MeCd^+$ was determined, which gives a good indication of a biologically medi-ated reaction for the formation of $MeCd^+$.

9.4   ORGANOTELLURIUM COMPOUNDS IN THE ENVIRONMENT

9.4.1   INTRODUCTION AND TERMINOLOGY

The chemistry of tellurium is very similar to the chemistry of selenium (see Section 9.10) and is often discussed in the same way. However, very little information is available for tellurium due to the fact that tellurium is less abundant in the earth's crust (5000-fold less abundant than arsenic; 10^{-7} mass-%) and is not established to be an essential element for life [55]. These facts make tellurium less attractive for scientists to investigate the occurrence of organotellurium compounds in the environment. The most stable oxidation state is Te(IV) but it also occurs as Te(VI), Te(II) and as elemental tellurium. The natural occurrence of tellurium as tellurite [Te(IV)], tellurate [Te(VI)] and Te(0) has been established [56] and the presence of organotellurium compounds has been hypothesized in seawater on the basis of discrepancy between the results of total tellurium concentration and the sum of Te(IV) and Te(VI), but this has not been confirmed [57]. The compound Me_2Te is dimethyl tel-luride (DMTe), and dimethylditelluride is $(CH_3)_2Te_2$, whereas Me_3Te^+ is the trimethyltelluronium ion.

The discussion of the essentially non-metallic selenium in Section 9.10 should be taken in conjunction with the present tellurium section, particularly in the context of biomethylation.

9.4.2 USES AND ENVIRONMENTAL DISTRIBUTION

In the Nineteenth century tellurium was used for the treatment of syphilis, leprosy and tuberculosis. Only 500 tonnes of tellurium were produced in 1978 [58] and this has not increased much since. This by itself shows the limited use of tellurium, but important amounts are used in the steel and glass industries. Diethyltelluride (Et_2Te) is used in the semiconductor industry [59]

9.4.3 TOXICOLOGY OF ORGANOTELLURIUM COMPOUNDS

There are no reports of tellurium toxicity to plants and lower animals. However, it seems that higher developed animals show significant effects when exposed to tellurite [Te(IV)]and tellurate [Te(VI)], whereas in general elemental tellurium is considered to be non-toxic. Tellurite is one order of magnitude more toxic than tellurate (4 and 47 mg kg^{-1} LD_{75} for rats) [60]. Levels of toxicity have not been investigated for Me_2Te or any other methylated tellurium compounds. LD_{50} values for Et_2Te in air are given for rats, guinea pigs and mice (55, 45 and 154 μg m^{-3}, respectively [61]) suggesting that volatile organotellurium compounds are much more toxic to mammals than their selenium counterparts; LD_{50} for dimethyl selenide (Me_2Se) in air cannot be given because more than 8043 ppm (39 mg Me_2Se L^{-1}) appeared to be non-active for rats [62]. Since the importance of tellurium has grown in the last two decades toxicological assessments become relevant. Taylor [63] summarized the biochemistry of tellurium in a fairly recent review. If tellurium is orally administered, animal studies suggest that 25 % of tellurium is absorbed in the gut and rapidly excreted ($t_{1/2}$ 0.81 days). Residual tellurium will be metabolized to Me_2Te, which is excreted into excreta, viscera and breath [64]. The intense smell of Me_2Te is noticeable if 0.5 μg of tellurium is consumed by humans [65]. The garlic smell can continue for several months depending on intake of tellurium. Uptake of 10 mg tellurium(IV) oxide (TeO_2) resulted in a smell, which was noticeable for more than a half year [66]. Interestingly, the threshold level for tellurium in workplace air (0.1 mg m^{-3}) is one order of magnitude too high to attenuate the garlic smell in the breath of workers [66]. Here, it has been reported that miners, who had been exposed to inorganic tellurium, excreted approximately 0.1 % of the tellurium as Me_2Te. However, no analysis could show conclusive Me_2Te in the breath.

On the one hand it seems that inorganic tellurium is biotransformed to volatile Me_2Te in mammals, although nothing is known about the mechanism.

On the other hand there are some indications that significant amounts of tellurite are also transformed into elemental tellurium in addition to Me_2Te production when it is injected directly into the blood [66].

9.4.4 SYNTHESIS OF ORGANOTELLURIUM STANDARDS

For the synthesis of methylated tellurium compounds it has to be considered that these compounds are some of the smelliest compounds known! Dimethyltellurium diiodide (Me_2TeI_2) is synthesized with a yield of 50% by the reaction of elemental tellurium with methyliodide for 39 h at 80 °C (Equation 9.5) [67].

$$Te + 2MeI \longrightarrow Me_2TeI_2 \qquad (9.5)$$

The product can be dissolved in chloroform to separate it from the residue. It has a m.p. of 127 °C. Me_2TeI_2 decomposes to Me_2Te and iodine at higher temperature. This can be mediated in an aqueous sodium sulfite (Na_2SO_3) solution (Equation 9.6).

$$Me_2TeI_2 + Na_2SO_3 + H_2O \longrightarrow Me_2Te + 2HI + Na_2SO_4 \qquad (9.6)$$

The mass spectrum shows the characteristic series of eight tellurium isotopes m/z 120 (0.10%), 122 (2.60%), 123 (0.91%), 124 (4.82%), 125 (7.14%), 126 (19.0%), 128 (31.7%), and 130 (33.8%) as Te^+, $(CH_3)Te^+$ and $(CH)_2Te^+$.

9.4.5 ANALYTICAL METHODS FOR ENVIRONMENTAL TELLURIUM SPECIATION

Volatile tellurium compounds (R_2Te) have been separated using a $30\,m \times 0.25\,mm$ i.d. DB5 column of 5% methylphenylsiloxane. The GC was coupled to a fluorine-induced chemiluminescence detector [68]. The fluorine-induced chemiluminescence detector is sensitive to a number of volatile organometalloid compounds. It has been used for organosulfur, organoselenium, organoantimony but also for organotellurium compounds [69]. The other technique used for volatile tellurium compounds (R_2Te) employed a packed column system filled with 10% SP2100 on Supelcoport coupled to an ICP-MS [30]. Only a few approaches have been made to specify non-volatile tellurium compounds in water or sediments. A hydride generation methodology was used to generate non-volatile tellurium to the volatile reduced counterparts; however, no systematic description of these methods has been published yet. Tellurate has been satisfactorily analysed by using anion exchange chromatography (IonPak AS 14, Dionex $250 \times 4\,mm$) and when linked to ICP-MS, a detection limit of $0.41\,\mu g\,L^{-1}$ could be achieved [70, 71]. However, tellurate has produced a sharp single peak, whereas tellurite gave a broad one, so the value of this method for tellurium speciation has still to be proven [72].

9.4.6 OCCURRENCE IN THE ENVIRONMENT

Since two organotellurium compounds are rather volatile (Me_2Te; b.p. 82 °C and MeTeH; b.p. 57 °C), process gases from municipal waste deposits and from fermentation tanks on municipal sewage treatment plants have been subject to the determination of volatile tellurium compounds. Me_2Te was the only species identified in the gas samples using various types of cryotrapping GC-ICP-MS. The identification of Me_2Te has been done by the comparison of the retention time with a Me_2Te standard [73]. The standard was synthesized and characterized by GC-MS. In landfill and sewage gases Me_2Te occurs in concentrations between 1.2 and 1760 ng Te m^{-3} [24].

Water condensate in a pipeline (in which landfill gas is transported from the deposits to the furnace) showed significant amounts of Me_2Te after acidification to pH 2 and hydride generation using $NaBH_4$ [74]. Me_2Te was detected when suspensions of freshwater river sediments were acidified to pH 2 and derivatized using $NaBH_4$. The trapped volatile compounds were analyzed by packed column GC coupled to ICP-MS and Me_2Te was identified using boiling point correlations only [75]. This approach has been performed for soil samples and in addition to Me_2Te, a less volatile tellurium compound was released from the derivatization vessel [32]. This is the only evidence so far that methylated tellurium compounds occur in sediment and soil.

The use of tellurium in heterogeneous catalysis is widespread in industry and the production of tellurium-containing wastewater can be expected. Klingenberg et al. [76] describe a situation in which the wastewater contains between 0.5–1.5 mg Te L^{-1}. This water is discharged to a sewage treatment plant in which the methylation of tellurium was suspected (noticeable by the unpleasant smell of Me_2Te). It was suggested that tellurium be removed with iron, and for this procedure tellurite and tellurate had to be monitored using liquid chromatography [76].

9.4.7 BIOLOGICAL PRODUCTION OF ORGANOTELLURIUM COMPOUNDS IN THE LABORATORY

Gadd and coworkers [77] have found that Penicillium citrinum did not produced any volatile tellurium from tellurite but only elemental tellurium, whereas the fungus Fusarium sp. has produced volatile as well as elemental tellurium after 48 h. Although no identification of the volatile tellurium was performed, the pungent garlic smell was a good indicator that Me_2Te had been produced. In contrast Chasteen and coworkers [68] inoculated different bacteria and fungi and amended with inorganic tellurium (tellurite). The bacteria Pseudomonas fluorescens generated Me_2Te, whereas fungi Acremonium falciforme and also Penicillium citrinum generated Me_2Te and Me_2Te_2. The compounds were determined by their retention times and their mass spectra (electron impact).

This confirmed early experiments from the Challenger group [78, 79] with several strains of *Pencillium* sp. which showed that inorganic tellurium could be methylated and volatilzed as Me₂Te. Cultures of *Pseudomonas fluorescens* K27 are able to transform almost 80% of added sodium tellurate and sodium tellurite (both at 0.2 mM or 2.0 mM) into Me₂Te [80].

Cultures known to methylate selenium *(Pencillium* sp.) show decreased production of Me₂Se if inorganic tellurium has been added. This might be a case of competition between selenium and tellurium in terms of methylation, i.e. tellurium was methylated and Me₂Te was produced. On the other hand it was shown that more Me₂Te can be generated if selenium was added to the culture [69]. Biofilms containing *Pencillium chrysogenum,* which forms during the electrolysis of copper, release Me₂Te [81]. It has been speculated that the resistance of purple non-sulfur bacteria *(Rhodocyclus tenius, Rhodobacter sphaeroides* 2.4.1, *Rhodobacter capsulantus, Rhodospirillum rubrum* Si, G9, *Rhodopseudomonas blastica)* to metalloid oxyanions are based on their ability to reduce and methylate metalloids like selenium and tellurium [69]. Some phototrophic bacteria (*R. tenuis, R. rubrum* Si and G9) release Me₂Te when tellurate *and* selenate were added to the culture, but all of the tested purple non-sulfur bacteria produced elemental tellurium. It is claimed that these synergetic effects are not caused by transmethylation of tellurium from organoselenium compounds. Sterile controls did not produce Me₂Te.

The addition of elemental tellurium, which remains as an insoluble powder on the bottom of the media flask, to *R. tenius, R. rubrum* Si and G4 did produce Me₂Te and Me₂Se when elemental selenium was added. This is most astonishing since elemental tellurium and selenium have been considered as non-bioavailable and relatively inert. Chasteen's [69] results oppose the mechanism for biomethylation of selenium and tellurium proposed by Challenger [78] (Figure 9.3)

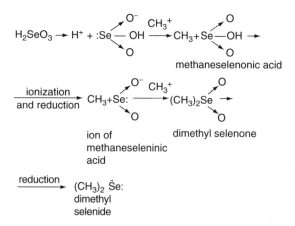

Figure 9.3   Challenger mechanism of the methylation of selenium or tellurium
Reproduced from this work, first edition, with permission

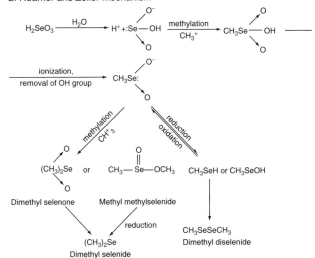

Figure 9.4　Tellurium methylation after (1) Lewis and (2) Reamer and Zoller Mechanisms for Selenium
Reproduced from this work, first edition, with permission

and later Lewis, and Reamer and Zoller (Figure 9.4) [82], which was based on reductive methylation with dimethylselenone (Me_2SeO_2) or dimethyltellurone (Me_2TeO_2) respectively as intermediates. The volatilization of tellurium or selenium to their dimethyl compounds supports Doran's mechanism [83] which has the elemental form as an intermediate (see also Section 9.9).

Inoculation of anaerobic sludge containing tellurium with sludge from contaminated-site bacteria have shown production of Me_2Te and two other unidentified volatile compounds [84]. The amounts, produced within 2 weeks, are between 1.2 and 161 μg Te kg^{-1} sludge. However, it was not been established how much biomass was established and which kind of bacteria had developed.

Furthermore, there is evidence available that tellurium-tolerant fungi strains are able to biosynthesize tellurium-amino acids such as telluromethionine and tellurocysteine, and also apparently a tellurium-containing protein [85, 86].

9.5 ORGANOMANGANESE COMPOUNDS IN THE ENVIRONMENT

9.5.1 INTRODUCTION AND TERMINOLOGY

Manganese is a classical transition element that does not form stable Mn—C bonds to alkyls unless they are stabilized by complexes such as tetrapyrrole rings or carboxyl ligands. However, stabilized by ligands, this d^7 element can form rather stable organomanganese complexes. No natural stable methylated manganese compounds are known and no studies of methylation or biomethylation of manganese have been reported. So far no natural existing organomanganese compounds have been characterized. However, organomanganese compounds have been synthesized and used extensively.

In this chapter, MMT is used for methylcyclopentadienyl manganese tricarbonyl or Me—(C_5H_4)—$Mn(CO)_3$, and CMT for cyclopentadienyl manganese tricarbonyl (C_5H_5—$Mn(CO)_3$). No other organomanganese compounds are discussed here.

9.5.2 USES AND ENVIRONMENTAL DISTRIBUTION

MMT is so far the only organometallic product other than the organolead compounds, produced in large quantities as a gasoline additive and is therefore distributed widely in the environment. MMT has been used in Canada since 1976 where it successfully replaced tetraalkyl lead in gasoline, which was phased out in 1990. Canada is the only country in the world to have authorized the replacement of lead with MMT. In 1994 the US Environmental Protection Agency (EPA) denied a petition from the Ethyl Corporation to allow the use of MMT in unleaded petrol because of health concerns related to the inhalation of manganese particulate emissions from combusted MMT, although no conclusive research had been done at that time [87]. Even after several years of use of MMT in Canada, many uncertainties remain, but MMT is now approved for use in the following countries: Argentina, Australia, Bulgaria, USA, France, Russia and (conditionally) New Zealand [88]. The MMT level in unleaded gasoline varied up to a maximum level of 18 mg L^{-1}, in leaded petrol the concentration was even higher [89].

9.5.3 TOXICOLOGY OF ORGANOMANGANESE COMPOUNDS

The toxicity of MMT has been tested in acute, chronic and subchronic tests. The LD_{50} values for orally administrated MMT to rats and mice ranged from 50–58 mg and 230 mg kg^{-1} body weight respectively [90, 91]. The primary site of action appears to be the lung for all species tested. Whether MMT itself or a metabolite causes the pneumotoxic effect is not clear since both inhibited and enhanced toxicity have been observed when MMT was mixed with P_{450} oxidase [92, 93]. Longer-term subacute inhalation tests on mice, rats and monkeys up to a concentration of 30 mg MMT m^{-3} have been carried out [89]. Microscopic changes in lung tissue were noted in both mice and rats but not in monkeys. There was a decline in body weight for mice and limited loss for rats but not for monkeys. The only pathological change observed in monkeys was a slight vacuolation of some types of brain tissue. In several cases of human exposure to accidental spills of MMT transpiratory distress, which disappeared within 24–48 hours, has been reported [89]. In general the inhalation hazard from vaporization is minimal since MMT has a higher boiling point than the hydrocarbons. Chronic inhalation studies of manganese(VI) oxide (MnO_2) by burning MMT showed no MMT related adverse health effects for rats or squirrels and monkeys (more than 1000 µg m^{-3} Mn for 22 hours per day for 9 months).

9.5.4 SYNTHESIS OF ORGANOMANGANESE COMPOUNDS

Both MMT and the non-methylated CMT are commercially available and laboratory synthesis is not required.

9.5.5 ANALYTICAL METHODS FOR ORGANOMANGANESE COMPOUNDS

Liquid chromatographic methods can be used to separate the different organomanganese compounds such as MMT and CMT. Recently a method that coupled diode-laser AAS with HPLC was reported by Niemax and coworkers [94]. They achieved detection limits of about $2\,\mu g\,L^{-1}$ as Mn and a linear range of five orders of magnitude. They spiked gasoline, human urine and tap water and achieved good recoveries. Another method was used to determine MMT in water samples. Here gas chromatographic separation was coupled to an atomic emission detector (GC-AED) to achieve good detection limits. MMT can be preconcentrated using liquid–liquid extraction or solid-phase microextraction (SPME) [95, 96]. Using SPME followed by GC-AED, ultratrace concentrations of MMT could be detected. With a linear range between 1 and $1000\,pg\,L^{-1}$ a detection limit of $0.3\,pg$ Mn as MMT per litre has been achieved [96], whereas if SPME is thermally desorbed and Mn is determined by graphite furnace AAS [95], a detection limit of only $25\,\mu g$ Mn per litre has been accomplished.

9.5.6 OCCURRENCE IN THE ENVIRONMENT

MMT is relatively unstable in air and is assumed to be totally oxidized during combustion in the car (99.9 %) [89] and subsequently emitted as MnO_2/Mn_3O_4 on suspended particles into the air (0.2 to 0.4 μm) [97]. Levels of airborne manganese (on particles, not specified)) were measured to be between $33–70\,ng\,m^{-3}$ in the US and in Toronto (Canada) $40\,ng\,m^{-3}$. These levels increase dramatically to values up to $300\,ng\,m^{-3}$ when inorganic manganese (not specified) is emitted from point sources such as the ferrous metal industries.

However, spills and vaporization of gasoline could emit large quantities of MMT directly into the air. Therefore it has been of interest to determine the half-life of MMT under atmospheric conditions. MMT absorbs light between 210 and 400 nm, σ (333 nm) $= 1.9 \pm 0.2.10^{-18}\,cm^2$ molecule^{-1}. Wallington et al. [98] have determined the photolysis rate of MMT at a typical summer day at a latitude of $40°$ N to be $1.3 \pm 0.1.10^{-2}\,s^{-1}$. Photolysis cleaves one CO from MMT to give CH_3—C_5H_4—$Mn(CO)_2$ in high yield. Furthermore, reactions with OH radicals and ozone take place in the atmosphere, resulting in the production of non-volatile manganese compounds. Reactions with OH radicals are slower than those with ozone. Overall however, MMT is entirely converted to inorganic manganese (presumably to MnO_2) within 1 day.

A multimedia exposure study was conducted in downtown Montreal, which considered air, food and water [99]. The daily average environmental exposure dose to manganese was estimated to be $0.010\,\mu g\,kg^{-1}\,d^{-1}$. This contribution to the multimedia exposure is considered to be very low. The mean atmospheric level of MMT was $5\,ng\,m^{-3}$, whereas respiratory manganese was seven-times higher ($36\,ng\,m^{-3}$) and total manganese in the atmosphere was determined to be $103\,ng\,m^{-3}$. The ratio of MMT to total atmospheric manganese was therefore determined to be approximately 5 %. Despite the short atmospheric half-life in the presence of UV light, highway runoffs and sewage samples showed water concentration in pg Mn as MMT per liter [95]. In addition to MMT, CMT (which is also used as an additive) was also identified in two out of three air samples taken in an underground car park [100]. Approximately one third of the MMT (8.8, 18.7 and 22.9 ng Mn m^{-3}) concentration was measured for CMT (5.5 and 7.4 ng Mn m^{-3}). Another study gave an estimate that less than 1 % of the daily manganese uptake is derived from inhalation by all age groups living in Canadian cities [101]. A recent study conducted by US-EPA and other standard setting bodies around the world on personal exposure in Canadian cities to manganese has determined no elevated manganese levels [102].

Since there is a high background level of manganese oxide, elevated levels of manganese are not easy to identify in an urban atmosphere. Therefore numerous studies have been done in order to determine the effect of MMT addition to gasoline on the environment. A very comprehensive study from Loranger and Zayed [103] shows how the manganese and lead levels behave in ambient air as well as on totally suspended particles in automobile emissions during the successive introductory process of MMT in Canada from 1981 to 1992. K-edge X-ray absorption near-edge fine structure spectrometry (XANES) has revealed that particles from automobile exhaust contain mainly Mn$_3$O$_4$, MnSO$_4$.H$_2$O, Mn$_5$(PO$_4$)[PO$_3$(OH)]$_2$.4H$_2$O, when MMT-containing gasoline was used [104].

Lead levels in the upper soil have often been analyzed against distance to the road to show the effect of tetraalkyl lead. This approach was also followed with MMT. However, since manganese occurs in large quantities as its oxides, and some of the manganese compounds emitted on particles in the automobile exhaust are very soluble in contrast to the soil derived manganese oxides, only the exchangeable fraction of the soil was studied against distance to roads. Although the different manganese species have been appreciated in the analysis no change of the manganese level in the exchangeable fraction of soil was identified [105].

An estimated increase in atmospheric manganese from MMT usage was calculated. With a concentration of 33 mg MMT L^{-1} of gasoline, the atmospheric manganese concentration would double to approximately 50 ng m^{-3} in an urban environment. Although the daily uptake of airborne manganese would double to about 0.2 μg, this would count only for a small amount of the daily uptake from water and food of about 120 μg per day [89]. This seems insignificant but it is not clear whether doubling of the daily intake rate has a significant impact on humans or not.

Biological production of organomanganese compounds in the laboratory appears not to have been investigated. Nothing is currently known about the biological production of organomanganese compounds in laboratory experiments, or even directly in the environment. It may not occur.

9.6 ORGANOGERMANIUM COMPOUNDS IN THE ENVIRONMENT

9.6.1 INTRODUCTION AND TERMINOLOGY

Germanium is a trace element in the earth's crust (5.6×10^{-4} mass-%) and is chemically very similar to silicon. Their similar atomic radii allow the geochemistry of germanium (at least at low temperatures) to be linked closely to that of silicon [106]. Germanium exists in oxidation state IV except under very unusual laboratory conditions; Ge(IV) is used to symbolize all purely inorganic germanium species in this chapter. Methylated compounds have the following abbreviations: for monomethylgermanium ($MeGe-X_3$), dimethylgermanium (Me_2Ge-X_2), trimethylgermanium (Me_3Ge-X) and tetramethylgermane (Me_4Ge) respectively.

9.6.2 USES AND ENVIRONMENTAL DISTRIBUTION

Germanium and, in particular organogermanium compounds are extensively used in the semiconductor industry and with it the distribution of germanium in anthropogenic influenced areas have been increased. Many different organogermanium products, like carboxyethylgermanium sesquioxide or citrate-lactate germanium have been synthesized and tested for their antitumour activities [107].

9.6.3 TOXICOLOGY OF ORGANOGERMANIUM COMPOUNDS

Embryonic teratogenic effects have been observed when Me_2GeO oxide was administered to chickens [108]. Triethylgermanium acetate (Et_3GeOAc) showed an LD_{50} value of 125–250 mg kg^{-1} body weight, when it was administrated to rats [109].

9.6.4 SYNTHESIS OF ORGANOGERMANIUM STANDARDS

Tetramethylgermanium (Me_4Ge) is commercially available and can be produced using the Grignard reaction: germanium tetrachloride ($GeCl_4$) reacts with methylmagnesium iodide in n-butyl ether [110, 111]. Me_4Ge was distilled off and washed with cold concentrated sulfuric acid and fractionated in a microcolumn with a yield of 40% (b.p. 44.3 °C), whereas diethyl ether as a solvent gave only 10% yield. The low yield can be explained by the highly volatile product, Me_4Ge. Quantitatively, Me_4Ge can be synthesized under vaccum in an ampoule [112], when stoichiometric amounts of dimethylzinc (Me_2Zn) and $GeCl_4$ are mixed and distilled *in vacuo* into a third ampoule.

Me_3Ge^+ can be purchased commercially and an aqueous solution is stable for 6 months when it is stored at 4°C [113]. Me_3Ge^+ is expected to be rather stable in air similar to Me_2Se, since the metal-carbon bond energies are very similar (see Chapter 1). No data on air stability have been found in the literature.

9.6.5 ANALYTICAL METHODS FOR ENVIRONMENTAL GERMANIUM SPECIATION

For the determination of inorganic and methylated germanium compounds in water samples, hydride generation with cold trapping using a water trap was linked to a graphite-furnace tube in order to atomize the methylated germanium hydrides to elemental germanium. Atomic absorption spectrometric detection guarantees a very specific detection of volatile germanium compounds [114]. An alternative approach is to make use of the Grignard reaction with pentylmagnesium iodide in order to pentylate the organogermanium species directly in the water sample. Subsequent extraction followed by gas chromatographic separation and flame photometric detection (GC-FPD) gave detection limits of 150, 50, 70 and 150 pg for Ge(IV), $MeGe^{3+}$, Me_2Ge^{2+} and Me_3Ge^+ respectively [113]. Liquid chromatography was also linked to element specific detection. Since this technique suffers from relatively high detection limits, post-column on-line hydride generation was introduced by using ICP-AES as the Ge-specifc detector [115]. A very specific technique for Me_3Ge^+ was developed by Jiang and Adams [113, 116]. A quartz surface-induced germanium emission for Me_3GeCl was used in a modified GC-FPD. Prior to the analysis, the column was pretreated with 2% HCl in methanol and then the water samples were injected. Detection limits of 3 pg Ge(IV), using 100 mL water sample a limit of detection of 15 ng L^{-1} could be achieved, with a recovery rate of Me_3GeCl of about 86–111%. No Me_3GeCl was found in water samples collected [113]. Even better detection limits can be achieved if hydride generation gas chromatography is coupled to an ICP-MS. Due to the high efficiency of hydride generation of 97% a detection limit of 0.08 pg Ge(IV), 0.1 pg $MeGe^{3+}$ and Me_2Ge^{2+} and 0.09 pg for Me_3Ge^+ has been achieved by a dynamic range or four orders of magnitude; germanium was monitored on m/z 74 [117].

9.6.6 OCCURRENCE IN THE ENVIRONMENT

A very good review of the biogeochemistry of methylated germanium species was published in 1988 [118], in which the behaviour of methylated species in the airborne and freshwater environments were discussed. In estuaries and surface water Ge(IV) displays a nutrient-like behavior similar to silicon. Siliceous organisms take up Ge(IV) as if it were a super-heavy isotope of silicon. The concentration of germanium increases steadily in deeper waters to concentrations of about 10 ng L^{-1}. For example, in deep waters ($>$ 200 m), the concentration of inorganic germanium increased from the Southern Tasman Sea towards the deeper water in the north, similar to arsenic. Analogous to arsenic which is linearly correlated to the phosphate ($\Delta As/\Delta P = 4.53 \times 10^{-3}$), Ge(IV) shows a linear concentration correlation to silicon ($\Delta Ge/\Delta Si = 0.73 \times 10^{-6}$) [119]. Another study established a very similar Ge/Si ratio in natural ocean waters (0.56×10^{-6}) [120]. This ratio would vary especially in organic rich waters by the stronger binding of Ge(IV) to organochelates [121].

In contrast to organoarsenic compounds such as Me_2As^+ or Me_3As^{2+}, $MeGe^{3+}$ is uniformly distributed in a vertical profile in the Pacific Ocean in a concentration of 16 ng L^{-1}. This indicates that $MeGe^{3+}$ concentration is independent of the distribution of biota. Thus $MeGe^{3+}$ is not involved in the biogeochemical cycling of Ge(IV) [119]. Furthermore, $MeGe^{3+}$ and Me_2Ge^{2+} show a linear behaviour to salinity in estuaries and oceans. The end-member (open ocean) data do not vary by location, in contrast to Ge(IV): a fairly constant concentration of about 133 $\pm$ 15 pM $MeGe^{3+}$ and 120 $\pm$ 20 pM Me_2Ge^{2+} occurs [122]. Methylated germaniums are barely detectable in pristine rivers ($<$ 3 pM $MeGe^{3+}$ and $<$ 7 pM Me_2Ge^{2+}) in contrast to contaminated rivers, where concentrations of about 15 $\pm$ 8 pM $MeGe^{3+}$ and 15 $\pm$ 6 pM Me_2Ge^{2+} have been determined [123]. From this behaviour in the open ocean it can be concluded that methylated germanium is entirely decoupled from the biogeochemical cycle of Ge(IV) and neither production nor sinks of $MeGe^{3+}$ and Me_2Ge^{2+} could be identified.

By applying hydride generation cryotrapping GC-ICP-MS directly to wet sediments, methylated germanium $MeGe^{3+}$ has been identified in freshwater sediments at a maximum concentration of 0.3 μg kg^{-1} [75]. Waters from New Zealand geothermal sources showed a considerable amount of organogermanium species ($MeGe^{3+}$, Me_2Ge^{2+} and also Me_3Ge^{3+}) [124]. From 11 different samples the total germanium concentration varied between 6 and 137 μg kg^{-1} in the water. The overall proportion of organogermanium counted only for $<$ 0.01 to 0.19 % of the total germanium; with $MeGe^{3+}$ ($<$ 0.01–0.06 μg kg^{-1}), Me_2Ge^{2+} (0.01–0.03 μg kg^{-1}) and Me_3Ge^+ ($<$ 0.01–0.17 μg kg^{-1}) [124].

9.6.7 BIOLOGICAL PRODUCTION OF ORGANOGERMANIUM COMPOUNDS IN THE LABORATORY

Methyl iodide (MeI) and methylcobalamin ($MeCoB_{12}$) were added to an aqueous solution of germanium at a concentration of about 100 μg L^{-1} at various pH. All experiments gave only $MeGe^{3+}$, with the best yields for MeI of 1.6 % and 6 % at pH $<$ 1 and pH $>$ 6, respectively. $MeCoB_{12}$ however, produced $MeGe^{3+}$ only at low pH (best yield at pH 1 with 1.3 %) [125]. It was assumed that the chemical methylation with MeI is based on an oxidative addition, although Ge(II) does not exist in the environment, whereas a free-radical mechanism is assumed for $MeCoB_{12}$ (Equations 9.7 and 9.8).

$$CH_3I + Ge^0 \longrightarrow CH_3Ge^{II}I \qquad (9.7)$$

$$CH_3I + CH_3Ge^{II}I \longrightarrow (CH_3)_2Ge^{IV}I_2 \qquad (9.8)$$

More likely is an SN_2 reaction of $MeCoB_{12}$ and Ge(IV). A carbanion (CH_3^-) could attack Ge(IV)as a strong nucleophile and replace a coordinating ligand to give a methylated CH_3—Ge(IV) compound. This would fit to the observation that no Ge(II) species exist under normal conditions in the environment.

Biomethylation of Ge(IV) to $MeGe^{3+}$, Me_2Ge^{2+} and Me_3Ge^+ has been observed under anoxic conditions in the presence of anaerobic bacteria whereas marine algae do not methylate germanium [126]. However, this could not be confirmed by the analysis of the hot spring water [124]. This study showed that oxic as well as anoxic geothermal waters could produce Me_3Ge^+ compounds.

9.7 ORGANOCOBALT COMPOUNDS IN THE ENVIRONMENT

9.7.1 INTRODUCTION AND TERMINOLOGY

Methylcobalamin ($MeCoB_{12}$) is an important enzyme cofactor required for methionine synthase activity. Methionine synthase (MS) catalyses the methylation of homocysteine to methionine and requires $MeCoB_{12}$ as a cofactor. The sixth axial metal compounds coordination site features a primary methyl group, which makes this complex the only fully established example of a natural organometallic compound in biochemistry, with a detailed function in bio-methylation of heavy metals plus adenosyl CoB_{12}. The importance of $MeCoB_{12}$ can be illustrated in this sense, that a methyltransfer as a radical or as a nucleophile is only possible via this stabilized cobalt species [127]. Most biomethylations use an electrophilic methyl group, readily available from S-adenosylmethionine (SAM) or through 5-methyltetrafolic acid (SMeTHF). This is important since easy polarizable and soft elements such as selenium and mercury can only be methylated by a carbanion (CH_3) via an SN_2 reaction. This may be extended to bismuth (see Section 9.1 above).

9.7.2 USES AND ENVIRONMENTAL DISTRIBUTION

Methylcobalamin is produced widely by microorganisms, but it also occurs in higher life forms, for example in mammals. In contrast to $5'$-desoxyadenosylCoB$_{12}$ (coenzyme B$_{12}$), $MeCoB_{12}$ is found in the circulatory system of mammals rather than in the liver [128]. This chapter does not discuss the biochemistry of this complex further but refers the reader to biochemistry literature for more detailed studies [129–131]. $MeCoB_{12}$ can be used for the synthesis of methylated metal compounds. For example, is it possible to synthesize fast and clean methylmercury ($MeHg^+$) within less than 4 hours, which is very attractive for radioactive labeled $Me^{203}Hg$ [132].

Main group metals of the groups (13 to 17) tend to be methylated in nature. The key question of whether the methylation is a spontaneous transmethylation or an enzymatically catalysed reaction has not been answered yet, although some findings suggest that methylation can take place without the involvement of biota, but to a lesser extent [34]. It has been shown that mercury can be methylated by *Desulfovibrio desulfuricans* by $MeCoB_{12}$ [133]. It is likely that the *in-vivo* methylation of mercury(II) is an enzymatically catalyzed process, since the reaction rate of methylation of the cell extract of *D. desulfuricans* is 600-times higher compared with the transmethylation by the free $MeCoB_{12}$ [134]. Berman *et al.* [135] have shown that radiocarbon from pyruvate and serine was incorporated into methylmercury. Tetrahydrofolate 4–35 µg, and 58–161 ng cobalamin were found per gram of cell protein. In a later study [136] they found that 95 % of the labeled cobalt (^{57}Co) were associated with macromolecules rather than with free cobalamin. Gel permeation and classical electrophoresis identified this macromolecule as a 40 kDa corrinoid protein, which might be the methyl donor. Summarizing, $MeCoB_{12}$ can transfer the methyl group to metals such as mercury(II) chemically, but it is more likely that *in vivo* this process is enzymatically catalysed.

In the mammalian organism, methylation of inorganic arsenic through arsenic methyltransferases is well known but the detection of the enzyme has not been reported. Zakharyan and Aposhian [137] have demonstrated in their experiments that arsenite can easily be converted to methylarsonic acid and small amounts of dimethylarsinic acid in the presence of thiols and $MeCoB_{12}$ in the absence of enzymes. This clearly is a study that provides information that the methylation of arsenic might be a non-enzymatic reaction. The reaction rate was enormously increased in the presence of dimercaptopropanesulfonate (DMPS) and selenite, and the reaction rate was not influenced by the addition of liver cytosol. On the other hand, it was reported that selenite inhibits the methylation of arsenite, but the addition of potential methylgroups, L-methionine and Me—B_{12} into caecal contents significantly increased the rate of methylation, especially of arsenate [138].

9.7.3 SYNTHESIS OF ORGANOCOBALT STANDARDS

$MeCoB_{12}$ itself can easily be synthesized using aquocobalamin and methyliodine in a yield of more than 90 % [139]. The aquocobalamin(III) is reduced by formate and subsequently methylated by methyliodide to give Me—Co(III) —B12 (Equation 9.9).

$$H_2O-Co(III)B_{12} \xrightarrow{+e} Co(II)B_{12} \xrightarrow{+CH_3^+} CH_3Co(III)B_{12} \qquad (9.9)$$

9.7.4 ANALYTICAL METHODS FOR ENVIRONMENTAL COBALT SPECIATION

For the determination and separation of organocobalt compounds, a microborereverse-phase HPLC has been coupled to UV (278 nm), ion-spray mass spectrometry (IS-MS) and ICP-MS as a detector. The following compounds have been separated: cyanocobalamin, 5'-desoxyadenosylcobalamin (coenzyme B_{12}), methylcobalamin (MeCoB$_{12}$), hydroxocobalamin, aquocobalamin, and cobinamide dicyanide. Detection limits of about $10\,ngmL^{-1}$ have been achieved [140, 141].

9.7.5 BIOLOGICAL PRODUCTION OF ORGANOCOBALT COMPOUNDS IN THE LABORATORY

The growth of *Chlorella vulgaris* is stimulated by the addition of aquocobalamin. In an aquocobalamin-free medium, *Chlorella* contained the cobalamin coenzymes, 5'-desoxyriboadenosylcobalamin (AdenCoB$_{12}$) and MeCoB$_{12}$, This is a strong indication that the algae have the ability to take up aquocobalamin and to synthesize the coenzyme B_{12} [142]. Clearly, the existence of MeCoB$_{12}$ as a natural product shows that the biochemistry of cobalt includes biomethylation.

9.8 ORGANOPOLONIUM COMPOUNDS IN THE ENVIRONMENT

Polonium only has radioactive isotopes (more than 27 such isotopes are known) but the isotope 210 amu is the most interesting one, since it can easily be generated in an atomic reactor when bismuth (209 amu) is radiated with neutrons. This isotope has a half-life of about 138 days. Polonium is generated when uranium decays, so that uranium minerals contain small amounts of polonium. Its chemistry is very similar to that of tellurium and bismuth. Since those two elements form rather stable organometallic compounds which can be found in the environment, it may be expected that methylated polonium compounds will be discovered to occur in the near future.

Recently Momoshima and coworkers [143] noticed the generation of a volatile polonium-containing product if sea sediments were inoculated with polonium. Sterilized sediments did not generate this not-yet-identified volatile polonium species. Furthermore, the same group found that polonium reacts with MeCoB$_{12}$ to form this volatile polonium species. It is quite likely that dimethyl polonium (Me_2Po) was generated but the positive identification has to be shown.

9.9 CARBONYL COMPLEX FORMING TRANSITION METALS

Since the metal–carbon bonds of the transition metals are very unstable unless stabilized by ligand bonding, methylated compounds of those metals can only occur in biological or environmental samples when the metal is stabilized in a chelating complex, e.g. carbonyls or the tetrapyrrole ligand in methylcobalamin. These elements do form strong complexes with carbon monoxide (CO) due to the synergetic effect of d_π–p_π backbonding. Although these compounds are not always considered as organometallic compounds, they should be mentioned here as some have been detected in the environment.

Recently, it has been fund that volatile carbonyl compounds of molybdenum, tungsten and nickel occur in landfill and sewage gases [144, 145]. The gases were cryotrapped and determined by packed GC-ICP-MS. In the landfill gas all isotopes of molybdenum and tungsten showed sharp peaks in the ICP-MS traces in the proportion which is expected by the natural abundance of those metals. The identifications were done by retention time comparison and standard addition of $Mo(CO)_6$ and $W(CO)_6$ to the sample. That these findings are not artifacts of the method but also not typical for Canadian landfill sites has been shown recently [29]. Krupp has found volatile molybdenum and tungsten hexacarbonyls ($M(CO)_6$) in gases from German landfill sites using capillary GC-ICP-MS. Whether these compounds are generated by an entirely abiotic pathway or by enzymatically catalyzed reactions cannot be answered yet. However, the release of industrially produced hexacarbonyl compounds from waste on the landfill site can be ruled out, since these volatile compounds could also be identified in sewage gas produced by fermentation of a sewage sludge at a municipal water treatment plant [145]; in addition to $Mo(CO)_6$, $W(CO)_6$, and $Ni(CO)_4$ were also identified in the headspace of the fermenter. It is perhaps not surprising that $Fe(CO)_5$ has not been detected yet, since this compound showed the least stability of all four metal carbonyls discussed here under a carbon monoxide atmosphere in the presence of water [145]. So if $Fe(CO)_5$ is also generated, the expected concentration may be rather small due to its low stability. Furthermore, there are some reports about volatile nickel compounds (not filterable) in urban air and in the exhaust of automobiles, which might suggest that $Ni(CO)_4$ is formed in automobiles [146]. As stated before, the generation of these compounds is not clear yet, but a purely inorganic process is most probable. If nickel can be reduced to elemental nickel (which might happen in an anaerobic fermenter) than this reactive nickel might react with the small concentration of CO in the fermenter. It has been known for half a century that NiS reacts under alkaline conditions in a CO atmosphere to give $Ni(CO)_4$ [147]. So a purely abiotic reaction is possible.

9.10 REFERENCES

1. Ojebuoboh, F.K., *J. Min. Met. Mater. Soc.*, 1992, **44**, 46.
2. Saager, P. In *Metallic Raw Material from Antimony to Zirconium (Bismuth)*, Bank von Tobel, Zürich, 1984, 95–98, cited in Thomas, D.W., In *Metals and Their Compounds in the Environment*, E. Merian (Ed.), VCH, Weinheim, 1991, pp. 789–800.
3. Menge, H., Gregor, M., Brosius, B., Hopert, R., and Lang, A., *European J. Gastroenterol. Hepatol.*, 1992, **4**, 41.
4. Banie, P.J., Djuran, M.I., Mazid, M.A., McPartlin, M., Sadler, P.J., Scowen, I.J., and Sun, H., *J.Chem. Soc. Dalton Trans.*, 1996, 2417.
5. Sandha, G.S., *et al.*, *Digest. Diseases Sci.*, 1998, **43**, 2727.
6. Alarcon, T., Domingo, D., and Lopez Brea, M., *Intern. J. Antimicrob. Agents*, 1999, **12**, 19.
7. Sadler, P.J., Sun, H., and Li, H. *Chem. Eur. J.*, 1999, **2**, 701.
8. Roosen, N., Doz, F., Yeomans, K.L., Dougherty, D.V., and Rosenblum, M.L., *Cancer Chemother. Pharmacol.*, 1994, **34**, 385.
9. Merkel, D., Matter, Y., and Appuhn, H., *Korrespondenz Abwasser*, 1994, **41**, 264.
10. Sun, H.Z., Li, H.Y., and Sadler, P.J., *Chem. Ber.-Recueil*, 1997, **130**, 669.
11. Giani, A. *et al.*, *Thin Solid Films*, 1998, **315**, 99.
12. *Minerals Yearbook 1986*, Vol. 1, Bureau of Mines, US Department of the Interior Washington, DC, 1988, Vol. I, p.164, cited in Thomas, D.W., In *Metals and Their Compounds in the Environment*, E. Merian (Ed.), VCH, Weinheim, 1991, pp. 789–800.
13. Erkens, L.J.H., and Vos, L.J., *Bulletin of the Bismuth Institute*, 1997, 70.
14. Arena, J.M. In *Poisoning; Toxicology Symptoms, Treatments*, 2nd edn, C.C. Thomas (Ed.), Springfield III., 1970, p.73.
15. Sollmann, T., and Seifter, J., *J. Phamacol. Environm. Therapeutics*, 1939, **67**, 17.
16. Charch, W.C., Mack E. Jr., and Boord, C.L., *Ind. Enging Chem.*, 1926, **18**, 334.
17. Thomas, D.W. In *Metals and Their Compounds in the Environment*, E. Merian (Ed.), VCH, Weinheim, 1991, pp. 789–800.
18. Knijff, R., Paton, G.I., and Feldmann, J., unpublished results, University of Aberdeen, 1999.
19. Akiba, K.Y., and Yamamoto, Y. In *The Chemistry of Organic Arsenic, Antimony and Bismuth Compounds*, S. Patai (Ed.), Wiley, Chichester, 1994, p. 761.
20. Haas, K., and Feldmann, J., *Anal. Chem.*, 2000, **72**, 4205.
21. Halt G.E.M., MacLaurin, A.I., Pelchat, J.C., and Gauthier, G., *Chem. Geol.*, 1997, **137**, 79.
22. Mota, P.V., Fernandez de la Campa, M.R., and Sanz-Medel, A., *J. Anal. At. Spectrom.*, 1998, **13**, 431.
23. Feldmann, J., Koch, I., and Cullen, W.R., *Analyst*, 1998, **123**, 815.
24. Feldmann, J., Krupp, E.M., Glindemann, D., Hirner, A.V., and Cullen, W.R., *Appl. Organomet. Chem.*, 1999, **13**, 739.
25. Müller, L.M., MSc thesis, University of Essen, 1994.
26. Koch, I., Feldmann, J., Lintschinger, J., Serves, S.V., Cullen, W.R., and Reimer, K.J., *Appl. Organomet. Chem.*, 1998, **12**, 129.
27. Desvyatykh, G.G., Nikishin, A.S., Moiseev, A.N., and Votintsev, V.N., *Vysokochistye Veshchestva English*, 1992, **5–6**, 133.
28. Skinner, H.A. In *Advances in Organometallic Chemistry*, Vol. II, F.G.A. Stone and R. West (Eds.), Academic Press, New York, 1964, p. 49.
29. Krupp, E., PhD thesis, University of Essen, Cuivillier, Göttingen, 1999.
30. Feldmann, J., Grümping, R., and Hirner, A.V., *Fresenius J. Anal. Chem.*, 1994, **350**, 228.

31. Hirner, A.V., Grüter, U.M., and Kresimon, J., *Fresenius J. Anal. Chem.*, 2000, **368**, 263.
32. Grüter, U.M., Kresimon, J., and Hirner, A.V., *Fresenius J. Anal. Chem.*, 2000, **368**, 67.
33. Feldmann, J., unpublished results, 2000.
34. Michalke, K., Meyer, J., Hirner, A.V., and Hensel, R., *Appl. Organomet. Chem.*, 2002, **16**, 221.
35. Kemper, F.H., and Bertram, H.P. In *Metals and Their Compounds in the Environment*, E. Merian (Ed.), VCH, Weinheim, 1991, pp. 1227–1241.
36. Gilman, H., and Jones, R.G., *J Am. Chem. Soc.*, 1946, **46**, 517.
37. Mareko, I.E., and Southern, J.M., *J. Org. Chem.*, 1990, **55**, 3368.
38. Schedlbauer, O.F., and Heumann, K.G., *Anal. Chem.*, 1999, **71**, 5459.
39. Schedlbauer, O.F., and Heumann, K.G, *Appl. Organomet. Chem.*, 2000, **14**, 330.
40. Hirner, A.V., Feldmann, J., Goguel, R., Rapsomanikis, S., Fischer, R., and Andreae, M.O., *Appl. Organomet. Chem.*, 1994, **8**, 65.
41. Huber, F., and Kotulla P., *Chemosphere*, 1982, **11**, N6.
42. Huber, F., and Kirchmann, H., *Inorg. Chim. Acta.* 1978, **29**, L1249.
43. Nesmeyanov, A.N., and Kocheshkov, K.A., *Methods of Elemento-Organic Chemistry*, Vol. 3, North Holland, Amsterdam, 1967, Chapter 1.
44. Chodkiewicz, W., Guillerm, D., Jore, D., Mathieu, E., and Wodzki, W., *J. Organomet. Chem.*, 1984, **269**, 107.
45. Cavanagh, K., and Evans, D.F., *J. Chem. Soc. A*, 1969, 2890.
46. Pongratz, R., and Heumann, K.G., *Anal. Chem.*, 1996, **68**, 1262.
47. Feldmann, J., and Hirner, A.V., *Int. J. Environm. Anal. Chem.*, 1995, **60**, 339.
48. Land, T., and Morel, F., *Proc. Nat. Acad. Sci.*, 2000, **97**, 4627.
49. Heumann, K.G. In *Environmental Contamination in Antarctica—A Challenge to Analytical Chemistry*, S. Caroli, P. Crescon and D.W.H. Walton (Eds), 2000 (in press).
50. Pongratz, R., and Heumann, K.G., *Chemosphere*, 1999, **39**, 89.
51. Pongratz, R., and Heumann, K.G., *Chemosphere*, 1999, **36**, 1935.
52. Robinson, J.W., and Kiesel E.L., *J. Environm. Sci. Health*, 1981, **A16**, 341.
53. Huey, C., Brinckman, F.E., Iverson, W.P., and Grim, S.O. In *Progress in Water Technology*, P.A. Krenkel (Ed.), 1976, p. 7.
54. Panichev, N.A., Diakov, A.O., and Kvitko, K.V., *Canadian J. Anal. Sci. Spectros.*, 1997, **42**, 116.
55. Sadeh, T. In *The Chemistry of Organic Selenium and Tellurium Compounds*, S. Patai (Ed.), Wiley, Chichester, Vol. 2, 1987, pp. 367–376.
56. Muangnoicharoen, S., Chiou, K.Y., and Manuel, O.K., *Talanta*, 1988, **58**, 679.
57. Yoon, B.M., Shim, S.C., Pyun, H.C., and Lee, D.S., *Anal. Sci.*, 1990, **6**, 561.
58. Beliles, R.P. In *The Lesser Metals*, F.W. Oehme (Ed.), Marcel Dekker, New York, 1979.
59. Giant A., Boulouz, A., Pascal-Delannoy, F., Foucaran, A., and Boyer, A., *Thin Solid Films*, 1998, **315**, 99.
60. Franke, K.W., and Moxon, A.L., *J. Pharm. Exp. Therapeut.*, 1937, **61**, 89.
61. Kozik, I.V., Novikova, N.P., Sedova, L.A., and Stepanova, E.N., *Gig. Tr. Prof. Zabol.*, 1981, 51.
62. Albayati, M.A., Raabe, O.G., and Teague, S.V., *J. Toxicol. Environm. Health*, 1992, **37**, 549.
63. Taylor, A., *Biol. Trace Elem. Res.*, 1996, **55**, 231.
64. Schroeder, HA., Buckman, J., and Bahlassa, J., *.J. Chronic Dis.*, 1967, **20**, 147.
65. Oettel, H. In *Ullmanns Encylcopadie der technischen Chemie*, VCH, Weinheim, 1965, p. 16.

66. Einbrodt, H.J., and Michels J. In *Metalle in der Umwelt*, E. Merian, (Ed.), VCH, Weinheim, 1984, pp. 561–569.
67. Rheinboldt, H. In *Methoden der organischen Chemie*, E. Müller (Ed.), Thieme, Stuttgart, 1967, pp. 1047–1085.
68. Chasteen T.G., Silver, G.M., Birks, J.W., and Fall, R, *Chromatographia*, 1990, **30**, 181.
69. Van Fleet-Stadler, V., and Chasteen, T.G., *J. Photochem. Photobiol.*, 1998, **43**, 193.
70. Lindemann, T., Prange, A., Danneker, W., and Neidhart, B., *Fresenius J. Anal. Chem.*, 1999, **364**, 462.
71. Lindemann, T., Prange, A., Danneker, W., and Neidhart, B., *Fresenius J. Anal. Chem.*, 2000, **368**, 214.
72. Lindemann, T, personal communication, 2000.
73. Riechmann, T., MSc thesis, University of Essen, 1995.
74. Feldmann, J., PhD thesis, University of Essen, Cuvillier, Göttingen, 1995.
75. Krupp, E.M., Grümping, R., Furchtbar, U.R.R., and Hirner, A.V., *Fresenius J. Anal. Chem.*, 1996, **354**, 546.
76. Klingenberg, H., van der Wal, S., de Koster, C., and Bart, J., *J. Chromatography A*, 1998, **794**, 219.
77. Gharieb, M.M., Kierans, M.L, and Gadd, G.M., *Mycol. Res.*, 1999 **103**, 299.
78. Challenger, F.E., *Adv. Enzymol.*, 1951, **12**, 429.
79. Bird, M.L., and Challenger, F., 1939, *J. Chem. Soc.*, 1939, 163.
80. Akpolat, O.M., MSc thesis, Sam Houston State University, Huntsville, Texas, 1999.
81. Solozhenkin, P.M., Lyubavina, L.L., Buyanova, N.N., and Kirshenina, L.V., *Chem. Abstr.*, 1993, **119**, 229275s.
82. Reamer, D.C., and Zoller,W.H., *Science*, 1980, **208**, 500; Lewis, B. A. G. In *Environmental Biochemistry*, Vol. 1, *Carbon, Nitrogen, Phosphorus, Sulfur and Selenium Cycles*, J.O. Nriagu (Ed.), Ann Arbor Science, Michigan, 1976, pp. 389–409.
83. Doran, J.W., *Adv. Microbiol. Ecol.*, 1982, **6**, 1.
84. Glindemann, D., Bergman, A., Hirner, A.V., Krupp, E.M., and Kuschk, P., 2002, in preparation.
85. Faulkner, D.J. In *Handbook of Environmental Chemistry*, Vol. 1A, O. Hutzinger (Ed.), Springer, Berlin, 1980, pp. 229–252.
86. Boles, J.O., *Diss. Abstr. Int.*, 1993, **53**, 4077B.
87. Davis, J.M., *Neutrotoxicol.*, 1999, **20**, 511.
88. Zayed, J., *Am. J. Ind. Med.*, 2001, **39**, 426.
89. Lynam, D.R., Pfeiffer, G.D., Fort, B.F., and Gelbcke, A.A., *Sci. Tot. Environm.*, 1990, **93**, 107.
90. Hysell, D.K., Moore, W., Stara, J.F., Miller, R., and Cambell, K.I., *Env. Res.*, 1974, **7**, 158.
91. Hinderer, R.K., *Am. Ind. Hyg. Assoc. J.*, 1979, **40**, 164; Hanzlik, R.P., Stitt, R., and Traiger, G.J., *Toxicol. Appl. Pharmacol.*, 1980, **56**, 353; Hinderer, R.K., *Am. Ind. Hyg. Assoc. J.*, 1979, **40**, 164. All cited in Lynam D.R., *et al.*, *Sci. Tot. Environm.*, 1990, **93**, 107.
92. Hakkinen, P.J., and Haschek, W.M., *Toxicol. Appl. Pharmacol.*, 1982, **65**, 11.
93. Clay, R.J., and Morris, J.B., *Toxicol. Appl. Pharmacol.*, 1989, **98**, 434.
94. Butcher, D.J., Zybin, A., Bolshov, M.A., and Niemax, K., *Anal. Chem.*, 1999, **71**, 5379.
95. Fragueiro, M.S., Alava-Moreno, F., Lavilla, I., and Bendicho, C., *Spectrochim. Acta*, 2001, **56B**, 215.
96. Fan, Y., and Chau, Y.K., *Analyst*, 1999, **124**, 71.
97. Ter Haar, G.L., Griffing, M.E., Brandt, M., Oberding, D.G., and Kapron, M., *J. Air Pollut. Control Assoc.*, 1975, **25**, 858.

98. Wallington, T.J., *et al.*, *Env. Sci. Technol.*, 1999, **33**, 4232.
99. Zayed, J., Thibault, C., Gareau, L., and Kennedy, G., *Neutrotoxicol.*, 1999, **20**, 151.
100. Chau, Y.K., Yang, F., and Brown, M., *Appl. Organomet. Chem.*, 1997, **11**, 31.
101. Egyed, M., and Wood, G.C., *Sci. Tot. Envi.*, 1996, **190**, 11.
102. Lynam, D.R., Roos, J.W., Pfeiffer, G.D., Fort, B.F., and Pullin, T.G., *Neurotoxicol.*, 1999, **20**, 145.
103. Loranger, S., and Zayed, J., *Atmosph. Env.*, 1994, **28**, 1645.
104. Ressler, T., Wong, J., Ross, J., and Smith, I.L., *Env. Sci. Technol.*, 2000, **34**, 950.
105. Normandin, L., Kennedy, G., and Zayed, J., *Sci. Tot. Environm.*, 1999, **239**, 165.
106. Goldschmidt, V.M., *Naturwissenschaften*, 1926, **14**, 295.
107. Schauss, A.G., *Ren. Fail.*, 1991, **13**, 1.
108. Caujolle, F., Caujolle, D., Cros, S., Giao, M.D.H., Moulas, F., Tollon, Y., and Caylus, J., *Chem. Abstr.*, 1966, **65**, 19054c.
109. Cremer, J.E., and Aldridge, W.N., *Br. J. Ind. Med.*, 1965, **21**, 214.
110. Lippincott, E.R., and Tobin, M.C., *J. Am. Chem. Soc.*, 1953, **75**, 4141.
111. Young, C.W., Koehler, J.S., and McKinney, D.S., *J. Am. Chem. Soc.*, 1947, **69**, 1411.
112. Lengel, J.H., and Dibeler, V.H., *J. Am. Chem. Soc.*, 1952, **74**, 2683.
113. Jiang, G.B., and Adams, F.C., *J. Chromatography A*, 1997, **759**, 119.
114. Hambrick, G.A., Froehlich, P.N., Andreae, M.O., and Lewis, B.L., *Anal. Chem.*, 1984, **56**, 421.
115. Padro, A., Rubio, R., and Rauret, G., *Fresenius J. Anal. Chem.*, 1995, **351**, 449.
116. Jiang, G.B., and Adams, F.C., *Anal. Chim. Acta*, 1997, **337**, 83.
117. Jin, K., Shibata, Y., and Morita, M., *Anal. Chem.*, 1991, **63**, 986.
118. Lewis, B.L., Andreae, M.O., and Froehlich, P.N. In *The Biological Alkylation of Heavy Metals*, P.J. Craig, P.J., and F. Glocking (Eds), 1988, pp. 77–91.
119. Santosa, S.J., Wada, S., Mokudai, H., and Tanaka, S., *Appl. Organomet. Chem.*, 1997, **11**, 403.
120. Froehlich, P.M., Hambrick, G.A., Andreae, M.O., Mortlock, R.A., and Edmond, J.M., *J. Geophys. Res.*, 1985, **90**, 1122.
121. Pokrovski, G.S., Martin, F, Hazemann, J.L., and Schott, L., *Chem. Geol.*, 2000, **163**, 151.
122. Lewis, B.L., Andreae, M.O., Froehlich, P.N., and Mortlock, R.A., *Sci Tot. Environm.*, 1988, **73**, 107.
123. Lewis, B.L, Andreae, M.O., and Foehlich P.N., *Mar. Chem.*, 1988, **27**, 179.
124. Hirner, A.V., Feldmann, J., Krupp, E., Grümping, R., Goguel, R., and Cullen, W.R., *Org. Geochem.*, 1998, **29**, 1765.
125. Mayer, H.P., and Rapsomanikis, S., *Appl. Organomet. Chem.*, 1992, **6**, 173.
126. Lewis, B.L., and Mayer, H.P. In *Metal Ions in Biological Systems*, Vol. 29, H. Sigel and A. Sigel (Eds), Marcel Dekker, NY, 1993, pp. 79–99.
127. Craig, P.J., and Glocking, F., *The Biological Alkylation of Heavy Elements*, Royal Society of Chemistry, London, 1988.
128. Kaim, W., and Schwederski, B., *Bioinorganic Chemistry: Inorganic Elements in the Chemistry of Life*, Wiley, Chichester, 1994.
129. Dolphin, D., B_{12}, Wiley, New York, 1982.
130. Schneider, Z., and Stroinski, A., *Comprehensive B_{12}*, de Gruyter, Berlin, 1987.
131. Toscano, P.J., and Marzilli, L.G., *Prog. Inorg. Chem.*, 1984, **31**, 105.
132. Rouleau, C., and Block, M., *Appl. Organomet. Chem.*, 1997, **11**, 751.
133. Choi, S.C., and Bartha, R., *Appl. Environm. Microbiol.*, 1993, **59**, 290.
134. Choi, S.C., Chase, T., and Bartha, R., *Appl. Environm. Microbiol.*, 1994, **60**, 1342.
135. Berman, M., Chase, T., and Bartha, R., *Appl. Environm. Microbiol.*, 1990, **56**, 298.

136. SungChang, C., Chase, T., and Bartha, R., *Appl. Environm. Micrcobiol.*, 1994, **60**, 1342.
137. Zakharyan, R.A., and Aposhian, H.V., *Toxicol. Appl. Pharmacol.*, 1999, **154**, 287.
138. Hall, L.L., George, S.E., Kohan, M.J., Styblo, M., and Thomas, D.J., *Toxicol. Appl. Pharmacol.*, 1997, **147**, 101.
139. Tollinger, M., Derer, T., Konrat, R., and Kraeutler, B., *J. Molecular Catalysis*, 1997, **116**, 147.
140. Chassaigne, H., and Lobinski, R., *Anal. Chim. Acta*, 1998, **359** 227.
141. Chassaigne, H., and Lobinski, R., *Analyst*, **123**, 131.
142. Watanabe, F., Abe, K., Takenaka, S., Tamura, Y., Maruyama, I., and Nakano., Y., *Biosci. Biotechnol. Biochem.*, 1997, **61**, 896.
143. Momoshima, N., Song, L.X., Osaki, S., and Maeda, Y., *Environm. Sci. Technol.*, 2001, **35**, 2956.
144. Feldmann, J., and Cullen, W.R., *Environm. Sci. Technol.*, 1997, **31**, 2125.
145. Feldmann, J., *J. Environm. Monit.*, 1999, **1**, 33.
146. Filkova, L., and Jäger, J., *C. Hygenia*, 1986, **31**, 255.
147. Behrens, H., and Eisenmann, E., *Z. Anorg. Allgem. Chem.*, 1955, **269**, 292.
148. Feldmann, J., and Kleimann, J., *Korrespondenz Abwasser*, 1997, **44**, 99.

10 Organoselenium Compounds in the Environment

P.J. CRAIG

Department of Molecular Sciences, De Montfort University, Leicester, UK

and

W.A. MAHER

Ecochemistry Laboratory, University of Canberra, Australia

10.1 INTRODUCTION

There is a large body of information and other material on the toxicology and role of selenium as a trace element, not all of it science based. The speciation of selenium in food and other matrices is discussed here, but the role of this element in toxicology and health is not discussed. It has been well covered in a very recent work, which in turn refers copiously to previous work on this subject [1]. Similarly, the role of selenium as a trace element in agriculture and animal husbandry will not be much discussed here [2].

Selenium has a similar oxidation state chemistry to the true metal tellurium (see Chapter 9), and the Se(II), Se(IV) and Se(VI) states are of most importance. In addition, dimethyl selenide (Me_2Se) and dimethyl diselenide ($MeSe_2$) are of considerable environmental importance (see Section 10.4.3). The atomic number of selenium is 34, and it occurs below sulphur in group 16 of the periodic table. Although there is a trend in the periodic table towards greater metallic behaviour as it is descended, selenium is mainly a non-metal. It is discussed in the present work because of its near metalloidal status and properties and because of its similarity to tellurium. It can be both oxidized and reduced in its environmental and biological chemistry.

10.2 USES AND DISTRIBUTION

Selenium is an essential trace element in animals and humans [3, 4]. The enzymes glutathione peroxidase and iodothyronine deiodinase contain selenium, and there is also a number of proteins which contain this element. Upper and

Organometallic Compounds in the Environment
Edited by P.J. Craig © 2003 John Wiley & Sons Ltd

lower limits of selenium in food have been established [5, 6]. The US RDA for selenium is 70 μg day^{-1} for men and 55 μg day^{-1} for women [7]. The EU has set a maximum level of intake at 450 μg day^{-1} [8]. The speciation of selenium in food and other matrices is discussed in Section 10.4.1 below. Interestingly, despite the health warnings, nuts generally contain a lot of selenium. The role of selenium in cancer prevention is currently of considerable interest, but is outside the scope of this work (but see Ref. 1 for source material).

The unique photochemical and semiconducting properties of selenium have resulted in the extensive use (20%) of this element, mostly as inorganic compounds, in photocell devices (20%) and in xerography (25%), solar batteries, transformers and rectifiers. Its major use is in glass manufacture (35%) for decorative purposes and to reduce glare and heat loss. Other uses include pigments and some small scale use, for example food additives, land supplementation, fungicides, metallurgy, radioactive tracers (^{75}Se selenomethionine) (20%) [9].

Selenium and its salts may be lost to the environment during processing, or be volatilized in the photocopying industry or during melting in the glass industry, leading to losses to atmosphere (about 200–300 tonnes per year in the USA in the 1970s). Burning of fossil fuels also leads to losses of selenium to the atmosphere. About several hundred tonnes from coal combustion escapes to atmosphere each year in the USA, with about 2500 tonnes being lost to land [9].

Taken together with the biomethylation processes that operate on natural trace element selenium, there is also microbial transformation of pollutant selenium into methyl forms. In general terms, these mechanisms have been considered as part of the section on tellurium (Chapter 9). These methylation processes ultimately lead to the species found in food, plants, soils, organisms, etc., and additionally to Me_2Se, Me_2Se_2 and Me_2Se_2O in the atmosphere. In waters, dissolved species as above also occur, together with DMSeP, dimethylselenopropiothetin, the selenium analogue of arsenobetaine, viz. $((CH_3)_2SeCH_2COO$.

Several genera of fungi produce Me_2Se from inorganic selenium species. Rats fed with selenate or selenite salts exhale Me_2Se and excrete Me_3Se in the urine. Various plants grown in selenite-enriched soils produce Me_2Se and Me_2Se_2. Soils incubated with selenite can produce Me_2Se. A strain of penicillium from raw sewage has been shown to produce Me_2Se. Volatile Me_2SeO_2 has been produced from soils and sewage sludge. Volatilization of selenium in sediments has been observed in both aerobic and anaerobic sediments. Methyl selenium species are therefore widely found in the environment.

10.3   ANALYTICAL CHEMISTRY

The analytical chemistry of selenium is discussed in detail in Reference [1] and has recently been reviewed [10–12]. The normal battery of modern techniques has been applied to the speciation analysis of organic (i.e. for this purpose, organometallic) selenium [13–17]. Coupling of HPLC or hydride generation to ICP-MS is the most effective method [18]. Capillary zone electrophoresis and capillary isoelectric focusing ICP-MS has been used to speciate organic and inorganic selenium in body fluids and HPLC-ICP-MS [19] has been used for speciation of selenium in biological samples [20]. Methodology for speciation in mammalian organisms has recently been reviewed [19, 20].

Selenium is normally incorporated into proteins and is present at low concentrations in animal and plant tissues. Thus selenium must be extracted from tissues by using a detergent and enzymatic or acid hydrolysis [21–23] to release the selenium moiety for identification. Commonly, selenoproteins are isolated and purified by size exclusion and anion chromatography or by capillary zone or isoelectric electrophoresis [24–26]. In animal tissues selenium has been identified as the selenocysteine unit. This moeity is not stable and must be derivatized (to carboxymethylselenocysteine [22]) before hydrolysis. Selenium species are identified by GC-MS [27–29], GC-FID [30], GC-ICPMS, HPLC or electrophoresis coupled to ICPMS, AAS or atomic fluorescence spectrometers [31–33]. Recently, selenium compounds in yeast (containing $1000\,s$ of μg^{-1} g Se) [34–35], mustards [36], garlic [37] and urine [38] have been identified by mass spectrometry; however, mass spectrometry is not widely used due to the low selenium concentrations present in tissues. Volatile selenium species (Me_2Se, Me_2Se_2) can be determined by purging, cryofocusing and detection by atomic fluorescence, GCFPD/AED or ICPMS [39, 40].

10.4 OCCURRENCE

10.4.1 FOOD PLANTS AND OTHER ORGANISMS

The organic compounds of selenium found in food and plants are given in Table 10.1. Compounds and amounts vary considerably, especially in less well characterized dietary supplements. Many organoselenium compounds have been reported in the literature. With improved understanding of selenium chemistry and improved analytical techniques, the identification of some of the compounds is in question. For example, selenocystine is reported by many authors but is unlikely to be present in animals or plants.

A detailed discussion of selenium in food is found in Reference 1. Numerous other recent references to naturally occurring organoselenium species occur [42–49].

10.4.2 BODY FLUIDS AND TISSUES

Selenium containing proteins ranging from 2–100 kDa have been identified in a variety of species. Selenoprotein P and glutathionine peroxidase are the most

Table 10.1   Selenium compounds in food, plants and other organisms

1.	Selenocysteine, $HSeCH_2CH(NH_2)COOH$
2.	Selenocysteine, $HOOC(NH_2)CHCH_2SeSeCH_2CH(NH_2)COOH$
3.	Selenomethionine, $CH_3SeCH_2CH_2CH(NH_2)COOH$
4.	Triphenylselenonium ion, $(C_6H_5)_3Se^+$
5.	Diphenyl selenide, $(C_6H_5)_2Se$
6.	Phenylmethyl selenide, $(C_6H_5)CH_3Se$
7.	Se-methylselenomethionine, $((CH_3)_2SeCH_2CH_2CH(NH_2)COOH$
8.	Selenocystathione, $Se(CH_2CH(NH_2)COOH)(CH_2CH(NH_2)COOH,$
9.	Se-methylselenocysteine, $CH_3SeCH_2CH(NH_2)COOH,$
10.	Se-methylselenocysteine selenoxide, $CH_3Se(O)CH_2CH(NH_2)COOH,$
11.	Selenocysteine selenic acid, $HOOCSeCH_2CH(NH_2)COOH$
12.	Se-propylselenocysteine selenoxide, $PrSe(O)CH_2CH(NH_2)COOH$
13.	Selenohomocystine, $HSeCH_2CH_2CH(NH_2)COOH$
14.	γ-L-Glutamyl-Se-methylselenocysteine, $GlSe(CH_3)CH_2CH(NH_2)COOH$
15.	Dimethylselenide, $(CH_3)_2Se$
16.	Dimethyldiselenide, $(CH_3)_2Se_2$
17.	Se-adenosylselenomethionine, $CH_3Se(Aden)CH_2CH_2CH(NH_2)COOH$
17.	Se-adenosylhomocysteine, $AdenSeCH_2CH_2CH(NH_2)COOH$
18.	Dimethylselenopropionate, $(CH_3)_2SeCH_2COO^-$
19.	Se-allyselenocysteine, $AuylSeCH_2CH(NH_2)COOH$

See also Ref. 41

widely reported selenoproteins [50, 51]. Possible transformation of selenium are shown in Figure 10.1 [52]. Selenocysteine and selenomethionine have been identified in bacteria, animal and plant proteins [50, 53, 54]. Trimethylselenonium ion is the major metabolite in human urine [55] while dimethyl selenide and dimethyldiselenide are produced in the breath of individuals with a high intake of selenium. The incorporation of selenium into amino acids is not fully understood. Two pathways may exist, the first pathway involves conversion of selenium obtained from dietary sources as selenomethionine to selenocysteine in the liver [56]; a specific tRNA molecule is responsible for incorporation of selenocysteine into proteins [57]. A second pathway involves the reduction of selenate/selenite with glutathionine as a cofactor [58], aminoacylation with serine and incorporation into proteins [53, 59].

In conclusion, most selenium in tissue is bound to protein and is determined as SeCyst and SeMet from the various selenium proteins. Me_3Se^+ is found in urine and Me_2Se and Me_2Se_2 in breath [60, 61].

10.4.3   ENVIRONMENT

Organoselenium compounds have been found in the natural environment on many occasions. Me_2Se, Me_2Se_2 and Me_2SeO_2 have been found in the atmosphere, lake sediments and silts and natural waters [60–64].

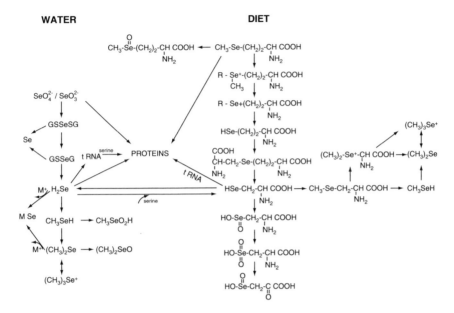

Figure 10.1 Pathways of Selenium transformations in marine organisms (After Reference 67, with permission)

Many organoselenium compounds have been found in the environment and the common ones are shown in Figure 10.2. Transformation and formation of organoselenium compounds in the environment occurs primarily by bacteria or by incorporation of inorganic selenium into plants and animals. Bacteria produce a range of volatile selenium and selenium–sulfur compounds [65, 66]. Animals and plants take up selenium and produce non-volatile organoselenium compounds, most of which have still not been identified [67, 68].

10.5 TOXICITY

Selenium toxicity and beneficial effects have been discussed and reviewed recently and will not be discussed here. They vary, of course, between selenium species, with the organic forms not appearing to be much more toxic than the inorganic forms. Apart from gross artificial aspects (overdosing) or inorganic selenium poisoning, as naturally occurring species in many organisms, methyl-selenium should be considered more as a natural product, with a neutral impact on toxicology if found at normal levels. The matter of selenium in food has been discussed [1]. The role of selenium in reducing mercury toxicity is well known, and this appears to operate by releasing mercury or methylmercury from its former linkage to protein sulfhydryl groups by competitive binding and removal to locations having less deleterious effects, without elimination of the

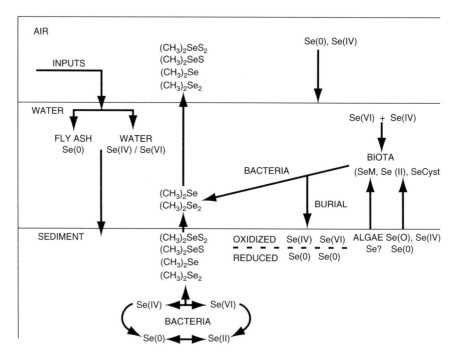

Figure 10.2   Biogeochemical cycling of selenium in aquatic ecosystems

mercury from the animal. Sometimes however, there is a synergism of selenium and mercury toxicity, not just antagonism.

10.6   REFERENCES

1. Ebdon, L., Pitts, L., Cornellis, R., Crews, H., Donard, O.F.X., and Quevauviller, Ph. (Eds), *Trace Element Speciation for Environment, Food and Health*, Royal Society of Chemistry, London, 2001.
2. World Health Organisation, *Trace Elements in Human Nutrition and Health*, Geneva, 1996.
3. Zachara, B.A., *J. Trace Elem. Health. Dis.*, 1993, **6**, 137.
4. *The Merck Index*, Merck and Co., Rahway, New Jersey, 1989, 1337.
5. National Research Council, *Dietary Recommendations*, 10th edn, National Research Council, Washington, DC, 1989.
6. European Commission, *Scientific Report on Human Nutrition and Health*, 31st Series, Luxembourg, 1994.
7. National Research Council, *Recommended Dietary Allowances*, National Academy Press, Washington, DC, 1989.
8. European Community, *Reports of the Scientific Committee for Food* (31st Series), Luxemburg, 1993, 204.
9. Chau, Y.K., and Wong, P.T.S. In *Organometallic Compounds in the Environment*, P.J. Craig (Ed.), Longman, London, 1986, pp. 254–278.

10. Olivas, R.M., Donard, O.F.X., Camara, C., and Quevauviller, Ph., *Anal. Chim. Acta*, 1994, **286**, 257.
11. Pyzyynska, *Analyst*, 1996, **121**, 77R.
12. Wang, Z., Xie, S., and Peng, A., *J. Agric. Food Chem.*, 1996, **44**, 2754.
13. Puskel, E., Mester, Z., and Fodor, P., *J. Anal. At. Spectrom.*, 1999, **14**, 973.
14. Michalke, B., and Schramel, P., *Electrophoresis*, 1998, **19**, 270.
15. Suzuki, K.T., Itoh, M., and Ohmichi, M., *J. Chromatogr. B*, 1995, **666**, 13.
16. Bird, S.M., Ge, H., Uden, P.C., Tyson, J.F., Block, E., and Denoyer, E., *J. Chromatogr. A*, 1997, **789**, 349.
17. Ge, H., Cai, X.-J., Tyson, J.F., Uden, P.C., Denoyer, E.R., and Block, E., *Anal. Comm.*, 1996, **33**, 279.
18. Michlalke, B., and Schramel, P., *J. Chromatogr.*, 1998, **807**, 71.
19. Behne, D., Hammel, C., Pfiefer, H., Rothlein, D., Gessner, H., and Kyriakopoulos, A. *Analyst*, 1998, **123**, 871.
20. Shibata, Y., Morita, M., and Fuwa, K., *Adv. Biophys.*, 1992, **28**, 31.
21. Beilstein, M., Whanger, P.D., and Yang, G.Q., *Biomedical and Environmental Sciences*, Vol. 4, 1991, p. 392.
22. Hammel, C., Kyriakopoulous, A., Rosick, U., and Behne, D., *Analyst*, 1997, **122**, 1359.
23. Larsen, E.H., Hansen, M., Fan, T., and Vahl, M., *J. Anal. At. Spectrom.*, 2001, **16**, 1403.
24. Onning, G., and Bergdahl, I.A., *Analyst*, 1999, **124**, 1435.
25. Sathe, S.K., Mason, A.P., Rodibaugh, R., and Weaver, C.M., *J. Agric. Food Chem.*, 1992, **40**, 2084.
26. Michalke, B.J., *J. Chromatog.*, 1995, 323.
27. Yasumoto, K., Suzuki, K.T., and Yoshida, M., *J. Agric. Food Chem.*, 1998, **36**, 463.
28. De La Calle-Guntinas, M.B., Brunori, C., Scerbo, R., Chiavarni, S., Quevauviller, P., Adams, F., and Morabito, R., *J. Anal. At. Spectrom.*, 1997, **12**, 1041.
29. Ansede, J.H., Pellechia, P.J., and Yoch, D.C., *Environ. Sci. Technol.*, 1999, **33**, 2064.
30. Janak, J., Billiet, H.A.H., Frank, J., Luyben, K.C.A.M., and Husek, P.J., *Chromatography*, 1994, **677**, 192.
31. Palez, M.V., Bayon, M.M., Alonso, J.I.G., and Snz-Medel, A., *J. Anal. At. Spectrom.*, 2000, **15**, 1217.
32. Bird, S.M., Uden, P.C., Tyson, J.F., Block, E., and Denoyer, E., *J. Anal. At. Spectrom.*, 1997, **12**, 785.
33. Gammlgaard, B., Jessen, K.D., Kristensen, F.H., and Jons, O., *Anal. Chim. Acta*, 2000, **404**, 47.
34. McSheehy, S., Pohl, P., Szpunar, J., Potin-Gautier, M., and Lobinski, R., *J. Anal. At. Spectrom.*, 2001, **16**, 68.
35. McSheehy, S., Szpunar, J., Haldys, V., and Tortajada, J., *J. Anal. At. Spectrom.*, 2002, **17**, 507.
36. Montes-Bayon, M., Yanes, E.G., Ponce de Leon, C., Jayasimhulu, K., Stalcup, A., Shann, J., and Caruso, J.A., *Anal. Chem.*, 2002, **74**, 107.
37. McSheehy, S., Yang, W., Pannier, F., Szpunar, J., Lobinski, R., Auger, J., and Potin-Gautier, M., *Anal. Chim. Acta*, 2000, **421**, 147.
38. Cao, T.H., Cooney, R.A., Wozichak, M.M., May, S.W., and Browner, R.F., *Anal. Chem.*, 2001, **73**, 2898.
39. Yanzer, D. and Heumann, K.G., *Atmospheric Environ.*, 1990, **24A**, 3099.
40. Pecheyran, C., Amouroux, D., and Donard, O.F.X., *J. Anal. At. Spectrom.*, 1998, **13**, 615.
41. Moesgaard, S., and Morril, R., in Ref 1, 261.

42. Shrift, A., *Selenium Compounds in Nature and Medicine*, Chapter 13, pp. 763–814.
43. McSheehy, S., Yang, W., Pannier, F., Szpunar, J., Lobinski, R., Auger, J., and Potin-Gautier, M., *Anal. Chim. Acta*, 2000, **421**, 147.
44. McSheehy, S., Szpunar, K., Haldys, V., and Tortajada, J., *J. Anal. At. Spectrom.*, 2002, **17**, 507.
45. McSheehy, S., Pohl, P., Szpunar, J., Potin-Gautier, M., and Lobinski, R., *J. Anal. At. Spectrom.*, 2001, **16**, 68.
46. B'Hymer, C., and Caruso, J., *J. Anal. At. Spectrom.*, 2000, **15**, 1531.
47. Ansede, J.H., Pellechia, P., and Yoch, D.C., *Environ. Sci. Technol.*, 1999, **33**, 2064.
48. Larsen, E.H., Hansen, M., Fan, T., and Vahl, M., *J. Anal. At. Spectrom.*, 2001, **16**, 1403.
49. Beilstein, M., Whanger, P.D. and Yang, G.Q., *Biomedical and Environmental Sciences*, 1991, **4**, pp. 392–298.
50. Burk, R.F., *FASEB J*, 1991, **5**, 2274.
51. Yang, J.G., Morrison-Plummer, J. and Burke, R.F., *J. Biol. Chem.*, 1987, **262**, 13372.
52. Maher, W., Deaker, M., Jolley, D., Krikowa, F., and Roberts, B., *Appl. Organomet. Chem.*, 1997, **11**, 313.
53. Yasumoto, K., Suzuki, K.T., and Yoshida, M., *J. Agric. Food Chem.*, 1988, **36**, 463.
54. De La Calle-Guntinas, M.B., Brunori, C., Scerbo, R., Chiavarni, S., Quevauiller, P., Adams, F., and Gammlgaard, B., Jessen, K.D., Kristensen, F.H., and Jons, O., *Anal. Chim. Acta*, 2000, **404**, 47.
56. Hawkes, W.C., Wilhelmsen, E.C., and Tappel, A.L., *J. Inorg. Chem.*, 1985, **23**, 77.
57. Alexander, J., Hogberg, J., Thomassen, Y., and Aaseth, J. In H.G. Seiler, H.G. Sigel and A. Sigel (Eds), *Handbook on Toxicity of Inorganic Compounds*, Marcel Dekker, 1988, pp. 581–594.
58. Hatfield, D., Lee, B.J., Hampton, L., and Diamond, A., *Nucleic Acid Research*, 1991, **9**, 939.
59. Lee, B.J., Rajagopalan, M., Kim, Y.S., You, K.H., Jacobson, K.B., and Hatfield, D., *Molec. Cell Biol.*, 1990, **10**, 1940.
60. Byard, J.L., *Arch. Biochem. Biophys. B*, 1969, 556.
61. Fleming, R.W., and Alexander, M., *Applied Microbiol.*, 1972, **24**, 424.
62. McConnell, K.P., and Portman, O.W., *J. Biol. Chem.*, 1952, 195.
63. Chai, Y.K., Wong, P.T.S., Silverberg, B.A., Luxon, P.L., and Bengert, G.A., *Science*, 1976, **192**, 1130.
64. Jiang, S., *Atm. Environ.*, 1983, **17**, 111.
65. Doran, J.W., In *Advances in Microbiology and Ecology*, Vol. 6, K.C. Marshall (Ed.), 1982, Chapter 1, pp. 1–32.
66. Peters, G.M., Maher, W.A., Jolley, D., Carroll, B.I., Gomes, V.G., Jenkinson, A.V., and Moist, G.D., *Org. Geochem.*, 1999, **30**, 1287.
67. Maher, W., Deaker, M., Jolley, D., Krikowa, F., and Roberts, B., *Appl. Organomet. Chem.*, 1997, **11**, 313.
68. Amouroux, D., Pecheyran, C., and Donard, O.F.X., *Appl. Organomet. Chem.*, 2000, **14**, 236.

Some Standard Reference Sources on Organometallic and Environmental Organometallic Chemistry

Dictionary of Organometallic Compounds, 2nd Edition, Vols. 1–5, Chapman and Hall, London 1995 (gives details of individual compounds).

Comprehensive Organometallic Chemistry II. A Review of the Literature 1982–1994. E.W. Abel, F.G.A. Stone and G. Wilkinson (Eds), Vols. 1–14, Pergamon (Elsevier), Oxford 1995. Pre-1982 information is given in *Comprehensive Organometallic Chemistry I*, 1982, same editors, publishers. These have to be used together.

The Open University, *Science: a Third Level Course.* S343 Inorganic Chemistry Block 6 Organometallic Chemistry (Course Team Chair, S. Bennett) 2nd Edition, 1994, Open University Press (excellent general introduction to organometallic chemistry).

Principles of Organometallic Chemistry, 2nd edition, P. Powell, Chapman and Hall, London. 1998. (Excellent discussion on stabilities of organometallic compounds.)

Determination of Metals in Natural and Treated Waters, T.R. Crompton, Spon Press (Taylor and Francis), London, 2002.

Trace Element Speciation for Environment, Food and Health, L. Ebdon, L. Pitts, R. Cornelis, H. Crews, O.F.X Donard and Ph. Quevauviller (Eds), Royal Society of Chemistry, London, 2001.

Organometallics, A Concise Introduction, 2nd Edition, Ch. Elschenbroich and A. Salzer, VCH Press Weinheim, 1992.

Metal Ions in Biological Systems, Vol. 29, *Biological Properties of Metal Alkyl Derivatives*, H. Sigel and A. Sigel (Eds), Marcel Dekker, New York, 1993 (Useful survey.)

Thermochemical Processes, Principles and Models, C.B. Hock, Butterworth Heinemann, Oxford, 2001 (good theoretical discussion).

Chemical Speciation in the Environment, A.M. Ure and C.M. Davidson (Eds), Blackie Academic and Professional, Glasgow, 1995 (primarily covers non-organometallic, analysis, methodology).

Occurrence and Analysis of Organometallic Compounds in the Environment, T.R. Crompton, John Wiley and Sons, Chichester, UK, 1998.

INDEX